Volatiles and Metabolites of Microbes

Volatiles and Metabolites of Microbes

Edited by

Ajay Kumar
Agriculture Research Organization, Volcani Center, Rishon, Lezziyon, Israel

Joginder Singh
Department of Biotechnology, Lovely Professional University, India

Jastin Samuel
Lovely Professional University, Phagwara, India

ELSEVIER

ACADEMIC PRESS
An imprint of Elsevier

Academic Press is an imprint of Elsevier
125 London Wall, London EC2Y 5AS, United Kingdom
525 B Street, Suite 1650, San Diego, CA 92101, United States
50 Hampshire Street, 5th Floor, Cambridge, MA 02139, United States
The Boulevard, Langford Lane, Kidlington, Oxford OX5 1GB, United Kingdom

British Library Cataloguing-in-Publication Data
A catalogue record for this book is available from the British Library

Library of Congress Cataloging-in-Publication Data
A catalog record for this book is available from the Library of Congress

ISBN: 978-0-12-824523-1

For Information on all Academic Press publications
visit our website at https://www.elsevier.com/books-and-journals

Publisher: Andre Gerhard Wolff
Acquisitions Editor: Linda Versteeg-buschman
Editorial Project Manager: Susan Ikeda
Production Project Manager: Niranjan Bhaskaran
Cover Designer: Christian J. Bilbow

Typeset by MPS Limited, Chennai, India

Contents

List of contributors

Nahid Akhtar Department of Molecular Biology and Genetic Engineering, School of Bioengineering and Biosciences, Lovely Professional University, Jalandhar, India

M. Amin-ul Mannan Department of Molecular Biology and Genetic Engineering, School of Bioengineering and Biosciences, Lovely Professional University, Jalandhar, India

R.M. Banik School of Biochemical Engineering, IIT (BHU), Varanasi, India

Ömür Baysal Faculty of Science, Department of Molecular Biology and Genetics, Molecular Microbiology Unit, Muğla Sıtkı Koçman University, Muğla, Turkey

Surojit Bera Department of Microbiology, School of Bioengineering and Biosciences, Lovely Professional University, Jalandhar, India

Protha Biswas Department of Life Sciences, Presidency University, Kolkata, India

Evangeline Christina Department of Molecular Biology and Genetic Engineering, School of Bioengineering and Biosciences, Lovely Professional University, Jalandhar, India

Tuyelee Das Department of Life Sciences, Presidency University, Kolkata, India

Banhishikha Datta Department of Molecular Biology and Genetic Engineering, School of Bioengineering and Biosciences, Lovely Professional University, Jalandhar, India

Priyanka Devi Department of Agronomy, School of Agriculture, Lovely Professional University, Jalandhar, India

Abhijit Dey Department of Life Sciences, Presidency University, Kolkata, India

Shipa Rani Dey Department of Agronomy, School of Agriculture, Lovely Professional University, Jalandhar, India

Padmanabh Dwivedi Department of Plant Physiology, Institute of Agriculture Sciences, BHU, Varanasi, India

Jing Han State Key Laboratory of Microbial Resources, Institute of Microbiology, Chinese Academy of Sciences, Beijing, P.R. China; College of Life Science, University of Chinese Academy of Sciences, Beijing, P.R. China

Chelliah Jayabaskaran Department of Biochemistry, Indian Institute of Science, Bengaluru, India

Subban Kamalraj Department of Biochemistry, Indian Institute of Science, Bengaluru, India

Prabhjot Kaur Department of Biotechnology, Lovely Faculty of Technology and Sciences, Lovely Professional University, Jalandhar, India

Prasann Kumar Department of Agronomy, School of Agriculture, Lovely Professional University, Jalandhar, India

Vijay Kumar Department of Biotechnology, Lovely Faculty of Technology and Sciences, Lovely Professional University, Jalandhar, India

Upasana Mangrolia (Deepak) School of Biosciences and Technology, Vellore Institute of Technology University, Vellore, India; Biomolecules lab, TT635, VIT, Vellore, Tamil Nadu, India

Ruchira Mitra State Key Laboratory of Microbial Resources, Institute of Microbiology, Chinese Academy of Sciences, Beijing, P.R. China; International College, University of Chinese Academy of Sciences, Beijing, P.R. China

Shreya Sikdar Mitra Department of Life Sciences, Presidency University, Kolkata, India

Anuradha Mukherjee MMHS, Joynagar, India

Samapika Nandy Department of Life Sciences, Presidency University, Kolkata, India

Santhanam Narmadha Biomolecules Lab, School of Bio Sciences and Technology, Vellore Institute of Technology, Vellore, India

Potshanghbam Nongdam Department of Biotechnology, Manipur University, Imphal, India

W. Jabez Osborne Biomolecules Lab, Department of Bio-Sciences, School of Bio Sciences and Technology, Vellore Institute of Technology, Vellore, India

Devendra Kumar Pandey Department of Biotechnology, Lovely Faculty of Technology and Sciences, Lovely Professional University, Jalandhar, India

Janmejay Pandey Department of Biotechnology, Central University of Rajasthan, Ajmer, India

Ramalingam Parthasarathy Department of Biochemistry, Indian Institute of Science, Bengaluru, India

T. Ramachandira Prabu Department of Microbiology, Prist Deemed University, Puducherry Campus, Puducherry, India

T.C. Prathna Department of Irrigation and Flood Control, Government of NCT of Delhi, New Delhi, India

Swarnkumar Reddy Biomolecules Lab, School of Bio Sciences and Technology, Vellore Institute of Technology, Vellore, India

Lalit Saini Department of Agronomy, School of Agriculture, Lovely Professional University, Phagwara, India

Khushbu Sharma Department of Agronomy, School of Agriculture, Lovely Professional University, Phagwara, India

Ragıp Soner Silme Center of Biotechnology and Genetic Engineering Research and Practice, Istanbul University, Istanbul, Turkey

Bishal Singh School of Medical Science and Technology, Indian Institute of Technology, Kharagpur, India

Monika Singh Centre of Advanced Study in Botany, Banaras Hindu University, Varanasi, India

Ranjit Singh Department of Chemistry, Saroop Rani Government College for Women, Amritsar, India

Astha Sinha Biomolecules Lab, School of Bio Sciences and Technology, Vellore Institute of Technology, Vellore, India

Saravanan Sivarajan VIT School of Agricultural Innovations & Advanced Learning (VAIAL)

Navnita Srivastava Department of Life Science, School of Science, Sardar Patel University, Balaghat, India

D. Sruthi Department of Biochemistry, Indian Institute of Science, Bengaluru, India

Kothandapani Sundar Department of Microbiology, Prist Deemed University, Puducherry Campus, Puducherry, India; Department of Microbiology, Sri Lakshmi Narayana Institute of Medical Sciences, Puducherry, India

Champa Keya Tudu Department of Life Sciences, Presidency University, Kolkata, India

Jeevanandam Vaishnavi Biomolecules Lab, Department of Bio-Sciences, School of Bio Sciences and Technology, Vellore Institute of Technology, Vellore, India

Pratibha Vyas Department of Microbiology, College of Basic Sciences and Humanities, Punjab Agricultural University, Ludhiana, India

Mrunal S. Wagh School of BioSciences and Technology, VIT, Vellore, India

Atif Khurshid Wani Department of Molecular Biology and Genetic Engineering, School of Bioengineering and Biosciences, Lovely Professional University, Jalandhar, India

Hua Xiang State Key Laboratory of Microbial Resources, Institute of Microbiology, Chinese Academy of Sciences, Beijing, P.R. China; College of Life Science, University of Chinese Academy of Sciences, Beijing, P.R. China

H.C. Yashavantha Rao Department of Biochemistry, Indian Institute of Science, Bengaluru, India

Foreword

There is increasing interest in how microbes interact or communicate with individuals within and between species. Microbial volatile organic compounds (VOCs) are part of the language microbes used to communicate at a distance with other individuals or populations. An example of a VOC that functions in this kind of communication is ethylene. Microbes produce ethylene through the activity of microbial ethylene synthase enzyme that uses the amino acid arginine as a substrate to generate ethylene. Ethylene is also a growth hormone produced by plants to stimulate growth. When microbes that are endophytic or epiphytic in plant roots produce ethylene, they trigger growth in plant root cells, and the root cells increase root exudation of nutrients (including carbohydrates, organic acids, and amino acids like arginine) that feed the endophytic or epiphytic microbes and enable them to grow and produce additional ethylene. Through ethylene generation, microbes modulate plant growth and increase nutrients that they have access to growth. VOCs comprise several chemical classes, including alkenes, alcohols, ketones, benzenoids, pyrazines, sulfides, and terpenes. Meta-analysis studies provided a comprehensive overview of VOCs derived from soilborne microbes. These secondary metabolites have been reported to be involved in microbial interactions produced by soil- and plant-associated microorganisms. Several studies have reported that the production of certain microbial VOCs could be induced or suppressed during interspecific microbial interactions. Microbial VOCs are often considered to be by-products of primary metabolism; however, recent findings have revealed that they also demonstrate biological activity. Compared to soluble metabolites, VOCs may be considered to be long-distance signal molecules. There are many different interactions in soils, including bacteria–bacteria, fungi–fungi, fungi–bacteria, bacteria–protists, fungi–plant, bacteria–plant, and bacteria–fungi–plant interactions.

We now live in a time in which more environmentally safe methods must replace the old methods of using chemicals to protect crops. The traditional synthetic fungicides currently used could be replaced with the so far underexplored microbial VOC-producing organisms for which significant proof of plant growth-promoting effects and plant protection ability already exists. Despite the apparent potential of microbial VOCs in agriculture, the field suffers from the typical "translational gap" because of the lack of studies evaluating other unexpected effects of these bioactive molecules on nontarget beneficial soil organisms.

The increasing global population and climate change are a significant concern for researchers, agronomists, environmentalists, and policy makers. The rising global human population creates pressure on agriculturists due to limited land resources, low soil productivity, pests and pathogens, and changing climate. Thus there is an urgent need to explore new biological methods for sustainable agriculture and health industries. Structural elucidation of volatile natural products with specific action potential would increase the impact of VOCs.

Deciphering how microbes use VOCs to communicate with and control other organisms could be vital in understanding how microbes may be used in agriculture and medicine to improve crop yields and health and to control diseases. This book unifies the knowledge of VOCs and their effects and utility in agriculture and health and is thus a step toward a more sustainable future for agriculture and medicine.

James Francis White
Department of Plant Biology, Rutgers University, New Brunswick, NJ, United States

Preface

A rising global population puts extra pressure on agriculture production and farmers due to limited land resources, low soil productivity, regular and extra use of chemical pesticides, and increased attacks from pests and pathogens due to changing climatic conditions. Global population increases and climate change are the major concerns for environmentalists, agronomists, researchers, and policymakers. There is an urgent need to explore new and advanced biological methods for sustainable agriculture and health industries to protect the environment from environmental pollution or contaminants, global warming and control the health of human beings from the side effects of various pharmaceutical products. While focusing on all the aforementioned factors, this book explores new and valuable aspects of microorganisms in terms of volatiles, enzymes, bioactive compounds synthesized from them, and their potential applications in the field of sustainable agriculture and health-related issues. In agriculture, various microbial volatile organic compounds are now being used to facilitate crop performance and production by acting as biocontrol agents to inhibit the growth and development of numerous plant pathogens and elicit induced systemic resistance in the plants or mitigating effects caused by stress factors. However, these volatile organic compounds are broadly utilized to synthesize antibiotics, bioactive compounds, etc. The ultimate goal is to support the scientific community, professionals, and enterprises that aspire to understand the latest developments and advancements about exploiting these volatile organic compounds, including their application, traditional uses, modern practices, and designing strategies to harness their potential. In this book, we compiled the latest research and advancement of volatiles, metabolites synthesized from microbial strains such as actinomycetes, bacteria, cyanobacteria, fungal species, and their potential applications in the field of sustainable agriculture and healthcare issues.

The purpose of editing *Volatiles and Metabolites of Microbes* is to present details of various systems approaches and provide a means to share the latest developments and advancements about the volatiles and metabolites of microbes, and their spectrum of applications, with academicians, industrialists and policymakers as well. With 21 chapters contributed by an exclusive group of researchers working at the forefront of microbiology, biotechnology, natural products, and development practices, the book covers advanced perspectives, latest research, and advancements in the field of volatiles, secondary metabolites, and bioactive compounds synthesized from the different microbial strains and their diverse potential applications. This book serves as a helpful reference book for academics, scientists, members of pharmaceutical and nutraceutical industries, plant pathologists, professionals, and strategy developers working in the environmental microbiology and food-beverage industries.

Ajay Kumar[1], Joginder Singh[2] and Jastin Samuel[3]
[1]Agriculture Research Organization, Volcani Center, Rishon, Lezziyon, Israel, [2]Department of Biotechnology, Lovely Professional University, India, [3]Lovely Professional University, Phagwara, India

Chapter 1

Microbial secondary metabolites: recent developments and technological challenges

Swarnkumar Reddy[1], Astha Sinha[1] and W. Jabez Osborne[2]
[1]Biomolecules Lab, School of Bio Sciences and Technology, Vellore Institute of Technology, Vellore, India, [2]Biomolecules Lab, Department of Bio-Sciences, School of Bio Sciences and Technology, Vellore Institute of Technology, Vellore, India

Chapter Outline

1.1 Introduction

Microbes play an immense significance in healthcare and sustainable environment and also serves as a primary source of essential nutrients of all life forms. The sum of various biochemical reactions are involved in the process of

Volatiles and Metabolites of Microbes. DOI: https://doi.org/10.1016/B978-0-12-824523-1.00007-9

microbial metabolism; these metabolites involve in various cellular process (Marinelli, 2009; Ahmed and El-Refai, 2010; Reddy and Osborne, 2020b). Metabolites produced in early idiophase phase of microbial growth are involved in cellular growth and development and are termed as primary metabolites (Poltronieri and Reca, 2020). The metabolites in the late idiophase was termed as secondary metabolites and coined by A. Kossel in 1891 (Thirumurugan et al., 2018). Primary metabolites are present in all living cells and involved in cell division, where secondary metabolite incidental metabolomic compounds, which does not have any significance in organism's life. However, these metabolites have various significant roles; for example, it provides defense against pathogens, facilitates strong symbiotic relationships with other organisms, enhances signaling molecules for communicating among microbial population, assists in adapting to adverse environmental conditions, and makes microbes the chief recycler in an environment (Prakash et al., 2019; Boyer-Joubert et al., 2003; Choudhary and Dhar, 2015). Microbial synthesis of secondary metabolites is highly facilitated with limiting the availability of key nutrient sources such as carbon, nitrogen, or phosphate (Ruiz et al., 2010) (Fig. 1.1).

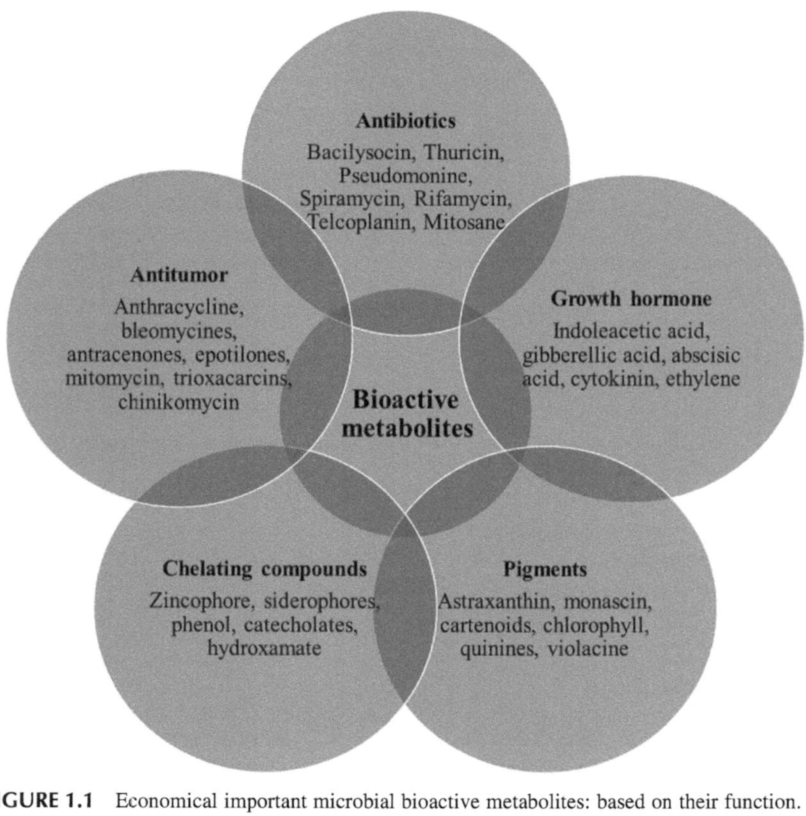

FIGURE 1.1 Economical important microbial bioactive metabolites: based on their function.

Secondary metabolites are special small bioactive molecules with major advantages over human health, nutrition, plant health, and also on society's economy. It was estimated that the global market on microbes and microbial metabolites was estimated to be nearly 186.3 billion USD in 2018 and was expected to reach 302.4 billion USD with an annual growth rate of 10.2% by 2023 (Bennett and Bentley, 1989; Singh et al., 2017). Antibiotics are the most known and widely produced microbial metabolites apart from antibiotics antiviral, antitumor, enzyme inhibitors, and various amino acids are other notable secondary metabolites. Microbial biosynthesis of secondary metabolites is mediated by certain pathways such as, polyketide synthase pathway, β-lactam synthesis, peptide synthesis, shikimate synthesis, and nonribosomal polypeptide synthesis (Butler, 2008; Kingston, 2011). Apart from the advantages over human health, secondary metabolites also assist in environmental and plant health by improving nutrient availability by producing metal-chelating agents, such as siderophore, provides mechanical strength against environmental stress by producing osmoprotectants and pigments, enhances competitive interactions with other organisms by synthesis of antibiotics or signaling molecules like quorum sensing molecules, and also provides metabolic defense mechanism through flavonoids and alkaloid toxins (Zhou et al., 2008; Bentley, 1997; Keller, 2019; Reddy and Osborne, 2020a).

The widespread diversity of secondary metabolites comprises over 35,000 terpenoids and steroid compounds, 12,000 known alkaloids, and over 10,000 fatty acids. The multitude ability of the secondary metabolites in application as immunomodulators, antitumor agents, and conventional antibiotics are well known (Pan et al., 2019; Park et al., 2019). The current chapter on "Microbial Secondary Metabolites: Recent Developments and Technological Challenges" comprises various aspects on exploitation and utilization of microbial metabolites in agriculture productivity by microbial-derived pesticides, insecticides, etc. The current report also deals with the recent advancements on synthetic biological approach of metabolite production, i.e., metabolomics as a key tool in bioprospecting new generation of secondary metabolites.

1.2 Microbial metabolites: definitions and diversity

Microbial metabolites include various organic compounds, amino acids, nucleotides, fermented alcohols, enzyme inhibitors, signally molecules, growth factors, antimicrobial agents, etc. (Vallianou et al., 2019; Aden et al., 2019). The origin of bioactive microbial metabolites are most important characteristic, specifically their interaction with environment, specific biological activities, and also based on chemical structures (Kallscheuer et al., 2019).

1.3 Taxonomic diversity

The ability of producing secondary metabolites are widespread and uneven among various species of life forms. Antibiotics and its derived natural products are the secondary metabolites commonly produced by all living organism including all higher forms of life (Cabello, 2020; Verma et al., 2020). A diverse group of organisms such as unicellular bacteria, eukaryotic fungi, and most particularly filamentous actinomycetes are the most abundant and versatile producers of bioactive secondary metabolites (Kaaniche et al., 2019). Among prokaryotes unicellular *Bacillus* and *Pseudomonas* species are the primary producers of various bioactive compounds. As result of intense research in microbial bioactive compounds two novel prolific species of *Myxo* and *Cyanobacteria* has been added in this distinguished category. Apart from these prokaryotes spirotheces, mycobacterium, and mycoplasmatales are other notable producers of bioactive compounds (Horak et al., 2019; Martínez et al., 2019). Eukaryotes hosts for both microscopic as well as multicellular produces. Among the microscopic eukaryotes, various filamentous and endophytic fungal species such as *Aspergillus, Colletotrichum*, and *Edenia* species, etc. (Rana et al., 2019). The basidiomycetes are the most frequently reported among the multicellular fungus. Yeast, phycomycetes, slime molds are other lower form eukaryotes with reported secondary metabolites (Nalini et al., 2019).

Over microbial secondary metabolites various higher forms plants and animals also produce various bioactive metabolites (Win and Laatsch, 2020). Among plant kingdom algae, lichens, and many vascular plants are main producers' antimicrobial and antiviral compounds. Marine and terrestrial animals together where reported to produce more than 7000 bioactive compounds (Tang et al., 2020).

1.4 Biological diversity

Secondary metabolites are widespread diverse compounds with various biological activity. The specific biological activity is the guiding line which distinguishes the "bioactive" metabolites from inactive metabolites. The biological activity of metabolites is highly specified by various uncommon factors like their specific chemical structures, and structural elements. Specific chemical groups like macrolactose, cyclopeptide skeleton, and unique functional groups also influence the specific biological activity of metabolites (Matsumoto and Takahashi, 2017; Vallesi et al., 2020).

Antimicrobial activity is the most primary function of wide range of secondary metabolites, it has been reported in various scientific literature against wide class of pathogenic microbes (Milshteyn et al., 2018). The second most abundant metabolite is antiviral compounds, which can be effective against viral infection in the host by inhibiting viral enzymes and activities connecting with neoplastic diseases. Antiviral activity can be monitored through simple cytotoxicity assay methods and angiogenesis inhibition (Ab Mutalib et al., 2020; Dhakal et al., 2017).

Apart from antimicrobial activity secondary metabolites exhibit various other activities generally termed as nonantimicrobial activity with respect to biological activity. Recent advances in microbial research has developed novel screening techniques of various bioactivates of secondary metabolite (Bills and Gloer, 2017; Newman, 2017). Cell-based-receptor binding and enzymatic assay methods are the target-based detection methods for various nonantimicrobial metabolites.

- Pharmacological-biochemical activity
- Biomedical activity
- Agricultural activity
- Regulatory, biophysical, and other inhibitory activities are the presently known nonantimicrobial bioactivities

1.5 Microbial metabolites: classifications

1.5.1 Primary metabolites

Metabolites, which are essential for the normal cell growth and division, are termed as primary metabolites. These are a conserved group of metabolites among closely related species. Primary metabolites are produced under typical physiochemical conditions of cell during early idiophase (Yadav et al., 2020; Sheflin et al., 2019). Various microbial species share similar group of primary metabolites including amino acids, flavor nucleotides, organic acids, polyols, polysaccharides, and vitamins. Primary metabolites are considered to have various vital functions in active phase of microbial growth.

1.5.2 Secondary metabolites

Microbial bioactive secondary metabolites are low molecular organic heterogeneous group of compounds which are not involved in any essential functions like cell division or vegetative growth of microbes. Although they have extremely important function in human health, nutrition, and also in sustainable environment (Lancini and Lorenzetti, 1993). This is because the secondary metabolites include a wide range of bioactive compounds like antibiotics, toxins, biopesticides, animal and plant growth factors, enzyme inhibitors, etc., which has tremendous economic significance (Sunazuka et al., 2008). Secondary metabolites are highly differentiated from primary metabolites in mode of synthesis and mode of action. Secondary metabolites are known to synthesis in late idiophase in lack of any essential nutrition and also it was known to have no function on cell growth and development where primary metabolites are involved in basic cellular metabolisms (Liu et al., 2020). Secondary metabolites are synthesized as mixtures of closely related members of chemical families and limited to certain restricted taxonomic groups of various microbes (Vallianou et al., 2019) (Table 1.1).

Secondary metabolites can be defined as

- Compounds produced in idiophase phase of batch culture
- Compounds with no cellular function
- Distributed among specific taxonomical group of microbes
- Exhibit varied and unusual chemical diversity
- Synthesized as mixture of closely related chemical compounds

Secondary metabolites are widely classified based on their structure, function, and its biosynthesis. Presently more than 2,00,000 known bioactive secondary metabolites are in various applications, and these bioactive compounds widely fall under five major classes, namely peptides, polyketides, volatile compounds, terpenes and steroids, and growth regulators (Jia-Xi et al., 2019; Singh et al., 2019a,b, Demain and Zhang, 2005).

1.5.3 Peptides

Peptides are soluble microbial metabolites often occur as cyclic compounds and rarely co-occur with diketopiperazines (amino acids found in the peptides). Microbial peptides are widely involved in antimicrobial activity most commonly as bacteriocins (Demain and Lancini, 2001; Sanchez and Demain, 2017). Microbial peptides are driven by enzymatically controlled condensations and also as ring extensions of diketopiperazine units. Microbial peptides are involved in various functions like antimicrobial activity, peptide vaccines, and also in degradation of various bacterial toxins (Stone and Williams, 1992). Ovine antimicrobial peptides are anionic surfactant-associated peptides has significance antibacterial activity against several Gram-negative bacteria. Dermaseptins are peptides with a variable C terminal antimicrobial domain (Salwan and Sharma, 2020; Ullah et al., 2020). Dermaseptins and derived peptides exhibit extensive sequence similarity with its precursor such as dermorphin, dermenkephalin, and adenregulin. Malaria CSP peptide a short fragment of approximately 400 amino acids derived from CircumSporozoite Protein with a previous malarial infection can be used as control peptide for melanoma vaccine studies. A similar PvMSP-1 peptide was known to induce lymphoproliferative response in individuals with prior history of *P. vivax* infection.

1.5.4 Polyketides

Polyketides structurally diverse group of bioactive microbial metabolites with diverse biological and pharmaceutical activities. Polyketides, biosynthesized by a similar process as fatty acid biosynthesis by decarboxylative

TABLE 1.1 Bioactive metabolites synthesized by various microbes.

Bioactive metabolite	Source of secondary metabolite	Activity	References
Thuricin S	*B. thuringenesis*	Bactericidal	Lisboa et al. (2006)
Halobacillin 5b	*B. licheniformis*	Cytotoxin, antitumor agent	Kalinovskaya et al. (2002)
Bacillomycin	*B. subtilis*	Antibacterial	Tamehiro et al. (2002)
Coaglin	*B. coagulans*	Bactericidal, Bacteriolytic	Le Marrec et al. (2000)
Oxohexaene and Cephalaxine	*Streptomyces* sp. *RM17*	Antibacterial	Vijayakumar and Malathi (2014)
Hydrogen cyanide	*P. pseudoalcaligenes P4*	Antifungal	Ayyadurai et al. (2007)
Pseudomonine	*P. stutzeri KC*	Competitive inhibition of phytopathogens	Lewis et al. (2000)
Bonactin	*Streptomyces* sp. *BD21–2*	Antibacterial	Schumacher et al. (2003)
Puromycin	*Streptomyces. alboniger*	*O*-methyltransferase (R)	Sankaran and Pogell (1975)
Limonene and guaiol	*Trichoderma viride*	Antimicrobial	Awad et al. (2018)
Tuberculariols	*Tubercularia* sp. *TF5*	Anticancer	Liu et al. (2012)
Roquefortine C	*P. roqueforti; P. crustosum*	Neurotoxin	Kim et al. (2004)
Pravastatin	*Penicillium citrinum*	Anticholesterolemics	Gonzalez et al. (2003)
Napyradiomycin (C-16 stereoisomers)	*S. antimycoticus*	Antibacterial	Motohashi et al. (2008)
Trioxacarcins	*S. ochraceus and S. bottropensis*	Antitumor and antimalarial	Maskey et al. (2003)
Megacin	*B. megaterium*	Bactericidal, bacteriolytic	Bizani et al. (2005)

condensation of malonyl CoA. Microbial biosynthesis of polyketides are guides by multifunctional, proteins called polyketide synthases. Polyketides are widely classified into three classes: type I: polyketides are mostly macrolides synthesized by multimodular megasynthases; type II: polyketides are aromatic compounds produced by reiterative action of dissociated enzymes; and type III: polyketides are usually small aromatic compounds widely found in fungal species. Pikromycin, the first-known macrolide antibiotic belongs to type I polyketide isolated from *S. venezuelae*. It is not clinically effective antibiotic but can be used as raw material for synthesis of other ketolide compounds (Weissman, 2009; Bills and Gloer, 2017). Apart from antibiotics polyketides can also be used as immunosuppressant and cell growth inhibitors. Radicicol and pochonin are monorden which bind to Hsp90 and alters its activity. So, this can be used to regulate cell growth, survival efficiency apoptosis, and oncogenesis. Tacrolimus, a macrolide lactone synthesized by *S. tsukubaensis*, can be used as immunosuppressive drug (Zaynab et al., 2018).

1.5.5 Volatile compounds

Among many reported secondary metabolites volatile organic compounds (VOC) are one group of bioactive microbial compounds highly by soil and plant-associated microorganisms. VOCs are comprising diverse group of low molecular VOC with high vapor pressure and low boiling point (Piechulla et al., 2017). High vapor pressure and low boiling point makes these molecules to evaporate and diffuse throughout the pores in soil and rhizosphere. VOC belongs to diverse class of chemical compounds such as alkenes, alcohols, ketones, benzenoids, sulfides, indole, and pyrazines (Hammerbacher et al., 2019).

Pyrazine 1,4-diazabenzene are major class of VOC widely distributes among a wide variety of plants and also in a few bacterial species. Microbial pyrazine is well known for their antimicrobial action. Pseudomonas, Bacillus, Streptomyces, and chondromyces are known producers of pyrazine. Indole and its derivates synthesized by ribosomal bacteria are known to enhance plant growth and inhibit bacterial pathogens. Sulfur-containing VOC are known to play a vital role in plant microbe interaction and also in interspecific microbe interactions. Sulfur VOC has a wide structural diversity ranging from small compounds like dimethylsulfide, dimethyldisulfide to complex structures like 2 methyltetrahydrothiophen-3-one which are synthesized by homocysteine bacteria.

1.5.6 Terpenoids and steroids

Terpenoids and steroids are biologically active organic compounds derived biosynthetically from isopentenyl diphosphate. Terpenoids are large and structurally diverse organic compounds derived from isoprene and isoprene polymers, where steroids are biologically active molecules with common tetracyclic carbon backbone (Lemfack et al., 2018). Endophytes are the major source of terpenoids such as sesquiterpenes, diterpenoids, and triterpenoids. Ergosterol and 5α,8α-epidioxyergosterol are the major antimicrobial steroids isolated from the endophytic fungus *Nodulisporium* sp. Terpenoids and Steroids from endophytes are known to show limited antimicrobial activity, which limits its application as effective drugs (Misztal et al., 2018).

1.5.7 Growth regulators

Plant-associated microbes are capable of producing several plant growth promoting molecules such as indole acetic acid, phosphate solubilizing low molecular weight organic acids, metal chelators like siderophore, etc. (Maheshwari et al., 2019; Gosal et al., 2017). Pathogenicity was the basic function of phytohormones by their elevated levels of various phytochemicals. Rhizo bacteria are known to stimulate the growth of plants by producing various plant growth promoting organic compounds and also by immobilizing nutrients. Plant resistance against microbial pathogens can also be induced by a phenomenon called "induced systemic resistance" (Sindhu and Sharma, 2020). Plant growth promotion is the direct mechanism of phytostimulation induces by various phytohormones including auxins, cytokinins, and gibberellins (Hassan et al., 2019). *Pseudomonas*, *Bacillus*, and *Azospirillum* are known producers of well-characterized phytohormones or growth regulators.

1.6 Secondary microbial metabolites: industrial significance

Microbial communities are promising sources of an enormous number of bioactive metabolites with various significant applications in healthcare and agriculture. Biological activity of microbial metabolites are highly dependent on operational and detection (Modolon et al., 2020). Secondary metabolites are highly reported with activities like plant growth stimulants, antimicrobial activity, and herbicides, but apart from these activities these bioactive metabolites also exhibit various antiinflammatory, anticoagulant, antitumor, anabolic, and vasodilating effects (Ranghar et al., 2019). These bioactive metabolites have potential source of food and feed supplements, bioinsecticides, and bioinsecticides in plant health, and also as antitumor and other therapeutic agents in human health (Fig. 1.2).

1.6.1 Nutraceutical industries

Nutraceuticals are the natural health promoting substances derived from various sources, which can exert various physiological benefits against

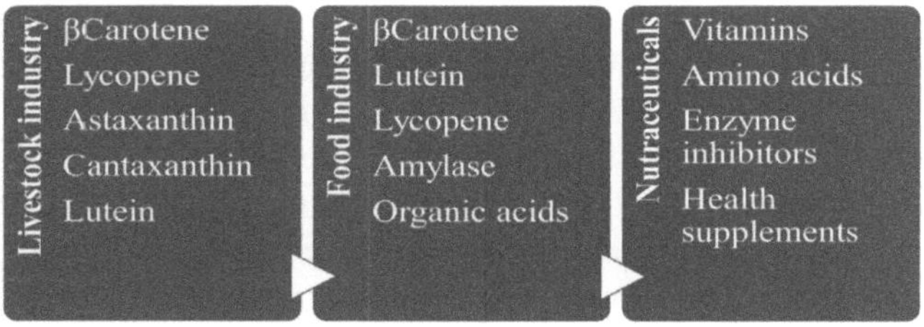

FIGURE 1.2 Industrial applications of microbial metabolites.

ageing-associated disorders, depression, inflammation, gastrointestinal diseases, and also in diabetes (Benkendorff, 2009). These molecules can be isolated from various sources like plants, microbes, and also from marine sources. Microbial nutraceuticals are widely used and has occupied an global market of \$230 billion. Microbial nutraceuticals are low cost and easily synthesized without the requirement of high pressure and heat. Amino acids, vitamins, enzymes, organic acids, and polyphenols are the widely synthesized commercial nutraceuticals (Bragazzi et al., 2017; Becker and Wittmann, 2012).

Amino acids, building blocks of proteins and also involved in various functions such as antioxidants, nutritional supplements, cattle feed additives, and also in various cosmetics. *Corynebacterium glutamicum* and *Escherichia coli* are the two major bacterial strains involved in the commercial production of various amino acids like L-glutamic acid, L-aspartic acid, L-lysine, L-threonine, and L-methionine. *Brevibacterium, Cornebacterium, Micrococcus,* and *Microbacterium* are also involved in the production of amino acids involved in animal feed additives like lysine, methionine, threonine, and tryptophan. The livestock industry was the largest consumer of amino acids, particularly lysine. Aromatic amino acids such as L-tryosine, L-phenylalanine, and L-tryptophan are the vital amino acids acts as important precursors and also in human diet (Mahmood, 2014). Recent advances in microbial fermentation technologies has made significant advancement in the large-scale synthesis of amino acids to meet the increasing demands of microbial metabolites.

1.6.2 Healthcare industries

Bioactive microbial metabolites are the most reliable source of drug development and antibiotics are the one of the greatest inventions of microbial metabolites with pharmaceutical applications (Berdy. 2005). In spite of accidental discovery of antibiotics, still its invention revolutionized the world of microbial metabolites and therapeutics. Apart from antimicrobial activity, recent research advancement in microbial metabolites were explored and proved various therapeutic activities such as antidiabetic, anticancerous, and various immunosuppressive. Diabetes, the most common metabolic disorder, was effectively treated through microbial-mediated therapeutics by genetically modified *E. coli*

(Raimundo et al., 2018). The genetically modified bacteria was able to produce recombinant insulin called Humulin, which was found to interact with human insulin receptors. *Micromonospora marina* is known to produce most effective anticancerous agent thiocoraline (Schwartsmann et al., 2001).

Bioactive immunosuppressant are most effective in treating autoimmune diseases and also in patients with organ transplantations. Cyclosporine A derived from *Tolypcladium nivenum* and sirolimus from *Streptomyces hygroscopicus* are the most effective and commercially available immune suppressant drugs (Malatesti et al., 2017). Enzyme inhibitors are the most recent advancements in microbial bioactive metabolites with increasing attention for their wide application in medicine. A many number of enzyme inhibitors with different purpose has been isolated; among all, amylase inhibitor was the most clinically used enzyme inhibitor in the treatment of carbohydrate related diseases (Sharma et al., 2020).

1.6.3 Agriculture industries

The extensive chemical pesticides have serious deleterious effects on soil fertility. In order to overcome this ecological problem, there have been carious alternates for the chemical pesticides like plant derive biopesticides and also bacterial mediated biopesticides and insecticides. The microbial-mediated biopesticides are considerable alternative for chemical pesticides

TABLE 1.2 Microbial phytohormones in plant growth promotion.

Bioactive phytohormone	Source of phytohormone	Plant inducer	References
Gibberellin 4	*Serratia nematodiphila*	*Arabidopsis thaliana*	Cohen et al. (2015)
Indole acetic acid	*Bacillus pumilus, Bacillus subtillis*	Tomato	Damodaran et al. (2011)
	Pseudomonas putida, Pseudomonas chlororaphis	Cotton	Egamberdieva et al. (2015)
ACC deaminase	*Curtobacterium albidum*	*Oryza sativa*	Vimal et al. (2019)
Auxins	*Bacillus amyloliquefaciens*	*O. sativa*	Shahzad et al. (2016)
Strigolactones	*Arbuscular mycorrhiza*	Lettuce, tomato	Ruiz et al. (2010)
Abscisic acid	*Azospirillum brasilense*	*A. thaliana*	Cohen et al. (2015)

(Vurukonda et al., 2018). The two most important bacteria: *Agrobacterium tumefaciens* and *Bacillus thuringinesis*, insecticidal bacteria widely used in various crops. The microbes associated in rhizosphere and nodular region of root are known to induce plant growth by production of various growth promotors, solubilization of nutrients, and facilitating efficient nitrogen fixation (Etesami and Adl, 2020).

Genetically modified bacterium is the widely used microbial biopesticides and insecticides, various species of Bt are employed in real time farming for their different function as bioinsecticide in order to control various pests including beetle larvae, caterpillars, etc. *Trichoderma* sp. and *Beauveria bassiana* are the most commonly available fungal biopesticides. Both biofungicides can be applied on both phylloplane and rhizosphere in order to control the fungal infection (Azizoglu, 2019) (Table 1.2).

1.7 Microbial secondary metabolites: recent advances

This part of the chapter details with recent technological advances in microbial-derived metabolites. Since there was growing interest in the application of microbial bioactive compounds in health and environment, exploring unique bioactive microbial metabolites from various sources has received enormous attention in past decades. This section of the chapter narrates the recent research advancements based on research and review papers.

1.7.1 Microbial antitumor agents

Being the deadliest diseases, anticancer drug discovery has witnessed a huge technological advancement in present decade. Microbes with enormous species diversity makes them as potential sources of novel antitumor agents (Kingston, 2009). *Salinispora arenicola*, a novel marine actinomycetes was reported to produce a novel polyketide called Arenicolide belonging to type I polyketides was proven to show cytotoxic effects on human colon adenocarcinoma cells (Kinghorn et al., 2009). Arenicolide was found to be effective at an IC50 of around 30 μg/mL. Various species of *Streptomyces* was found to produce a novel anticancerous compound Chalcomycin. This belong to the class of macrolide; other species of *Streptomyces* was also found to produce similar compounds. Saliniketal A and B are the most identified polyketides with antitumor potential. Both salinketal are known to source from same species of actinomycetes acts on tumor cells by inhibiting ornithine decarboxylase (Selvakumar et al., 2020) (Table 1.3).

Macrolactin-A produced by symbiotic bacteria *Noctiluca scintillans* was found to suppress B16-F10 melanoma murine cancer cell lines. It was also known to protect T-lymphocyte attack and inhibits the proliferation of

TABLE 1.3 Pharmaceutically important microbial bioactive compounds.

Bioactive metabolite	Pharmaceutical application
Ampicillin	Antibiotic and semisynthetic penicillin
Azithromycin	Antibiotic, semisynthetic erythromycin
Benzylpenicillin	Antibiotic, natural penicillin
Bialaphos	Herbicide, natural peptide
Cefaclor	Semisynthetic cephalosporin, Antibiotic
Cephalexin	Antibiotic, semisynthetic cephalosporin
Daunorubicin	Antitumor, natural anthracyclin
Erythromycin A	Antibiotic, natural macrolide
Kanamycin sulfate	Antibiotic, natural aminoglycoside
Mitomycin C Antitumor, natural mitosane	Antitumor, natural mitosane
Oxytetracycline	Antibiotic, natural polyketide, feed additive
Rifampin	Antibiotic, semisynthetic rifamycin, ansamycin
Spiramycin	Antibiotic, natural macrolide
Streptomycin sulfate	Antibiotic, natural aminoglycoside
Tacrolimus (FK506)	Natural immunosuppressant, macrolide
Tylosin phosphate	Natural macrolide, growth promotant
Vancomycin	HCl Antibiotic, natural glycopeptide
Pravastatin	Hypocholesterolemic polyketide, statin made by bioconversion

human immunodeficiency virus. Bryostatins, sarcodictyin, eleutherobin aand discodermolide are the other potent antitumor peptides and polyketides from microbial source (Nirmala and Zyju, 2017).

1.7.2 Microbial immunosuppressive agents

Immunosuppressive drugs have drawn wide importance with advancements in organ transplantation. Immunosuppressive drugs are used against autoimmune disorders and to prevent graft rejection in transplants (Scheuplein et al., 2020). Cyclosporin was the first microbial metabolite to exhibit immunosuppressive property, which was isolated from a mold *T. nivenum*. Apart from cyclosporin, two other important immunosuppressive drugs where isolated from actinomycetes namely Tacrolimus and sirolimus.

These metabolites belong to polyketide macrolactoes synthesized by *S. tsukubaensis* and *S. hygroscopicus*, they both exhibited dual function of antifungal activity and also inhibits T-cell activation and proliferation involved in IL-2 and other cytokines synthesis.

1.7.3 Microbial novel antimicrobials

The search for novel antimicrobial compounds constantly increases with the battle of humanity with pathogenic microbes. Plant endophytic microbes are the preliminary source of a wide range of novel microbes. Ecomycins, xiamycins, pseudomycins, and munumbicins are the most novel endophyte-derived microbial bioactive compounds. *Streptomyces* sp. are the most novel group of microbes known to produce wide spectrum of antibiotics, which was proven to show significant activity against HIV (Arnau et al., 2016).

Extremophiles, the microbes exist in extreme habitats are the abundant source of novel microbial species. Pyochelin, an antifungal metabolite was isolated from the extracts of thermophilic *Pseudomonas*. Microcin was another proteinaceous toxin isolated from psychrophilic bacteria. Microcin was found to exhibit a broad spectrum of antibacterial activity against various pathogens (Giudice and Fani, 2016).

Marine actinobacteria are the known source of various bioactive compounds of economic importance. Marine cyanobacteria are known to produce small bioactive peptides called cyanobactins, which are proven to show wide potential antimalarial activity. Viridamide, a low molecular organic compound obtained from a blue green alga of *Oscillatoria* genes, showed antileishmanial and antitrypanosomal activity (Rule and Cheeptham, 2013) (Table 1.4).

1.7.4 Microbial enzyme inhibitors

Enzyme inhibitors also plays a crucial role in treatment of various metabolic disorders. Enzyme inhibitors are often administrated in combination therapy. Clavulanic acid and β-lactum obtained from *Streptomyces clavuligerus* was shown to exhibit inhibitory action of penicillinase with poor antibiotic activity. Penicillins was administrated as combination therapy for penicillin resistant bacterial infections. *Streptomyces pilosus* was screened to produce desferal an siderophore and it was effective against high metal binding activity in iron loaded diseases and aluminum overload in dialysis patients (Takizawa and Yamasaki, 2018). Acarbose, an inhibitor of α-glucosidase obtained from *Actinoplanes* sp., was used in the treatment of diabetes and hyperlipoproteinemia by inhibiting α-glucosidase.

TABLE 1.4 Screening assays for various pharmaceutical bioactive molecules.

Type of bioactive metabolite	Assay system	Mode of action
Immunosuppressant	Cell-free: inhibition of receptor-ligand binding Cell-dependent: inhibition of immune response	Inhibition of interleukin production in cell lines Suppression of lymphocyte rection in cell lines Immunoassay based on the cyclosporin binding to cyclophilin
Insecticide	In vivo in insects	Microbial solid cultures were administered orally as food to larvae
Antihyperglycemic	Enzyme inhibition, genetic screening	Inhibition of α-glucosidases Polymerase chain reaction based method to detect sedoheptulose 7- phosphate cyclase
Antitumor	Cell-based: cytotoxicity vs. rapidly proliferating cells or tumor cells	Cytotoxicity, growth inhibition Microbial prescreening
Antiholesterolemic	Enzyme inhibition	Inhibition of 3-hydroxy-3-methylglutaryl-coenzyme A reductase
Antiinflammatory	Inhibition of receptor-ligand binding	Immobilized-ligand IL-1 receptor binding assay
Antiparasitic	In vivo in animals	Activity in mice against nematode
Antiviral	Enzyme inhibition of viral enzymes	Inhibition of HIV-1 integrase

1.7.5 Microbial plant growth promotors

Genetically modified bacterium is the widely used microbial biopesticides and insecticides, various species of Bt are employed in real time farming for their different function as bioinsecticide in order to control various pests including beetle larvae, caterpillars, etc. *Trichoderma* sp. and *B. bassiana* are the most commonly available fungal biopesticides. Both biofungicides can be applied on both phylloplane and rhizosphere in order to control the fungal infection (Ancheeva et al., 2020).

Microbial bioactive compounds can also be employed as biopesticides and insecticides. The microbial-mediated biopesticides are considerable alternative for chemical pesticides. The two most important bacteria: *A. tumefaciens* and *B. thuringinesis*, insecticidal bacteria widely used in various crops. The microbes associated in rhizosphere and nodular region of root are known to induce plant growth by production of various growth promotors, solubilization of nutrients and facilitating efficient nitrogen fixation.

1.7.6 Omics approach in biosynthesis of microbial metabolites

Bacteria and actinomycetes are some of the most prolific sources of natural bioactive materials, with a wide variety of biologicals. Several popular bioactive molecules, widely used in medicine, are used to treat most of the infectious diseases (Palazzotto and Weber, 2018; Mohite et al., 2019). With development of advanced sequencing technologies coupled with efficient bioinformatic tools the wide range of untapped metabolic potential of microbial metabolites can be explored. The advancements in omics such as genomics, metabolomics, and transcriptomics enhances the biosynthetic potential of various gene clusters and discovering new natural bioactive compounds. Multiomics analysis enables the analysis of multitremendous evolution of both diversity and distribution of biosynthetic gene clusters. Most insight advancements in the omics technologies with combined use of biosynthetic gene clusters gives wide knowledge on the various microbial communities and their bioactive metabolites.

1.8 Conclusion

The microbial secondary metabolites with recent developments and technological advancements show a spectrum of promising applications driving the revenue and market for several industries while contributing to sustainable ecofriendly approaches. Further research on genetically engineering and target specific approaches will gain momentum attracting investors across the globe.

References

Ab Mutalib, N.S., Wong, S.H., Ser, H.L., Duangjai, A., Law, J.W.F., Ratnakomala, S., et al., 2020. Bioprospecting of microbes for valuable compounds to mankind. Prog. Microbes Mol. Biol. 3 (1).

Aden, K., Rehman, A., Waschina, S., Pan, W.H., Walker, A., Lucio, M., et al., 2019. Metabolic functions of gut microbes associate with efficacy of tumor necrosis factor antagonists in patients with inflammatory bowel diseases. Gastroenterology 157 (5), 1279–1292.

Ahmed, E.M., El-Refai, H.A., 2010. Cyclodextrin glucosyltransferase production by Bacillus megaterium NCR: evaluation and optimization of culture conditions using factorial design. Indian. J. Microbiol 50 (3), 303–308.

Ancheeva, E., Daletos, G., Proksch, P., 2020. Bioactive secondary metabolites from endophytic fungi. Curr. Med Chem. 27 (11), 1836–1854.

Arnau, G.V., Danilovich, M.E., Sánchez, L.A., Acosta, F., Delgado, O., 2016. Novel sources of antimicrobials from pristine and poorly explored environments. The patagonia microbiota case. Biology and Biotechnology of Patagonian Microorganisms. Springer, Cham, pp. 127–146.

Awad, N.E., Kassem, H.A., Hamed, M.A., El-Feky, A.M., Elnaggar, M.A.A., Mahmoud, K., et al., 2018. Isolation and characterization of the bioactive metabolites from the soil derived fungus *Trichoderma viride*. Mycology 9 (1), 70–80.

Ayyadurai, N., Ravindra Naik, P., Sakthivel, N., 2007. Functional characterization of antagonistic fluorescent pseudomonads associated with rhizospheric soil of rice (*Oryza sativa* L.). J. Microbiol. Biotechnol. 17, 919–927.

Azizoglu, U., 2019. Bacillus thuringiensis as a biofertilizer and biostimulator: a mini-review of the little-known plant growth-promoting properties of Bt. Curr. Microbiol 1–7.

Becker, J., Wittmann, C., 2012. Systems and synthetic metabolic engineering for amino acid production–the heartbeat of industrial strain development. Curr. Opin. Biotechnol. 23 (5), 718–726.

Benkendorff, K., 2009. Aquaculture and the production of pharmaceuticals and nutraceuticals. New Technologies in Aquaculture. Woodhead Publishing, pp. 866–891.

Bennett, J.W., Bentley, R., 1989. What's in a name?—microbial secondary metabolism, Advances in Applied Microbiology, vol. 34. Academic Press, pp. 1–28.

Bentley, R., 1997. Microbial secondary metabolites play important roles in medicine; prospects for discovery of new drugs. Perspect. Biol. Med. 40 (3), 364–394.

Berdy, J., 2005. Bioactive microbial metabolites. J. Antibiotics 58 (1), 1–26.

Bills, G.F., Gloer, J.B., 2017. Biologically active secondary metabolites from the fungi. Fungal Kingd. 1087–1119.

Bizani, D., Dominguez, A.P.M., Brandelli, A., 2005. Purification and partial chemical characterization of the antimicrobial peptide cerein 8A. Lett. Appl. Microbiol. 41, 269–273.

Boyer-Joubert, C., Lorthiois, E., Moreau, F., 2003. Annual reports in medicinal chemistry, vol. 38, Doherty, A.M., (Ed.).

Bragazzi, N.L., Martini, M., Saporita, T.C., Nucci, D., Gianfredi, V., Maddalo, F., et al., 2017. Nutraceutical and functional food regulations in the European Union. Developing new functional food and nutraceutical products. Academic Press, pp. 309–322.

Butler, M.S., 2008. Natural products to drugs: natural product-derived compounds in clinical trials. Nat. product. Rep. 25 (3), 475–516.

Cabello, L.O., 2020. Microorganisms as alternative sources of new natural products. Natural Products-From Bioactive Molecules to Human Health. IntechOpen.

Choudhary, K.K., Dhar, D.W., 2015. Microbes in Soil and their Agricultural Prospects. Nova Publishers.

Cohen, A.C., Bottini, R., Piccoli, P., 2015. Role of abscisic acid producing PGPR in sustainable agriculture. In: Bacterial Metabolites in Sustainable Agroecosystem, Springer, Cham, pp. 259–282.

Damodaran, T., Rai, R.B., Mishra, V.K., Sharma, D.K., Ram, R.A., Rai, S., 2011. Integrated farming system and livelihood security: an approach. Karnal, Central Soil Salinity Research Institute (CSSRI), India, pp. 1–108.

Demain, A.L., Lancini, G., 2001. Bacterial pharmaceutical products. The Prokaryotes: An Evolving Electronic Resource for the Microbiological Community. Springer-Verlag, New-York.

Demain, A.L., Zhang, L., 2005. Natural products and drug discovery. Natural Products. Humana Press, pp. 3–29.

Dhakal, D., Pokhrel, A.R., Shrestha, B., Sohng, J.K., 2017. Marine rare actinobacteria: isolation, characterization, and strategies for harnessing bioactive compounds. Front. Microbiol 8, 1106.

Egamberdieva, D., Jabborova, D., Hashem, A., 2015. Pseudomonas induces salinity tolerance in cotton (Gossypium hirsutum) and resistance to Fusarium root rot through the modulation of indole-3-acetic acid. Saudi journal of biological sciences 22 (6), 773–779.

Etesami, H., Adl, S.M., 2020. Plant growth-promoting rhizobacteria (PGPR) and their action mechanisms in availability of nutrients to plants. Phyto-Microbiome in Stress Regulation. Springer, Singapore, pp. 147–203.

Giudice, A.L., Fani, R., 2016. Antimicrobial potential of cold-adapted bacteria and fungi from Polar regions. Biotechnology of Extremophiles:. Springer, Cham, pp. 83–115.

Gonzalez, J.B., Fernandez, F.J., Tomasini, A., 2003. Microbial secondary metabolites production and strain improvement. Indian J. Biotechnol. 2, 322–333.

Gosal, S.K., Kaur, J., Kaur, J., 2017. Plant growth-promoting rhizobacteria: a probiotic for plant health and productivity. Probiotics and Plant Health. Springer, Singapore, pp. 589–600.

Hammerbacher, A., Coutinho, T.A., Gershenzon, J., 2019. Roles of plant volatiles in defence against microbial pathogens and microbial exploitation of volatiles. Plant, Cell Environ. 42 (10), 2827–2843.

Hassan, M.K., McInroy, J.A., Kloepper, J.W., 2019. The interactions of rhizodeposits with plant growth-promoting rhizobacteria in the rhizosphere: a review. Agriculture 9 (7), 142.

Horak, I., Engelbrecht, G., van Rensburg, P.J., Claassens, S., 2019. Microbial metabolomics: essential definitions and the importance of cultivation conditions for utilizing *Bacillus* species as bionematicides. J. Appl. Microbiol 127 (2), 326–343.

Jia-Xi, L., Chun-Xia, Z., Ying, H., Meng-Han, Z., Ya-Nan, W., Yue-Xin, Q., et al., 2019. Application of multiple chemical and biological approaches for quality assessment of *Carthamus tinctorius* L.(safflower) by determining both the primary and secondary metabolites. Phytomed 58, 152826.

Kaaniche, F., Hamed, A., Abdel-Razek, A.S., Wibberg, D., Abdissa, N., El Euch, I.Z., et al., 2019. Bioactive secondary metabolites from new endophytic fungus *Curvularia.* sp isolated from *Rauwolfia macrophylla.* PLoS One 14 (6), e0217627.

Kalinovskaya, N., Kuznetsova, T.A., Ivanova, E.P., Romanenko, L.A., Voinov, V.G., Huth, F., et al., 2002. Characterization of surfactin-like cyclic depsipeptides synthesized by *Bacillus pumilus* from ascidian *Halocynthia aurantium.* Marine Biotechnol 4, 179–189.

Kallscheuer, N., Classen, T., Drepper, T., Marienhagen, J., 2019. Production of plant metabolites with applications in the food industry using engineered microorganisms. Curr. Opin. Biotechnol. 56, 7–17.

Keller, N.P., 2019. Fungal secondary metabolism: regulation, function and drug discovery. Nat. Rev. Microbiol 17 (3), 167–180.

Kinghorn, A.D., Chin, Y.W., Swanson, S.M., 2009. Discovery of natural product anticancer agents from biodiverse organisms. Curr. Opin. Drug. Discov Dev. 12 (2), 189.

Kingston, D.G., 2009. Tubulin-interactive natural products as anticancer agents. J. Nat. Products 72 (3), 507–515.

Kingston, D.G., 2011. Modern natural products drug discovery and its relevance to biodiversity conservation. J. Nat. Products 74 (3), 496–511.

Kim, S., Shin, D.S., Lee, T., Oh, K.B., 2004. Periconicins, two new fusicoccane diterpenes produced by an endophytic fungus *Periconia* sp. with antibacterial activity. J. Nat. Prod. 67, 448–450.

Lancini, G., Lorenzetti, R., 1993. Biotechnology of Antibiotics and Other Bioactive Microbial Metabolites. Springer Science & Business Media.

Lemfack, M.C., Gohlke, B.O., Toguem, S.M.T., Preissner, S., Piechulla, B., Preissner, R., 2018. mVOC 2.0: a database of microbial volatiles. Nucleic Acids Res. 46 (D1), D1261–D1265.

Le Marrec, C., Hyronimus, B., Bressollier, P., Verneuil, B., Urdaci, M.C., 2000. Biochemical and genetic characterization of coagulin, a new antilisterial bacteriocin in the pediocin family of bacteriocins, produced by *Bacillus coagulans* I4. Appl. Environ. Microbiol. 66, 5213–5220.

Lewis, T.A., Cortese, M.S., Sebat, J.L., Green, T.L., Lee, C.H., Crawford, R.L., 2000. A *Pseudomonas stutzeri* gene cluster encoding biosynthesis of the CCl4-dechlorination agent pyridine-2,6-bis(thiocarboxylic acid). Environ. Microbiol. 2, 407–416.

Lisboa, M.P., Bonatto, D., Bizani, D., Henriques, J.A.P., Brandelli, A., 2006. Characterization of a bacteriocin-like substance produced by *Bacillus amyloliquefaciens* isolated from the Brazilian Atlantic forest. Int. J. Microbiol. 9, 111–118.

Liu, L., Xu, Y., Han, Z., Li, Y., Lu, L., Lai, P., et al., 2012. Four new antibacterial xanthones from the marine derived actinomycetes *Streptomyces caelestis*. Mar. Drugs. 10 (11), 2571–2583.

Liu, Y., Teng, K., Wang, T., Dong, E., Zhang, M., Tao, Y., et al., 2020. Antimicrobial *Bacillus velezensis* HC6: production of three kinds of lipopeptides and biocontrol potential in maize. J. Appl. Microbiol 128 (1), 242–254.

Maheshwari, D.K., Saraf, M., Dheeman, S., 2019. Plant growth-promoting rhizobacteria (PGPR) as protagonists of ever-sustained agriculture: an introduction. Field Crops: Sustainable Management by PGPR. Springer, Cham, pp. 1–10.

Mahmood, Z.A., 2014. Microbial amino acids production. Microb. Biotechnol Prog. Trends 187.

Malatesti, N., Munitic, I., Jurak, I., 2017. Porphyrin-based cationic amphiphilic photosensitisers as potential anticancer, antimicrobial and immunosuppressive agents. Biophys Rev. 9 (2), 149–168.

Marinelli, F., 2009. From microbial products to novel drugs that target a multitude of disease indications. Methods Enzymol 458, 29–58.

Martínez, L., Bastida, P., Castillo, J., Ros, G., Nieto, G., 2019. Green alternatives to synthetic antioxidants, antimicrobials, nitrates, and nitrites in clean label Spanish chorizo. Antioxidants 8 (6), 184.

Maskey, R.P., Helmke, E., Laatsch, H., 2003. Himalomycin A and B: isolation and structure elucidation of a new fridamycin type antibiotics from a marine *Streptomyces isolate*. J. Antibiot. 56 (11), 942–949.

Matsumoto, A., Takahashi, Y., 2017. Endophytic actinomycetes: promising source of novel bioactive compounds. J. Antibiotics 70 (5), 514–519.

Milshteyn, A., Colosimo, D.A., Brady, S.F., 2018. Accessing bioactive natural products from the human microbiome. Cell Host Microbe 23 (6), 725–736.

Misztal, P.K., Lymperopoulou, D.S., Adams, R.I., Scott, R.A., Lindow, S.E., Bruns, T., et al., 2018. Emission factors of microbial volatile organic compounds from environmental bacteria and fungi. Environ. Sci. Technol. 52 (15), 8272–8282.

Modolon, F., Barno, A.R., Villela, H.D., Peixoto, R.S., 2020. Ecological and biotechnological importance of secondary metabolites produced by coral-associated bacteria. J. Appl. Microbiol. 129 (6), 1441–1457. Available from: https://doi.org/10.1111/jam.14766.

Mohite, O.S., Weber, T., Kim, H.U., Lee, S.Y., 2019. Genome-scale metabolic reconstruction of actinomycetes for antibiotics production. Biotechnol. J. 14 (1), 1800377.

Motohashi, K., Sue, M., Furihata, K., Ito, S., Seto, H., 2008. Terpenoids produced by actinomycetes: napyradiomycins from *Streptomyces antimycoticus* NT17. J. Nat. Prod. 71 (4), 595–601.

Nalini, M.S., Prakash, H.S., Tejesvi, M.V., 2019. Bioactive potentials of novel molecules from the endophytes of medicinal plants. Medically Important Plant Biomes: Source of Secondary Metabolites. Springer, Singapore, pp. 293–351.

Newman, D., 2017. Screening and identification of novel biologically active natural compounds. F1000Research 6.

Nirmala, L., Zyju, D.P., 2017. Novel sources of antimicrobials. Bioresources and Bioprocess in Biotechnology. Springer, Singapore, pp. 327–349.

Palazzotto, E., Weber, T., 2018. Omics and multi-omics approaches to study the biosynthesis of secondary metabolites in microorganisms. Curr. Opin. Microbiol 45, 109–116.

Pan, R., Bai, X., Chen, J., Zhang, H., Wang, H., 2019. Exploring structural diversity of microbe secondary metabolites using OSMAC strategy: a literature review. Front. Microbiol 10, 294.

Park, S.R., Yoon, Y.J., Pham, J.V., Yilma, M.A., Feliz, A., Majid, M.T., et al., 2019. A review of the microbial production of bioactive natural products and biologics. Front. Microbiol 10, 1404.

Piechulla, B., Lemfack, M.C., Kai, M., 2017. Effects of discrete bioactive microbial volatiles on plants and fungi. Plant, Cell Environ. 40 (10), 2042–2067.

Poltronieri, P., Reca, I.B., 2020. Microbial products and secondary metabolites in plant health. Applied Plant Biotechnology for Improving Resistance to Biotic Stress. Academic Press, pp. 189–202.

Prakash, B., Singh, P.P., Kumar, A., Das, S., Chaudhari, A.K., 2019. Microbes as a novel source of secondary metabolite products of industrial significance. Role of Plant Growth Promoting Microorganisms in Sustainable Agriculture and Nanotechnology. Woodhead Publishing, pp. 21–37.

Raimundo, I., Silva, S.G., Costa, R., Keller-Costa, T., 2018. Bioactive secondary metabolites from octocoral-associated microbes—new chances for blue growth. Mar. Drugs 16 (12), 485.

Rana, K.L., Kour, D., Sheikh, I., Dhiman, A., Yadav, N., Yadav, A.N., et al., 2019. Endophytic fungi: biodiversity, ecological significance, and potential industrial applications. Recent Advancement in White Biotechnology through Fungi. Springer, Cham, pp. 1–62.

Ranghar, S., Agrawal, S., Agrawal, P.K., 2019. Microbial products: protein, enzyme, secondary metabolites and chemicals. Microbial Interventions in Agriculture and Environment. Springer, Singapore, pp. 347–384.

Reddy, S., Osborne, W.J., 2020a. Heavy metal determination and aquatic toxicity evaluation of textile dyes and effluents using Artemia salina. Biocatalysis and Agricultural Biotechnology 25, 101574.

Reddy, S., Osborne, J.W., 2020b. Biodegradation and biosorption of Reactive Red 120 dye by immobilized Pseudomonas guariconensis: kinetic and toxicity study. Water Environment Research 92 (8), 1230–1241.

Ruiz, B., Chávez, A., Forero, A., García-Huante, Y., Romero, A., Sánchez, M., et al., 2010. Production of microbial secondary metabolites: regulation by the carbon source. Crit. Rev. Microbiol 36 (2), 146–167.

Rule, D., Cheeptham, N., 2013. The effects of UV light on the antimicrobial activities of cave actinomycetes. Int. J. Speleol 42 (2), 7.

Salwan, R., Sharma, V., 2020. Molecular and biotechnological aspects of secondary metabolites in actinobacteria. Microbiol Res. 231, 126374.

Sanchez, S., Demain, A.L., 2017. Bioactive products from fungi. Food Bioactives. Springer, Cham, pp. 59–87.

Sankaran, L., Pogell, B.M., 1975. Biosynthesis of puromycin in Streptomyces alboniger: regulation and properties of O-demethylpuromycin O-methyltransferase. Antimicrobial agents and chemotherapy 8 (6), 721–732.

Scheuplein, N., Bzdyl, N., Kibble, E.A., Lohr, T., Holzgrabe, U., Sarkar-Tyson, M., 2020. Targeting protein folding; a novel approach for the treatment of pathogenic bacteria. J. Med Chem.

Schumacher, R.W., Talmage, S.C., Miller, S.A., Sarris, K.E., Davidson, B.S., Goldberg, A., 2003. Isolation and structure determination of an antimicrobial ester from a marine-derived bacterium. J. Nat. Prod. 66, 1291–1293.

Schwartsmann, G., da Rocha, A.B., Berlinck, R.G., Jimeno, J., 2001. Marine organisms as a source of new anticancer agents. Lancet Oncol. 2 (4), 221–225.

Selvakumar, V., Singh, S., Kannan, K., Annamalai, P., 2020. Pharmaceutically important metabolites from marine fungi. Bioactive Natural products in Drug Discovery. Springer, Singapore, pp. 411–425.

Shahzad, R., Waqas, M., Khan, A.L., Asaf, S., Khan, M.A., Kang, S.M., Yun, B.W., Lee, I.J., 2016. Seed-borne endophytic Bacillus amyloliquefaciens RWL-1 produces gibberellins and regulates endogenous phytohormones of Oryza sativa. Plant Physiology and Biochemistry 106, 236–243.

Sharma, D., Piplani, M., Sharma, A., Dhamija, M., Sharma, P.C., 2020. Application of microbes in pharmaceutical industries. Microbes Humankind Appl. 165.

Sheflin, A.M., Chiniquy, D., Yuan, C., Goren, E., Kumar, I., Braud, M., et al., 2019. Metabolomics of sorghum roots during nitrogen stress reveals compromised metabolic capacity for salicylic acid biosynthesis. Plant. direct 3 (3), e00122.

Sindhu, S.S., Sharma, R., 2020. Plant growth promoting rhizobacteria (PGPR): a sustainable approach for managing soil fertility and crop productivity. Microbes Humankind Appl. 97.

Singh, R., Kumar, M., Mittal, A., Mehta, P.K., 2017. Microbial metabolites in nutrition, healthcare and agriculture. 3 Biotech. 7 (1), 15.

Singh, B.P., Rateb, M.E., Rodriguez-Couto, S., Polizeli, M.D.L.T.D.M., Li, W.J., 2019a. Microbial secondary metabolites: Recent developments and technological challenges. Front. Microbiol 10, 914.

Singh, H.B., Keswani, C., Reddy, M.S., Sansinenea, E., García-Estrada, C. (Eds.), 2019b. Secondary Metabolites of Plant Growth Promoting Rhizomicroorganisms: Discovery and Applications. Springer.

Stone, M.J., Williams, D.H., 1992. On the evolution of functional secondary metabolites (natural products). Mol. Microbiol 6 (1), 29–34.

Sunazuka, T., Hirose, T., Omura, S., 2008. Efficient total synthesis of novel bioactive microbial metabolites. Acc. Chem. Res. 41 (2), 302–314.

Takizawa, N., Yamasaki, M., 2018. Current landscape and future prospects of antiviral drugs derived from microbial products. J. Antibiotics 71 (1), 45–52.

Tamehiro, N., Okamoto-Hosoya, Y., Okamoto, S., Ubukata, M., Hamada, M., Naganawa, H., et al., 2002. Bacilysocin, a novel phospholipid antibiotic produced by *Bacillus subtilis* 168. Antimicrob. Agents Chemother. 46, 315–320.

Tang, D.Y.Y., Khoo, K.S., Chew, K.W., Tao, Y., Ho, S.H., Show, P.L., 2020. Potential utilization of bioproducts from microalgae for the quality enhancement of natural products. Bioresour. Technol. 304, 122997.

Thirumurugan, D., Cholarajan, A., Raja, S.S., Vijayakumar, R., 2018. An introductory chapter: secondary metabolites. Second. Metab—Sources Appl. 1–21.

Ullah, A., Bano, A., Janjua, H.T., 2020. Microbial secondary metabolites and defense of plant stress. Microbial Services in Restoration Ecology. Elsevier, pp. 37–46.

Vallesi, A., Pucciarelli, S., Buonanno, F., Fontana, A., Mangiagalli, M., 2020. Bioactive molecules from protists: perspectives in biotechnology. Eur. J. Protistol 125720.

Vallianou, N., Stratigou, T., Christodoulatos, G.S., Dalamaga, M., 2019. Understanding the role of the gut microbiome and microbial metabolites in obesity and obesity-associated metabolic disorders: current evidence and perspectives. Curr. Obes. Rep. 8 (3), 317–332.

Verma, M.L., Kumar, S., Jeslin, J., Dubey, N.K., 2020. Microbial production of biopolymers with potential biotechnological applications. Biopolymer-Based Formulations. Elsevier, pp. 105–137.

Vijayakumar, R., Malathi, R., 2014. Isolation, characterization and antibacterial activity of actinobacteria from dye polluted soils of Tirupur. FU Med. Biol. 16, 43–48.

Vimal, S.R., Patel, V.K., Singh, J.S., 2019. Plant growth promoting Curtobacterium albidum strain SRV4: an agriculturally important microbe to alleviate salinity stress in paddy plants. Ecological Indicators 105, 553–562.

Vurukonda, S.S.K.P., Giovanardi, D., Stefani, E., 2018. Plant growth promoting and biocontrol activity of *Streptomyces* spp. as endophytes. Int. J. Mol. Sci. 19 (4), 952.

Weissman, K.J., 2009. Introduction to polyketide biosynthesis. Methods Enzymol 459, 3–16.

Win, H.Y., Laatsch, H., 2020. Macrolide antibiotics and indole alkaloid produced from marine and terrestrial bacteria. J. Myanmar Acad. Arts Sci. Vol. XVIII. No.1A.

Yadav, V.K., Khan, S.H., Malik, P., Thappa, A., Suriyaprabha, R., Ravi, R.K., et al., 2020. Microbial synthesis of nanoparticles and their applications for wastewater treatment. Microbial Biotechnology: Basic Research and Applications. Springer, Singapore, pp. 147–187.

Zaynab, M., Fatima, M., Abbas, S., Sharif, Y., Umair, M., Zafar, M.H., et al., 2018. Role of secondary metabolites in plant defense against pathogens. Microb. Pathogenesis 124, 198–202.

Zhou, H., Xie, X., Tang, Y., 2008. Engineering natural products using combinatorial biosynthesis and biocatalysis. Curr. Opin. Biotechnol. 19 (6), 590–596.

Chapter 2

Bacterial volatile organic compounds and gene-induced host-defense pathways

Ragıp Soner Silme[1] and Ömür Baysal[2]

[1]*Center of Biotechnology and Genetic Engineering Research and Practice, Istanbul University, Istanbul, Turkey,* [2]*Faculty of Science, Department of Molecular Biology and Genetics, Molecular Microbiology Unit, Muğla Sıtkı Koçman University, Muğla, Turkey*

Chapter Outline

2.1 Introduction

Microbial volatile organic compounds (mVOCs) play a key role in the antagonism and mutualism among competitive populations. Moreover, microbial volatiles are responsible for various intra- and interspecies cellular and developmental processes. Micro/macroorganisms produce enzymes, hormones, proteins, and volatile compounds that these compounds provide advantages for their surviving in nature besides significant roles related to metabolism, nutrition, establishment, and conservation. Volatiles are the compounds with high vapor pressure mainly classified into two different groups as organic and inorganic phase. Microbes emit volatile compounds that lead to forming of colonization on the plant roots and rhizospheres (Mendes et al., 2013). Scientists have found more than 1000 bacterial volatile organic compounds (VOCs; Lemfack et al., 2014, http://bioinformatics.charite.de/mvoc) playing a role in

Volatiles and Metabolites of Microbes. DOI: https://doi.org/10.1016/B978-0-12-824523-1.00002-X

communication signals with other microorganisms, which in turn influence interacting of microbial communities (Romoli et al., 2011). Bacterial VOCs such as 3-butanediol and acetoin have a negative effect on fungal pathogens and affect plant growth. Abiotic stress resistance occurrence of the bacteria, belonging to a different genus associated with production of volatile compounds and biofilm formation, has been reported (Wenke et al., 2012a). In general, bacterial volatiles have a positive effect on growth, differentiation, and stress resistance (Davis et al., 2013).

2.2 Types of bacterial volatile organic compounds and their biological role

Bacterial volatile compounds are organic molecules such as fatty acid derivatives (sulfur and nitrogen-containing compounds, and terpenes). The mVOCs, different chemical compounds, are generally related to catabolic pathways involving glycolysis, protein, and lipid degradation pathways (Penuelas et al., 2014). Bacterial volatiles possibly assist microbial communities as signals during inter- and intra-cell and/or cell-to-cell communication (Fig. 2.1). They have a role in growth-promoting or growth-inhibiting of the plants. Volatiles also play an important role in population dynamism on bacterial species for development of different communities living in specific rhizospheres (Fig. 2.1). The biological and ecological functions of mVOCs have been investigated more in detail on reactions and adaptations at the physiological, transcriptional, protein, and metabolic levels of the target organisms (Bailly and Weisskopf, 2012; Wenke et al., 2012b).

2.3 Bioconversion

Anaerobic and aerobic microbial communities produce CO_2 and water (Owen et al., 2007). The most part of anaerobic microorganisms produce formic or acetic acid and degrades VOCs (Krzycki and Zeikus, 1984; Guyot and Brauman, 1986). This metabolic pathway from production of methane belongs to Archaea. In soil, methanogenic bacteria degrade hydrocarbons, saturated/unsaturated fatty acids, and alcohols, and encompass an acetogenic partner that they are not able to grow alone on a certain organic compound (Stams and Plugge, 2009). In fact, VOCs are inaccessible to microorganisms since they stay in the gas phase. The degradation process occurring after their adsorption to humic acid or clay mineral surfaces (polar and apolar interactions) work as biofilters (Malhautier et al., 2005). These natural biofilters consist of various microorganism species and environmental conditions. Abiotic VOC degradation processes play an important role in soils (Wilson and Jones, 1996). Atkinson and Arey (2003) showed physical–chemical degradation of VOCs in the atmosphere by photolysis and spontaneous reactions with OH radicals and NO_3 radicals, O_3, and Cl atoms depending on light exposure.

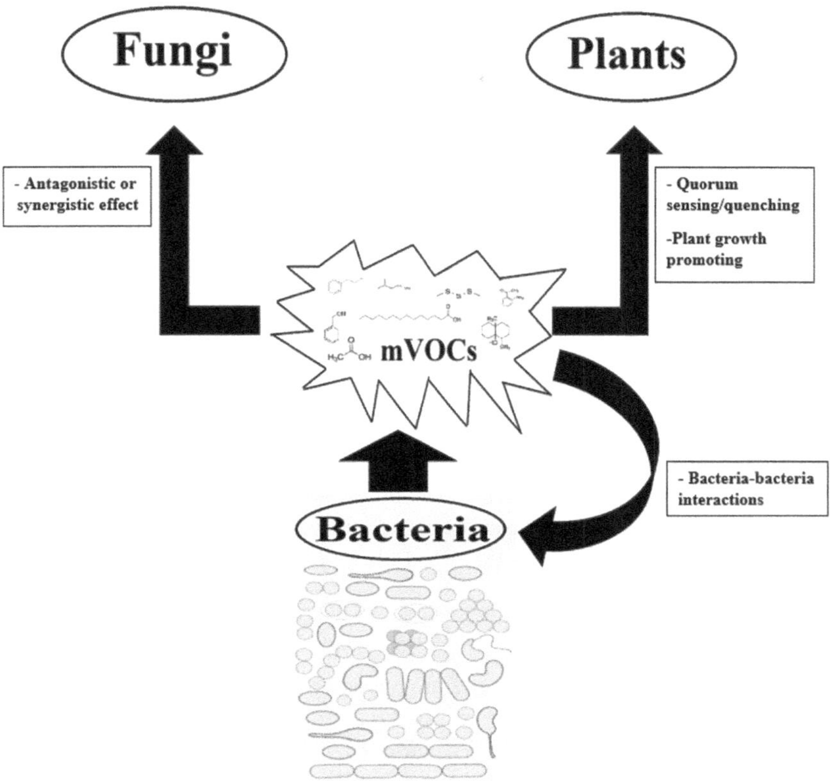

FIGURE 2.1 Overview of mVOC interactions.

They are the essential processes for VOC degradation by abiotic factors. Similarly, other powerful oxidants (for instance, hydrogen peroxide) have been formed by microorganisms by reacting with humic acids or phospholipids.

2.4 Biomineralization of elements

Biomineralization by prokaryotes has a major impact in biology, microbiology, and studies related to evolution. Sorption and precipitation properties of bacteria are useful for bioremediation of radionuclide and heavy metal-contaminated lands and waters (Lovley, 2013). The growth of iron- and manganese-oxidizing bacteria remove iron and manganese from wastewater that renders elimination to possible for the problems due to sticking minerals in pipelines (Mouchet, 1992). Biomineralization is a mineralization that occurs in biological environments involving organic matrix or soluble biomolecules, along with biological-induced local environments that helps for the crystallization of minerals and determines their morphologies and locations of nucleation (Mann, 2001). Bacterially formed minerals are useful as biomarkers in various situations. These biomarkers are useful not only in determining evolutionary processes of prokaryotes on Earth but also give information of the beginning of life in extraterrestrial materials. Biomolecules play important roles in controlling the kinetics of the biomineralization process (George and Veis, 2008; Suzuki et al., 2009). The capability of some microbial species is to adsorb some heavy metals on their surface (Kumar et al., 2009) or accumulate them within their structure (Atici et al., 2010; Saravanan et al., 2011). The studies on the formation of the insoluble gold-amine complexes attributed to the presence of some volatile biogenic metabolites in the biogas related to growth of the aerobic microorganism showed the efficiency of VOCs. The detoxification of heavy metals by microorganisms converts VOCs into other forms to reduce their toxicity in process of chelation, which occurs in metals inside or outside. Lefebvre et al. (2006) studying on cyanobacterial strains demonstrated to mineralize the metals ions (Hg^{2+}, Cd^{2+}, Pb^{2+}, and Cu^{2+}) using the VOCs found in the cultural biogas produced with the aerobic growth of bacteria. However, further detailed studies are required to identify the cyanobacterial biogases and to understand their role in the metal chelation process.

2.5 Quorum sensing/quenching

VOCs produced by rhizospheric strains belonging to *Pseudomonas fluorescens* and *Serratia plymuthica* inhibits communication. This case can be attributed as a kind of quorum-sensing (QS) mediated by *N*-acyl-homoserine lactone (AHL) among the gram-negative bacterial cells. These signal communication molecules are produced by various bacteria, including strains of *Agrobacterium*, *Chromobacterium*, *Pectobacterium*, and *Pseudomonas* (Chernin et al., 1998). Inhibition of this signaling, called the quorum-quenching effect, occurred when AHL-producing bacteria treated with VOCs emitted by these strains led to drastically decrease the amount of AHL production. Along with AHLs, bacterial volatiles can be considered as another type of signal molecule involved in microbial communication in the rhizosphere.

S. plymuthica suppresses disease symptoms caused by soil-borne pathogens and stimulates the growth of plants. Research findings about the underlying mechanisms and regulation are important for the application in terms of biocontrol strategies. The regulatory function of AHLs in the synthesis of the plant growth hormone indole-3-acetic acid has shown decreasing by AHL regulation in vitro assays. However, production of extracellular hydrolytic enzymes positively affected AHL regulation. A broad spectrum of VOCs involved in antifungal activity is influenced by QS. To improve the performance of any biocontrol agent additional studies on its regulatory mechanisms are necessary. Antagonistic activity occurs by antibiosis, competition for colonization, nutrients, minerals, parasitism, and lytic enzymes (Whipps, 2001). The interaction with the host plant and microbes plays an important role in the production of phytohormones, rhizosphere competence, enhancement of the availability of nutrients, and induction of systemic resistance (Baysal et al., 2013; Baysal and Tor, 2014). However, the studies have shown the production of antagonistic compounds by rhizobacteria regulated by AHL-dependent QS systems (Liu et al., 2007). Although several QS-regulated traits identified in bacterial genus belonging to *Serratia*, there is no study analyzing its role in biocontrol of plant pathogens, as reported for *Pseudomonas* (Wei and Zhang, 2006). Our recent unpublished data has clarified the genomic property of *Serratia* spp. involving gene clusters encoding proteins which helps for understanding the sequence profiles playing role in biocontrol. The VOCs produced by the rhizospheric strains *P. fluorescens* B-4117 and *S. plymuthica* IC1270 showed inhibition in cell–cell communication and QS network mediated by AHL signal molecules produced by various bacteria (Chernin et al., 2011). VOCs promoting QS quenching has a significant impact on the producer bacteria and neighboring bacteria in the rhizosphere, including plant pathogenic and beneficial strains in view of plant growth promoting and biocontrol effect (Chernin et al., 2013).

2.6 Communication signals and defense induction

Bacterial volatiles known for their ecological and biological functions have many beneficial properties. Although complete volatile spectra of bacteria and their relevant concentration are not exactly known, the studies on a model plant *Arabidopsis thaliana* developed normally in coculture with *Bacillus subtilis*, *Burkholderia cepacia*, *Staphylococcus epidermidis*, and *Escherichia coli* showed growth recession with *P. fluorescens*, *Pseudomonas trivialis*, *Serratia odorifera*, *S. plymuthica*, *Stenotrophomonas maltophilia*, and *Stenotrophomonas rhizophila* alone, but phenotypical changes has obtained by cocultivation with *S. plymuthica* and *S. maltophilia* resulted in significant effects on foliage parts of the plant and roots (Wenke et al., 2012a). Bacterial volatiles, which are the reason for biochemical alterations, give rise to changes at the cellular and physiological levels. Changing in gene expression due to pathogen response followed using synthetic plant promoter/GUS (β-glucuronidase) constructed regulatory elements (e.g., S-box; AGCCACC, GCC-box; AGCCGCC) has been followed for observing expression of the genes related to VOCs. Rushton et al. (2002) have demonstrated the existence of the promoter regions of defense genes (Ohme-Takagi and Shinsi, 1995). This region confers gene expression upon fungal elicitor effect and can be manipulated by recombinant DNA technology in vitro (Kirsch et al., 2001). To detect the gene activation postemission course for bacterial volatiles, the S-box and the GCC-box promoter/GUS constructed sequences using reporter gene led to a non-specific expression of the ethylene-inducible GCC-box with *S. plymuthica* on cocultivated seedlings. These studies showed that the volatiles of both bacteria have the capability to activate genes in plants via stress responsive promoters (Vespermann et al., 2007).

2.7 Plant growth promoting effect of bacterial volatiles in agriculture

The plant growth-promoting rhizobacteria support plant growth by mechanisms that cause (1) synthesis of plant hormones by bacteria present in rhizosphere, (2) increasing the content of soil minerals, (3) fixation of N_2, and (4) antibiotic production (Raaijmakers et al., 2002). Bacterial volatiles have positive growth effects for plant growth promoting, which affected the auxin metabolism, photosynthetic capacity, chloroplast intensity, chlorophyll content, and iron uptake; increased tolerance to abiotic stress, reduced severity of disease symptoms caused by plant pathogens; and increased resistance of the model plant *A. thaliana*. For instance, *B. subtilis* (GB03) emits 38 different VOCs (Farag et al., 2006), they have the potential to influence cellular or molecular processes. 2,3-butanediol production induces leaf growth and decreases of disease symptoms; however, 2,3-butanediol did not improve any photosynthetic efficiency. The bacterial (*B. subtilis*) VOCs mixture have shown inhibitory effects on plant growth (Farag et al., 2006). The antagonistic effect of co-inoculation of two different strains observed on plant trials has shown negatively regulation on the *A. thaliana* growth. We should consider that the emitted bacterial volatiles have inhibitory effects on growth of the plant with co-cultivation of *Bacillus amyloliquefaciens* (IN937a) and *B. subtilis* (GB03), though they have known with growth-promoting effects.

Microbiota with microenvironments is forming in the soil, which cause colonizing and create microecosystems. The interaction occurring within microbial communities by mutually exploiting, antagonism, and competition have given positive and/or negative impacts, which is useful if the pathogen is weakened and the strengthened one is the host plant. Therefore, the effect of biological agent is not only with its competitors' property related to antibiosis and antagonism. The dominant volatile compounds play a role in suppression of the pathogen with multilayered defense adaptations including morphological and physiological alterations by the activation of signaling pathways to withstand this environmental influence. For instance, microbial volatiles have the potential for a biocontrol agent in agriculture. In another study, *B. subtilis* strain colonizing on the root, and the rhizoplane is able to supresses cucumber wilt disease caused by *Fusarium oxysporum* f.sp. *cucumerinum* (Cao et al., 2011). The *B. subtilis* strain, which is effective to control of *Botrytis mali* and *Phytophthora sojae*, emits bacterial volatiles playing a role in biocontrol (Jamalizadeh et al., 2010). Many *Bacillus* species synthesizing plant growth-promoting VOCs such as 2,3-butandiol and acetoin emitted by *B. subtilis* GB013 and *B. amyloliquefaciens* IN937 triggered induced systemic resistance against *Pectobacterium carotovora* in *A. thaliana* (Ryu et al., 2004). In our previous study, 2,3-butanediol and acetoin production emitted by *B. amyloliquefaciens* (EU07) has correlated well with its suppressive effect on *F. oxysporum* f.sp. *radicis-lycopersici* growth

(Baysal et al., 2013). In the findings of the studies obtained in the last decades reported that mVOC (dimethyl disulfide) emitted by bacteria, registered as a novel soil fumigant, plays a key role in pathogen inhibition as an alternative to the usage of the methyl bromide for fumigation of soils infected by soil-borne fungal pathogens (Insam and Seewald, 2010).

Moreover, the role of VOCs produced by microorganisms (fungi and bacteria) in soil has also positive effect to control of the plant-parasitic nematodes. The evidence of VOCs effect has attracted the attention of the researchers. Noteworthy further studies are needed to understand the utility of the VOC knowledge and its positive effects, which suppresses the plant pathogens. Major VOC molecules produced by *B. subtillis* and *B. amyloliquefaciens* with fungicidal activity showed chemical structure similarity with acetamide, benzaldehyde, benzothiazole, 1-butamine, methanamine, phenylacetaldehyde and 1-decene due to their common ketone groups (Arrebola et al., 2010). Additionally, VOCs from bacterial isolates on the nematode antagonism tested in vitro for evaluating inhibitory effect on second stage juvenile (J2) immobility, mortality, and egg-hatching in infected plant have given positive results, which could be put into practice after expanded field trials. *Bacillus megaterium* VOCs have resulted in nearly 100% death of *Meloidogyne incognita* J2 and strong suppression on egg-hatching on tomato roots (Huang et al., 2010).

2.8 The role of bacterial VOCs in diary production

Microorganisms play an important role in the development of dairy product (cheese, yogurt, curd, etc.) flavor. Volatiles from lipolysis of the diary product by microbial transformation of lactose and citrate are emitted into these foods (Cheng, 2010). Most important VOCs are acetaldehyde, diacetyl, acetoin, acetone, and 2-butanone have positive effect on the extended shelf-life of the products cause of lipid oxidation (Penuelas et al., 2014). New strains isolated from dairy products tested for their production of odor-active volatile compounds to modulate product flavors are very important microorganisms in view of their characteristic property of them (Pogacic et al., 2015). mVOC profiles of typical fermenting bacteria (*Leuconostoc lactis*, *Lactobacillus* spp., *Brachybacterium* spp., *Brevibacterium* sp., and *Propionibacterium* sp.) determined by different analyses that 52 different mVOCs including ethyl esters, sulfur compounds, branched chain alcohols and acids, and the diacetyl with carbonyl compounds have been detected (Pogacic et al., 2015).

2.9 Outlook and future perspectives

We believe that mVOCs emitted are widespread and have a more complex structure than properties previously estimated. Emission profiles of bacteria under different growth conditions should be investigated more in detail, which will help for understanding biological and ecological effects of individual compounds as well as their mixtures to elucidate the communication ongoing as microbial–host interactions. Volatile syntheses in bacteria have a massive potential to provide sustainable and alternative methods, which may replace chemical pesticides and fertilizers in the plant cultivation and dairy products.

Scientists should find the potential microbial isolates and their VOCs for the control of different problems caused by plant pathogens and undesired contamination occurring in nature, considering environment protection and ecosystem balance. The further studies will be helpful for a better evaluation of the effects of VOCs from microorganism to effective strategies, aiming at benefits for human beings and environmental conservation.

References

Arrebola, E., Sivakumar, D., Korsten, L., 2010. Effect of volatile compounds produced by *Bacillus* strains on postharvest decay in citrus. Biol. Control: Theory Appl. Pest. Manag. 53 (1), 122–128. Available from: https://doi.org/10.1016/j.biocontrol.2009.11.010.

Atici, T., Obali, O., Altindag, A., Ahiska, S., Aydin, D., 2010. The accumulation of heavy metals (Cd, Pb, Hg, Cr) and their state in phytoplanktonic algae and zooplanktonic organisms in Beysehir Lake and Mogan Lake, Turkey. Afr. J. Biotechnol. 9 (4), 475–487.

Atkinson, R., Arey, J., 2003. Atmospheric degradation of volatile organic compounds. Chem. Rev. 103 (12), 4605–4638. Available from: https://doi.org/10.1021/cr0206420.

Bailly, A., Weisskopf, L., 2012. The modulating effect of bacterial volatiles on plant growth: current knowledge and future challenges. Plant Signal. Behav. 7 (1), 79–85. Available from: https://doi.org/10.4161/psb.7.1.18418.

Baysal, O., Tor, M., 2014. Smart biologics for crop protection in agricultural systems. Turk. J. Agric. For. 38 (5), 723–731. Available from: http://doi.org/10.3906/tar-1309-26.

Baysal, O., Lai, D., H.-H, X., Siragusa, M., Carimi, F., Silva, J.A., et al., 2013. A proteomic approach provides new insights into the control of soil-borne plant pathogens by *Bacillus* species. PLoS One 5 (1), 1–12. Available from: https://doi.org/10.1371/journal.pone.0053182.

Cao, Y., Zhang, Z., Ling, N., Yuan, Y., Zheng, X., Shen, B., et al., 2011. *Bacillus subtilis* SQR9 can control Fusarium wilt in cucumber by colonizing plant roots. Biol. Fertil. Soils 47, 495–506. Available from: https://doi.org/10.1007/s00374-011-0556-2.

Cheng, H., 2010. Volatile flavor compounds in yogurt: a review. Crit. Rev. Food Sci. Nutr. 50 (10), 938–950. Available from: https://doi.org/10.1080/10408390903044081.

Chernin, L.S., Winson, M.K., Thompson, J.M., Haran, S., Bycroft, B.W., Chet, I., et al., 1998. Chitinolytic activity in *Chromobacterium violaceum*: Substrate analysis and regulation by quorum sensing. J. Bacteriol. 180 (17), 4435–4441.

Chernin, L., Toklikishvili, N., Ovadis, M., Kim, S., Ben-Ari, J., Khmel, I., et al., 2011. Quorum-sensing quenching by rhizobacterial volatiles. Environ. Microbiol. Rep. 3 (6), 698–704. Available from: https://doi.org/10.1111/j.1758-2229.2011.00284.x.

Chernin, L., Toklikishvili, N., Ovadis, M., Khmel, I., 2013. Quorum-sensing quenching by volatile organic compounds emitted by rhizosphere bacteria. In: de Bruijn, F.J. (Ed.), Molecular Microbial Ecology of the Rhizosphere, first ed. John Wiley & Sons Ltd., pp. 791–800. Available from: http://doi.org/10.1002/9781118297674.ch75.

Davis, T.S., Crippen, T.L., Hofstetter, R.W., Tomberlin, J.K., 2013. Microbial volatile emissions as insect semiochemicals. J. Chem. Ecol. 39 (7), 840–859. Available from: https://doi.org/10.1007/s10886-013-0306-z.

Farag, M.A., Ryu, C.M., Summer, L.W., Pare, P.W., 2006. GC-MS SPME profiling of rhizobacterial volatiles reveals prospective inducers of growth promotion and induced systemic resistance in plants. Phytochemistry 67 (20), 2262–2268. Available from: https://doi.org/10.1016/j.phytochem.2006.07.021.

George, A., Veis, A., 2008. Phosphorylated proteins and control over apatite nucleation, crystal growth, and inhibition. Chem. Rev. 108 (11), 4670–4693. Available from: https://doi.org/10.1021/cr0782729.

Guyot, J.-P., Brauman, A., 1986. Methane production from formate by syntrophic association of *Methanobacterium bryantii* and *Desulfovibrio vulgaris*. Appl. Environ. Microbiol. 52 (6), 1436–1437. Available from: https://doi.org/10.1128/AEM.52.6.1436-1437.1986.

Huang, Y., Xu, C., Ma, L., Zhang, K., Duan, C., Mo, M., 2010. Characterization of volatiles produced from *Bacillus megaterium* YFM 3.25 and their nematicidal activity against *Meloidogyne incognita*. Eur. J. Plant Pathol. 126 (3), 417–422. Available from: https://doi.org/10.1007/s10658-009-9550-z.

Insam, H., Seewald, M.S.A., 2010. Volatile organic compounds (VOCs) in soils. Biol. Fertil. Soils 46, 199–213. Available from: https://doi.org/10.1007/s00374-010-0442-3.

Jamalizadeh, M., Etebarian, H.R., Aminian, H., Alizadeh, A., 2010. Biological control of *Botrytis mali* on apple fruit by use of *Bacillus* bacteria, isolated from the rhizosphere of wheat. Arch. Phytopathol. Plant Prot. 43 (18), 1836–1845. Available from: https://doi.org/10.1080/03235400902830960.

Kirsch, C., Logemann, E., Lippok, B., Schmelzer, E., Hahlbrock, K., 2001. A highly specific pathogen responsive promoter element from the immediate-early activated CMPG1 gene in *Petroselinum crispum*. Plant J. Cell Mol. Biol. 26 (2), 217–227. Available from: https://doi.org/10.1046/j.1365-313x.2001.01015.x.

Krzycki, J.A., Zeikus, J.G., 1984. Acetate catabolism by Me*th*anosarcin*a* bark*eri*: hydrogen-dependent methane production from acetate by a soluble cell protein fraction. FEMS Microbiol. Lett. 25 (1), 27–32. Available from: https://doi.org/10.1111/j.1574-6968.1984.tb01369.x.

Kumar, J.I.N., Oommen, C., Kumar, R.N., 2009. Biosorption of heavy metals from aqueous solution by green marine macroalgae from Okha port, Gulf of Kutch, India. American-Eurasian J. Agric. Environ. Sci. 6 (3), 317–323.

Lefebvre, D.D., Kelly, D., Budd, K., 2006. Biotransformation of Hg(II) by Cyanobacteria. Appl. Environ. Microbiol. 73 (1), 243–249. Available from: https://doi.org/10.1128/AEM.01794-06.

Lemfack, M.C., Nickel, J., Dunkel, M., Preissner, R., Piechulla, B., 2014. mVOC: a database of microbial volatiles. Nucleic Acids Res. 42, D744–D748. Available from: https://doi.org/10.1093/nar/gkt1250.

Liu, H., He, Y., Jiang, H., Peng, H., Huang, X., Zhang, X., et al., 2007. Characterization of a phenazine-producing strain *Pseudomonas chlororaphis* GP72 with broad-spectrum antifungal activity from green pepper rhizosphere. Curr. Microbiol. 54, 302–306. Available from: https://doi.org/10.1007/s00284-006-0444-4.

Lovley, D., 2013. Dissimilatory Fe(III)- and Mn(IV)-reducing prokaryotes. In: Rosenberg, E., DeLong, E.F., Lory, S., Stackebrandt, E., Thompson, F. (Eds.), The Prokaryotes. Springer, Berlin, Heidelberg, pp. 635–658. Available from: http://doi.org/10.1007/978-3-642-30141-4_69.

Malhautier, L., Khammar, N., Bayle, S., Fanlo, J.L., 2005. Biofiltration of volatile organic compounds. Appl. Microbiol. Biotechnol. 68 (1), 16–22. Available from: https://doi.org/10.1007/s00253-005-1960-z.

Mann, S., 2001. Biomineralization: Principles and Concepts in Bioinorganic Materials Chemistry. Oxford University Press.

Mendes, R., Garbeva, P., Raaijmakers, J.M., 2013. The rhizosphere microbiome: significance of plant beneficial, plant pathogenic, and human pathogenic microorganisms. FEMS Microbiol. Rev. 37 (5), 634–663. Available from: https://doi.org/10.1111/1574-6976.12028.

Mouchet, P., 1992. From conventional to biological removal of iron and manganese in France. J. Am. Water Work. Assoc. 84 (4), 158–167. Available from: https://doi.org/10.1002/j.1551-8833.1992.tb07342.x.

Ohme-Takagi, M., Shinsi, H., 1995. Ethylene-inducible DNA binding proteins that interact with an ethylene-responsive element. Plant Cell 7 (2), 173–182. Available from: https://doi.org/10.1105/tpc.7.2.173.

Owen, S., Clark, S., Pompe, M., Semple, K.T., 2007. Biogenic volatile organic compounds as potential carbon sources for microbial communities in soil form the rhizosphere of *Populus tremula*. FEMS Microbiol. Lett. 268 (1), 34–39. Available from: https://doi.org/10.1111/j.1574-6968.2006.00602.x.

Penuelas, J., Asensio, D., Tholl, D., Wenke, K., Rosenkranz, M., Piechulla, B., et al., 2014. Biogenic volatile emissions from the soil. Plant Cell Environ. 37 (8), 1866–1891. Available from: https://doi.org/10.1111/pce.12340.

Pogacic, T., Maillard, M.B., Leclerc, A., Herve, C., Chuat, V., Yee, A.L., et al., 2015. A methodological approach to screen diverse cheese-related bacteria for their ability to produce aroma compounds. Food Microbiol. 46, 145–153. Available from: https://doi.org/10.1016/j.fm.2014.07.018.

Raaijmakers, J.M., Vlami, M., de Sou, J.T., 2002. Antibiotic production by bacterial biocontrol agents. Antonie Van Leeuwenhoek 81, 537–547. Available from: https://doi.org/10.1023/A:1020501420831.

Romoli, R., Papaleo, M.C., de Pascale, D., Tutino, M.L., Michaud, L., LoGiudice, A., et al., 2011. Characterization of the volatile profile of Antarctic bacteria by using solid-phase microextraction-gas chromatography-mass spectrometry. J. Mass. Spectrom. 46, 1051–1059. Available from: https://doi.org/10.1002/jms.1987.

Rushton, P.J., Reinstädler, A., Lipka, V., Lippok, B., Somssich, I.E., 2002. Synthetic plant promoters containing defined regulatory elements provide novel insights into pathogen- and wound induced signaling. Plant Cell 14 (4), 749–762. Available from: https://doi.org/10.1105/tpc.010412.

Ryu, C.M., Farag, M.A., Hu, C.H., Reddy, M.S., Kloepper, J.W., Paré, P.W., 2004. Bacterial volatiles induce systemic resistance in *Arabidopsis*. Plant Physiol. 134 (3), 1017–1026. Available from: https://doi.org/10.1104/pp.103.026583.

Saravanan, A., Brindha, V., Krishnan, S., 2011. Characteristic study of the marine algae *Sargassum* sp. on metal adsorption. Am. J. Appl. Sci. 8 (7), 691–694. Available from: https://doi.org/10.3844/ajassp.2011.691.694.

Stams, A.J.M., Plugge, C.M., 2009. Electron transfer in syntrophic communities of anaerobic bacteria and archaea. Nat. Rev. Microbiol. 7 (8), 568–577. Available from: https://doi.org/10.1038/nrmicro2166.

Suzuki, M., Saruwatari, K., Kogure, T., Yamamoto, Y., Nishimura, T., Kato, T., et al., 2009. An acidic matrix protein, Pif, is a key macromolecule for nacre formation. Science 325 (5946), 1388–1390. Available from: https://doi.org/10.1126/science.1173793.

Vespermann, A., Kai, M., Piechulla, B., 2007. Rhizobacterial volatiles affect the growth of fungi and *Arabidopsis thaliana*. Appl. Environ. Microbiol. 73 (17), 5639–5641. Available from: https://doi.org/10.1128/AEM.01078-07.

Wei, H.-L., Zhang, L.-Q., 2006. Quorum-sensing system influences root colonization and biological control ability in *Pseudomonas fluorescens* 2P24. Antonie Van Leeuwenhoek 89 (2), 267–280. Available from: https://doi.org/10.1007/s10482-005-9028-8.

Wenke, K., Weise, T., Warnke, R., Valverde, C., Wanke, D., Kai, M., et al., 2012a. Bacterial volatiles mediating information between bacteria and plants. In: first ed. Witzany, G., Baluska, F. (Eds.), Biocommunication of Plants. Springer Signaling and Communication in Plants, vol. 14. Springer, Berlin, Heidelberg, pp. 327–347. Available from: https://doi.org/10.1007/978-3-642-23524-5_17.

Wenke, K., Wanke, D., Kilian, J., Berendzen, K., Harter, K., Piechulla, B., 2012b. Volatiles of two growth-inhibiting rhizobacteria commonly engage AtWRKY18 function. Plant J. 70 (3), 445–459. Available from: https://doi.org/10.1111/j.1365-313X.2011.04891.x.

Whipps, J., 2001. Microbial interactions and biocontrol in the rhizosphere. J. Exp. Bot. 52 (Suppl. 1), 487–511. Available from: https://doi.org/10.1093/jexbot/52.suppl_1.487.

Wilson, S., Jones, K.C., 1996. The fate and behavior of volatile aromatic hydrocarbons in sewage sludge-amended soil. In: Wang, W., Schnoor, J., Doi, J. (Eds.), Volatile Organic Compounds in the Environment. ASTM International, West Conshohocken, PA, pp. 119–123. 978-0-8031-5323-3.

Chapter 3

Microbial volatiles: small molecules with an important role in intra- and interbacterial genus interactions-quorum sensing

Jeevanandam Vaishnavi and W. Jabez Osborne

Biomolecules Lab, Department of Bio-Sciences, School of Bio Sciences and Technology, Vellore Institute of Technology, Vellore, India

Chapter Outline

3.1 Introduction

Secondary metabolites are regarded as the best example of microbial interactions. The microbial volatile organic compounds (MVOCs) are also the secondary metabolites, among which many are undiscovered. Frequently reported MVOC are described in Table 3.1. MVOC are widely characterized as odorless, lipophilic compounds with molecular weight less than 300 kDa (Effmert et al., 2012). MVOCs are produced during the oxidation of glucose

Volatiles and Metabolites of Microbes. DOI: https://doi.org/10.1016/B978-0-12-824523-1.00005-5

TABLE 3.1 Frequently reported microbial volatile organic compounds in the environment.

S. no	Microbial volatile organic compounds	References
1	3-Methyl 2-Butanol	Ström et al. (1994)
2	2-Methylisoborneol	Smedje et al. (1996)
3	2-Isopropyl-3-methoxy-pyrazine	
4	Geosmin	
5	3-Octanol	Carlson and Quraishi (1999)
6	2-Octen-1-ol	Morey et al. (2000)
7	3-Methyl-1-butanol	Mehrer and Lorenz (2005)
8	2-Pentanol	Wieslander et al. (2007)
9	1-Octen-3-ol	
10	3-Methylfuran	
11	2-Hexanone	
12	2-Heptanone	
13	3-Octanone	
14	Dimethyl disulphide	

using numerous substrates (Berry, 1988). The synthesis of MVOC is affected by pH, growth condition, and temperature. The bacteria which produce volatile organic compounds are called bacterial volatile organic compound (BVOC) as primary metabolites (acetic acid, acetone, ethanol, etc.) and secondary metabolites as signaling molecules produced in different combination based on the inter and intraspecies communication between bacteria (Zhu et al., 2010). As per literature, there are more than 200 compounds regarded as BVOC these compounds are not solely produced by bacterial metabolism but it also has other environmental metabolism. It can also be used as an indicator for the contaminants in the food processing industries by the unpleasant odor that indicates the presence of microbes during the storage of food (Wilkins and Scholl, 1989). From the food industry, it was diverted to microbial research to identify the separate strains of microbes (Karlshøj and Larsen, 2005).

Every microorganism has a signaling pathway to sense the environment and coordinates to give the proper response. The sensing occurs by recognizing the specific molecules, which play a key role in this mechanism. These molecules are produced when the microbes interact with close proximity, and they are called quorum sensing (QS). These microbes manipulate the signaling pathway by screening other specific molecules (Bednarek et al., 2010). Plant-associated microbes also secrete specific metabolites which affect the plant growth indirectly. These molecules scavenge the nutrients available in the environment for the benefit of microorganisms, and plants there are called siderophores. Pathogenic bacteria also produce BVOC, but these compounds have a virulence factor that affects the plant growth by attenuating the defense mechanism of the plants, which helps in the proliferation of virulence factor (Martin and Kamoun, 2012).

Nutritional imbalances trigger the production of BVOC and QS molecules (Korpi, 2001). The other research also states that these BVOC are the inhibitors of primary metabolites (Sunesson, 1995). The BVOC is converted into various products by environmental factors, for example, alcohol is converted to aldehyde and carboxylic acids (Wilkins et al., 1997). Diverse chemicals in the environment also lead to the production of BVOC, and looking beyond, this suggests that BVOC also has other resources by human activities such as smoking, trafficking, pollution, etc. (Helmig et al., 1999). The chapter contends with the bacterial release of BVOC and QS molecules as a boon for agriculture, which helps as an alternate to synthetic fertilizers and stimulants. It also describes the human infection with QS molecules.

3.2 Biosynthesis of bacterial volatile organic compounds

Bacteria has wide ability in producing diverse range of volatile organic compounds, which are organic and inorganic in nature. Organic BVOC are usually fatty acids derivatives such as ketones, terpenes, alcohols, acids and hydrocarbons.

3.2.1 Organic volatile compounds

3.2.1.1 Ketones

Ketones are derived from decarboxylation of fatty acids. According to Ryu et al. (2003), acetoin and 2,3, butanedione is derived during pyruvate formation in the anaerobic condition. The enzyme acetolactate synthase catalyzes two pyruvate molecules into acetolactate which in turn decarboxylated to acetoin and the oxidation leads to the formation of 2,3, butanedione.

3.2.1.2 Terpenes

The organic BVOC terpenes are produced either by mevalonate or deoxyxylulose phosphate pathway. According to Schulz and Dickschat (2007) monoterpenes and sesquiterpenes derivatives are the end products produced by the bacteria. Albaflavenone, the organic BVOC of terpenes group is only found in *Streptomyces* and was first isolated in *Streptomyces albidoflavus*.

3.2.1.3 Alcohols

Aliphatic alcohols with the long chain are synthesized by Enterobacteriaceae through α/β oxidation of fatty acid derivatives (Hamilton-Kemp et al., 2005) where short chain aliphatic alcohols are produced in low oxygen condition by proteobacteria and firmicutes (Farag et al., 2013). According to Marilley and Casey (2004), branched short chain alcohols like 2-methyl-1-butanol and 3-methyl-1-butanol are produced by the microbes via enzymatic conversion in the Ehrlich pathway.

3.2.1.4 Acids

According to Schulz and Dickschat (2007), when compared to ketones and alcohols, organic acids are less abundantly present in bacteria. However, short chain fatty acids such as propionic acid and acetic acid have been produced by bacteria. Anaerobic fermentation of carbohydrates leads to the production of various aliphatic organic acids as byproducts.

3.2.1.5 Hydrocarbons

Hydrocarbons with linear chains are most likely derived from fatty acid metabolism by two main pathway elongations: decarboxylation and head to head condensation, whereas short chain hydrocarbons such as decane to tetradecane are found in microorganism and long chain hydrocarbon hexadecane are abundant among cyanobacteria and also have the ability to produce branched hydrocarbons (Ladygina et al., 2006).

3.2.2 Inorganic volatile compounds

Bacteria also produce inorganic volatile compounds such as ammonia, nitric oxide, hydrogen sulfide, etc. Ammonia is produced by the metabolism of peptides and amino acids (Bernier et al., 2011). Nitric oxide is produced by nitric oxide synthase from L-arginine, which is homologous to mammalian enzymes (Mattila and Thomas, 2014). Hydrogen sulfide gas is produced when bacteria degrade cysteine, an orthologous of mammalian enzymes, and it is produced at the end of the exponential phase at low oxygen concentration. Hydrogen cyanide has been emitted by bacteria such as *Chromobacterium*, *Pseudomonas*, and *Rhizobium*. And it is catalyzed by hydrogen cyanide synthase and encoded by hcnABC genes, as a result it forms HCN and CO_2 from glycine. Cyanogenesis regulation through QS regulators which are appeared to be strain specific to *Pseudomonas aeruginosa* (PA01) and *Chromobacterium violaceum* (CV0) (Blom et al., 2011)

3.3 Bacterial volatile organic compounds in biofilm formation

The volatile compound secretion by bacteria gets influenced by the biofilm formation with different stages of biofilm, the type of bacteria involved, and their motility. Létoffé et al. (2014) reported 12 nontoxic volatile compounds were tested against *Escherichia coli* and *Pseudomonas aeruginosa*. The exposure to 1-butanol decreased the motility of *E. coli* and *P. aeruginosa*. Where the motility gets increased when both of the organism gets exposed to 2-butanone or acetoin. *Paenibacillus polymyxa, Burkholderia glumae, Pseudomonas aeruginosa* gets affected by the exposure of volatile compounds secreted by *Bacillus subtilis* (Kim et al., 2013). The swarming of *E. coli* is decreased when glyoxylic acid is aerially exposed.

According to Nijland and Burgess (2010) and Létoffé et al. (2014), BVOC influence the bacteria to form biofilm. Ammonia induces *Staphylococcus aureus, licheniform*, and *B. subtilis* to form biofilm. Indole is also said to induce biofilm formation of *P. aeruginosa, Vibrio cholerae*, and *P. fluorescence* but, has a negative effect on *E. coli*. When volatile signal was used which inhibits *P. aeruginosa* and *E. coli* to form biofilm formation and stimulates biofilm in *S. aureus*. Other BVOC such as ethanol, hexadecane, acetoin, and ammonia has more positive and negative influence in biofilm formation when tested, bacterial species like *S. aureus, B. subtilis, P. aeruginosa*, and *E. coli*. Barraud et al. (2009) reported that nitrogen oxide induces the dispersal of bacterial biofilm formation. The sublethal dose of antibiotics has positive and negative effects on biofilm formation with respect to several bacterial species. The exposure of BVOC affects the antibiotic resistant profile which could indirectly affect biofilm formation and affects the ability of the antibiotic (Bernier et al., 2011).

3.4 The role of bacterial volatile organic compounds in agriculture

Food crops have been adversely affected due to pathogens. Synthetic fertilizers aid in crop protection against the pests but it has an adverse effect on human health. To prevent this, researchers are continuously working on biofertilizers and biopesticides as an alternative to synthetic fertilizers. The exposure of BVOC to plants has a significant impact on modifying the metabolic and physiological changes which help in the confirmation that plants recognize and react to BVOC. Before a decade plant-BVOC interactions are laboratorial scale but now the field trials have been carried out (Song and Ryu, 2013; Cortes-Barco et al., 2010). BVOC plays a crucial role in promoting plant growth with no contact between plant and bacteria. Sharifi and Ryu (2018) stated that *B. subtilis* GB03 and *Bacillus amyloliquifaciens* IN937 have proved to stimulate plant growth by producing BVOC. The leaf surface got increased in ethylene insensitive (etr1), gibberellic acid insensitive (gia2), auxin transporter deficient, and ethylene insensitive and brassienosteroid insensitive (cbbi). These prove the necessities of ethylene, brassienosteroid, and gibberellic acid are not required for the plant growth when BVOC is used. *B. subtilis* GB03 proved that seeds interact with GB03 before germination used root exudates during germination and further multiplies a healthy progeny through plant–microbe interaction (Kloepper et al., 2004).

BVOC also helps in inducing systemic resistance toward abiotic and biotic stress including salinity, drought, and phytopathogens by PGPR (Yang et al., 2009). Under high salinity, crops undergo osmotic stress, which is mediated BVOC that helps in reducing the uptake of sodium in roots and increases the release of sodium ions in shoots. While the plants face drought, there is an increase in accumulation of osmoprotectants which in turn increases osmotic pressure to reduce the free water transport to cells. BVOC such as acetic acid induces biofilm formation when it contains a high amount of exopolysaccharide, which indirectly provides moisture and prevents drought (Chen et al., 2015). In order to compete, the biotic stress from BVOC helps in protecting the plant against phytopathogens. Globally, 13% of total crop loss occurs due to phytopathogens. BVOC first helps in the competition, where the rhizobacteria moved toward the root exudates after the first chemotactic movement and compete pathogen by acquiring the nutrients and specific niche and thereby reducing the pathogen population. BVOC next involves in antibiosis where the rhizobacteria produce antibacterial agents to inhibit the pathogen and BVOC finally involves increasing the plant immunization by colonizing the rhizobacteria which act as a defense system and responds against the pathogens (Panpatte et al., 2017).

3.5 Inter- and intraspecies communication between bacterial species

Inter- and intraspecies communication between the bacteria comes under the mechanism called QS. It is a process in which bacteria monitor cell density by measuring the signaling molecules also known as auto inducers (AIs). The individual bacteria secrete AIs into the extracellular environment and their concentration is measured and correlated with cell density. Target gene expression can be altered by QS when Gram-positive bacteria gets involved in oligopeptide based two component type of QS. Whereas Gram-negative bacteria get involved luxI/luxR type of QS. luxI is responsible in producing acyl homoserine lactone (AHL). Each species of Gram-negative bacteria produces unique AHL or the combination of AHL and only the same species recognizes and responds to it. The detection of AHL and the alteration in gene expression is mediated by luxR (Engebrecht et al., 1983). The autoinducer moves in and out of the cell spontaneously by luxI mediated AHL (Kaplan and Greenberg, 1985). Therefore, the external AHL secretion is equal to internal AHL secretion and it is increased by cell density (Fuqua

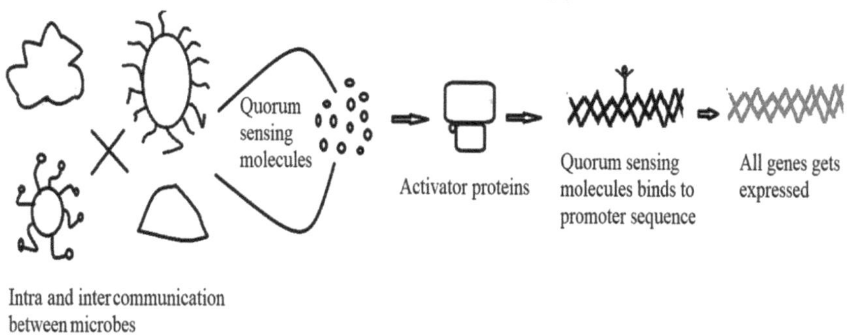

FIGURE 3.1 Intra and interspecies communication between microbes.

et al., 1994). After it reaches the threshold, the autoinducer gets bound to luxR protein.

Intraspecies communication is more specific in the interaction between signal and receptors. When the multiple signals are produced by the same bacteria it regulates with overall receptors and maximize the response as mentioned in Fig. 3.1 (Jayaraman and Wood, 2008). Autoinducer-mediated QS is the highest degree of affinity. The bacteria should categorize between the signals produced by inter- and intraspecies. It is an important criteria in the case of pathogenic bacteria where the virulence genes interaction with infection and the other signals, which impact on eliminating the infection. Boedicker et al. (2009) reported that *S. aureus* or *P. aeruginosa* has a characteristic of producing QS molecules when they are confined to a constructed place, which results in producing QS molecules as if they are in high cell density.

3.6 Bacterial quorum sensing in human infection

Infectious disease is usually caused by biofilm forming microbes. These bacteria have the capability to resist antibiotics. QS bacteria sense the neighboring cell density by releasing QS molecule, which in turn induces the expression of virulence factors. The pathogenic bacteria and the eukaryotic host interaction is the important outcome in human infection, and it is mediated by QS. The pathogens communicate the information to the high cell density through biofilm formation and simultaneously induces the virulence gene and develops resistance. The QS system allows the bacteria to interact as a group with high cell density to invade the immune system and establish the infection which is impossible for the individual cells.

P. aeruginosa strain is an important model organism in the QS system because of its highlighted presence in clinical strains (Castillo-Juárez et al., 2015). It communicates with the eukaryotic cells with QS signaling molecules AHL. *P. aeruginosa* has three QS systems, among these two are luxI/luxR and the third system is *Pseudomonas quinolone* signals (PQS) mediated by quinolones (2-heptyl-3-hydroxy-4-quinolone) produced by anthranilate. The luxI/luxR is AHL dependent. lasI is the homolog of luxI, which produces N-3-OXO-dodecanoyl-L-homoserine lactone (30-C_{12}-HSL), which is identified by lasR homolog of luxI (cytoplasmic receptor) (Parsek et al., 1999). RhlI the other homolog of luxI synthesis another AHL N-butyrl homoserine lactone (C4-HSL), which binds RhlR (cytoplasmic receptor) (Pearson et al., 1997). Both lasR and RhlR are transcriptional regulators. AHL-luxR together control the activation of >3000 genes of *P. aeruginosa.* maximum of these genes codes for virulence factors which promote infection (Toder et al., 1991; Gambello et al., 1993). The HSL was termed after the virulence gene which were found to be under their control. LAS upregulates the production of elastase by inducing the expression of elastase gene (lasB) (Gambello and Iglewski, 1991). The PQS system controls the formation of biofilm (De Kievit, 2009). The intracellular orphan receptor in bacteria which is a homology of luxR binds to 30-C12-HSL (Oinuma and Greenberg, 2011). These innate targets become dimer with LasR and RhlR which makes them inactive and further repress LasRI and RhlRI depending on genes which in turn prevents QS response before it reaches the high cell density community (Holm and Vikström, 2014; Ledgham et al., 2003). Table 3.2 represents the QS molecules produced by bacteria.

S. aureus has several virulence factors, and these virulence factors are regulated by QS. The QS regulation in Gram-positive bacteria is mediated by autoinducing peptides (AIP). In *S. aureus* QS controls the expression of virulence factors including leukocidins, exoenzymes, hemolysins, adhesion to the cell surface, and biofilm formation through AIP (Castillo-Juárez et al., 2015). The accessory gene regulator (agr) mediated QS upregulates the expression of virulence factors and decreases the cell surface proteins in exponential to stationary phase (Novick, 2003). The expression of agr contributes pathogenicity in *S. aureus*. Biofilm formation and QS in *S. aureus* suggest that it is important in the development and establishment of chronic infection (Yarwood and Schlievert, 2003).

According to Høiby et al., (2010) there are numerous reports on the resistance of antibiotics by biofilm formation, which ultimately causes difficulties in treating infection. These effects have been observed in *P. aeruginosa* (Strateva and Yordanov, 2009), *S. aureus* (Savage et al., 2013), and *Klebsiella pneumoniae* (Vuotto et al., 2014). The biofilm helps the bacteria to develop different kinds of defense mechanisms like physical barriers and modification in gene expression (De la Fuente-Núñez et al., 2013). He et al. (2015) reported the clinical isolate *Acinetobacter baumanii* was increasing the antibody resistance due to levofloxacin and meropenum antibiotic that triggers the release of AHL, which in turn stimulates QS-mediated biofilm and as a result, it increases the resistance. QS inhibition makes a potent way by increasing antibiotic sensitivity and decreasing the antibiotic dosage.

TABLE 3.2 Quorum sensing molecules produced by bacteria.

Quorum sensing molecules	Bacterial sp.	References
C_4-HSL	*P. aeruginosa*	Veliz-Vallejos et al. (2014)
3-Oxo-C_{12}-HSL		
C_6-HSL	*Sinorhizobium meliloti*	
C_8-HSL		
C_{10}-HSL		
C_{12}-HSL		
C_{14}-HSL		
C_{16}-HSL		
C_{16}-:1-9 cis-(L)-HSL		
C_{18}-HSL		
3-Oxo-C_8-HSL		
3-Oxo-C_{14}-HSL		
C_{14}-:1-9 cis-(L)-HSL	*Agrobacterium vitis*	
3-Oxo-C_{16}:1-11cis-(L)-HSL		
BHL & HHLg	*Aeromonas hydrophila* *Aeromonas salmonicida*	Defoirdt et al. (2010)
HHLg	*Burkholderia vietnamiensis*	
AI-2	*Erwinia amylovora* *Fusobacterium nucleatum* *Porphyromonas gingivalis* *Prevotella intermedia* *Vibrio vulnificus*	
OH-OHLg	*Photobacterium phosphoreum*	
AI-2, OH-BHLg, CAI-1	*Vibrio campbellii*	
CAI-1, AI-2, OH-BHLg	*Vibrio harveyi*	
C_8-HSL, C_{10}-HSL, OHC_8-HSL, OHC_{10}-HSL, $3OC_{14}$-HSL, C_8-HSL, C_{10}-HSL, $3OHC_8$-HSL, $3OHC_{10}$-HSL	*Burkholderia pseudomallei*	Kalia et al. (2014)
OHC_6-HSL, $3OC_6$-HSL, C_7-HSL, $3OHC_{14}$-HSL	*Rhizobium leguminosarum*	
C_{14}-HSL to C_{18}-HSL	*Sinorhizobium meliloti*	

3.7 Quorum sensing as a boon for plant growth promotion

QS mechanism supplies nitrogen and potassium, and it also induces the resistance against pathogenic bacteria in the rhizospheric region. Intercellular communication is regulated by AHL, a signaling molecule. AHL and AI production is mediated by Gram-positive and Gram-negative soil bacteria. Plant pathogens use QS for the regulation of virulence factors (Pandey et al., 2013). The rhizospheric region of the plant is a niche to favorable bacteria, which aids in the important biological activities in plants.

Rekadwad and Khobragade (2017) reported the QS mechanism in paddy. The root nodules in paddy plants are formed through legumes and rhizobia, which is nitrogen-fixing bacteria. These rhizobia synthesized AHL QS signals (Singh et al., 2015). The signal induces the exopolysaccharide, which increase the efficiency of nodes, transfer of plasmid DNA, *nif* gene regulation, and the swarming behavior (Mukherji and Prabhune, 2015). The carbon sequestration of N-3-Oxo-dodecanyl-HSL and AHL indicates QS molecules were produced by Gram-negative bacteria (Kalia, 2015).

Bacteria responds to the QS molecules produced by rhizobacteria and by plants. It also destroys the other QS molecules produced by other bacterial species (Dong et al., 2002). *Bacillus* sp. secretes the enzyme to degrade AHL produced by Gram-negative bacteria. Gene encoding AHL degrading enzyme is *aiia* in *Bacillus thuringenesis* and other subspecies (Lee et al., 2002). Bacillus strain breaks down lactone bond via hydrolysis and stimulates AI; in turn, these species struggles in competing with other Gram-negative bacteria. Therefore, the bacteria functions in the rhizospheric region which can be directly changed by the plants or by means of other microorganisms by QS molecules (Podile et al., 2014).

Burdman et al. (2001), reported the integral membrane protein of bacteria helps in the identification of host. *Azospirillum brasilense* outer-membrane protein has better adhesion to root exudates of cereals when compared to tomato and legumes, and it also helps in root adsorption in bacteria. Dekkers et al. (1998) reported the bacterial lipopolysaccharides help in root habitation. Endophytes have reported having QS mechanisms, which helps in the habituation of host plants and prevents the plant pathogen.

3.8 Quorum sensing detection technologies

The QS signaling molecules is detected by various biosensor strains. These strains are constructed and used to identify the QS molecules. The reporter strain *Agrobacterium tumifaciens* showed the other strains produced QS molecules, which were detected with different chromatographic patterns (Anbazhagan et al., 2012). Gonzalez et al. (2001) reported *Acinetobacter colcoaceticus* (BD413) culture supernatant was detected with four QS compounds, and their maximum activity was achieved in its stationary phase.

The induction test was performed for the identification of short and long chain AHL molecules. The strain, which produces AHL molecules, is expected with purple colonies. The induction inhibition test was carried out and produced colorless colonies by previously adding synthetic AHL and inhibit the QS molecules. The biosensor strain *C. violaceum* (CV026) and *A. tumefaciens* NT1 (pZLR4). These strain does not produce AHL, but responds to AHL. The test strains are streaked in LB plates against *C. violaceum* and 70 μg/mL X gal was supplemented for *A. tumifaciens*. Plates were incubated and checked for the purple colonies indicates AHL positive and colorless colonies indicates AHL negative (Pinto et al., 2007).

Kumar et al. (2016) reported the biofilm formation assay using microliter plate; 200 μL of LB broth was added into each well and 2 μL of overnight bacterial culture was added and incubated for 18 hours at 30°C. Hundred microliters of fresh LB broth was prepared and added into a fresh well containing 1 μL of 18 hours old culture from the titer plates and it was added in triplicates. It was further incubated for 24 hours at 30°C. The supernatant was removed and 150 μL of crystal violet (1%) was added into the well and was incubated at room temperature for 30 minutes. The stain was removed and washed twice with distilled water and to it, 175 μL of DMSO was added to dissolve crystal violet and was read in spectrophotometre at 570 nm. The higher degree of biofilm formation was subjected to NMR and quadrupole time-of-flight mass spectrometry (QTOF MS).

Wynendaele et al. (2013) reported the QS short chain amino acids are tested based on the biosensor system and approximately 300 QS peptides, and its structural analogs were performed using liquid chromatography–mass spectroscopy. Plasmid transfer demonstrates the presence of QS peptides, the genes present in plasmid encodes for bacteriocins, hemolysins, and antibiotic resistance; these genes are also transferable. The *Enterococcus faecalis* strain transfers plasmids by pheromones to a QS peptide (Dunny et al., 1979). A specific QS strain induces the other strains with conjugative plasmid to produce protein on the surface, which aids in forming aggregates and helps in the transfer of plasmids. QS peptides are also known as clumping inducing agents.

3.9 Conclusion

Over the last few decades, BVOC and QS molecules produced by different bacteria have been increased. These molecules are small in size with an odorless nature, which modifies the behavior and induces/inhibits the growth of neighboring strains by playing a vital role in intra- and interspecies interaction. QS molecules also help in preventing human infection by QSI, which increases the sensitivity to antibiotics. The QS molecules also act as a boon for the plant growth by supplying the required form of nitrogen, phosphorus, and potassium (NPK) and inducing the resistance against pathogens.

References

Anbazhagan, D., Mansor, M., Yan, G.O.S., Yusof, M.Y.M., Hassan, H., Sekaran, S.D., 2012. Detection of quorum sensing signal molecules and identification of an autoinducer synthase gene among biofilm forming clinical isolates of *Acinetobacter* spp. PLoS One 7 (7), e36696.

Barraud, N., Schleheck, D., Klebensberger, J., Webb, J.S., Hassett, D.J., Rice, S.A., et al., 2009. Nitric oxide signaling in *Pseudomonas aeruginosa* biofilms mediates phosphodiesterase activity, decreased cyclic di-GMP levels, and enhanced dispersal. J. Bacteriol. 191 (23), 7333–7342.

Bednarek, P., Kwon, C., Schulze-Lefert, P., 2010. Not a peripheral issue: secretion in plant–microbe interactions. Curr. Opin. Plant Biol. 13 (4), 378–387.

Bernier, S.P., Létoffé, S., Delepierre, M., Ghigo, J.M., 2011. Biogenic ammonia modifies antibiotic resistance at a distance in physically separated bacteria. Mol. Microbiol 81 (3), 705–716.

Berry, D.R., 1988. Physiology of Industrial Fungi. Blackwell Scientific Publications.

Blom, D., Fabbri, C., Connor, E.C., Schiestl, F.P., Klauser, D.R., Boller, T., et al., 2011. Production of plant growth modulating volatiles is widespread among rhizosphere bacteria and strongly depends on culture conditions. Environ. Microbiol 13 (11), 3047–3058.

Boedicker, J.Q., Vincent, M.E., Ismagilov, R.F., 2009. Supporting information for microfluidic confinement of single cells of bacteria in small volumes initiates high-density behavior of quorum sensing and growth and reveals its variability. Angew. Chem. Int. Ed. Engl. 121 (32), 6022–6025.

Burdman, S., Dulguerova, G., Okon, Y., Jurkevitch, E., 2001. Purification of the major outer membrane protein of *Azospirillum brasilense*, its affinity to plant roots, and its involvement in cell aggregation. Mol. Plant Microbe Interact. 14 (4), 555–561.

Carlson, N.E.I.L., Quraishi, A., 1999. Anatomy of a fungal problem. In: Proceedings of the Third International Conference: Bioaerosols, Fungi and Mycotoxins: Health Effects, Assessment, Prevention and Control. Fungal Research Group Foundation, Inc., Albany, NY, Boyd Printing, pp. 245–253.

Castillo-Juárez, I., Maeda, T., Mandujano-Tinoco, E.A., Tomás, M., Pérez-Eretza, B., García-Contreras, S.J., et al., 2015. Role of quorum sensing in bacterial infections. World J. Clin. Cases: WJCC 3 (7), 575.

Chen, Y., Gozzi, K., Yan, F., Chai, Y., 2015. Acetic acid acts as a volatile signal to stimulate bacterial biofilm formation. MBio 6 (3).

Cortes-Barco, A.M., Hsiang, T., Goodwin, P.H., 2010. Induced systemic resistance against three foliar diseases of *Agrostis stolonifera* by (2R, 3R)-butanediol or an isoparaffin mixture. Ann. Appl. Biol. 157 (2), 179–189.

De Kievit, T.R., 2009. Quorum sensing in *Pseudomonas aeruginosa* biofilms. Environ. Microbiol 11 (2), 279–288.

De la Fuente-Núñez, C., Reffuveille, F., Fernández, L., Hancock, R.E., 2013. Bacterial biofilm development as a multicellular adaptation: antibiotic resistance and new therapeutic strategies. Curr. Opin. Microbiol 16 (5), 580–589.

Defoirdt, T., Boon, N., Bossier, P., 2010. Can bacteria evolve resistance to quorum sensing disruption? PLoS Pathog. 6 (7), e1000989.

Dekkers, L.C., Phoelich, C.C., Van Der Fits, L., Lugtenberg, B.J., 1998. A site-specific recombinase is required for competitive root colonization by *Pseudomonas fluorescens* WCS365. Proc. Natl Acad. Sci. 95 (12), 7051–7056. Available from: https://doi.org/10.1007/s10886-012-0135-5.

Dong, Y.H., Gusti, A.R., Zhang, Q., Xu, J.L., Zhang, L.H., 2002. Identification of quorum-quenching N-acyl homoserine lactonases from *Bacillus* species. Appl. Environ. Microbiol 68 (4), 1754–1759.

Dunny, G.M., Craig, R.A., Carron, R.L., Clewell, D.B., 1979. Plasmid transfer in Streptococcus faecalis: production of multiple sex pheromones by recipients. Plasmid 2 (3), 454–465.

Effmert, U., Kalderas, J., Warnke, R., Piechulla, B., 2012. Volatile mediated interactions between bacteria and fungi in the soil. J. Chem. Ecol. 38, 665–703.

Engebrecht, J., Nealson, K., Silverman, M., 1983. Bacterial bioluminescence: isolation and genetic analysis of functions from *Vibrio fischeri*. Cell 32 (3), 773–781.

Farag, M.A., Zhang, H., Ryu, C.M., 2013. Dynamic chemical communication between plants and bacteria through airborne signals: induced resistance by bacterial volatiles. J. Chem. Ecol. 39 (7), 1007–1018.

Fuqua, W.C., Winans, S.C., Greenberg, E.P., 1994. Quorum sensing in bacteria: the LuxR-LuxI family of cell density-responsive transcriptional regulators. J. Bacteriol. 176 (2), 269.

Gambello, M.J., Iglewski, B.H., 1991. Cloning and characterization of the *Pseudomonas aeruginosa* lasR gene, a transcriptional activator of elastase expression. J. Bacteriol. 173 (9), 3000–3009.

Gambello, M.J., Kaye, S.U.S.A.N., Iglewski, B.H., 1993. LasR of *Pseudomonas aeruginosa* is a transcriptional activator of the alkaline protease gene (apr) and an enhancer of exotoxin A expression. Infect. Immun. 61 (4), 1180–1184.

Gonzalez, R.H., Nusblat, A., Nudel, B.C., 2001. Detection and characterization of quorum sensing signal molecules in *Acinetobacter* strains. Microbiol Res. 155 (4), 271–277.

Hamilton-Kemp, T., Newman, M., Collins, R., Elgaali, H., Yu, K., Archbold, D., 2005. Production of the long-chain alcohols octanol, decanol, and dodecanol by *Escherichia coli*. Curr. Microbiol 51 (2), 82–86.

He, X., Lu, F., Yuan, F., Jiang, D., Zhao, P., Zhu, J., et al., 2015. Biofilm formation caused by clinical *Acinetobacter baumannii* isolates is associated with overexpression of the AdeFGH efflux pump. Antimicrob Agents Chemother 59 (8), 4817–4825.

Helmig, D., Klinger, L.F., Guenther, A., Vierling, L., Geron, C., Zimmerman, P., 1999. Biogenic volatile organic compound emissions (BVOCs). I. Identifications from three continental sites in the U.S. Chemosphere 38 (9), 2163–2187.

Høiby, N., Bjarnsholt, T., Givskov, M., Molin, S., Ciofu, O., 2010. Antibiotic resistance of bacterial biofilms. Int. J. Antimicrob. Agents 35 (4), 322–332.

Holm, A., Vikström, E., 2014. Quorum sensing communication between bacteria and human cells: signals, targets, and functions. Front. Plant Sci. 5, 309.

Jayaraman, A., Wood, T.K., 2008. Bacterial quorum sensing: signals, circuits, and implications for biofilms and disease. Annu. Rev. Biomed. Eng. 10, 145–167.

Kalia, V.C., 2015. Microbes: the most friendly beings? Quorum Sensing vs Quorum Quenching: A Battle With No End in Sight. Springer, New Delhi, pp. 1–5.

Kalia, V.C., Wood, T.K., Kumar, P., 2014. Evolution of resistance to quorum-sensing inhibitors. Microb. Ecol. 68 (1), 13–23.

Kaplan, H.B., Greenberg, E.P., 1985. Diffusion of autoinducer is involved in regulation of the *Vibrio fischeri* luminescence system. J. Bacteriol. 163 (3), 1210–1214.

Karlshøj, K., Larsen, T.O., 2005. Differentiation of species from the *Penicillium roqueforti* group by volatile metabolite profiling. J. Agric. Food Chem. 53 (3), 708–715.

Kim, K.S., Lee, S., Ryu, C.M., 2013. Interspecific bacterial sensing through airborne signals modulates locomotion and drug resistance. Nat. Commun. 4, 1809.

Kloepper, J.W., Ryu, C.M., Zhang, S., 2004. Induced systemic resistance and promotion of plant growth by *Bacillus* spp. Phytopathology 94 (11), 1259–1266.

Korpi, A., 2001. Faculty of natural and environmental sciences C129. Fungal volatile metabolites and biological responses to fungal exposure. Thesis. Kuopio University, Finland, pp. 1–C177.

Kumar, J.S., Umesha, S., Prasad, K.S., Niranjana, P., 2016. Detection of quorum sensing molecules and biofilm formation in *Ralstonia solanacearum*. Curr. Microbiol 72 (3), 297–305.

Ladygina, N., Dedyukhina, E.G., Vainshtein, M.B., 2006. A review on microbial synthesis of hydrocarbons. Process Biochem. 41 (5), 1001–1014.

Ledgham, F., Ventre, I., Soscia, C., Foglino, M., Sturgis, J.N., Lazdunski, A., 2003. Interactions of the quorum sensing regulator QscR: interaction with itself and the other regulators of *Pseudomonas aeruginosa* LasR and RhlR. Mol. Microbiol. 48 (1), 199–210.

Lee, S.J., Park, S.Y., Lee, J.J., et al., 2002. Genes encoding the N-acyl homoserine lactone-degrading enzyme are widespread in many subspecies of *Bacillus thuringiensis*. Appl. Environ. Microbiol. 68, 3919–3924.

Létoffé, S., Audrain, B., Bernier, S.P., Delepierre, M., Ghigo, J.M., 2014. Aerial exposure to the bacterial volatile compound trimethylamine modifies antibiotic resistance of physically separated bacteria by raising culture medium pH. MBio 5 (1).

Marilley, L., Casey, M.G., 2004. Flavours of cheese products: metabolic pathways, analytical tools and identification of producing strains. Int. J. Food Microbiol 90 (2), 139–159.

Martin, F., Kamoun, S. (Eds.), 2012. Effectors in Plant-Microbe Interactions. Wiley-Blackwell, Oxford, UK.

Mattila, J.T., Thomas, A.C., 2014. Nitric oxide synthase: non-canonical expression patterns. Front. Immunol 5, 478.

Mehrer, A., Lorenz, W., 2005, Potential influences on MVOC measurements. In: Proceedings of the 10th International Conference on Indoor Air Quality and Climate, pp. 2537–2542.

Morey, P., Horner, E., Gareis, M., Johanning, E., Worthan, A., Lstiburek, J., et al., 2000. Microbial evaluation in a partially remediated residence. In: Proceedings of Healthy Buildings, pp. 385–390.

Mukherji, R., Prabhune, A., 2015. Enzyme purification and kinetic characterization of AHL lactonase from *Bacillus* sp. RM1 a novel and potent quorum quencher isolated from Fenugreek root nodule rhizosphere. Int. J. Curr. Microbiol. Appl. Sci. 4, 909–924.

Nijland, R., Burgess, J.G., 2010. Bacterial olfaction. Biotechnol. J. 5 (9), 974–977.

Novick, R.P., 2003. Autoinduction and signal transduction in the regulation of staphylococcal virulence. Mol. Microbiol 48 (6), 1429–1449.

Oinuma, K.I., Greenberg, E.P., 2011. Acyl-homoserine lactone binding to and stability of the orphan *Pseudomonas aeruginosa* quorum-sensing signal receptor QscR. J. Bacteriol. 193 (2), 421–428.

Pandey, R., Swamy, K.V., Khetmalas, M.B., 2013. Indole: a novel signaling molecule and its applications. Indian J. Biotechnol. 12 (3), 297–310.

Panpatte, D.G., Shukla, Y.M., Shelat, H.N., Vyas, R.V., Jhala, Y.K., 2017. Bacterial volatile organic compounds: a new insight for sustainable agriculture. Microorganisms for Green Revolution. Springer, Singapore, pp. 151–166.

Parsek, M.R., Val, D.L., Hanzelka, B.L., Cronan, J.E., Greenberg, E.P., 1999. Acyl homoserine-lactone quorum-sensing signal generation. Proc. Natl Acad. Sci. 96 (8), 4360–4365.

Pearson, J.P., Pesci, E.C., Iglewski, B.H., 1997. Roles of *Pseudomonas aeruginosa* las and rhl quorum-sensing systems in control of elastase and rhamnolipid biosynthesis genes. J. Bacteriol. 179 (18), 5756–5767.

Pinto, U.M., de Souza Viana, E., Martins, M.L., Vanetti, M.C.D., 2007. Detection of acylated homoserine lactones in gram-negative proteolytic psychrotrophic bacteria isolated from cooled raw milk. Food Control 18 (10), 1322–1327.

Podile, A.R., Vukanti, R.V.N.R., Sravani, A., Kalam, S., Dutta, S., Durgeshwar, P., et al., 2014. Root colonization and quorum sensing are the driving forces of plant growth promoting rhizobacteria (PGPR) for growth promotion. Proc. Indian. Natl Sci. Acad. 80 (2), 407–413.

Rekadwad, B., Khobragade, C., 2017. Bacterial quorum sensing (QS) in rhizosphere (paddy soil): understanding soil signaling and N-recycling for increased crop production, Microbial Applications, vol. 2. Springer, Cham, pp. 119–131.

Ryu, C.M., Farag, M.A., Hu, C.H., Reddy, M.S., Wei, H.X., Paré, P.W., et al., 2003. Bacterial volatiles promote growth in *Arabidopsis*. Proc. Natl Acad. Sci. 100 (8), 4927–4932.

Savage, V.J., Chopra, I., O'Neill, A.J., 2013. *Staphylococcus aureus* biofilms promote horizontal transfer of antibiotic resistance. Antimicrob Agents Chemother 57 (4), 1968–1970.

Schulz, S., Dickschat, J.S., 2007. Bacterial volatiles: the smell of small organisms. Nat. Product. Rep. 24 (4), 814–842.

Sharifi, R., Ryu, C.M., 2018. Revisiting bacterial volatile-mediated plant growth promotion: lessons from the past and objectives for the future. Ann. Bot. 122 (3), 349–358.

Singh, R.P., Baghel, R.S., Reddy, C.R.K., Jha, B., 2015. Effect of quorum sensing signals produced by seaweed-associated bacteria on carpospore liberation from *Gracilaria dura*. Front. Plant Sci. 6, 117.

Smedje, G., Norback, D., Wessen, B., Edling, C., 1996. Asthma among school employees in relation to the school environment. In: Proceedings from Indoor Air'96, vol. 1, Nagoya, Japan, p. 1.

Song, G.C., Ryu, C.M., 2013. Two volatile organic compounds trigger plant self-defense against a bacterial pathogen and a sucking insect in cucumber under open field conditions. Int. J. Mol. Sci. 14 (5), 9803–9819.

Strateva, T., Yordanov, D., 2009. *Pseudomonas aeruginosa*—a phenomenon of bacterial resistance. J. Med. Microbiol 58 (9), 1133–1148.

Ström, G., West, J., Wessen, B., Palmgren, U., 1994. Quantitative analysis of microbial volatiles in damp Swedish houses. Health Implic. Fungi Indoor Environ. 2, 291–305.

Sunesson, A.L., 1995. Volatile metabolites from microorganisms in indoor environments: sampling, analysis and identification. Doctoral Dissertation. Umeå Universitet.

Toder, D.S., Gambello, M.J., Iglewski, B.H., 1991. *Pseudomonas aeruginosa* LasA: a second elastase under the transcriptional control of LasR. Mol. Microbiol 5 (8), 2003–2010.

Veliz-Vallejos, D.F., van Noorden, G.E., Yuan, M., Mathesius, U., 2014. A *Sinorhizobium meliloti*-specific N-acyl homoserine lactone quorum-sensing signal increases nodule numbers in *Medicago truncatula* independent of autoregulation. Front. Plant Sci. 5, 551.

Vuotto, C., Longo, F., Balice, M.P., Donelli, G., Varaldo, P.E., 2014. Antibiotic resistance related to biofilm formation in *Klebsiella pneumoniae*. Pathogens 3 (3), 743–758.

Wieslander, G., Norbäck, D., Venge, P., 2007. Changes of symptoms, tear film stability and eosinophilic cationic protein in nasal lavage fluid after re-exposure to a damp office building with a history of flooding. Indoor Air 17 (1), 19–27.

Wilkins, C.K., Scholl, S., 1989. Volatile metabolites of some barley storage molds. Int. J. Food Microbiol 8 (1), 11–17.

Wilkins, K., Ntelsen, E.M., Wolkoff, P., 1997. Patterns in volatile organic compounds in dust from moldy buildings. Indoor Air 7 (2), 128–134.

Wynendaele, E., Bronselaer, A., Nielandt, J., D'Hondt, M., Stalmans, S., Bracke, N., et al., 2013. Quorumpeps database: chemical space, microbial origin and functionality of quorum sensing peptides. Nucleic Acids Res. 41 (D1), D655–D659.

Yang, J., Kloepper, J.W., Ryu, C.M., 2009. Rhizosphere bacteria help plants tolerate abiotic stress. Trends Plant Sci. 14 (1), 1–4.

Yarwood, J.M., Schlievert, P.M., 2003. Quorum sensing in *Staphylococcus* infections. J. Clin. Investig 112 (11), 1620–1625.

Zhu, J., Bean, H.D., Kuo, Y.M., Hill, J.E., 2010. Fast detection of volatile organic compounds from bacterial cultures by secondary electrospray ionization-mass spectrometry. J. Clin. Microbiol 48 (12), 4426–4431.

Further reading

Heeb, S., Fletcher, M.P., Chhabra, S.R., Diggle, S.P., Williams, P., Cámara, M., 2011. Quinolones: from antibiotics to autoinducers. FEMS Microbiol Rev. 35 (2), 247–274.

Høiby, N., Ciofu, O., Johansen, H.K., Song, Z.J., Moser, C., Jensen, P.Ø., et al., 2011. The clinical impact of bacterial biofilms. Int. J. Oral Sci. 3 (2), 55–65.

Ochsner, U.A., Koch, A.K., Fiechter, A., Reiser, J., 1994. Isolation and characterization of a regulatory gene affecting rhamnolipid biosurfactant synthesis in *Pseudomonas aeruginosa*. J. Bacteriol. 176 (7), 2044–2054.

Rutherford, S.T., Bassler, B.L., 2012. Bacterial quorum sensing: its role in virulence and possibilities for its control. Cold Spring Harb. Perspect. Med. 2 (11), a012427.

Ryu, C.M., Farag, M.A., Hu, C.H., Reddy, M.S., Kloepper, J.W., Paré, P.W., 2004. Bacterial volatiles induce systemic resistance in *Arabidopsis*. Plant Physiol. 134 (3), 1017–1026.

Xiao, G., He, J., Rahme, L.G., 2006. Mutation analysis of the *Pseudomonas aeruginosa* mvfR and pqsABCDE gene promoters demonstrates complex quorum-sensing circuitry. Microbiology 152 (6), 1679–1686.

Chapter 4

Detection and purification of microbial volatile organic compounds

Jeevanandam Vaishnavi[1], Swarnkumar Reddy[2], Santhanam Narmadha[2] and W. Jabez Osborne[1]

[1]*Biomolecules Lab, Department of Bio-Sciences, School of Bio Sciences and Technology, Vellore Institute of Technology, Vellore, India,* [2]*Biomolecules Lab, School of Bio Sciences and Technology, Vellore Institute of Technology, Vellore, India*

Chapter Outline

4.1 Introduction

Microbial volatile organic compounds (MVOCs) are the diverse compounds formed during bacterial and fungal metabolism. Of the more than 1200 compounds that have been identified as MVOC, none of them are specifically

Volatiles and Metabolites of Microbes. DOI: https://doi.org/10.1016/B978-0-12-824523-1.00021-3

among any microbial origin. The MVOC is usually produced in moisture conditions and during microbial damage (Korpi et al., 2009). MVOCs are formed during microbial metabolism, and they are formed as the byproducts (Berry, 1988). The imbalance of nutrition can also trigger the microorganism for the secretion of numerous MVOC. The MVOC can be converted into other compounds by the chemical reaction occurred in the environment, and these chemical reactions also produce volatile organic compounds (VOCs) in the atmosphere.

According to Fiedler et al. (2001), gas chromatography–mass spectroscopy combined with headspace solid-phase microextraction (GC-MS-HS-PME) was used to detect the MVOC released by different microbial species grown on various substrates. In the late 1980s and early 1990s, MVOC was detected using activated carbon/Tenax for sampling combined with chemical elution/thermal desorption in the gas chromatography unit (Larsen and Frisvad, 1994). In the late 1990s, MVOC detection and analysis can be carried out by HS-SPME. About 150 compounds were identified using the HS-SPME technique, and all the compounds were identified to be unique. MVOCs are synthesized as the byproducts by all the living organisms during the biological metabolism. These compounds are categorized as endogenous and exogenous based on the compounds emitted by external factors and by the normal biological function (Mochalski et al., 2014)

Due to the contact with an external source, living organisms produce exogenous metabolites of VOCs. These exogenous compounds enter into the organism by ingestion and inhalation, and contributes to the metabolites generated. The exogenous source includes the environment (including pollutant), drug, smoking habit, and even ionizing radiation and nutrition, among others (Montero-Montoya et al., 2018). Living species naturally produce endogenous VOC metabolites. Finding the same metabolites in different body fluids does not imply that they have an endogenous origin. In hemoglobin, the inhaled VOCs are dissolved and stored in the body compartments and then exhaled/excreted via urine. The release of VOCs from plants plays a significant role in plant–plant interactions, plant–microbial interactions, and herbivores-induced VOCs. Depending on the stage of growth or medium, the cell or microbial interactions produce VOCs. Relevant VOC release from animals has been recorded recently in dolphins and whales (Cumeras and Correig, 2018).

A range of sophisticated techniques is used to study MVOC. The reference techniques are based on chromatographic separation and mass spectrometry (GC/MS or LC/MS). Those applications in metabolomics are less complex in the instrumentation dependent on ion mobility spectrometry [IMS and high field asymmetric ion mobility spectrometry (FAIMA)] and chemical sensor arrays, as in e-noses, which ensure full integration and minimal cost. Metabolomics is researching the metabolic small molecular (the metabolome) of an rRNA. These provide the key lead to emerging fields such as transcriptomic and proteomic. The evolvement of omics science also applies to the low-weight compounds, including carbohydrates, lipids, hormones, and vitamins, which perform a great deal of the cell's function. The collection of metabolites synthesized by the biological system constitute their "metabolome" in parallel to the terms "transcriptome and proteome." The analysis of the volatile metabolome part is called Volatilomics. A reductionist approach at every single omics provides an insight into what is happening at a precise level to create a complete model (Cumeras and Correig, 2018).

4.2 Classification of microbial volatile organic compounds

VOCs produced by bacteria, yeast, fungi, algae, and plants are often divided into two major categories based on the source of microbial origin of primary and secondary metabolites. Primary metabolites reflect a unit of biochemistry and are essential to the life of the organism. Intermediate Krebs cycle, lipids, nucleic acid, and amino acids are some examples (Keller et al., 2005). Secondary metabolites, on the other hand, represent a huge class of diverse natural products, which are often of very unusual chemical structure not necessary for growth and which are almost always constrained by taxonomic distribution. The documented example of fungal secondary metabolites is antibiotics and hallucinogens such as the ergot alkaloids, and mycotoxins such as the trichothecenes (Cole et al., 2003, Matysik et al., 2009). Specific pathways (e.g., polyketides, nonribosomal peptides, and isoprenoids) are involved in the biosynthesizing of secondary metabolites, which constitutes the bulk of a field known as natural products. Advances in genomics help to classify the clusters of their signature genes through bioinformatics (Fig. 4.1).

As a group, fungal VOCs are often known as secondary metabolites; these are characterized by their low molecular weight and their phase-dependent appearance. However, this simplistic categorization is well understood using the modern technologies available. Secondary metabolites are typically formed by only a limited number of species whereas the majority of VOCs can find in across a broad range of organisms. For instance, the toxic secondary metabolite aflatoxin is produced by certain species of the genus *Aspergillus* while the volatile 1-octen-3-ol is widely spread across many plants, animals, and fungi (Cole et al., 2003, Combet et al., 2006). In general, secondary metabolites are produced through complex metabolic pathways coded by linked gene clusters (Cumeras and Correig, 2018). While plenty is known about the pathways that produce VOCs, many of them are either metabolic transformation products of heterocyclic metabolites, lipids, proteins, and other components of living tissues, which are degradation of fungal catabolic pathways. 1-octen-3-ol is a typical example of the well-studied fungal VOC, and it comes from linoleic acid (Bennett et al., 2012, Claeson, 2006). Although 1-octen-3-ol was mostly referred to as a secondary metabolite, it is more commonly synthesized in lipid degradation. The

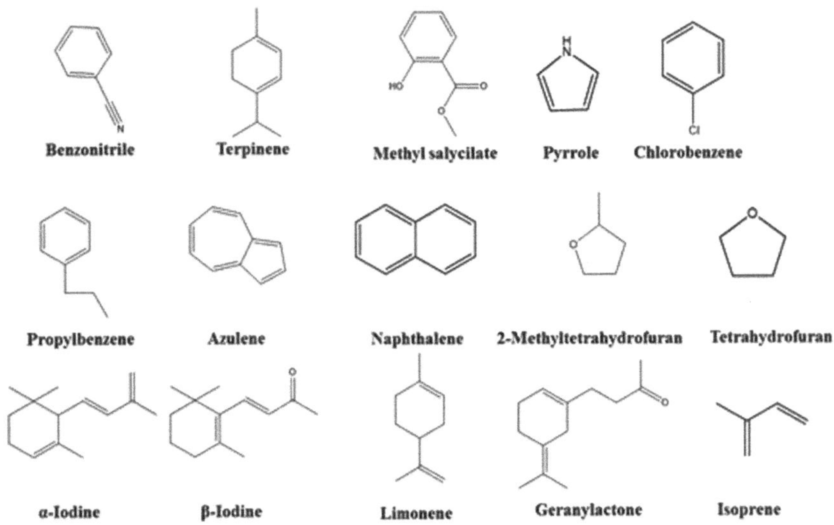

FIGURE 4.1 Common microbial volatile compounds.

enzyme oxidation and cleavage of linoleic and linolenic acids are responsible for the synthesis of 1-octen-3-ol and a less volatile 10-oxo-trans-8-decenoic acid. In particular, not all the "small molecules" outside of the central pathways of intermediary metabolism are secondary metabolites. Therefore, fungal VOCs can be categorized as primary or secondary metabolites. Rather we will describe VOCs according to their ring structures, their substituent group, their number of carbons, and as ketones, acids, terpenes, and aldehydes (Keller et al., 2005).

4.2.1 Overview of fungal volatiles

In a variety of various scientific disciplines, fungal VOCs are both theoretical and practical. They have been studied as indicators to detect the presence of fungal growth, as contributors to "sick building syndrome," as signals for fungal development, and also for their flavor properties (Ashrafi et al., 2017). Also, endophyte VOCs has emerged as a specific interest in recent years since some of them have demonstrated antibiotic activity, while others have the potential for possible use as fuel compounds or "biodiesel." All of this research has shown that fungal VOC profiles are both complex and dynamic: the compounds produced and their abundance differ with the species produced, the temperature, abundance of moisture, the age of the fungal colony, the type of substrate, and other environmental parameters (Achyuthan et al., 2017). The fungal species *Tuber indicum* is reported to produce 2-methylpentan-1-ol (Splivallo et al., 2007). There are many fungal

TABLE 4.1 Microbial volatile compounds.

Microbes	Microorganism	Volatile organic compounds	References
Bacteria	*Acinetobacter baumanii*	Acetic acid	Julák et al. (2003)
	Arthrobacter bgilis UMCV2		
	Bacillus pumilus ES4		Amavizca et al. (2017)
	Bacteroides bivius		Wiggins et al. (1985)
	Clostridium bifermentans		
	Enterococcus faecalis		
	Escherichia coli	Propionic acid	Bunge et al. (2008)
	Fusobacterium necrophorum		Julák et al. (2003)
	Lactobacillus acidophilus		
	Klebsiella pneumoniae		
	Staphylococcus aureus	3-hydroxybutan-2-one	Preti et al. (2009)
	Bacillus cereus B-569		Blom et al. (2011)
	Burkholderia tropica LMG 22274		
	Chondromyces crocatus CM C5	Phenylmethanol	Dickschat et al. (2005)
	Pseudomonas syringae S22		
	Serratia proteamaculans B5a	Butane-2,3-diol	Blom et al. (2011)
	Staphylococcus warneri CCM 2730		Lemfack et al. (2018)
	Marine *Streptomyces*	4-ethenyl-2-methoxyphenol	Stritzke et al. (2004)
Fungi	Laccaria bicolor	3-hydroxybutan-2-one	Müller et al. (2013)
	Trichoderma viride		
	Saccharomyces cerevisiae		
	Armillaria mellea	GEOSMIN	Müller et al. (2013)
	Aspergillus niger		
	Penicillium expansum		Mattheis and Roberts (1992)
Algae	*Prochlorococcus*	Isoprene, Monoterpenes	Meskhidze et al. (2015)
	Thalassiosira weissflogii		
	Thalassiosira pseudonana		
	Pleurochrysis carterae		
	Rhodomonas salina		

species producing MVOC, which is used in the hospital (pathogens), industrial, and environment (Table 4.1).

4.2.2 Overview of bacterial volatiles

Like fungi, bacterial species produce complex VOCs. They are produced as primary (ethanol, acetic acid) and secondary (signaling molecules) metabolites in various amounts and combinations based on the different bacterial species. Lee et al. (2012) reported the MVOC Pentan-1-ol produced by the bacteria *Bacillus amyloliquifaciens*. Using several molecular mechanisms, bacteria communicate with each other and have interactive effects on species with which they share ecological niches, including fungi, bacteria, plants, and animals (Bos et al., 2013). Bacterial volatiles was studied for (1) their production of "off" odors in food and water supplies, (2) other applied areas, and (3) their uses in foods, particularly in the dairy industry. In particular, rhizosphere-associated soil bacteria have been studied for their growth-promoting activities. The research of quorum sensing has transformed our view of "single-cell" organisms in basic science (Chen et al., 2017).

4.2.3 Overview of algae volatiles

Algal VOCs (AVOC or biogenic VOC, BVOC) are produced under various conditions as secondary metabolites (Jerković et al., 2018). The chemistry of VOC, rate of production, and the number of emissions depend on several biotic and abiotic factors, such as species/strain type, growth phase, water, stress (seasonal changes, temperature, light intensity, pH, salinity), aeration (mixing/turbulence) or static culture, gases (H_2O, CO_2, O_3), nutrients, and the presence of predators (Achyuthan et al., 2017). The algae VOCs and their respective organisms are listed in Table 4.1.

4.3 Analytical techniques used for the detection purification and analysis of microbial volatile organic compound

Maurer et al. (2019) reported the emission and detection of MVOC, and it can also be used as a detection tool in the diagnosis of cancer, diabetes, asthma, cholera, schizophrenia, chronic lung disease, etc., in humans and animals. The detection of MVOC from *Mycobacterium tuberculosis* is performed by closed-loop stripping analysis coupled with GCMS, and the detection of MVOC from *Mycobacterium bovis* is from select ion flow tube (SIFT) coupled with a mass spectrometer (SIFT-MS) (Nawrath et al., 2012, Sethi et al., 2013). Ammonia levels in the breath are used to detect *Helicobacter pylori* and also to detect the acetonitrile levels in the smoker's breath. Sethi et al. (2013) reported electronic e-nose technology for the detection of MVOC in the sputum samples. E-nose technology has the ability to differentiate *Mycobacterium tuberculosis* and *Mycobacterium bovis* from others. Laser spectroscopy, IMS, proton transfer reaction coupled with mass spectrometer (PTR-MS) can be used for the detection of MVOC with the diagnosis of tuberculosis (Krisher et al., 2014; Purkhart et al., 2011).

4.4 Chromatography techniques for microbial volatile organic compound identification

Tait et al. (2014) reported the VOC was identified using HS-SPME coupled with GC-MS. The effect of the medium was optimized with a gas chromatography column based on VOC produced. Gram-positive bacteria (*S. aureus*) and Gram-negative bacteria (*E. coli, K. pneumoniae*) were used and evaluated for VOC. Multivariate analysis is required to determine the VOC parameters. The solid-phase microextraction (SPME) is said to alter the detected VOC, and it was also varied with an increase in polarity in the GC column (Souza-Silva et al., 2015). Critical evaluation is required further for the VOC to be used in clinical diagnostics. SPME method is an effective protocol for the extraction of VOC. SPME comes with numerous varieties of coatings. The higher degree of selectivity was seen in two different types of SPME fibers such as carboxy-polydimethylsiloxane and carbowax divenyl benzene (Jia et al., 2009). The identification, purification, and analysis of VOC in the bacterial extract gets based on the culture medium and the column used for purification.

The VOC is also detected in the sediments. BITEX, aromatic compounds, alcohols, aldehydes, terpenes, and sulfur compounds are the wide range of VOC produced, and these VOCs are produced by both natural and anthropogenic manner. The VOC concentration depends upon the presence of contaminants. The types of sediments matter for the type of VOC produced. If there are muddy sediments, VOC could be dimethyl sulfide, methyl mercaptan, etc. If the sediments are sandy, they have very little possibility of VOC production (Kataoka et al., 2016).

Matysik et al. (2009) reported the separation, identification, and quantification of VOC depend on GC-MS. Several methods are in use depending upon the type of VOC. Air sampling for the VOC production was carried out by Tenax desorption tubes, which is accompanied by thermodesorption and which paves the way for the precise sampling in time and passive diffusion monitors on charcoal adsorbents used to determine VOC over a long period. The fungal species *Penicillium chrysogenum, Aspergillus fumigatus, Penicillium expansum, Aspergillus versicolor, Penicillium brevicompactum, Cladosporium cladosporoides*, and *Aspergillus niger* were grown on a dichloran glycerol agar, and passive sampling coupled with GC-MS was used to detect MVOC. It was found that this approach was especially appropriate in epidemiological studies attempting to correlate concentrations of specific VOCs and exposure to indoor molds. After the separation MVOC, mass spectroscopy distinguishes the individual constituents present in the MVOC. To confirm the identification, mass spectra were compared with library spectra, and the chromatographic retention indices are also used in combination with the parallel determination of standards (Reyes-Garcés et al., 2017).

4.4.1 Gas chromatography–mass spectroscopy for microbial volatile organic compound

Dewulf et al. (2002) reported MVOC detection and analysis using GC. The injection of samples into the GC column is carried out by sample loops and syringes. When the MVOC sample is of solvents, a large volume injection has been developed to load the samples (Adahchour et al., 2001). Pocurull et al. (2000) reported the special swing injection for a large number of water samples on water containing samples. Sample evaporation was carried out with the programmed temperature vaporization. Cryogenically preserved samples were subjected to thermal desorption to overcome the band thickening problem.

MVOC was separated on the capillary columns coated with silicone or alumina-based porous layers for the highly VOC. The multipurpose columns are designed and developed for the analysis of VOC according to the United States Environmental Protection Agency standards. The separating columns are coupled with computer modeling to reduce the analysis time with higher accuracy in peaks. Sixty-six samples were analyzed in 30 minutes using a wall coated open tubular column, and 15 samples were analyzed in 16 minutes using a porous layer open tubular column (PLOT). Ji et al. (1999) reported the extension PLOT column by improving thermal, chemical, and mechanical stability. Zeng et al. (2000) designed the crown ether-based column thrown sol-gel process and the silica gel coating. This column showed high selectivity in the separating of aromatic compounds. Chiral cyclodextrin-based stationary phase-based column was used for the analysis of aromatic VOC.

The VOC detection employs flame ionization detection, electron capture detection, mass spectrometry detection, flame photometry detection, photoionization detection, and nitrogen phosphorous detection (Hill et al., 2000). Based on the selectivity and sensitivity, the detectors are connected in the outlet of the GC column.

4.4.2 Ultra-performance liquid chromatography for microbial volatile organic compound

High-performance liquid chromatography (HPLC) is an important technique of liquid chromatography (LC) used in the mixture segregation of different components. It is also used in the process of drug development for the detection and quantification of compounds and has been used for decades around the world. A significant development in instrumentation and column technology (column particle size and column dimension) has been made to further achieve the dramatic increase in resolution, speed, and sensitivity in LC (Murray, 2012). In 2004 waters introduced and trademarked Ultra-Performance Liquid Chromatography (UPLC) which is based on small, porous particles (sub-2-micron particles), to achieve the above target. The principle behind this evolution is the Van Deemter equation, which correlates the relation between linear velocity and plate height. The role of UPLC, the small particles need high pressure, that is, 6000 psi, which is generally the upper limit of traditional HPLCs. It has been found that there is a significant increase in effectiveness when the particle size is reduced below 2.5 μm, and this effect does not decrease when the linear speed or flow rate is increased (Garrity et al., 2001). The tiny radius particles and the maximum number of resolvable peaks followed by flow rate understand efficiency along with the resolution. Compared to HPLC, this technique decreases the mobile phase volume consumption by at least 80% with a shorter runtime of about 1.5 minutes. The separation retention factors will be increased accordingly with the rise of pressure (up to 1000 bars or more) in small particles. For UPLC, the lower injection volume is required, which results in higher efficiency and an increase in resolution. The higher column temperature decreases the viscosity of the mobile phase resulting in the high diffusion coefficient and flow rate without significant loss in efficiency and increase in column back pressure. UPLC is a special version of HPLC that has the advantage of technological development in the performance of particle chemistry, system optimization, detector design, processing, and control of data. The vital increase in the efficiency, sensitivity, resolution with the rapid outcome, and less usage of solvents, which lends to cost-effectiveness and ecofriendliness (Murray, 2012).

4.4.3 Proton transfer reaction coupled with mass spectroscopy for microbial volatile organic compound

Detection of microbial VOC with GC-MS is highly limited because high time consuming and only discontinuous measurements are recorded. In order to overcome this mass spectroscopy can be coupled with the PTR. PTR enables both detection and purification of VOCs with the requirement of crude samples without any preprocessing. Proton ionization facilitates the chemical ionization of volatile compounds with hydronium ions along with real-time detection of VOCs. Franke et al. (2019) detection of ethanol, acetic acid, 3-hydroxy, and 2,3-butanedione using PTR-MS. Bunge et al. (2008) in their work on online monitoring of microbial volatiles reported the temporal emission of microbial volatiles in the growth phase of the bacteria, which was characterized and detected by PTR-MS. The use of PTR-MS was limited by the major drawback that is the VOCs with similar nominal numbers cannot be identified due to the use of low volume resolution of quadrupole mass spectrometer.

4.4.4 Matrix-assisted laser desorption/ionization time-of-flight mass spectrometry for microbial volatile organic compound

Matrix-assisted laser desorption/ionization time-of-flight mass spectrometry (MALDI-TOF mass spectrometry) was conventionally used for the detection of various peptides proteins and other higher molecular mass organic compounds. The application of MALDI-TOF mass spectrometry can be extended to a wide range of organic molecules because of its various advantages such as wide detection range, high sensitivity, and easy sample preparation. For the analysis of various organic molecules, a wide range of matrices has been developed based on the nature of the sample. Parylene-matrix with graphene, porous silicon, ionic liquid matrix, and titania are the most commonly used matrices for the detection of small volatile compounds. These matrices are limited for the use of small volatiles, due to the sample preparation in which the sample is dried in the parylene-matrix chip. The use of graphene as a matrix in MALDI-TOFMS adds the advantage of analyzing a wide range of volatiles.

4.4.5 Selected ion flow tube–mass spectrometry

Microbial volatiles is commonly detected using GC-MS, whereas to overcome the disadvantages like tedious sample processing and narrow detection range SIFT coupled with MS is a highly recommended alternate (Scotter et al., 2005). The selected ion flow tube detects the volatiles by chemical ionization of selected ions for the characterization of volatile molecules. SIFT-MS is more advantageous over others as it detects the extremely lower concentration of volatiles and also facilitates the real-time detection. Ethanol, methanol, acetaldehyde, crotonaldehyde, and acetone are the most common volatiles from various microbial origin detected using SIFT-MS (Kumar et al., 2013).

4.5 Conclusions

A general overview of some of the purification and identification techniques for MVOCs was provided in this chapter; also, the classification of MVOCs has been discussed, such as bacterial-produced MVOCs, fungal-produced MVOCs, and algae-produced MVOCs. Many MVOCs/VOCs are potential sensory irritants and can exceed individually or in combination irritation thresholds. Some of the emitted compounds can also react to the production of other, more irritating compounds. A large number of VOC are released during the microbial growth of building materials; ketone and alcohol are the most common production groups, depending on both species.

References

Achyuthan, K.E., Harper, J.C., Manginell, R.P., Moorman, M.W., 2017. Volatile metabolites emission by in vivo microalgae—an overlooked opportunity? Metabolites 7 (3), 39.

Adahchour, M., Kristenson, E.M., Vreuls, R.J.J., Brinkman, U.T., 2001. Investigations on analyte losses in systems suited for large-volume on-column injections. Chromatographia 53 (5-6), 237–243.

Amavizca, E., Bashan, Y., Ryu, C.M., Farag, M.A., Bebout, B.M., de-Bashan, L.E., 2017. Enhanced performance of the microalga *Chlorella sorokiniana* remotely induced by the plant growth-promoting bacteria *Azospirillum brasilense* and *Bacillus pumilus*. Sci. Rep. 7 (1), 1–11.

Ashrafi, S., Stadler, M., Dababat, A.A., Richert-Pöggeler, K.R., Finckh, M.R., Maier, W., 2017. *Monocillium gamsii* sp. nov. and *Monocillium bulbillosum*: two nematode-associated fungi parasitising the eggs of *Heterodera filipjevi*. MycoKeys 27, 21–38.

Bennett, J.W., Hung, R., Lee, S., Padhi, S., 2012. 18 Fungal and bacterial volatile organic compounds: an overview and their role as ecological signaling agents. Fungal Associations. Springer, Berlin, Heidelberg, pp. 373–393.

Berry, D.R., 1988. In: Berry, D.R. (Ed.), Products of Primary Metabolic Pathways, 130. Blackwell Scientific Publications, Oxford, p. 160.

Blom, D., Fabbri, C., Connor, E.C., Schiestl, F.P., Klauser, D.R., Boller, T., et al., 2011. Production of plant growth modulating volatiles is widespread among rhizosphere bacteria and strongly depends on culture conditions. Environ. Microbiol 13 (11), 3047–3058.

Bos, L.D., Sterk, P.J., Schultz, M.J., 2013. Volatile metabolites of pathogens: a systematic review. PLoS Pathog. 9 (5), e1003311.

Bunge, M., Araghipour, N., Mikoviny, T., Dunkl, J., Schnitzhofer, R., Hansel, A., et al., 2008. On-line monitoring of microbial volatile metabolites by proton transfer reaction-mass spectrometry. Appl. Environ. Microbiol 74 (7), 2179–2186.

Chen, J., Tang, J., Shi, H., Tang, C., Zhang, R., 2017. Characteristics of volatile organic compounds produced from five pathogenic bacteria by headspace-solid phase micro-extraction/gas chromatography-mass spectrometry. J. Basic Microbiol. 57 (3), 228–237.

Claeson, A.S., 2006. Volatile organic compounds from microorganisms: identification and health effects. Doctoral Dissertation, Folkhälsa och klinisk medicin.

Cole, R.J., Schweikert, M.A., Jarvis, B.B. (Eds.), 2003. Handbook of Secondary Fungal Metabolites, 3-Volume Set. Elsevier.

Combet, E., Henderson, J., Eastwood, D.C., Burton, K.S., 2006. Eight-carbon volatiles in mushrooms and fungi: properties, analysis, and biosynthesis. Mycoscience 47 (6), 317–326.

Cumeras, R., Correig, X. (Eds.), 2018. Volatile Organic Compound Analysis in Biomedical Diagnosis Applications. CRC Press.

Dewulf, J.O., Van Langenhove, H., Wittmann, G., 2002. Analysis of volatile organic compounds using gas chromatography. TrAC. Trends Anal. Chem. 21 (9-10), 637–646.

Dickschat, J.S., Helmke, E., Schulz, S., 2005. Volatile organic compounds from arctic bacteria of the cytophaga-flavobacterium-bacteroides group: a retrobiosynthetic approach in chemotaxonomic investigations. Chem. Biodivers. 2 (3), 318–353.

Fiedler, K., Schütz, E., Geh, S., 2001. Detection of microbial volatile organic compounds (MVOCs) produced by moulds on various materials. Int. J. Hyg. Environ. Health 204 (2-3), 111–121.

Franke, C., Hilgarth, M., Vogel, R.F., Petermeier, H., Langowski, H.C., 2019. Characterization of the dynamics of volatile organic compounds released by lactic acid bacteria on modified atmosphere packed beef by PTR-MS. Food Packaging Shelf Life 22, 100400.

Garrity, G.M., Holt, J.G., Whitman, W.B., Keswani, J., Boone, D.R., Koga, Y., et al., 2001. Phylum AII. Euryarchaeota phy. nov. Bergey's Manual® of Systematic Bacteriology. Springer, New York, NY, pp. 211–355.

Hill Jr., H.H., Gardea-Torresdey, J., Eiceman, G.A., 2000. Anal. Chem. 72 (12), 137Rr144R.

Jerković, I., Marijanović, Z., Roje, M., Kuś, P.M., Jokić, S., Čož-Rakovac, R., 2018. Phytochemical study of the headspace volatile organic compounds of fresh algae and seagrass from the Adriatic Sea (single point collection). PLoS One 13 (5), e0196462.

Ji, Z., Majors, R.E., Guthrie, E.J., 1999. Porous layer open-tubular capillary columns: preparations, applications and future directions. J. Chromatogr. A 842 (1-2), 115–142.

Jia, B., Sohnlein, B., Mortelmans, K., Coggiola, M., Oser, H., 2009. Distinguishing methicillin-resistant and sensitive *Staphylococcus aureus* using volatile headspace metabolites. IEEE Sens. J. 10 (1), 71–75.

Julák, J., Procházková-Francisci, E., Stránská, E., Rosová, V., 2003. Evaluation of exudates by solid phase microextraction–gas chromatography. J. Microbiol Methods 52 (1), 115–122.

Kataoka, H., Ishizaki, A., Saito, K., 2016. Recent progress in solid-phase microextraction and its pharmaceutical and biomedical applications. Anal. Methods 8 (29), 5773–5788.

Keller, N.P., Turner, G., Bennett, J.W., 2005. Fungal secondary metabolism—from biochemistry to genomics. Nat. Rev. Microbiol 3 (12), 937–947.

Korpi, A., Järnberg, J., Pasanen, A.L., 2009. Microbial volatile organic compounds. Crit. Rev. Toxicol. 39 (2), 139–193.

Krisher, S., Riley, A., Mehta, K., 2014. Designing breathalyser technology for the developing world: how a single breath can fight the double disease burden. J. Med. Eng. Technol. 38 (3), 156–163.

Kumar, S., Huang, J., Abbassi-Ghadi, N., Španěl, P., Smith, D., Hanna, G.B., 2013. Selected ion flow tube mass spectrometry analysis of exhaled breath for volatile organic compound profiling of esophago-gastric cancer. Anal. Chem. 85 (12), 6121–6128.

Larsen, T.O., Frisvad, J.C., 1994. Production of volatiles and presence of mycotoxins in conidia of common indoor Penicillia and Aspergillii. Health Implic. Fungi Indoor Air Environ. 251–279.

Lee, G.W., Ko, J.A., Oh, B.T., Choi, J.R., Lee, K.J., Chae, J.C., et al., 2012. Biological control of postharvest diseases of apples, peaches and nectarines by *Bacillus subtilis* S16 isolated from halophytes rhizosphere. Biocontrol Sci. Technol. 22 (3), 351–361.

Lemfack, M.C., Gohlke, B.O., Toguem, S.M.T., Preissner, S., Piechulla, B., Preissner, R., 2018. mVOC 2.0: a database of microbial volatiles. Nucleic Acids Res. 46 (D1), D1261–D1265.

Mattheis, J.P., Roberts, R.G., 1992. Identification of geosmin as a volatile metabolite of *Penicillium expansum*. Appl. Environ. Microbiol 58 (9), 3170–3172.

Matysik, S., Herbarth, O., Mueller, A., 2009. Determination of microbial volatile organic compounds (MVOCs) by passive sampling onto charcoal sorbents. Chemosphere 76 (1), 114–119.

Maurer, D.L., Ellis, C.K., Thacker, T.C., et al., 2019. Screening of microbial volatile organic compounds for detection of disease in cattle: development of lab-scale method. Sci. Rep. 9, 12103.

Meskhidze, N., Sabolis, A., Reed, R., Kamykowski, D., 2015. Quantifying environmental stress-induced emissions of algal isoprene and monoterpenes using laboratory measurements. Biogeosciences 12 (3).

Mochalski, P., Al-Zoairy, R., Niederwanger, A., Unterkofler, K., Amann, A., 2014. Quantitative analysis of volatile organic compounds released and consumed by rat L6 skeletal muscle cells in vitro. J. Breath Res. 8 (4), 046003.

Montero-Montoya, R., López-Vargas, R., Arellano-Aguilar, O., 2018. Volatile organic compounds in air: sources, distribution, exposure and associated illnesses in children. Ann. Glob. Health 84 (2), 225.

Müller, A., Faubert, P., Hagen, M., zu Castell, W., Polle, A., Schnitzler, J.P., et al., 2013. Volatile profiles of fungi–chemotyping of species and ecological functions. Fungal Genet. Biol. 54, 25–33.

Murray, P.R., 2012. What is new in clinical microbiology—microbial identification by MALDI-TOF mass spectrometry: a paper from the 2011 William Beaumont Hospital Symposium on molecular pathology. J. Mol. Diagn 14 (5), 419–423.

Nawrath, T., Mgode, G.F., Weetjens, B., Kaufmann, S.H., Schulz, S., 2012. The volatiles of pathogenic and nonpathogenic mycobacteria and related bacteria. Beilstein J. Org. Chem. 8 (1), 290–299.

Pocurull, E., Biedermann, M., Grob, K., 2000. Introduction of large volumes of water-containing samples into a gas chromatograph: improved retention of volatile solutes through the swing system. J. Chromatogr. A 876 (1-2), 135–145.

Preti, G., Thaler, E., Hanson, C.W., Troy, M., Eades, J., Gelperin, A., 2009. Volatile compounds characteristic of sinus-related bacteria and infected sinus mucus: analysis by solid-phase microextraction and gas chromatography–mass spectrometry. J. Chromatogr. B 877 (22), 2011–2018.

Purkhart, R., Köhler, H., Liebler-Tenorio, E., Meyer, M., Becher, G., Kikowatz, A., et al., 2011. Chronic intestinal Mycobacteria infection: discrimination via VOC analysis in exhaled breath and headspace of feces using differential ion mobility spectrometry. J. Breath Res. 5 (2), 027103.

Reyes-Garcés, N., Gionfriddo, E., Gómez-Ríos, G.A., Alam, M.N., Boyacı, E., Bojko, B., et al., 2017. Advances in solid phase microextraction and perspective on future directions. Anal. Chem. 90 (1), 302–360.

Scotter, J.M., Langford, V.S., Wilson, P.F., McEwan, M.J., Chambers, S.T., 2005. Real-time detection of common microbial volatile organic compounds from medically important fungi by selected ion flow tube-mass spectrometry (SIFT-MS). J. Microbiol Methods 63 (2), 127–134.

Sethi, S., Nanda, R., Chakraborty, T., 2013. Clinical application of volatile organic compound analysis for detecting infectious diseases. Clin. Microbiol Rev. 26 (3), 462–475.

Souza-Silva, É.A., Reyes-Garcés, N., Gómez-Ríos, G.A., Boyacı, E., Bojko, B., Pawliszyn, J.A., 2015. Critical review of the state of the art of solid-phase microextraction of complex matrices III. Bioanalytical and clinical applications. TrAC. Trends Anal. Chem. 71, 249–264.

Splivallo, R., Novero, M., Bertea, C.M., Bossi, S., Bonfante, P., 2007. Truffle volatiles inhibit growth and induce an oxidative burst in *Arabidopsis thaliana*. N. Phytol 175 (3), 417–424.

Stritzke, K., Schulz, S., Laatsch, H., Helmke, E., Beil, W., 2004. Novel caprolactones from a marine streptomycete. J. Nat. Products 67 (3), 395–401.

Tait, E., Perry, J.D., Stanforth, S.P., Dean, J.R., 2014. Identification of volatile organic compounds produced by bacteria using HS-SPME-GC–MS. J. Chromatograph Sci. 52 (4), 363–373.

Wiggins, R.J., Wilks, M., Tabaqchali, S., 1985. Analysis by gas liquid chromatography of production of volatile fatty acids by anaerobic bacteria grown on solid medium. J. Clin. Pathol. 38 (8), 933–936.

Zeng, Z.R., Qiu, W.L., Xing, H., Huang, Z.F., 2000. Anal. Sci. 16, 851.

Chapter 5

Microbial volatiles as new frontiers in antibiotic research

Upasana Mangrolia (Deepak)[1,2] and W. Jabez Osborne[3]
[1]School of Biosciences and Technology, Vellore Institute of Technology University, Vellore, India, [2]Biomolecules lab, TT635, VIT, Vellore, Tamil Nadu, India, [3]Biomolecules Lab, Department of Bio-Sciences, School of Bio Sciences and Technology, Vellore Institute of Technology, Vellore, India

Chapter Outline

5.1 Introduction

The environment comprises of biotic and abiotic components and their coexistence is the key to its sustainability. To adapt to the changing environmental conditions, change at the molecular as well as the genetic level is important, indicating evolution is inevitable for survival (Avalos et al., 2018). This is applicable to the prokaryotic microbial cells as much as it is applicable for the eukaryotes. Although the eukaryotes and prokaryotes thrive in synergy, there are disadvantages due to the pathogenic species, especially bacterial pathogenic population. Infections due to such bacterial strains have emerged in recent decades as an important threat, which needs further exploration. Given

Volatiles and Metabolites of Microbes. DOI: https://doi.org/10.1016/B978-0-12-824523-1.00016-X

the wide diversity in the bacterial phylum and its ever-evolving characteristic, the cells are observed to develop resistance against the administered antibiotics (Avalos et al., 2018). An important factor contributing to this adaption is antibiotic abuse. The antibiotic prescribed is of an advanced generation than required, or the patients often do not complete the dosage prescribed in the treatment designed, rendering bacteria to survive abruptly in the antibiotic-infused environment and adapt to it (Smith et al., 2006). Once the resistance develops, the need for successively more lethal drug molecules arises. This need is fulfilled primarily by mimicking the natural products, both microbial and plant based (Audrain et al., 2015). However, there is a need of bacterial populations producing advanced molecules capable of targeting physically distant microbial cells. The volatiles produced by the microbial cells have been recently gaining attention as potential antibiotic compounds (Kanchiswamy, 2015). The need for new antibiotics, the possible approaches of discovering such molecules, and some already known compounds reported for their action potential have been discussed in this chapter.

5.2 Antibiotic resistance in bacteria

Bacterial infections are prevented and eliminated with the aid of antibiotics. The discovery and subsequent implementation of antimicrobial agents in clinical practices revolutionized bacterial infection treatment sector of the medical field (Komolafe, 2003). As indicated by the World Health Organization (WHO), antibiotic resistance is a leading threat to global health as well as food security. The administration of antibiotic dose over a period of time forces the target microbe to either die or adapt to the antibiotic molecule. Antibiotic resistance is not only a natural evolutionary process, for example, vancomycin resistance in *Escherichia coli* (Tomasz and Munoz, 1995), but it can also be an acquired phenomenon. Genetic mutation to cope up with the elevated drug concentration is the most commonly known mechanism of resistance development. This acquired resistance occurs frequently and is a result of bacteria undergoing mutation within itself or by acquisition of resistant DNA (Hawkey, 1998). The surviving microbes carry the resistance gene that are directly transferable to other strains via transposons and plasmids during the inter- and intraspecies interactions (Wise et al., 1998). The resistant genes hence transferred to the neighboring cells eventually result in multidrug resistance (Komolafe, 2003). Additionally, antibiotic abuse in humans and animals is accelerating the development of antibiotic resistance. This leads to advanced challenges in treating the infections and hence raising the pursuit of novel antibiotics.

5.2.1 Mechanism of action of the antibiotics against the bacteria

In order to understand antibiotic resistance, it is important to study the mechanisms underlying the activity of the drug. Moreover, drugs can be either cidal or static in action. The drugs can be categorized based on their mode of attack on the bacterial cells as (1) inhibitors of cell wall synthesis, (2) protein synthesis inhibitors, (3) metabolic antagonists, and (4) nucleic acid synthesis inhibition.

1. Inhibitors of cell wall synthesis.
   Antibiotic groups: penicillins, cephalosporins, and vancomycins.
2. Protein synthesis inhibitors.
   Antibiotic groups: aminoglycosides, tetracyclins, macrolides, and chloramphenicol.
3. Metabolic antagonists.
   Antibiotic groups: sulfonamides, trimethoprim, dapsone, and isoniazid.
4. Nucleic acid synthesis inhibition.
   Antibiotic groups: quinolones and fluoroquinolones, rifampicin.
5. Cell wall disruption.

Antibiotic group: Polymyxin B

5.2.1.1 *Mechanisms of antibiotic resistance*

The clinically surfacing bacterial property of developing antibiotic resistance is well-illustrated in the ever-decreasing efficacy of the antibiotic therapy. To become refractory to a lethal antibiotic treatment, bacterial populations adopt different strategies, thus complicating the clinical practices. Survival strategies emerge as a result of vertical de novo mutation transmission or mobile genetic elements transferred horizontally. In comparison to vertical gene transfer, horizontal gene transfer is difficult to track in an evolution study set up in a laboratory. Minimum inhibitory concentration (MIC) is a routine measure for quantifying resistance; however, strains with low MIC may also express its ability to survive during the high dose, although unable to proliferate. This is attributed to its property of tolerance, level of which can be quantified as the minimum duration for killing at high concentration (Fridman et al., 2014; Wiegand et al., 2008). Tolerance of a fraction of bacterial population to the antibiotic is referred to as persistence (Brauner et al., 2016; Dewachter et al., 2019). Exposing a bacteria to a steady dose of antibiotics often results into an evolved resistance. Low-level resistance is developed as the selection pressure decline when a single resistance mutation is developed. Further, a gradual increase in the antibiotic concentration or prolonged time frame of exposure to antibiotics facilitates accumulation of multiple and highly resistant mutations (Lázár et al., 2013; Oz et al., 2014; Spagnolo et al., 2016; Toprak et al., 2012).

Bacteria that share an ecological niche with an antimicrobial-producing organism develop intrinsic resistance to withstand the harmful effects of the antibiotic molecule in order to thrive in its presence. Given the genetic plasticity in bacteria as an important tool to respond to a variety of environmental changes, it is equally important to understand the mechanism of resistance that it adapts. This knowledge would help in advancing with the drug development studies. Genetic mutations within the cells and acquired mutations are the two groups under which mechanisms of resistance are studied.

5.2.1.1.1 Mutational resistance

As mentioned earlier, a constant exposure to an antibiotic dose triggers antibiotic-resistance adaptation in the bacterial cell, most commonly via mutations. Upon the onset of mutation, the susceptible cells die, and the resistant bacterial cells predominate. With the limited genetic material, the mutations for resistance often decreases the fitness since they are expressed at the expense of cell homeostasis (Munita et al., 2016). The mutational resistance mechanisms include (1) modification of the drug target, (2) reduced drug uptake, (3) enhanced drug efflux mechanisms, and (4) modulation of regulatory pathways.

5.2.1.1.2 Horizontal gene transfer

Horizontal gene transfer, which enables acquisition of foreign DNA, is an important propulsion for bacterial evolution. Of the other environmental components, soil is a major source of bacterial cells and also for most of the antimicrobial agents used in current clinical practices. This supports the robust evidence that suggests existence of "environmental resistome" arising from a diverse bacterial population and supporting the acquisition of resistance genes in bacteria relevant clinically (Munita et al., 2016). The mechanisms underlying horizontal gene transfer include (1) transformation (incorporation of a naked DNA from the milieu), (2) transduction (mediated through phage), and (3) conjugation (bacterial "copulation"). Resistance genes initially found in a few species/pathogens have been now observed to be distributed among other species widely. One of the classic examples is presence of tetracycline resistance determinants, initially found to be present in *Streptococcus* spp. and *Campylobacter jejuni*, but is now widely distributed among different bacterial species. This can be explained on account of their presence in numerous plasmids and broad-range of conjugative transposons (Connell et al., 2003).

5.2.1.2 Antimicrobial resistance-mechanistic understanding

A comprehensive classification based on the biochemical route involved in resistance mechanism can be categorized as (1) drug molecule modification, (2) hindrance in reaching the target, (3) change in and bypass of the target site, and (4) changes in the regulatory networks. The mechanisms have been graphically described in Fig. 5.1.

1. *Drug molecule modification*

As an efficient strategy, bacteria-produced enzymes that either change the chemical structure of the antibiotic molecule or destroy the complete molecule, resulting in an inactive antibiotic.

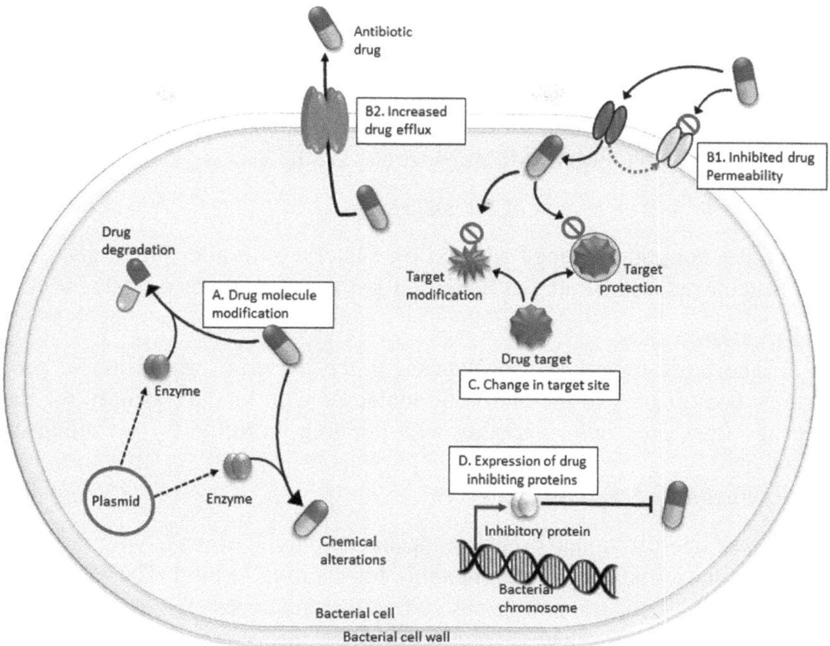

FIGURE 5.1 Mechanistic understanding of the antibiotic resistance in bacteria: (A) drug molecule modification; (B1) inhibited drug permeability; (B2) increased drug efflux; (C) change in target site; and (D) expression of drug-inhibiting proteins.

a. *Chemical alterations*

The primary targets of the resistant bacterial machinery are the drug molecules that are acted upon by inhibiting protein synthesis at the ribosomal level (Wilson, 2014).

b. *Destruction of the molecule*

Certain enzymes destroy the entire drug molecule; the most relevant example is B-lactamase enzyme. The B-lactamases render the molecule inactive by breaking the amide bond of the lactam ring (Abraham and Chain, 1940; Costa et al., 2011).

2. *Decreased antibiotic permeability and increased efflux*

a. *Decreased antibiotic permeability*

The prime targets of antibiotics are present usually intracellularly and in the inner membrane of the cytoplasmic membrane in case of gram-negative bacteria. Since the mode of activity requires the bacteria to penetrate the cell membrane as the primary step, bacteria prevent the influx of the molecules with the adapted mechanisms. Here the bacteria utilizes the outer membrane as the first line of defense (Pagès et al., 2008).

b. *Increased efflux action*

Bacteria producing complex efflux machineries are another efficient way of expressing antibiotic resistance. The earliest evidence of drug efflux can be noted from early 1980s, where in *E. coli* was reported to pump out tetracyclins from the cytoplasm (McMurry et al., 1980).

3. *Change in and/or bypass of the target site*

Another common strategy adapted by bacteria is to alter the target sites. This is achieved with strategies classified in the following broad categories.

a. *Target protection*
The basic underlying mechanism is the bacterial production of proteins homologues to the antibiotic molecule with binding affinity for the drug target site, thus competing with the drug molecule for executing its activity.

b. *Modification of the target site*

Another interesting and common mechanism involves modifying the drug target site, thus rendering it incompatible for the drug to bind. These changes may be (1) point mutations in genes coding the target sites, (2) enzymatically altering the target site, and (3) replacement of the active site of the target molecule.

4. *Changes in the regulatory networks*
To survive extreme hostile environments inclusive of the human body, bacteria adapt sophisticated mechanisms of survival. They not only need to compete with nutrients but also survive by protection from rival cells (eukaryotic and prokaryotic). While residing in the biological niches of an animal body, it is under constant stress induced by the immune system; bacteria develop resistance to these attacks over a series of reactions.

5.3 Approaches for discovering microbial volatiles as antibiotics

5.3.1 Culture-based approach

Screening for antagonism imbibes/utilizes a common principal of the inhibition of a test strain in the presence of a closely cultivated indicator strain, a systematic method first developed by Selman Waksman in 1940s for soil bacteria. Bacteria evolve in a way that also supports their existence in presence of bacteria-producing antimicrobial compounds; however, production of the antimicrobials is also a part of their defense strategy of survival in cases of foreign bacterial invasions. Bacterial strains are screened with techniques such as cross-streak method, spot-on-the-lawn method, and diffusion method (Durand et al., 2019). Bacteria as a source of antibiotic property is among the earliest as well as the well-known strategy for procuring antibiotic molecules.

5.3.2 Assessing the antimicrobial properties of the already known compounds

Looking into the available resources, such as chemical databases, and filtering potent drug molecules is another successful and progressive approach. These databases comprise of several millions of compounds awaiting an exploration of their biological significance, for instance, molecules such as fidaxomicin and linezolid were discovered several years before their use as antibiotics (Durand et al., 2019; Lewis, 2013).

5.3.3 Synthesis of new molecules and improvement of already known compounds

For a drug to be efficient it must display appropriate absorbance upon administration, target specificity, and nontoxicity in the host; a commonly recognized rule for industrial purposes is Lipinski's rule. The rule evaluates a compound's drug likeness, with the compounds to be having a molecular mass not greater than 500 Da, five or less hydrogen bond donors, 10 or less hydrogen bond acceptors, and an octanol-water partition coefficient lop P not greater than 5 (Lewis, 2013). In place of looking for volatile compounds that satiate these properties, rational drug designing takes the reverse approach. It is an empirical synthesis of a novel molecule satisfying the objective. Alternatively, modification of the known compound is carried out to specifically overcome resistance in bacteria; for example, for plazomicin, a modified sisomicin (aminoglycoside) molecule, the efficacy is comparable to levofloxacin in treating pyelonephritis as well as urinary tract infection (Armstrong and Miller, 2010; Connolly et al., 2018). Although there is an availability of several molecules and available expertise, the number of drugs reaching clinical trials are limited (Lewis, 2012).

5.3.4 Gene mining

Every molecule synthesized by a cell can be tracked down to genetic level. Enormous prokaryotic gene data is available online in the form of sequence databases; of these, the biosynthetic gene clusters (BGC) code for secondary metabolites can be the gene pool of interest for antibiotic discovery. In the Human Microbiome Project, 74 BGCs were identified of 59 genomes, and these BGCs majorly belonged to firmicutes, proteobacteria, and bacteroidetes. The putative volatile bacteriocins, hence coded by these BGCs, belonged to class III and IV (Walsh et al., 2015).

5.3.5 CRISPR-Cas9

Bacterial cells gain genetic material from phages; these are not always beneficial to the bacterial host, and a mechanism is required to prevent the bacterial cell from imbibing these genetic materials.

The latest addition to the discovery of the bacterial immune system is the CRISPR-Cas (clustered regularly interspaced short palindromic repeats) system. This bacterial defense system is being explored to target the resistance gene of highly resistant bacterial populations, thus making them sensitive to the antibiotics. The CRISPR-Cas9 system, transformed by plasmids and transduced by bacteriophages in the target populations of bacterial cells, acts on the resistance gene; for example, phagemid targets *Staphylococcus aureus* methicillin-resistant gene (Bikard et al., 2014). This approach targets not only the plasmidic gene but also those present on the chromosome (Citorik et al., 2014).

The role of compounds found in the form of microbial volatiles as a result of a culture-based approach has been discussed in detail.

5.4 Microbial volatiles and their action potential in the antibiotic research

As mentioned earlier, survival in a native environment diverse in microbial population requires the bacteria to develop defense mechanisms. Since these are not crucial products under a laboratory environment, they are termed as secondary metabolites. Thus, natural macromolecules (Proteins, DNA, and RNA) and their precursors are not included in the working definition of natural products (Katz and Baltz, 2016). Over the past 75 years, since the discovery of penicillin, more than 23,000 natural products have been characterized for their potential roles; bacteria belonging to the family *Actinomycetaceae* are found to be the major contributors (Bérdy, 2012). Natural products, their derivatives, and synthetic drugs (based on the pharmacophore of the natural products) contribute to about 50% to the total new chemical entities registered as new drugs (Bhanot et al., 2016; Demain, 2014; Giddings and Newman, 2013; Kinch et al., 2014). Antiinfection, especially antibacterial therapy, is the prominent use of the natural products and their derivatives of advanced generations.

Bacterial population survival in a given environmental niche is an outcome of their synchronized gene expression (Ryan and Dow, 2008). To achieve this, cell-to-cell communication becomes essential to monitor their population density (quorum sensing). The processes regulated in such coordinated fashion includes developmental as well as survival strategies by contributing to virulence, antibiotic production, biofilm formation, and so forth.

The signaling molecules produced by bacteria are chemically diverse and vary in size. Some of these with lower molecular weight might act as volatile compounds. Microbial volatiles can be considered as an important reservoir of molecular templates for designing drugs, especially antibiotics.

5.4.1 Bacterial volatile compounds and their antibacterial activity

With the advantage of traveling distances, the bacterial volatile organic compounds (VOCs) have a direct antagonist effect on other bacteria. The VOCs produced by various marine and terrestrial microbes are diverse in their chemical properties (Piechulla et al., 2017; Schulz et al., 2010). The various chemical classes include alkanes, alkenes, esters, alcohols, ketones, terpenoids, and other sulfur-containing compounds. The variations in these classes are enormous but yet are to be discovered (Klapschinski et al., 2016; von Reuß et al., 2010). Several bacterial volatiles have been identified for their antibacterial properties, and some are discussed here. The bacterial volatile metabolites produced are key players in promoting cross-kingdom interactions. Briefly, Gram-negative bacteria use cyclic peptides, fatty acid derivatives, or *N*-acylhomoserine lactones (quorum-sensing molecules); Gram-positive bacteria employ amino acids, modified peptides, or gamma butyrolactones for communication with their kind (Kai et al., 2009).

5.4.1.1 Bacterial volatile compounds with direct and indirect activity

Pseudomonas fluorescens and *Serratia plymuthica* are rhizobacteria reported to emit dimethyl sulfide having a bacteriostatic effect on plant pathogens—*Agrobacterium tumefaciens* and *Agrobacterium vitis* (Dandurishvili et al., 2011). Another strain, *P. fluorescens* WR-1, had bacteriostatic effects on tomato pathogen *Ralstonia solanacearum* via its volatiles benzothiazole and 1-methyl naphthalene (Raza et al., 2016a). The growth of *R. solanacearum* is also inhibited by volatiles, 1,2-benzisothiazol-3(2 H)-one and 1,3-butadiene, produced by *Bacillus* spp. (Tahir et al., 2017). The g-butyrolactones are hormone-like in structure with antifungal and antibacterial activity (Schulz et al., 2010). The compound albaflevenone is another terpenoid (sequestered ketone) produced by bacteria. It was first reported and isolated as a product from *Streptomyces albidoflavus* (Schulz and Dickschat, 2007). It has been reported to express bactericidal activity against *Bacillus subtilis* (Gürtler et al., 1994).

In addition to direct antagonistic effects, certain bacterial volatiles have been evaluated to influence other bacteria to produce antibiotic molecules. *Collimonas pratensis* and *S. plymuthica* volatiles induced growth of *P. fluorescens* Pf0-1 and provoked increased secondary metabolite production that had antagonistic activity against Bacillus (Garbeva et al., 2014). Paenibacillus when interacts with Burkholderia, emits 2,5-bis(i-methyl)-pyrazine that has been reported to exhibit activity against human pathogens such as *E. coli* and *S. aureus* (Barka et al., 2016; Bérdy, 2012). Pyrizines have become an important class of antimicrobials in recent times.

5.4.1.2 Molecular mechanisms involved in the action of bacterial volatiles

VOCs are also capable of altering the transcriptional expression levels of genes involved in pathogenecity and motility, such as type III secretion system, virulence regulator PhcA, and extracellular polysaccharide production. Schleiferons A and B, produced by skin-borne *Staphylococcus schleiferi*, interferes with prodiginine production in Gram-positive bacteria, thereby inhibiting their growth and modulating the skin microbiome (Lemfack et al., 2016). Furfuryl isovalerate acts as quorum quencher in Gram-negative bacteria but inhibits growth of both Gram-negative and positive bacteria (Schulz et al., 2010). About 15% of the total tested actinobacteria species emitted volatiles that exhibited inhibitory activity against *B. subtilis* or *E. coli* (Avalos et al., 2018). The volatiles acted on either Gram-negative bacteria or Gram-positive bacteria but not on both; suggesting a unique mode of action on both. Among others chemical VOCs, a well-studied molecule is the diffusible signal factor (cis-11-methyl-2-dodecenoic acid). It has been observed to modulate virulence factors in *Xanthomonas campestris* (Barber et al., 1997; Wang et al., 2003); it reduced the transcription of genes involved in drug efflux and biofilm formation in *B. subtilis* (Deng et al., 2014). Certain beneficial bacteria produce volatile metabolites that enhance their exploration properties; for example *Streptomyces* cells produce trimethylamine which increases the pH of the environment, favoring the exploration for itself and the neighboring exploration-competent streptomycetes (Jones and Elliot, 2017). On the contrary, the trimethylamine has been shown to influence the antibiotic resistance in pathogens such as *E. coli* (Létoffé et al., 2014). The bacterial volatiles impart immunity in indirect ways significantly. Inside the human host, the bacterial diversity is intriguing. The gut microbial load is often considered to be an organ in itself. It functions and self-maintains the microbial load by timely exclusion and/or elimination of pathogenic species from the gut environmental niche. For elimination, the primary method explored is its ability to produce antibacterial compounds. The human gut provides protection from the invading pathogens with the aid of short-chain fatty acids (SCFAs) such as butyrate, propionate, and acetate. These SCFAs are metabolites of the microbial homeostasis, that is, they are bacterial fermentation products of undigested starch and dietary fibers. Butyrate is the most abundant among the SCFAs in the gut; along with various antioxidative, antiinflammatory, and anticarcinogenic activities, it primes in synchronizing gastrointestinal immunity (Corrêa-Oliveira et al., 2016; Louis et al., 2014; Natarajan and Pluznick, 2014; Thorburn et al., 2014). It aids the immune system by playing a protective role against the enteric pathogens either directly or by mediating the process through the immune system. They are capable of directly diffusing through the bacterial membrane, entering the pathogenic cells, and reducing the pH, this imparts damage to the pathogens. On the other hand, the SCFAs maintain the integrity of the epithelial barrier function thereby inhibiting the pathogenic invasion (Ashida et al., 2012; Honda and Littman, 2012). Additionally they induce antimicrobial peptide production in enterocytes. The macrophages differentiating in the presence of butyrate have been shown to express increased antimicrobial properties (Schulthess et al., 2019).

5.4.1.3 Sensitizing effects on the drug-resistant bacterial cells

Although the volatilome is abundant in chemical diversity, these volatiles must be further studied to explore their activity-synergism with the soluble antibiotic molecules. Identifying molecules with properties to sensitize the resistant cells would be a boon for overcoming the increasing threat of bacterial infections. This effect has been reported for terpene eugenol that lowers the MIC for antibiotic-resistant bacteria, thereby conferring antibiotic sensitivity (Hemaiswarya and Doble, 2009). Also, phenylpropanoids, B-cinnamic acid and ferulic acid, have shown restored sensitivity towards erythromycin, amikazin, and vancomycin (Hemaiswarya and Doble, 2009). A combination of monoterpenes (G-terpinene, 1S-a-pinene, B-myrcene, and B-pinene) synthesized by *C. pratensis* showed inhibitory activity on *S. aureus* and *E. coli* (Song et al., 2015). The activity of antituberculosis drugs such as ethanbutol, isoniazid and rifampicin has been subjected to enhancement upon pretreatment with essential volatile oil components such as limonene. sabinene, thymol, a-pinene and even carvacrol (Sieniawska et al., 2017).

5.4.1.4 Involvement of microbial volatiles in promoting plant health

Beyond human pathogens, bacterial pathogens also affect the plant health; growth inhibition has been reported in tomato wilt pathogens *R. solanacearum* by *Bacillus amyloliquifaciens* strain SQR-9 volatilome (Raza et al., 2016b), although the exact compound and its mechanism of action remains unevaluated so far. Another activity of the bacterial volatilome has been observed in strains of *Pseudomonas chlororaphis, S. plymuthica*, and *Serratia proteamaculans*; their volatilomes showed bacteriostatic activity against the well-studied *A. tumefaciens* (Popova et al., 2014). *P. fluorescens* and *S. plymuthica* have been reported to be capable of emitting bacteriostatic compound dimethyl sulfide, active against *A. tumefaciens* and *A. vitis* (Dandurishvili et al., 2011). This is also indicative of efficiency in activity due to the diversity in chemical composition when tested collectively as a volatilome.

The activity of these VOCs are as diverse as their chemistry range. However, the reported cases of their antibacterial cases are limited.

5.4.2 Fungal volatile compounds and their antibacterial activity

Bacteria not only contribute to antibacterial activity but also influence the growth of fungal colonies in their niche of existence. On the other hand, fungal VOCs impact the bacterial cells in their coexistence phenotypically.

Fungal strains have the potential to attract mutualistic bacteria and repel the competitors by targeting the bacterial motility under the influence of the VOCs they produce (Piechulla et al., 2017). The antibacterial activity of fungal volatiles are poorly explored, however, the antifungal activities of fungal volatiles have been reported. The volatiles produced by *Fusarium oxysporum* acts antagonistically against *F. oxysporum lactucae* strain Fuslat10 by repressing its two putative virulent genes (Minerdi et al., 2009).

5.4.3 Volatile compounds from Cyanobacteria

Microalgae are extremely evolutionary and phylogenetically diverse. The micro-green algae and the cyanobacteria are widely known and recognized by their toxins; however these also hold the potential of developing antibiotics (Berry et al., 2004). The microalgal metabolites are chemically diverse (Senhorinho et al., 2015). The production of volatile compounds by *Spirulina platensis*, with heptadecan and tetradecan as major components, showed antibacterial activity but at low efficacy (Ozdemir et al., 2004). Although many compounds sourced from cyanobacteria have been investigated for antimicrobial activity, the data available on volatile microalgal compounds is limited.

5.5 Conclusion

The limited exploration of microbial volatiles, specifically for their antibiotic potential, indicates the existence of persistent challenges. These typically include the primary detection of the volatile compounds, and purification is further challenging. Appropriate methods need to be developed to identify its antibiotic activity since the diffusion pattern is often affected by the volatility of the compounds. Given together, although volatile compounds have the potency, their identification and role in an ecosystem needs to be mechanistically better understood.

References

Abraham, E.P., Chain, E., 1940. An enzyme from bacteria able to destroy Penicillin. Nature 146, 837–837. Available from: https://doi.org/10.1038/146837a0.

Armstrong, E.S., Miller, G.H., 2010. Combating evolution with intelligent design: the neoglycoside ACHN-490. Curr. Opin. Microbiol. 13, 565–573. Available from: https://doi.org/10.1016/j.mib.2010.09.004.

Ashida, H., Ogawa, M., Kim, M., Mimuro, H., Sasakawa, C., 2012. Bacteria and host interactions in the gut epithelial barrier. Nat. Chem. Biol. 8, 36–45. Available from: https://doi.org/10.1038/nchembio.741.

Audrain, B., Farag, M.A., Ryu, C.M., Ghigo, J.M., 2015. Role of bacterial volatile compounds in bacterial biology. FEMS Microbiol. Rev. 39, 222–233. Available from: https://doi.org/10.1093/femsre/fuu013.

Avalos, M., van Wezel, G.P., Raaijmakers, J.M., Garbeva, P., 2018. Healthy scents: microbial volatiles as new frontier in antibiotic research? Curr. Opin. Microbiol. 45, 84–91. Available from: https://doi.org/10.1016/j.mib.2018.02.011.

Barber, C.E., Tang, J.L., Feng, J.X., Pan, M.Q., Wilson, T.J.G., Slater, H., et al., 1997. A novel regulatory system required for pathogenicity of *Xanthomonas campestris* is mediated by a small diffusible signal molecule. Mol. Microbiol. 24, 555–566. Available from: https://doi.org/10.1046/j.1365-2958.1997.3721736.x.

Barka, E.A., Vatsa, P., Sanchez, L., Gaveau-Vaillant, N., Jacquard, C., Klenk, H.-P., et al., 2016. Taxonomy, physiology, and natural products of Actinobacteria. Microbiol. Mol. Biol. Rev. 80, 1–43. Available from: https://doi.org/10.1128/MMBR.00019-15.

Bérdy, J., 2012. Thoughts and facts about antibiotics: where we are now and where we are heading. J. Antibiot. (Tokyo) 65, 385–395. Available from: https://doi.org/10.1038/ja.2012.27.

Berry, J.P., Gantar, M., Gawley, R.E., Wang, M., Rein, K.S., 2004. Pharmacology and toxicology of pahayokolide A, a bioactive metabolite from a freshwater species of Lyngbya isolated from the Florida Everglades. Comp. Biochem. Physiol. Part C Toxicol. Pharmacol. 139, 231–238. Available from: https://doi.org/10.1016/j.cca.2004.11.005.

Bhanot, A., Sharma, R., Noolvi, M.N., 2016. Natural sources as potential anti-cancer agents: a review. Int. J. Phytomed. 3, 2011.

Bikard, D., Euler, C.W., Jiang, W., Nussenzweig, P.M., Goldberg, G.W., Duportet, X., et al., 2014. Exploiting CRISPR-Cas nucleases to produce sequence-specific antimicrobials. Nat. Biotechnol. 32, 1146–1150. Available from: https://doi.org/10.1038/nbt.3043.

Brauner, A., Fridman, O., Gefen, O., Balaban, N.Q., 2016. Distinguishing between resistance, tolerance and persistence to antibiotic treatment. Nat. Publ. Gr. 14, 320–330. Available from: https://doi.org/10.1038/nrmicro.2016.34.

Citorik, R.J., Mimee, M., Lu, T.K., 2014. Sequence-specific antimicrobials using efficiently delivered RNA-guided nucleases. Nat. Biotechnol. 32, 1141–1145. Available from: https://doi.org/10.1038/nbt.3011.

Connell, S.R., Tracz, D.M., Nierhaus, K.H., Taylor, D.E., 2003. Ribosomal protection proteins and their mechanism of tetracycline resistance. Antimicrob. Agents Chemother. 47, 3675–3681. Available from: https://doi.org/10.1128/AAC.47.12.3675.

Connolly, L.E., Riddle, V., Cebrik, D., Armstrong, E.S., Miller, L.G., 2018. A multicenter, randomized, double-blind, phase 2 study of the efficacy and safety of plazomicin compared with levofloxacin in the treatment of complicated urinary tract infection and acute pyelonephritis. Antimicrob. Agents Chemother. 62. Available from: https://doi.org/10.1128/AAC.01989-17.

Corrêa-Oliveira, R., Fachi, J.L., Vieira, A., Sato, F.T., Vinolo, M.A.R., 2016. Regulation of immune cell function by short-chain fatty acids. Clin. Transl. Immunol. 5, e73. Available from: https://doi.org/10.1038/cti.2016.17.

Costa, V.M.D., King, C.E., Kalan, L., Morar, M., Sung, W.W.L., Schwarz, C., et al., 2011. Antibiotic resistance is ancient. Nature 477, 457–461. Available from: https://doi.org/10.1038/nature10388.

Dandurishvili, N., Toklikishvili, N., Ovadis, M., Eliashvili, P., Giorgobiani, N., Keshelava, R., et al., 2011. Broad-range antagonistic rhizobacteria *Pseudomonas fluorescens* and *Serratia plymuthica* suppress Agrobacterium crown gall tumours on tomato plants. J. Appl. Microbiol. 110, 341–352. Available from: https://doi.org/10.1111/j.1365-2672.2010.04891.x.

Demain, A.L., 2014. Importance of microbial natural products and the need to revitalize their discovery. J. Ind. Microbiol. Biotechnol. 41, 185–201. Available from: https://doi.org/10.1007/s10295-013-1325-z.

Deng, Y., Lim, A., Lee, J., Chen, S., An, S., Dong, Y.-H., et al., 2014. Diffusible signal factor (DSF) quorum sensing signal and structurally related molecules enhance the antimicrobial efficacy of antibiotics against some bacterial pathogens. BMC Microbiol. 14, 51. Available from: https://doi.org/10.1186/1471-2180-14-51.

Dewachter, L., Fauvart, M., Michiels, J., 2019. Bacterial heterogeneity and antibiotic survival: understanding and combatting persistence and heteroresistance. Mol. Cell 76, 255–267. Available from: https://doi.org/10.1016/j.molcel.2019.09.028.

Durand, G.A., Raoult, D., Dubourg, G., 2019. Antibiotic discovery: history, methods and perspectives. Int. J. Antimicrob. Agents 53, 371–382. Available from: https://doi.org/10.1016/j.ijantimicag.2018.11.010.

Fridman, O., Goldberg, A., Ronin, I., Shoresh, N., Balaban, N.Q., 2014. Optimization of lag time underlies antibiotic tolerance in evolved bacterial populations. Nature 99. Available from: https://doi.org/10.1038/nature13469.

Garbeva, P., Hordijk, C., Gerards, S., de Boer, W., 2014. Volatile-mediated interactions between phylogenetically different soil bacteria. Front. Microbiol. 5. Available from: https://doi.org/10.3389/fmicb.2014.00289.

Giddings, L.-A., Newman, D.J., 2013. Microbial natural products: molecular blueprints for antitumor drugs. J. Ind. Microbiol. Biotechnol. 40, 1181–1210. Available from: https://doi.org/10.1007/s10295-013-1331-1.

Gürtler, H., Pedersen, R., Anthoni, U., Christophersen, C., Nielsen, P.H., Wellington, E.M.H., et al., 1994. Albaflavenone, a sesquiterpene ketone with a zizaene skeleton produced by a streptomycete with a new rope morphology. J. Antibiot. (Tokyo) 47, 434–439. Available from: https://doi.org/10.7164/antibiotics.47.434.

Hawkey, P.M., 1998. The origins and molecular basis of antibiotic resistance. BMJ 317 (September), 657–660. Available from: https://doi.org/10.1136/bmj.317.7159.657.

Hemaiswarya, S., Doble, M., 2009. Synergistic interaction of eugenol with antibiotics against Gram negative bacteria. Phytomedicine 16, 997–1005. Available from: https://doi.org/10.1016/j.phymed.2009.04.006.

Honda, K., Littman, D.R., 2012. The microbiome in infectious disease and inflammation. Annu. Rev. Immunol. 30, 759–795. Available from: https://doi.org/10.1146/annurev-immunol-020711-074937.

Jones, S.E., Elliot, M.A., 2017. Streptomyces exploration: competition, volatile communication and new bacterial behaviours. Trends Microbiol. 25, 522–531. Available from: https://doi.org/10.1016/j.tim.2017.02.001.

Kai, M., Haustein, M., Molina, F., Petri, A., Scholz, B., Piechulla, B., 2009. Bacterial volatiles and their action potential. Appl. Microbiol. Biotechnol. 81, 1001–1012. Available from: https://doi.org/10.1007/s00253-008-1760-3.

Kanchiswamy, C.N., 2015. Chemical diversity of microbial volatiles and their potential for plant growth and productivity. 6. https://doi.org/10.3389/fpls.2015.00151.

Katz, L., Baltz, R.H., 2016. Natural product discovery: past, present, and future. J. Ind. Microbiol. Biotechnol. Available from: https://doi.org/10.1007/s10295-015-1723-5.

Kinch, M.S., Haynesworth, A., Kinch, S.L., Hoyer, D., 2014. An overview of FDA-approved new molecular entities: 1827–2013. Drug. Discov. Today 19, 1033–1039. Available from: https://doi.org/10.1016/j.drudis.2014.03.018.

Klapschinski, T.A., Rabe, P., Dickschat, J.S., 2016. Pristinol, a Sesquiterpene Alcohol with an Unusual Skeleton from Streptomyces pristinaespiralis. Angew. Chem. Int. Ed. 55, 10141–10144. Available from: https://doi.org/10.1002/anie.201605425.

Komolafe, O., 2003. antibiotic resistance in bacteria - an emerging public health problem. Malawi Med. J. 15, 63–67.

Lázár, V., Pal Singh, G., Spohn, R., Nagy, I., Horváth, B., Hrtyan, M., et al., 2013. Bacterial evolution of antibiotic hypersensitivity. Mol. Syst. Biol. 9, 700. Available from: https://doi.org/10.1038/msb.2013.57.

Lemfack, M.C., Ravella, S.R., Lorenz, N., Kai, M., Jung, K., Schulz, S., et al., 2016. Novel volatiles of skin-borne bacteria inhibit the growth of Gram-positive bacteria and affect quorum-sensing controlled phenotypes of Gram-negative bacteria. Syst. Appl. Microbiol. 39, 503–515. Available from: https://doi.org/10.1016/j.syapm.2016.08.008.

Létoffé, S., Audrain, B., Bernier, S.P., 2014. Raising culture medium pH aerial exposure to the bacterial volatile compound trimethylamine. 5, 1–12. https://doi.org/10.1128/mBio.00944-13.Editor.

Lewis, K., 2012. Recover the lost art of drug discovery. Nature 485, 439–440. Available from: https://doi.org/10.1038/485439a.

Lewis, K., 2013. Platforms for antibiotic discovery. Nat. Rev. Drug Discov. 12, 371–387. Available from: https://doi.org/10.1038/nrd3975.

Louis, P., Hold, G.L., Flint, H.J., 2014. The gut microbiota, bacterial metabolites and colorectal cancer. Nat. Rev. Microbiol. 12, 661–672. Available from: https://doi.org/10.1038/nrmicro3344.

McMurry, L., Petrucci, R.E., Levy, S.B., 1980. Active efflux of tetracycline encoded by four genetically different tetracycline resistance determinants in *Escherichia coli*. Proc. Natl. Acad. Sci. 77, 3974–3977. Available from: https://doi.org/10.1073/pnas.77.7.3974.

Minerdi, D., Bossi, S., Gullino, M.L., Garibaldi, A., 2009. Volatile organic compounds: a potential direct long-distance mechanism for antagonistic action of *Fusarium oxysporum* strain MSA 35. Environ. Microbiol. 11, 844–854. Available from: https://doi.org/10.1111/j.1462-2920.2008.01805.x.

Munita, J.M., Arias, C.A., Unit, A.R., De Santiago, A., 2016. Mechanisms of antibiotic resistance. Microbiol. Spectr. 4, 1–37. Available from: https://doi.org/10.1128/microbiolspec.VMBF-0016-2015.Mechanisms.

Natarajan, N., Pluznick, J.L., 2014. From microbe to man: the role of microbial short chain fatty acid metabolites in host cell biology. Am. J. Physiol. Physiol. 307, C979–C985. Available from: https://doi.org/10.1152/ajpcell.00228.2014.

Oz, T., Guvenek, A., Yildiz, S., Karaboga, E., Tamer, Y.T., Mumcuyan, N., et al., 2014. Strength of selection pressure is an important parameter contributing to the complexity of antibiotic resistance evolution. Mol. Biol. Evol. 31, 2387–2401. Available from: https://doi.org/10.1093/molbev/msu191.

Ozdemir, G., Karabay, N.U., Dalay, M.C., Pazarbasi, B., 2004. Antibacterial activity of volatile component and various extracts of *Spirulina platensis*. Phyther. Res. 18, 754–757. Available from: https://doi.org/10.1002/ptr.1541.

Pagès, J., James, C.E., Winterhalter, M., 2008. The porin and the permeating antibiotic: a selective diffusion barrier in Gram-negative bacteria. Nat. Rev. Microbiol. 6, 893–903. Available from: https://doi.org/10.1038/nrmicro1994.

Piechulla, B., Lemfack, M.C., Kai, M., 2017. Effects of discrete bioactive microbial volatiles on plants and fungi. Plant Cell Environ. 40, 2042–2067. Available from: https://doi.org/10.1111/pce.13011.

Popova, A.A., Koksharova, O.A., Lipasova, V.A., Zaitseva, J.V., Katkova-Zhukotskaya, O.A., Eremina, S.I., et al., 2014. Inhibitory and toxic effects of volatiles emitted by strains of Pseudomonas and Serratia on growth and survival of selected microorganisms, Caenorhabditis elegans, and Drosophila melanogaster. Biomed. Res. Int. 2014, 1–11. Available from: https://doi.org/10.1155/2014/125704.

Raza, W., Ling, N., Liu, D., Wei, Z., Huang, Q., Shen, Q., 2016a. Volatile organic compounds produced by *Pseudomonas fluorescens* WR-1 restrict the growth and virulence traits of *Ralstonia solanacearum*. Microbiol. Res. 192, 103–113. Available from: https://doi.org/10.1016/j.micres.2016.05.014.

Raza, W., Ling, N., Yang, L., Huang, Q., Shen, Q., 2016b. Response of tomato wilt pathogen *Ralstonia solanacearum* to the volatile organic compounds produced by a biocontrol strain Bacillus amyloliquefaciens SQR-9. Sci. Rep. 6, 24856. Available from: https://doi.org/10.1038/srep24856.

Ryan, R.P., Dow, J.M., 2008. Diffusible signals and interspecies communication in bacteria. Microbiology 154, 1845–1858. Available from: https://doi.org/10.1099/mic.0.2008/017871-0.

Schulthess, J., Pandey, S., Capitani, M., Rue-Albrecht, K.C., Arnold, I., Franchini, F., et al., 2019. The short chain fatty acid butyrate imprints an antimicrobial program in macrophages. Immunity 50, 432–445. Available from: https://doi.org/10.1016/j.immuni.2018.12.018. e7.

Schulz, S., Dickschat, J.S., 2007. Bacterial volatiles: the smell of small organisms. Nat. Prod. Rep. 24, 814–842. Available from: https://doi.org/10.1039/b507392h.

Schulz, S., Dickschat, J.S., Kunze, B., Wagner-Dobler, I., Diestel, R., Sasse, F., 2010. Biological activity of volatiles from marine and terrestrial bacteria. Mar. Drugs 8, 2976–2987. Available from: https://doi.org/10.3390/md8122976.

Senhorinho, G.N.A., Ross, G.M., Scott, J.A., 2015. Cyanobacteria and eukaryotic microalgae as potential sources of antibiotics. Phycologia 54, 271–282. Available from: https://doi.org/10.2216/14-092.1.

Sieniawska, E., Swatko-Ossor, M., Sawicki, R., Skalicka-Woźniak, K., Ginalska, G., 2017. Natural terpenes influence the activity of antibiotics against isolated *Mycobacterium tuberculosis*. Med. Princ. Pract. 26, 108–112. Available from: https://doi.org/10.1159/000454680.

Smith, C.J., Rocha, E.R., Paster, B., 2006. The prokaryotes: a handbook on the biology of bacteria - symbiotic associations. Biotechnol. Appl. Microbiol. Appl. Microbiol. Available from: https://doi.org/10.1007/0-387-30746-x_22.

Song, C., Schmidt, R., de Jager, V., Krzyzanowska, D., Jongedijk, E., Cankar, K., et al., 2015. Exploring the genomic traits of fungus-feeding bacterial genus Collimonas. BMC Genomics 16, 1103. Available from: https://doi.org/10.1186/s12864-015-2289-3.

Spagnolo, F., Rinaldi, C., Sajorda, D.R., Dykhuizen, D.E., 2016. Evolution of resistance to continuously increasing streptomycin concentrations in populations of *Escherichia coli*. Antimicrob. Agents Chemother. 60, 1336–1342. Available from: https://doi.org/10.1128/AAC.01359-15.

Tahir, H.A.S., Gu, Q., Wu, H., Niu, Y., Huo, R., Gao, X., 2017. Bacillus volatiles adversely affect the physiology and ultra-structure of *Ralstonia solanacearum* and induce systemic resistance in tobacco against bacterial wilt. Sci. Rep. 7, 40481. Available from: https://doi.org/10.1038/srep40481.

Thorburn, A.N., Macia, L., Mackay, C.R., 2014. Diet, metabolites, and "Western-Lifestyle" inflammatory diseases. Immunity 40, 833–842. Available from: https://doi.org/10.1016/j.immuni.2014.05.014.

Tomasz, A., Munoz, R., 1995. b-Lactam antibiotic resistance in gram-positive bacterial pathogens of the upper respiratory tract: a brief overview of mechanisms. Microb. Drug. Resist. 1, 103–109.

Toprak, E., Veres, A., Michel, J.-B., Chait, R., Hartl, D.L., Kishony, R., 2012. Evolutionary paths to antibiotic resistance under dynamically sustained drug selection. Nat. Genet. 44, 101–105. Available from: https://doi.org/10.1038/ng.1034.

von Reuß, S.H., Kai, M., Piechulla, B., Francke, W., 2010. Octamethylbicyclo[3.2.1]octadienes from the Rhizobacterium Serratia odorifera. Angew. Chem. Int. Ed. 49, 2009–2010. Available from: https://doi.org/10.1002/anie.200905680.

Walsh, C.J., Guinane, C.M., Hill, C., Ross, R.P., O'Toole, P.W., Cotter, P.D., 2015. In silico identification of bacteriocin gene clusters in the gastrointestinal tract, based on the Human Microbiome Project's reference genome database. BMC Microbiol. 15, 183. Available from: https://doi.org/10.1186/s12866-015-0515-4.

Wang, L.-H., He, Y., Gao, Y., Wu, J.E., Dong, Y.-H., He, C., et al., 2003. A bacterial cell-cell communication signal with cross-kingdom structural analogues. Mol. Microbiol. 51, 903–912. Available from: https://doi.org/10.1046/j.1365-2958.2003.03883.x.

Wiegand, I., Hilpert, K., Hancock, R.E.W., 2008. Agar and broth dilution methods to determine the minimal inhibitory concentration (MIC) of antimicrobial substances. Nat. Protoc. 3, 163–175. Available from: https://doi.org/10.1038/nprot.2007.521.

Wilson, D.N., 2014. Ribosome-targeting antibiotics and mechanisms of bacterial resistance. Nat. Publ. Gr. 12, 35–48. Available from: https://doi.org/10.1038/nrmicro3155.

Wise, R., Hart, T., Otto, C., et al., 1998. Antimicrobial resistance. BMJ 609–610.

Chapter 6

Fungal volatile compounds: a source of novel in plant protection agents

Prasann Kumar, Priyanka Devi and Shipa Rani Dey
Department of Agronomy, School of Agriculture, Lovely Professional University, Jalandhar, India

Chapter Outline

Volatiles and Metabolites of Microbes. DOI: https://doi.org/10.1016/B978-0-12-824523-1.00001-8

6.1 Introduction

Endophytes may be based on theory, and defined by Hallmann et al. (1997) as microbes that could be either placed apart from the surface-infected plant tissue or extracted between the plant body as well as not damaging the plant. The definition may not be appropriate because it does not include endophytes that are uncultivable and difficult to access phytopathogenic and distinguish hidden pathogens from endophytes, in particular for unculturable fungi related to the microbiome family (Hyde and Soytong, 2008; Hardoim et al., 2015; Mercado-Blanco, 2015; Card et al., 2015). Fungi can manage the symbiotic relationship with the host along with several lifestyles, such as mutual interaction with only one plant (Hardoim et al., 2015). Scientific endophytes are assumed to be microbes that can be detected within plant tissue and generally asymptomatic. To examine the endophytic lifestyle of microbes first, it must be shown that a healthy seedling study with the help of a microscope could be reintroduced into the Koch postulate (Hyde and Soytong, 2008).

6.2 Main objective of chapter

Because we know the basic features of fungal volatile endophytes, the main objective of the chapter is to gain insight into how fungal endophytes are further used for positive effects as enlightening the development and growth of plant health along with adaptation to different stress conditions. Fungal endophytes used in agriculture and horticulture will be replaced as an example, although many reviews have interpreted several characteristics of fungal endophytes (Strobel and Daisy, 2003; Aly et al., 2011; Saikkonen et al., 1998; Friesen, 2013; Johnson et al., 2013b). More of them have focused on the combined aspect of:

1. Some selected endophytes with their secondary metabolite improve the healthy development and implementation of plant adaptation.
2. The product and commercialization of fungal endophytes themselves can be achieved.
3. Loss of beneficial fungal endophyte activity due to cultivated plant breeding and overuse of fungicide prevention of these properties in the future.

Although most plants remain unstudied in terms of their endophytic diversity or function, as not all plants contain endophytes. Although the exponential increase in plant adaptation by the use of endophytes represented in various studies (Johnson et al., 2013b; Hardoim et al., 2015; Card et al., 2016; Kivlin et al., 2013) shows a growing appreciation of the limited scientific community for crop production by exploitative endophytes.

6.3 Grouping of endophytic volatile fungi by functional group

The clarification of the particular symbiotic and environmental function of endophytic fungi requires a proper classification of fungal endophytes according to a functional group based on several criteria, which provides an essential framework and provides scholars and researchers with important information on the fundamental and biological queries of almost these microbes. Rodriguez and Redman (2008) state that there are different endophytic fungi functional groups based on their phylogenetic and lifetime characteristics. Class 1 endophytes are represented as claviciptaceous endophytes (including balansia species and epic hole species). Epic hole classes are considered to be one of the most important examples of the relationship of plant endophytes as they form symbiotic interaction and association with grass tissue of above ground parts (Johnson et al., 2013b). The nonclaviciptaceous species are further divided into three subclasses 2, 3, and 4, where Class 2 encompasses both ascomycetes and several basidiomycetes. The most distinct character of colonizing root stems and leaves is the extensive infection of plants (2009). Class 3 endophytes are very diverse, forming a major infection in overground tissue, for example, tropical trees and vascular and nonvascular plants (Rodriguez and Redman, 2008). Class 4 endophytes are also known as shadow septate endophytes, and their functioning biotrophic fungi have a distinctive characteristic of dark septate hyphae (Jumpponen, 2001) and plant root colonization. They suggest that an additional class is essential for identifying and distinguishing entomopathogenic endophyte fungi, which are considered to be symptomless endophytes of a plant with unit characteristics of infecting and colonizing insects (Quesada Moraga et al., 2014; Vidal and Jaber, 2015).

6.4 Occurrence, spread and biosynthesis of fungal volatile

There is a large microbial community of bacterial, fungal archaeal, and protistic microbes in every plant (Hardoim et al., 2015). The different identification of the classification, as well as the discovery of fungi along with the estimation of the diversity of fungi present on earth, varies by about 1.5 million species (Hawksworth, 2001) to the current estimate of 5.1 million, based on the type of high throughput sequence Blackwell (2011). Researchers estimate that at least 1 million endophytic fungal species are still present (Strobel and Daisy, 2003; Ganley et al., 2004). Such a figure indicates the importance of fungal microbes as an essential endophyte for fungi biodiversity. They are mainly ascomycetes with a ubiquitous appearance in biodiversity due to their presence in large ecosystems along with hot deserts, for example, the plant adopted mangroves of the arctic tundra. Temperate and tropical forests have a significant presence of fungi, including grassland, cropland, and savannah (Arnold, 2008). Along with major terrestrial lineages, Fungi is commonly present in mosses and other nonvascular plants, both ferns and flowering plants (Arnold, 2008); while appreciative of the fungi variety across a wide geographic area, it examined the fact that the plant's endophytic diversity has decreased from tropical to northern boreal forests (Arnold, 2008). Although some endophyte groups that are characterized by comparatively few fungal species, different endophytic species that are large in number compressed within less class dominate tropical endophytes (Arnold, 2008). The fungi range in the plant fluctuates according to time, space, and leaves, and is colonized by several fungi resulting in a different range of secondary metabolism.

In parallel transmission situations, proliferation is generally dependent on the reproductive structures of the endophyte, which carry spores that can be blown along with wind or dispersed by rain, as well as by vector-like insects from one plant to another plant in the soil, by air and vector movements. The different class of endophytes, such as 2 and 3, and the colonization of plants occurs through the infected structure, which directly penetrates the hyphae of plant tissue (Ernst et al., 2003; Gao and Mendgen, 2006). The spread of the sexual species of *Epichlöe* (Class 1) horizontally through well-known ascospores (reviewed in Leuchtmann et al., 2014). If mating partners are compatible ascospores, they are produced and expelled into the air and the wind that mediates transmittable infections (Brem and Leuchtmann, 1999; Chung and Schardl, 1997). The choice of a spread path chosen by the sexual *Epichlöe* classes for transmission amusingly influences the advantageous or antagonistic symbiotic result for the host plant due to the direct effect of the endophyte life cycle on the host reproduction (Fig. 6.1).

6.4.1 Role in the ecosystem

It plays an important role in protecting plants against various stresses such as abiotic and biotic stress and increasing flexibility, as well as helping them to adapt to new habitats, and fungal endophytes play critical roles in ecosystems (Strobel and Daisy, 2003; Aly et al., 2011; Friesen 2013). In return, plants provide a spacious structure, protection from parchment and nutrients. In the case of vertical distribution, it provides broadcasting to the next generation of hosts (Schulz, 2006). It also plays a role in the ecosystem as antagonistic fungal interfaces affect plant growth. Examples of interactions between endophytes and pathogens in host plants of maize are *Ustilago maydis* and *Fusarium verticillioides*, where metabolites are secreted by endophytes that break down plant compounds that limit Ustilago growth. This may result in a reduction in the growth rate of pathogens. Some fungal endophytes initiate biological deprivation of dead and decaying host plants that help to recycle nutrients in the ecosystem (Strobel and Daisy, 2003; Aly et al., 2011; Boberg et al., 2011). In a study conducted by Vázquez de Aldana et al. (2013) on grass, endophytes dominate endophyte grasses that are found to be airborne fungi. It was assumed that some of these species require sporulation after host senescence to complete

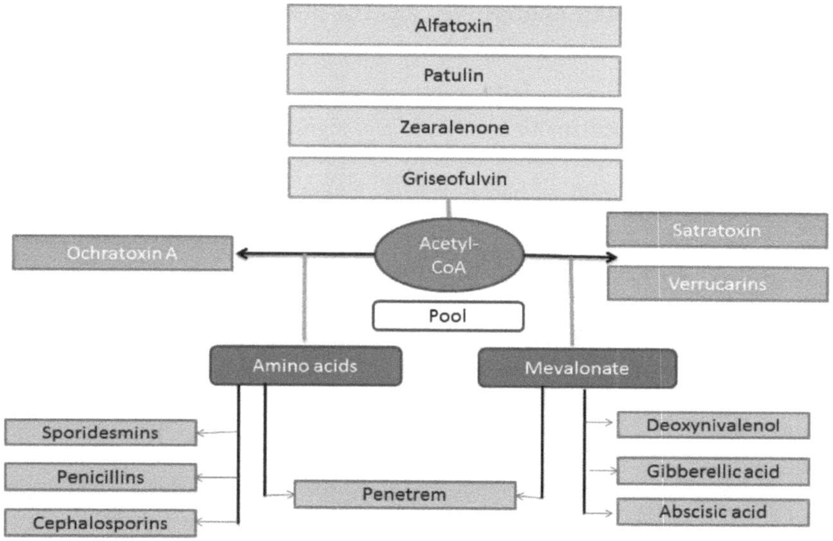

FIGURE 6.1 Common metabolites generated from secondary biosynthesis pathway in fungi.

their life cycle from endophyte to saprophyte. A well-known presence in tissues, endophytes show instant contact with nutrients in plants accessible during plant senescence (Aly et al., 2011; Rodriguez et al., 2008).

6.5 Improvement in plant performance with volatile fungal endophytes

Fungal endophytes have a significant contribution to the health benefits of the plant, providing plant host adaptation for different stresses (Rodriguez and Redman, 2008; Franken, 2012; Aly et al., 2011; Johnson et al., 2013b, 2014; Card et al., 2016). Varied secondary metabolites of various types of fungi (Gunatilaka, 2006; Tan and Zou, 2001), some of which are bioactive compounds articulated as defensive mechanisms to protect the host plant from both pest and disease. Habitat-adapted symbiosis, a term coined by Rodriguez et al. (2008), defines how, under different environmental conditions, some fungal endophytes (Class 2 types) adapt to stress in a habitat-specific manner. The different molecular and biochemical mechanisms behind this result are unknown with habitat symbiotic interactions, resulting in high-stress tolerance in the plant. Another beneficial feature of conferred is the promotion of endophyte-enhanced plant growth. For root colonization, endophyte is most commonly achieved through improved translocation as well as nutrient uptake and intonation of phytohormones involved in growth and development (Johnson et al., 2014).

6.5.1 Fungal volatile endophytes against pest and disease resistance

Epichlöe endophytes (Class 1), which are found in a few temperate grass species along with their major symbiotic and significant influence in the host plant result in the production of bioactive metabolites, which can act as preventions to pests as well as herbivores. *Epichlöe festucae* var. *lolii*, a common toxic endophyte, produces three major secondary metabolites:

1. Lolitrem B: a neurotoxin well-known to cause ryegrass staggers (Gallagher et al., 1981, 1984; Fletcher, 1993; Fletcher et al., 1999).
2. Ergovaline: a metabolite causing vasoconstriction (Oliver, 2005) resulting in heat stress (Fletcher, 1993; Fletcher et al., 1999) and in tall fescue it causes the condition referred to as fescue toxicosis (Hoveland, 1993; Bacon, 1977).
3. Peramine: an insect feeding deterrent (Rowan et al., 1990; Rowan, 1993).

Endophyte type of free grass is nontoxic for fodder purpose of animals and having prone to insect or pests. A novel (ARI) strain of Epichlöe which was familiarized 15 years ago did not produce ergovaline and lolitrem B, but forming peramine which provided resistant to crucial insect pest such as *Listronotus bonariensis* (Aegentine stem weevil, ASW) (Johnson et al., 2013b). After that, the existence of additional alkaloids was found in endophytes, A novel strain of *Epichlöe* consisting epoxy-janthitrems is an indole diterpene type, which is only present in perennial type ryegrass infected by AR37 strain. In New Zealand, five main pasture pests are presents out of six against those grasses, which are infected by AR37 strain having the side effect of pesticides. Yet it has not been properly explained that the compounds epoxy-janthitrem role against the resistance form insects (Johnson et al., 2013a). There is a brisk acceptance by farmers for these novel endophytes introduced by improving ryegrass preserving without causing any harmful effect to animals. Johnson et al. (2013a) examined that different munificent stories that the strain of Epichlöe contains important secondary metabolites and it also establishes their application. For the solution of economic and sustainable agriculture, the endophytes of *Epichlöe* received worldwide acceptance for its significance in grassland areas and has also contributed in providing the best model for studying numerous facet related to entophyte biology in accumulation to explore both the symbiosis of the molecular basis and ecological implication. *E. festucae* is acquiescent to molecular function studies and above all can be civilized on axenic media. Increasing numbers in mutants *E. festucae* to characterizing studies have divorced that maintaining both iron homeostasis and native oxygen by (Johnson et al., 2013b) that are a crucial factor for keeping good mutualistic cooperation in between grasses (Takemoto et al., 2011; Tanaka et al., 2006, 2008). The facultative root colonizing *Piriformospora indica* endophyte forms favorable symbiosis by roots of plants and also has potential to use in floriculture, horticulture, and agriculture (Gill et al., 2016; Johnson et al., 2014; Varma et al., 2013; Oberwinkler et al., 2013; Ansari et al., 2013; Qiang et al., 2012; Franken, 2012; Oelmüller et al., 2009). Verma et al. (1998) were the first who isolated *P. indica* from the xerophytic plant's rhizosphere in the Thar Desert of India. *P. indica* filamentous fungus generally belongs to Sebacinales order of Basidiomycota (Weiss et al., 2004) and provides large numbers of host benefits for example nutrient stress condition, plant growth promoter, and also provided tolerance to abiotic stress (heavy metals, salinity, water, extreme temperature, drought) and biotic stresses (foliar and root pathogens). Original inoculation experiment showed that *P. indica* has an ability to colonizing the roots of the plant (Verma et al., 1998). *P. indica* making symbiotic interaction of roots with both dicotyledonous and monocotyledonous plants also included model plants, important crops as barley, *Arabidopsis thaliana*, and *Nicotiana tabacum* (Johnson et al., 2014).

Plants showed enhanced tolerance to various root and foliar pathogens when colonized with *P. indica* (Johnson et al., 2014). *P. indica* can potentially be used as a biocontrol solution to control various root diseases in different crop species such as maize, tomato, wheat, and barley (Kumar et al., 2009; Fakhro et al., 2010; Serfling et al., 2007; Rabiey et al., 2015; Waller et al., 2005). Systemic resistance to numerous foliar pathogens via *P. indica* has also been demonstrated by Qiang et al. (2012). Fakhro et al. (2010) noticed the effects of inoculation of tomato with *P. indica* in a study and observed that *P. indica* colonizes the roots of tomato relatively increased the biomass of leaves up to 20% and reducing the disease severity of Verticillium wilt caused by *Verticilliumdahliae* by more than 30%, and in hydroponics it increased biomass of fresh tomato fruit up to 100%, resulting in higher fruit yield and increased the dry matter content up to 20%. Moreover, at light densities, it repressed Pepino Mosaic Virus a widely found disease of tomato found in greenhouses of many European countries, Morocco, South and North America, and in China (Fakhro et al., 2010). Many other fungal endophytes have also been found to guard host plants against pathogenic fungi. *Epicoccum nigrum* is an endophyte isolated from sugarcane particularly known for its biocontrol activity against pathogens, like *Sclerotinia sclerotiorum* in sunflower, *Pythium* in cotton, phytoplasma bacteria in apple and *Monilinia* spp. in nectarines and peaches (de L Favaro et al., 2012).

6.5.2 The effect of the pathogenic fungi in host plants protected by most of the fungal endophytes

The endophyte, which is isolated from sugarcane (*E. nigrum*) is particularly used as the various acts of biocontrol against many pathogens like in sunflower (*S. sclerotiorum*), in cotton (*Pithyum*), phytoplasma bacteria in apple, and *Monilinia spp.* in peaches and nectarines (de L Favaro et al., 2012). The fungal endophytes, which are isolated from the wild plants *Hordeum murinum* subsp. *Murinum* L (Murphy et al., 2014) and verified the seed-borne fungal infections activity on barley seeds, which were successfully able to hold back the activity of the seed-borne fungal infections. The tested seeds protect the most barely deteriorating pathogens as well as species of *Fusarium*, *Pyrenophora*, *Cochiolobus*, and *Rhyncosporium*. The suppression of the infections by fungal endophytic isolates includes strain results in 100% suppression. Besides destroying seed-borne pathogens, *Penicillium vicompactum* strain also suppressed the growth of the soil-borne pathogen *Gaeumannomyces graminis vartritici* (Murphy et al., 2014). From *Theobroma cacao* tissue, the fungal endophytes are isolated and have been screened for in vitro resistance against main pathogens of cacao, which may include *Moniliopthora roreri* (causal organism of frosty pod rot), *Phytophthora palnivora* (causal organism of black rod pot), and *M. perniciosa* (causal organism of witches broom). The endophytic fungi in field trials have exposed the treatment with *Colletotrichum gloeosporioides* decreased the pod loss due to black pod and the treatment with *Clonostachys rosea* reduced sporulation lesions on cacao poda (Mejía et al., 2008). The fungal entomopathogens are considered as an important class of endophytes having properties of biocontrol for insects. The capability to infect along with colonizing the insects result in affecting survival as well as a reproduction of insects is unique not all have the ability to colonize plant (Vidal and Jaber, 2015). Interestingly the species of *Beauveria*, *Metarrhizium*, *Lecanicillium*, and *Isaria* are the most commercially produced entomopathogenic fungi and have the characteristics of endophytes in the parts of their life cycle (Vidal and Jaber, 2015). The mode of action of the commercial endophytic products have not been used, however, at least able to colonize plants as such products endophytically (reviewed in Vidal and Jaber, 2015). The penetration as well as colonization of various plant tissues by the entomopathogenic fungi *Beauveria bassiana* was first described Wagner and Lewis (2000). The root colonization of plants by entomopathogenic fungi is further reviewed by Vidal and Jaber (2015) and discussed the variability as well as the effectiveness of colonization rates while using with different combinations of strains/cultivator. One of the most distressing pests of coffee throughout the world is coffee berry borer (*Hypothenemus hampei*) and the fungal endophytes of the coffee plants are collected from different places of the world like Hawaii, Colombia, Mexico. The major role played by the fungal entomopathogens is nitrogen fixation in soil. The capability of *Metarhizium robertsii* is the movement of insect-derived nitrogen to plants was tested by Behie et al. (2012). The wax moth larva is injected with 15-N labeled nitrogen, which placed the labeled insect on haricot bean (*Phaseolus vulgaris*) and switch grass (*Panicum virgatum*) in the presence of *Me. Roberts* and observed the incorporation of 15-N into amino acid of plants.

6.6 Fungal volatile for deliberating the abiotic stress along with climate change

Mutual fungal endophytes provide fitness benefits and are responsible for plant adaptation, not only to biotic stress but also to abiotic stress by increasing acceptance of drought and water stress as well as acceptance of high temperature and salinity (Aly et al., 2011). Plants with compelling observable facts, such as habitat, adapted the stress tolerance of the plant to produce habitat-specific symbionts and the symbiotic process, and are subject to plant growth in a stressful environment (Rodriguez et al., 2008). A large number of Class 2 endophytes have habitat adaptation properties that contribute to the tolerance of habitat-specific selective stress such as pH, temperature, and salinity (Rodriguez and Redman, 2008). Rodriguez et al. (2008) studied and determined that grass species can be adapted to coastal and geothermal habitat locations by maintaining salinity and heat tolerance, respectively, with the help of symbiotic fungal endophytes. Dunegrass plants from coastal beach habitats in the United States have shown a symbiotic relationship with the endophyte *Fusarium culmorum.* Some experiments have been conducted to demonstrate endophytic fungi against heat tolerance. Tropical panic grasses, *Dichanthelium lanuginosum*, which grow in geothermal soils in Yellowstone National Park, have been found to form mutualistic symbiotic interactions with Class 2 endophytes, *Curvularia protuberates*, which provide heat tolerance (Redman et al., 2011). Symbiotic plants inoculated with endophytes tolerate and survive a temperature of 65°C in the root areas of the plants, whereas individual plants or endophytes cannot tolerate a temperature of more than 38°C (Márquez et al., 2007). However, the relationship between the plant and the endophyte is compounded, and it is discovered that the endophyte raises the virus in a mutually symbiotic manner and that the virus is an essential partner for the heat tolerance of the host plant, whereas the virus-free endophyte is not capable of providing tolerance to the plant (Márquez et al., 2007).

Moreover, water consumption by endophytes reduced by 20%–30% with increasing reproductive yield, growth rate, and biomass of greenhouse-grown plants, and they also provide cold tolerance to greenhouse-grown plants. The incorporation of fungal symbionts may be a useful strategy in reducing the effects of climate change on major crops and expanding agricultural production on marginal lands (Redman et al., 2011) (Fig. 6.2).

Turning toward endophytes has the ability to increase the growth of the plant. *P. indica* is known for the encouragement of early flowering, germination of seeds, vegetative growth, and setting of seeds has been found repeatedly with different families of the plant by their species (Franken, 2012). Networks of signaling phytohormones and the set of extensive phytohormones involved in decreasing growth activities of plants, which are responsible for increasing the early growth of roots promotion, and lastly to raising

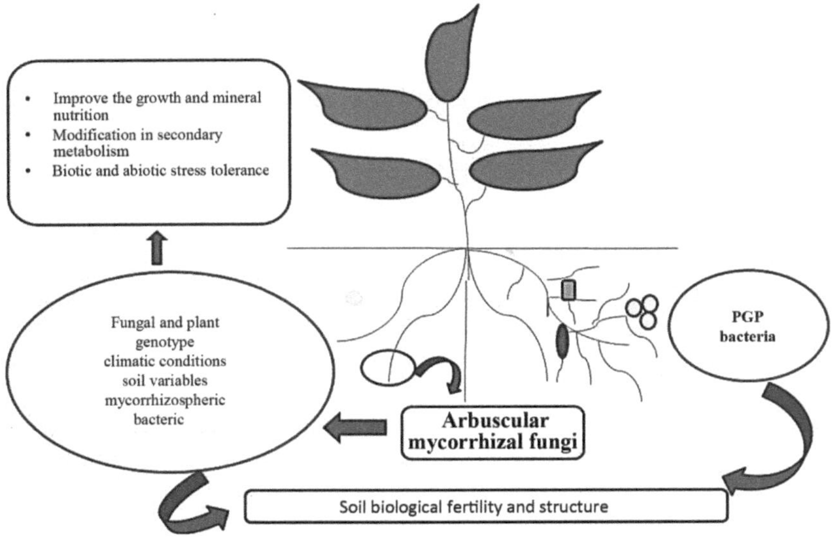

FIGURE 6.2 Role of fungal volatiles endophytes in the improvement of plant growth.

the biomass. Around 50.0% in the promotion of growth is extended, yet there are major differences in this because the growth of a plant is done in the environmental variability and various condition of experiments. In plants the mechanism of tolerance due to stress of salt colonized by *P. indica* has been considerably researched, and the same trait also has been used for demonstration in tobacco and rice as well as barley and wheat and the species of reactive oxygen are detoxified by the induction of a higher amount of environmental antioxidants, which increases the efficiency of photosynthesis (Johnson et al., 2014).

The broad host benefits of *P. indica* as well as its convenience for the study of fundamentals of biology, are tempting properties for research purposes and potential for sustainably improving agricultural crop production is normally exciting. Interaction of *P. indica* with barley shows increases in numbers of grains and yield (Murphy et al., 2014; Waller et al., 2005). Improving grain yield and earlier flowering of the trigger appeared due to *P. indica*, and increases the input of nutrients with lower temperature. The result shows that *P. indica* grows effectively in the treatment of crop under low-stressed temperature barley, and it has the potential to improve the yield of the crop under an environment of the temperate region to provide with a proper supply of nutrients (Murphy et al., 2014). Franken (2012) informs that *P. indica* is difficult for supplying in market or agricultural shops because (1) India was the first country to isolate and patent the fungus and in different countries (number of international publication-WO 99/29177) forming any reliant mercantile vision on the owner of a patent to licensing and manufacturing it; (2) sometimes it has an enigmatic synergetic effect on the growth of a plant. In India it has been deployed for trials in the field, and *P. indica* is developed in the formulation of powder under the ROOTONIC trade name (Varma et al., 2013).

There are various types of barley endophytes that can be advantageous in an agricultural setting. The wild variety of barley form isolated type of endophytic fungi (*Murinum L., H. murinum* subsp.) and previously explained the different types of biotic benefits. Among these, of all the isolates few were able to raise the numbers of grains and yield in a cultivar of barley having nutrient deficiency (up to 28%). The higher impact of isolated endophytes on yield comes from grains, and the weight of dry shoot was cognizable under the lowest input of nutrients. Keeping as it is the theme of increasing the performance of the plant, the class (IV) and Dark septate endophytic (DSE) fungi having root colonizing, characterized by bright septa of melanize, enthralling endophytes group that are existing in a broad range of terrestrial (subastral) ecosystem, but they are popular especially in the alpine and polar ecosystems (Rodriguez and Redman, 2008). In the environment, the stress of cold water (arbuscular mycorrhizal fungi, AMF) mutualists in the roots of grass at lesser latitudes along with altitudes are virtually absent (Newsham, 2011) and it also informs that these species of DSE might behave as surrogate fungus in those habitats. Dicotyledonous and monocotyledonous species of plants inoculated with that fungi and increases the biomass of shoot and root, nitrogen of shoots, and content of phosphorous. The inorganic form of nitrogen is readily accessed by plants roots when species of DSE does not form benefits by apparent plants (Newsham, 2011).

6.7 The various secondary metabolites which are produced by fungal endophytes

6.7.1 Fungal endophytes are considered as a source of bioactive natural products

Due to the presence of incredible endophytes of fungal diversity, we get several opportunities to discover natural products having distinctive chemical structures. The present-day progress of showing technologies has disclosed capability of endophytes for the production of naturally active complexes in fields such as medicine and agriculture (Zhang et al., 2006; Aly et al., 2011; Wu et al., 2015b). The process of communication between organisms such as protection of plant as well as habitat adaptation according to environmental changes are carried by these types of molecules. Many synthetic products of agriculture have been and will be removed from the common market due to the reasons of environmental problems and safety issues. The pest and pathogens can be controlled by the fungal endophytes produced by secondary metabolites (Strobel and Daisy, 2003).

For the better understanding of endophyte-produced bioactive, we have to know that some are known for antioxidants, anticancer agents, antibiotics, and biofuels (Strobel and Daisy, 2003). In the first step, we have to discover plants which bear these microbes having a capability for production novel bioactive metabolites.

6.7.2 Microbial bioactive metabolites discovered by plants

Some microbial metabolites have characters of biotopes at both the taxonomic and environmental level (Schulz, 2001). The study showed that the organisms reside on distinctive biotopes as well as habitats which were exposed to constant metabolic as well as environmental interactions results in increased in yield by secondary metabolites than the organisms which do not reside (Schulz, 2001). This is the reason that search mainly held on organisms that inhibit unique biotopes for novel secondary metabolites. There is a collection of successful plants harboring endophytes that are producing novel and unique natural bioactive, which requires the plant documentation (Strobel and Daisy, 2003).

- from the distinctive settings of environment, especially with infrequent biology characters and possessing new approaches for survival;
- that has an ethnobotanical history, which relates to specific uses or requests of interest;
- that is endemic having infrequent longevity and acquires certain ancient land masses; and
- that grow and develop in areas with greater biodiversity.

Secondary metabolites which are produced by endophytic fungi are discovered in Amazonian rain forest expeditions.

6.7.3 Antimicrobial secondary metabolites

There are several examples of secondary metabolites, which are produced from endophytic fungi and are used in agriculture and horticulture for controlling pest and pathogens. This involves endophytes of fungal, which produce auspicious but not so identified complexes through the well understood and described antimicrobial, considered as secondary metabolites. Gunatilaka (2006) published the 230 metabolites, which are produced by a plant that is associated with microbial stains includes many endophytes of fungal. Some of the examples which are well-characterized as antimicrobial secondary metabolites from fungal of endophytic are given below. Numerous compounds are described by Strobel and Daisy (2003). This chemical structure of selected complexes is shown in the figures.

Some of the antimicrobial complexes produced by the sugarcane endophyte *E. nigrum* have been characterized as (Brown et al., 1987):

- Epicorazines A-B by Baute et al. (1978) where flavipin (Bamford et al., 1961), epirodines A-B by Ikawa et al. (1987), epipiridones and epicocarines by Wangun and Hertweck (2007), and the last one was epicoccines A-D by Zhang et al. (2007).

Particularly the compounds like flavipin and epicorazines A-B was associated with *E. nigrum*, which is further considered as a biocontrol activity (Madrigal and Melgarejo, 1995; Brown et al., 1987).

We obtain the antifungal and antioomycete acid called ambuic acid from an endophytic fungus named as *Pestalotiopsis microspora*. Several of the world's tropical rain forest have isolates of it. It is highly active against many species of *Fusarium* as well as *Phythium ultimum* (Li et al., 2001). The appearance looks like quorum-sensing (Fuqua et al., 1994) inhibitor (Gary Strobel, pers. comm.). *Colletotrichum* sp., which is isolated from *Artemisia annua* produces Colletonoic acid (Bills et al., 2002). This can be used as a good antibacterial, antialgal as well as antifungal (Hussain et al., 2014). Colletotric acid is considered as one of the metabolites of *C. gloeosporioides* species, which is an endophytic fungus obtained from Artemisia *mongolica*. It further acts as an antimicrobial agent against bacteria and some fungi like *Helminthosporium sativum* (Zou et al., 2000). An endophytic fungus *Cordyceps dipterigena* produces Cordycepsidone A, which acts strongly as an antifungal agent against some plant pathogenic fungi like *Gibberella fujikuroi*. A lipopeptide cryptocandin is obtained from fungus *Cryptosporiopsis quercina*, which is usually related to the hardwood species in Europe. It also acts strongly against much plant pathogenic fungi involves *S. sclerotiorum* and *Botrytis cinerea*. Stated that it also acts strongly in opposite to various fungus of plant pathogen involving *B. cinerea* and *S. sclerotiorum*. These antimycotic type of compounds are also called as *penumocandins* and the *echinocandins* (Walsh, 1992). Li et al. (2001) also obtained Cryptocin from *C. quercina*, which shows the highest activity of potent against the fungus of *Pyricularia oryzae*. It may have energetic activities against various types of pathogenic fungi named as *Fusarium oxysporum, S. sclerotiorum, G. candidum*, and *Rhizoctonia solani*, along these; it is effective for oomycetes of plant pathogens such as *Phytophthora cinnamon, Ph. Citrophthora*, and *Py. Ultimum* (Li et al., 2001). Papua New Guinea has *P. Jester* and evolves jesterone from it, which is further isolated from the highly moist areas where there the plant pathogenic oomycetes usually grows. These compounds have the antioomycete type of activity as anticipated. The *P. microspora* evolved by the antioxidants of isopestacin and pestacin, which is obtained by *Terminalia morobensis* (Harper et al., 2003). Antimicrobial activity is performed by both antioxidants (Daisy et al., 2002). The secondary type of metabolites Phomopsichalasin obtained by species of *Phomopsis* and have some antibacterial activities in

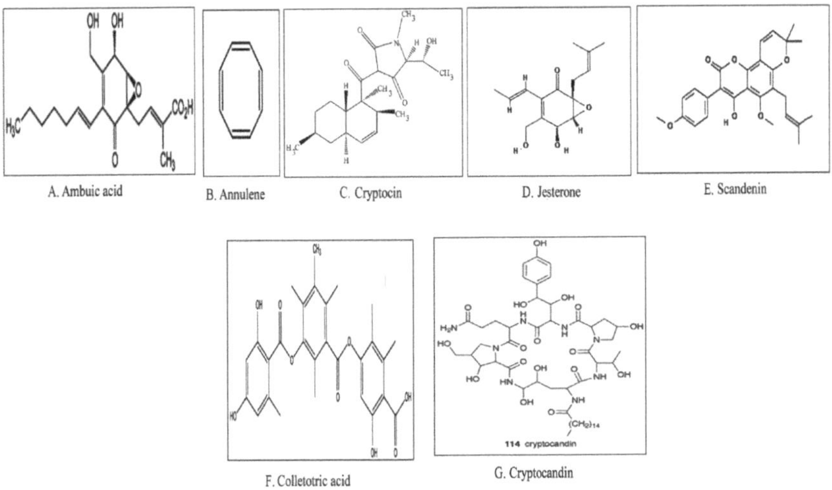

FIGURE 6.3 (A-G) Examples of antimicrobial secondary metabolites from endophytic fungi.

opposite to *Staphylococcus Bacillus subtilis* and *Salmonella enteric* (Horn et al., 1995). Scandenin is taken by the plant named *Derris scandens*, which is present in Pakistan and strongly behaves as a representative antibacterial in opposite to *B. megaterium* and also having good properties of antialgal and antifungal (Hussain et al., 2015). Various diseases of the plant caused by microbes controlled with bacteria's present in the fungi of endophytic, none of which are produced by the secondary metabolites of an antimicrobial (Figs. 6.3 and 6.4).

6.8 Antivolatile organic compounds

Even though the volatile antibiotics topic was disused in the previous section, they are described in a different heading because special methods and techniques are involved in the detection of these compounds. There are many ways to detect the antimicrobial activity of VOCs such as by creating air contact between the fungus (producing volatile compound) and the marked pathogen. The detection and separation of volatile compounds can be subsequently done by using many methods but liquid chromatography–high resolution mass spectrometry is one of most prominent method another most adopted method used for separation is nuclear magnetic resonance and multivariate data analysis (Wu et al., 2015a). It has been observed that the endophytic fungus *Muscodor albus* is producing almost 28 volatile compounds, which are isolated from little branches of *Cinnamomum zeylanicum* (cinnamon tree) and have the ability to kill other bacteria and fungus (Worapong et al., 2001) (Strobel and Daisy, 2003). Some individual volatile

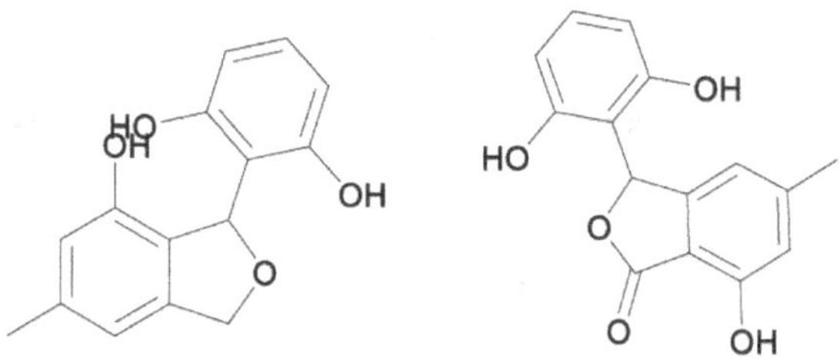

1: Pestacin 2: Isopestacin

FIGURE 6.4 Some examples of antioxidant secondary metabolites from endophytic fungi.

compounds had several inhibitory effects against the test fungi and some bacteria, but none of them was considered lethal. Although the effect of these compounds has been found more efficient collectively as compared to an individual, collectively they generate synergistic interaction and cause strong damage to a wider range of plants/human pathogenic bacteria or fungus. There are five-known classes of volatile compounds such as lipids, alcohols, ketones, esters, and acids. Among all five classes ester was considered as the most effective and only class of producing inhibitory and biologically active compounds like isoamyl acetate. (Strobel and Daisy, 2003). Later on many strains of *M. albus* strains were developed to produce a variety of volatile compounds but from isolated strains, not a single one had ester or any biologically active compounds.

6.9 Fungal volatile compounds: reduction in their beneficial activities

We all know that there are very few wild plants species have fungal endophytes, which are not enough to fulfill their promising target. The most challenging requirement of endophyte fungus is to maintain a desirable environment within the plant, endophyte fungus within the plants required specific environment to continue their physiological activity. Because of the inadequate use of fungicides and pesticides along with synthetic fertilizer plants does not get the benefit of endophytes and presence of these chemical compounds within the plant cell inhibits the activity of fungal volatile compounds. Under the situation where low selective pressure, the effect of endophytes were found less because due to the absence of secondary metabolites or because they preferred mutation. Finally, the plant does not get expected benefit from an application of endophytes.

Here, some experimental proofs to understand the behavior of endophytes within the plants following:

1. Weese et al. (2015) rhizobium was collected from two different fields where the clover was cultivated and one field was taken as control, where

no nitrogenous fertilizers were applied, compared to another field where nitrogenous fertilizers were applied regularly for 22 years. When a comparison was made, there was a frequent reduction in N-fixing activity of rhizobacteria, lesser chlorophyll content, and reduction in the crop by 16%–30% in N fertilizer applied field.

2. Redman et al. (2011) show that the activity of endophyte fungus in the rice field was increased under salt and drought stress was found more prominent and resulted in generating tolerance to these two abiotic stresses. It was observed that the colonization of endophytes fungus during the absence of abiotic stresses was decreased by 100% at planting to 65%. Plants grown under drought and salt stress were strongly maintained at 100% colonization.

6.10 Conclusion and future scopes for increasing production, use, and maintenance of fungal volatile compounds for raising their standards to commercial levels

Previously, in this chapter, we were discussing fungal (endophyte) volatile compounds, and our main focus of discussion was targeted on fungus endophytes, but this study is not limited to fungus; it is also provisional for bacterial VOCs and their uses. Our observation has been targeted toward fungal volatile endophytes, but in many cases, it will apply for many bacterial endophytes. Many germplasm libraries should be appreciated for maintaining the probability that the seeds having many fungal endophytes have scientific and economic importance. In seed storage houses, seeds were stored at low temperatures and low humidity that will promote the durability of seed along with its viability to any of the fungal endophyte they may contain. The treatment of seed with fungicides and placing at higher temperature should be avoided. Our second request is toward our plant breeders that they should realize the benefits of their crop importance for fungal endophytes, and the following suggestions should be taken:

1. They have to check the loss of recognized valuable microbes had been lost during breeding.
2. For an objective to incorporate the microbiomes, they should adopt the protocols.
3. They should sequence and collect the DNA from many plant seeds, mainly from their center of origin before the propagation so that there can be an identification of involvement of potential endophytes, which are beneficial.
4. Check the loss of valuable microbes as well as mutation of genes from the origin center while treating the seeds or during the propagation process.

5. Seeds companies should also check the killing of microbes or weakness the useful microbes, which are present in or on the seeds while heating the seeds and using of artificial as well as natural fungicides for seed coating processes.
6. The substances used during the process of seed coating should be checked so that the beneficial microbes should not be lost during storage and germination.
7. The seeds, having beneficial fungal endophytes should be stored at low temperature and humidity.
8. Request the microbiologists try to arrange or preserve all microbiomes of plants from their center of origin.
9. Identify that the collection of microbes are beneficial for agricultural and horticultural point of view.
10. A search should be done to find out the beneficial microbes and secondary metabolites from the center of origin and from tropical rain forests.
11. Some suggestions are there for the politicians, public and other opinion makers that know the position of the microbiomes of plants for plant health as well as food production.
12. Wild plants should be preserved, and also the microbiomes in their habitats. There are lots of natural habitats, plant species, and their fungal endophytes by various anthropogenic activities like clearing, harvesting, fire, agricultural developments, mining, or the other human-oriented activities.

Since microbes hold the crucial role to give prospects for the attainment of sustainable agriculture, we should search for the profits of these beneficial microbes with some urgency.

Acknowledgment

P.D. and P.K gratefully acknowledge the support provided by Department of Agronomy, School of Agriculture, Lovely Professional University, Punjab, India.

References

Aly, A.H., Debbab, A., Proksch, P., 2011. Fungal endophytes: unique plant inhabitants with great promises. Appl. Microbiol. Biot. 90, 1829–1845.

Ansari, M.W., Trivedi, D.K., Sahoo, R.K., 2013. A critical review on fungi mediated plant responses with special emphasis to *Piriformospora indica* on improved production and protection of crops. Plant Physiol. Biochem. 70, 403–410.

Arnold, A., 2008. Hidden Within Our Botanical Richness, a Treasure Trove of Fungal Endophytes. Plant Press, pp. 13–15.

Bacon, C.W., 1977-present. Grass-endophyte research, an ecological miasma clarified through history. In: Hume, D.E., Latch, G.C.M., Easton, H.S. (Eds.). Proceedings of the Second International Symposium on Acremonium/Grass Interactions. Palmerston North, NZ: Research, Grasslands Research Centre, pp. 3–10.

Bamford, P.C., Norris, G.L.F., Ward, G., 1961. Flavipin production by *Epicoccum* spp. T Brit Mycol. Soc. 44, 354–356.

Baute, M.A., Deffieux, G., Baute, R., 1978. New antibiotics from the fungus *Epicoccum nigrum*. I. Fermentation, isolation and antibacterial properties. J. Antibiot. 31, 1099–1101.

Behie, S.W., Zelisko, P.M., Bidochka, M.J., 2012. Endophytic insect-parasitic fungi translocate nitrogen directly from insects to plants. Science 336, 1576–1577.

Bills, G., Dombrowski, A., Peleaz, F., 2002. Recent and future discoveries of pharmacologically active metabolites from tropical fungi. In: Watling, R., Frankland, J.C., Ainsworth, A.M., et al.,Tropical Mycology: Micromycetes, vol. 2. CABI Publishing, Wallingford, Oxon, pp. 165–194.

Blackwell, M., 2011. The fungi: 1, 2, 3 ... 5.1 million species? Am. J. Bot. 98, 426–438.

Boberg, J.B., Ihrmark, K., Lindahl, B.D., 2011. Decomposing capacity of fungi commonly detected in *Pinus sylvestris* needle litter. Fungal Ecol. 4, 110–114.

Brem, D., Leuchtmann, A., 1999. High prevalence of horizontal transmission of the fungal endophyte *Epichlöe sylvatica*. Bull, Geobotan., Inst. ETH 65, 3–12.

Brown, A.E., Finlay, R., Ward, J.S., 1987. Antifungal compounds produced by *Epicoccum purpurascens* against soil-bone plant pathogenic fungi. Soil Biol. Biochem. 19, 657–664.

Card, S.D., Hume, D.E., Davood, R., 2015. Beneficial endophytic microorganisms of *Brassica*—a review. Biol. Control. 90, 102–112.

Card, S.D., Johnson, L.J., Teasdale, S., 2016. Deciphering microbial behaviour—the link between endophyte biology and efficacious biological control agents. FEMS Microbiol. Ecol. 92. Available from: https://doi.org/10.1093/femsec/fiw114.

Chung, K.-R., Schardl, C.L., 1997. Sexual cycle and horizontal transmission of the grass symbiont, *Epichlöe typhina*. Mycol. Res. 101, 295–301.

Daisy, B.H., Strobel, G.A., Castillo, U., 2002. Naphthalene, an insect repellent, is produced by *Muscodor vitigenus*, a novel endophytic fungus. Microbiology 148, 3737–3741.

de L Favaro, L.C., de S Sebastianes, F.L., Aráujo, W.L., 2012. *Epicoccum nigrum* P16, a sugarcane endophyte, produces antifungal compounds and induces root growthLiles MR (Ed.) PLoS One 7, e36826.

Ernst, M., Mendgen, K.W., Wirsel, S.G.R., 2003. Endophytic fungal mutualists: seed-borne *Stagonospora* spp. enhance reed biomass production in axenic microcosms. Mol Plant-Microbe Interact 16, 580–587.

Fakhro, A., Andrade-Linares, D.R., von Bargen, S., et al., 2010. Impact of *Piriformospora indica* on tomato growth and on interaction with fungal and viral pathogens. Mycorrhiza 20, 191–200.

Fletcher, L.R., 1993. Grazing ryegrass/endophyte associations and their effect on animal health and performance. In: Hume, D.E., Latch, G.C.M., Easton, H.S. (Eds.), Proceedings 2nd International Symposium on Acremonium/Grass Interactions: Plenary Papers. Agresearch, Palmerston North, pp. 115–120.

Fletcher, L.R., Sutherland, B.L., Fletcher, C.G., 1999. The impact of endophyte on the health and productivity of sheep grazing ryegrass-based pastures. In: Woodfield, D.R., Matthew, C. (Eds.), Ryegrass Endophyte: an Essential New Zealand Symbiosis. Research and Practice Series No. 7. New Zealand Grassland Association, Palmerston North, New Zealand, pp. 11–17.

Franken, P., 2012. The plant strengthening root endophyte *Piriformospora indica*: potential application and the biology behind. Appl. Microbiol. Biot. 96, 1455–1464.

Friesen, M.L., 2013. Microbially mediated plant functional traits, Molecular Microbial Ecology of the Rhizosphere, vol. 1. John Wiley & Sons, Inc, Hoboken, NJ, pp. 87–102.

Fuqua, W.C., Winans, S.C., Greenberg, E.P., 1994. Quorum sensing in bacteria: the LuxR-LuxI family of cell density-responsive transcriptional regulators. J. Bacteriol. 176, 269–275.

Gallagher, R.T., White, E.P., Mortimer, P.H., 1981. Ryegrass staggers: isolation of potent neurotoxins lolitrem a and lolitrem B from staggers-producing pastures. N. Zeal Vet. J. 29, 189–190.

Gallagher, R.T., Hawkes, A.D., Steyn, P.S., et al., 1984. Tremorgenic neurotoxins from perennial ryegrass causing ryegrass staggers disorder of livestock: structure elucidation of lolitrem B. J. Chem. Soc. 9, 614–616.

Ganley, R.J., Brunsfeld, S.J., Newcombe, G., 2004. A community of unknown, endophytic fungi in western white pine. Proc Nat. Acad. Sci. USA 101, 10107–10112.

Gao, K., Mendgen, K., 2006. Seed-transmitted beneficial endophytic Stagonospora sp. can penetrate the walls of the root epidermis, but does not proliferate in the cortex, of *Phragmites australis*. Can. J. Bot. 84, 981–988.

Gill, S.S., Gill, R., Trivedi, D.K., et al., 2016. *Piriformospora indica*: potential and significance inplant stress tolerance. Front. Microbiol. 7, 1–20.

Gunatilaka, A.A.L., 2006. Natural products from plant-associated microorganisms: distribution, structural diversity, bioactivity, and implications of their occurrence. J. Nat. Prod. 69, 509–526.

Hallmann, J., Quadt-Hallmann, A., Mahaffee, W.F., et al., 1997. Bacterial endophytes in agricultural crops. Can. J. Microbiol. 43, 895–914.

Hardoim, P.R., van Overbeek, L.S., Berg, G., et al., 2015. The hidden world within plants: ecological and evolutionary considerations for defining functioning of microbial endophytes. Microbiol. Mol. Biol. R. 79, 293–320.

Harper, J.K., Arif, A.M., Ford, E.J., et al., 2003. Pestacin: a 1,3-dihydro isobenzofuran from *Pestalotiopsis microspora* possessing antioxidant and antimycotic activities. Tetrahedron 59, 2471–2476.

Hawksworth, D.L., 2001. The magnitude of fungal diversity: the 1.5 million species estimate revisited. Mycol. Res. 105, 1422–1432.

Horn, W.S., Simmonds, M.S.J., Schwartz, R.E., et al., 1995. *Phomopsichalasin*, a novel antimicrobial agent from an endophytic *Phomopsis* sp. Tetrahedron 51, 3969–3978.

Hoveland, C.S., 1993. Importance and economic significance of the *Acremonium* endophytes to performance of animals and grass plant. Agr. Ecosyst. Environ. 44, 3–12.

Hussain, H., Root, N., Jabeen, F., et al., 2014. Seimatoric acid and colletonoic acid: two new compounds from the endophytic fungi, *Seimatosporium* sp. and *Colletotrichum* sp. Chin. Chem. Lett. 25, 1577–1579.

Hussain, H., Al-Harrasi, A., Krohn, K., et al., 2015. Phytochemical investigation and antimicrobial activity of *Derris scandens*. J. King Saud. Univ. 27, 375–378.

Hyde, K.D., Soytong, K., 2008. The fungal endophyte dilemma. Fungal Divers. 33, 163–173.

Ikawa, M., McGratten, C.J., Burge, W.R., et al., 1987. Epirodin, a polyene antibiotic from the mold *Epipoccum nigrum*. J. Antibiot. 31, 159–161.

Johnson, L.J., De Bonth, A.C.M., Briggs, L.R., et al., 2013a. The exploitation of epichloae endophytes for agricultural benefit. Fungal Divers. 60, 171–188.

Johnson, L.J., Koulman, A., Christensen, M., et al., 2013b. An extracellular siderophore is required to maintain the mutualistic interaction of *Epichlöe festucae* with *Lolium perenne*. PLoS Pathog. 9, e1003332.

Johnson, J.M., Alex, T., Oelmüller, R., 2014. *Piriformospora indica*: the versatile and multifunctional root endophytic fungus for enhanced yield and tolerance to biotic and abiotic stress in crop plants. J. Trop. Agric. 52, 103–122.

Jumpponen, A., 2001. Dark septate endophytes are they mycorrhizal? Mycorrhiza 11, 207–211.

Kivlin, S.N., Emery, S.M., Rudgers, J.A., 2013. Fungal symbionts alter plant responses to global change. Am. J. Bot 100, 1445–1457.

Kumar, M., Yadav, V., Tuteja, N., et al., 2009. Antioxidant enzyme activities in maize plants colonized with *Piriformospora indica*. Microbiology 155, 780–790.

Leuchtmann, A., Bacon, C.W., Schardl, C.L., et al., 2014. Nomenclatural realignment of Neotyphodium species with genus *Epichöe*. Mycologia 106, 202–215.

Li, J.Y., Harper, J.K., Grant, D.M., et al., 2001. Ambuic acid, a highly functionalized cyclohexenone with antifungal activity from *Pestalotiopsis* spp. and *Monochaetia* sp. Phytochemistry 56, 463–468.

Madrigal, C., Melgarejo, P., 1995. Morphological effects of *Epiccocum nigrum* and its antibiotic flavipin on *Monilinia laxa*. Can. J. Bot. 73, 425–431.

Márquez, L.M., Redman, R.S., Rodriguez, R.J., et al., 2007. A virus in a fungus in a plant: three-way symbiosis required for thermal tolerance. Science 315, 513–515.

Mejía, L.C., Rojas, E.I., Maynard, Z., et al., 2008. Endophytic fungi as biocontrol agents of *Theobroma cacao* pathogens. Biol. Control. 46, 4–14.

Mercado-Blanco, J., 2015. Life of microbes inside the plant. In: Lugtenberg, B. (Ed.), Principles of Plant-Microbe Interactions- Microbes for Sustainable Agriculture. Springer, Dordrecht, The Netherlands, pp. 25–32.

Murphy, B.R., Doohan, F.M., Hodkinson, T.R., 2014. Yield increase induced by the fungal root endophyte *Piriformospora indica* in barley grown at low temperature is nutrient-limited. Symbiosis 62, 29–39.

Newsham, K.K., 2011. A meta-analysis of plant responses to dark septate root endophytes. N. Phytol. 190, 783–793.

Oberwinkler, F., Riess, K., Bauer, R., et al., 2013. Enigmatic *Sebacinales*. Mycol. Prog. 12, 1–27.

Oelmüller, R., Sherameti, I., Tripathi, S., et al., 2009. *Piriformospora indica*, a cultivable root endophyte with multiple biotechnological applications. Symbiosis 49, 1–17.

Oliver, J.W., 2005. Pathophysiologic response to endophyte toxins. Neotyphodium in Cool-Season Grasses. Blackwell Publishing Ltd, Oxford, UK, pp. 291–304.

Qiang, X., Weiss, M., Kogel, K.H., et al., 2012. *Piriformospora indica*-a mutualistic basidiomycete with an exceptionally large plant host range. Mol. Plant Pathol. 13, 508–518.

Quesada Moraga, E., Herrero, N., Zabalgogeazcoa, I., 2014. Entomopathogenic and nematophagous fungal endophytes. In: Verma, V.C., Gange, A.C. (Eds.), Advances in Endophytic Research. Springer, Heidelberg, Germany, pp. 85–99. , Chapter 19.

Rabiey, M., Ullah, I., Shaw, M.W., 2015. The endophytic fungus *Piriformospora indica* protects wheat from fusarium crown rot disease in simulated UK autumn conditions. Plant Pathol. 64, 1029–1040.

Redman, R.S., Kim, Y.O., Woodward, C.J.D.A., et al., 2011. Increased fitness of rice plants to abiotic stress via habitat-adapted symbiosis: a strategy for mitigating impacts of climate change. PLoS One 6, e14823.

Rodriguez, R., Redman, R., 2008. More than 400 million years of evolution and some plants still can't make it on their own: plant stress tolerance via fungal symbiosis. J. Exp. Bot. 59, 1109–1114.

Rodriguez, R.J., Henson, J., Van Volkenburgh, E., et al., 2008. Stress tolerance in plants via habitat-adapted symbiosis. ISME J. 2, 404–416.

Rowan, D.D., 1993. Lolitrems, parasite and paxilline: mycotoxins of the ryegrass/endophyte interaction. Agric. Ecosyst. Environ. 44, 103–122.

Rowan, D.D., Dymock, J.J., Brimble, M.A., 1990. Effect of fungal metabolite per amine and analogs on feeding and development of Argentine stem weevil (*Listronotus bonariensis*). J. Chem. Ecol. 16, 1683–1695.

Saikkonen, K., Faeth, S.H., Helander, M., et al., 1998. Fungal endophytes: a continuum of interactions with host plants. Annu. Rev. Ecol. Syst. 29, 319–343.

Schulz B., 2001. Endophytic fungi: a source of novel biologically active secondary metabolites. In: British Mycological Society, International Symposium Proceedings: Bioactive Fungal Metabolites—Impact and Exploitation. University of Wales, Swansea, Wales, p. 20.

Schulz, B.J.E., 2006. Mutualistic interactions with fungal root endophytes. In: Schulz, B.J.E., Boyle, C.J.C., Sieber, T.N. (Eds.), Microbial Root Endophytes, vol. 9. Springer, Berlin, Heidelberg, pp. 261–279.

Serfling, A., Wirsel, S.G.R., Lind, V., et al., 2007. Performance of the biocontrol fungus Piriformospora indica on wheat under greenhouse and field conditions. Phytopathology 97, 523–531.

Strobel, G., Daisy, B., 2003. Bioprospecting for microbial endophytes and their natural products bioprospecting for microbial endophytes and their natural products. Microbiol. Mol. Biol. R. 67, 491–502.

Takemoto, D., Kamakura, S., Saikia, S., et al., 2011. Polarity proteins Bem1 and Cdc24 are components of the filamentous fungal NADPH oxidase complex. Proc Natl. Acad. Sci. USA 108, 2861–2866.

Tan, R.X., Zou, W.X., 2001. Endophytes: a rich source of functional metabolites. Nat. Prod. Rep. 18, 448–459.

Tanaka, A., Christensen, M.J., Takemoto, D., et al., 2006. Reactive oxygen species play a role in regulating a fungus-perennial ryegrass mutualistic interaction. Plant Cell 18, 1052–1066.

Tanaka, A., Takemoto, D., Hyon, G.S., et al., 2008. NoxA activation by the small GTPase RacA is required to maintain a mutualistic symbiotic association between *Epichlöe festucae* and perennial ryegrass. Mol. Microbiol. 68, 1165–1178.

Varma, A., Bajaj, R., Agarwal, A., et al., 2013. Memoirs of rootonic-themagic fungus: promotes agriculture. Horticulture and Forest Productivity. Amity University, Noida, India.

Vázquez de Aldana, B.R., Bills, G., Zabalgogeazcoa, I., 2013. Are endophytes an important link between airborne spores and allergen exposure? Fungal Divers. 60, 33–42.

Verma, S., Varma, A., Rexer, K.-H., et al., 1998. *Piriformospora indica*, gen. et sp. nov., a new root-colonizing fungus. Mycologia 90, 896.

Vidal, S., Jaber, L.R., 2015. Entomopathogenic fungi as endophytes: plant–endophyte–herbivore interactions and prospects for use in biological control. Curr. Sci. 109, 46–54.

Wagner, B.L., Lewis, L.C., 2000. Colonization of corn, *Zea mays*, by the entomopathogenic fungus *Beauveria bassiana*. Appl. Environ. Microbiol. 66, 3468–3473.

Waller, F., Achatz, B., Baltruschat, H., et al., 2005. The endophytic fungus *Piriformospora indica* reprograms barley to salt-stress tolerance, disease resistance, and higher yield. Proc. Natl Acad. Sci. USA 102, 13386–13391.

Walsh, T.A., 1992. Inhibitors of β-glucan synthesis. In: Sutcliffe, J., Georgopapadakou, N.H. (Eds.), Emerging Targets in Antibacterial and Antifungal Chemotherapy. Champman & Hall, London, pp. 349–373.

Wangun, H.V.K., Hertweck, C., 2007. Epicoccarines A, B and epipyridone: tetramic acids and pyridone alkaloids from an *Epicoccum* sp. associated with the tree fungus *Pholiota squarrosa*. Org. Biomol. Chem. 5, 1702–1705.

Weese, D.J., Heath, K.D., Dentinger, B.T.M., et al., 2015. Long-term nitrogen addition causes the evolution of less-cooperative mutualists. Evolution 69, 631–642.

Weiss, M., Selosse, M.A., Rexer, K.H., et al., 2004. *Sebacinales*: a hitherto overlooked cosm of heterobasidiomycetes with a broad mycorrhizal potential. Mycol. Res. 108, 1003–1010.

Worapong, J., Strobel, G., Ford, E.J., et al., 2001. *Muscodor albus* anam. gen. et sp. nov., and endophyte from *Cinnamomum zeylanicum*. Mycotaxon 79, 67–79.

Wu, C., Kim, H.K., Van Wezel, G.P., et al., 2015a. Metabolomics in the natural products field - a gateway to novel antibiotics. Drug. Discov. Today Technol. 13, 11–17.

Wu, C., Zacchetti, B., Ram, A.F.J., et al., 2015b. Expanding the chemical space for natural products by *Aspergillus-Streptomyces* co-cultivation and biotransformation. Sci. Rep. B 5, 10868.

Zhang, H.W., Song, Y.C., Tan, R.X., 2006. Biology and chemistry of endophytes. Nat. Prod. Rep. 23, 753–771.

Zhang, Y., Liu, S., Che, Y., et al., 2007. Epicoccins A-D, epipolythiodioxopiperazines from a Cordyceps-colonizing isolate of *Epicoccum nigrum*. J. Nat. Prod. 70, 1522–1525.

Zou, W.X., Meng, J.C., Lu, H., et al., 2000. Metabolites of *Colletotrichum gloeosporioides*, an endophytic fungus in *Artemisia mongolica*. J. Nat. Prod. 63, 1529–1530.

Chapter 7

Endophytic microbes: an array of organic volatiles and secondary metabolites

Pratibha Vyas

Department of Microbiology, College of Basic Sciences and Humanities, Punjab Agricultural University, Ludhiana, India

Chapter Outline

7.1 Endophytes

Microorganisms colonizing and residing in the internal tissues of plants without causing any diseases are termed as endophytes. They can be isolated from a woody tree to herbaceous plant, monocotyledonous to dicotyledonous plant (Ryan et al., 2008). These microorganisms are known to make a mutualistic association with almost all plants, including herbaceous, woody, monocotyledons, and dicotyledons. They can be isolated from all plant parts like roots, shoots, leaves, bark, flowers, fruits, etc. During the association, the endophytic microorganisms obtain nutrients and minerals from their host for their survival. In return, these microorganisms produce metabolites helping the host plant to resist environmental stress. The association between the host plant and microbial population of endophytes is dependent upon different components including type and density of microbes, plant species and development, and environmental factors (Khare et al., 2018).

Volatiles and Metabolites of Microbes. DOI: https://doi.org/10.1016/B978-0-12-824523-1.00004-3

Endophytes can be grouped into obligate or facultative categories depending upon their dependence on plants, whereby the former are exclusively dependent on plants for their survival. They are transmitted along with their host by vectors or vertically. While the endophytes falling under the group "facultative" spend some of their life cycle stages outside host plant and are connected with their host through the soil. It has been reported that plant–endophyte relationship is highly specific with respect to plant specificity, transmission mode, and abundance (Andreote et al., 2014). It has also been proposed that endophytic association is more likely to be exhibited by microorganisms with a small genome (Mitter et al., 2013).

There has been an increased interest in developing and discovering new drugs due to the threat caused by drug-resistant microbial pathogens. New and novel solutions are required to overcome the challenge of emerging drug resistance. Endophytes have a vast potential to be used in medicine and agriculture due to their ability to produce a large number of secondary metabolites and organic volatiles. The ability of these metabolites to be used in a wide range of fields has grabbed the attention for several years in view of their antimicrobial, insecticidal, cytotoxic, anticancer, antidiabetic, and antioxidant properties. They are known to produce metabolites belonging to different classes like alkaloids, steroids, terpenes, terpenoids, diterpenes, coumarins, isocumarins, flavonoids, polyketides, quinones, lactones phenylpropanoids, peptides, lignans, tannins, and phenolics. Many bioactive metabolites including paclitaxel, camptothecin, isopectacin, cytosporone, oocydin, etc. have been extracted from various endophytes with anticancer, antibacterial, antifungal, and antioxidant activities (Zhao et al., 2005; Elsebai et al., 2014). Additionally, organic volatiles including 1,8-cineole, acetic acid, ethyl ester, dimethyl disulfide, dibenzofuran, methane-thiol, ketones, propanoic acid, 2-methyl-methyl ester and acetic acid, 2-methyl propyl ester, etc. isolated from endophytes have inhibited numerous plant and human pathogens (Banerjee et al., 2014; Chen et al., 2016; Wang et al., 2017). In many cases, metabolites extracted from these endophytes are similar to those produced by their host plants. In addition, endophytes are also reported to improve plant growth, soil fertility, and crop production due to the production of many plant growth-promoting metabolites.

7.2 Secondary metabolites: diversity and significance

Endophytes have immense potential for synthesizing secondary metabolites with different applications and biological activities. Efforts have been made recently to explore the chemical nature of compounds produced by endophytes. Endophytes including bacteria, fungi, and actinomycetes have been reported to produce antibiotics, anticholesterol, anticancer, and immunosuppressants, highlighting their importance in pharmaceutical industries. For example, a fungus *Penicillium* sp. LDL4.4 isolated from *Huperzia serrata* produced Huperzine A (HupA) with the potential to treat Alzheimer's disease (Le et al., 2019).

Apart from their importance in medicine, these endophytes are equally important in agriculture. An endophytic bacterium *Pseudomonas aeruginosa* isolated from rice produced N-acylhomoserine lactone, pyochellin, phloroglucinol, lahorenoic acid, diketopiperazines, hydroxy-2-alkylquinolines, and rhamnolipids with antimicrobial activity and biosurfactant potential (Yasmin et al., 2017). Many of the compounds have been grouped as alkaloids, steroids, peptides, terpenoids, flavonoids, etc. with diverse functions. Biologically active secondary metabolites from *Alternaria alternata* isolated from *Maytenus hookeri* were identified as alternariol, alternariol monomethyl ether, altenuene, adenosine, 5-epialtenuene, uridine, ACTG toxin-E, and ergosta-4,6,8,22-tetraen-3-one by nuclear magnetic resonance (Ma et al., 2010). Recently, peptides, including acrepeptins A-D, guangomide A and destruxin B with antiinflammatory activity is reported from *Acremonium* sp. isolated from marine microalgae (Hsiao et al., 2020).

7.3 Metabolites with anticancer activity

Many important anticancer compounds belonging to different chemical classes have been isolated from endophytes. Taxol or Paclitaxel, a diterpenoid compound used to cure ovarian, breast, and lung cancer, was isolated for the first time from *Taxus brevifolia* (Stierle et al., 1993). It was subsequently isolated from various endophytes including actinomycetes and fungi associated with *Taxus* species and other plants. Taxol has been isolated from various endophytes: *Pestalotiopis microspore* from *Taxus wallichiana* (Strobel et al., 1996); *Colletotrichum gloeosporioides* from the leaves of *Justicia gendarussa* (Gangadevi and Muthumary, 2008); *Metarhizium anisopliae* isolated from *Taxus chinensis* (Liu et al., 2009a); *Pestalotiopsis* species from *Taxus cuspidata* (Kumaran et al., 2010); *Lasiodiplodia theobromae* from *Morinda citrifolia* (Pandi et al., 2011); *Taxomyces andreanae* from the inner bark of *T. brevifolia* (Selim et al., 2017); *Fusarium redolens* from Himalayan Yew (Garyali et al., 2013); *Acremonium, Colletotrichum*, and *Fusarium* spp. from *Taxus baccata* (El-Bialy and El-Bastawisy, 2020); and *Epicoccum nigrum* from *T. baccata* (El-Sayed et al., 2020a).

Apart from taxol, other important anticancer compounds like camptothecin, vincristine and podophyllotoxin have been isolated from endophytes (Table 7.1). *Cordyceps militaris* extract inhibited the growth of MCF-7 human breast cancer cells in a dose and time-dependent manner (Xu, 2008). The fungal extract reduced methylation of DNA through the suppression of methyltransferase transcripts leading to the recovery of tumor-suppressor genes and eventually inhibiting tumor cell growth. Deoxy-podophyllotoxin production is reported from the fungi associated with *Juniperus communis* L. Horstmann (Kusari et al., 2009). *Altersolanol* A with anticancer potential was isolated from endophyte *Stemphylium globuliferum* (Teiten et al., 2013). A quinoline alkaloid- camptothecine extracted from endophytic bacteria associated with *Miquelia dentata* Bedd. has been shown to be a powerful

TABLE 7.1 **Endophytes as source of secondary metabolites.**

Endophyte	Host plant	Secondary metabolite	Activity	References
Petalotiopsis jesteri	*Fragraea bodenii*	Jesterone, Hydroxyjesterone	Against the oomycetous fungi	Li and Strobel (2001)
Pestalotiopsis microspora	*Torreya taxifolia*	Torreyanic acid	Cytotoxic activity	Li et al. (2001)
Penicillium sp.	*Melia azedarach*	Preaustinoid A, B	Bacteriostatic effect	dos Santos and Rodrigues-Fo (2003)
Streptomyces sp.	*Grevillea pteridifolia*	Kakadumycin A (chemically related to echinomycin*)	Antimalarial and antibiotic activity	Castillo et al. (2003)
Aspergillus fumigatus CY 018	*Cynodon dactylon*	Asperfumoid, Asperfumin and Physcion	Antifungal against *Candida albicans*	Liu et al. (2004)
Aspergillus niger	*C. dactylon*	Rubrofusarin B, Aurasperone A	Inhibitors on xantin oxidase, colon cancer cell and pathogens	Song et al. (2004)
Streptomyces sp.	*Monstera* sp.	Coronamycin	Against *Cryptococcus neoformans* and *Plasmodium falciparum*	Ezra et al. (2004)
Xylaria sp.	*Abies holophylla*	Griseofulvin, Dechlorogriseofulvin	Antifungal antibiotic	Park et al. (2005)
Streptomyces sp.	*Maytenus hookeri*	Dimeric dinactin, Dimeric nonactin, Cyclohomononactic acid, Cyclononactic acid	Strong antineoplastic activity and antibacterial activity	Zhao et al. (2005)
Trametes hirsuta	*Podophyllm hexandrum*	Podophyllotoxin, Aryl Tetralin Lignans	Anticancer activity	Puri et al. (2006)

Xylaria sp.	Mangrove plant	Xylopyridine A	DNA-binding affinity	Xu et al. (2009)
Alternaria sp.	*Sonneratia alba*	Alternariol, perylene quinones	Cytotoxic activity	Kjer et al. (2010)
Penicillium sp.	*Kandelia candel*	7-Epiaustdiol, stemphyperylenol, secalonic acid A	Antibacterial activity	Liu et al. (2010)
Aspergillus ochraceus	*A.* ochraceus	Steroidal derivative, 3β,11α-dihydroxy ergosta- 8,24(28)- dien-7-one, 7-nor-ergosterolide,	Anticancer activity	Cui et al. (2010)
Aspergillus terreus	Tree hole	Butenolides, butyrolactones VI and VII	Antimycobacterial activity, antiplasmodial activity	Haritakun et al. (2010)
Xylaria sp.	*Licuala spinosa*	Eremophilane type sesquiterpenoids	Cytotoxic activity	Isaka et al. (2010)
Aspergillus tubingensis	*Pongamia pinnata*	Monomeric and Dimeric naphtho-γ-pyrones	Cytotoxic activity	Huang et al. (2010)
Chaetomium globosum	*C. dactylon*	Chaetoglocins A–D	Antibacterial activity	Ge et al. (2011)
Alternaria sp.	*Aegiceras corniculatum*	Alterporriol K–M	Cytotoxic activity	Huang et al. (2011)
Emericella sp.	*A. corniculatum*	Isoindolones emerimidine A and B	Cytopathogenic effect	Zhang et al. (2011)
Pseudonocardia sp.	Artemisia plant	Artemisinin	Antimalarial drugs	Li et al. (2012)
Cordyceps dipterigena	*Desmotes incomparabilis*	Cordycepsidone A	Activity against *Gibberella fujikuroi*	Varughese et al. (2012)
Bionectria ochroleuca	*Sonneratia caseolaris*	Pullularins E and F, verticillin D	Cytotoxic activity	Ebrahim et al. (2012a)

(*Continued*)

TABLE 7.1 (Continued)

Endophyte	Host plant	Secondary metabolite	Activity	References
Corynespora cassiicola	*Laguncularia racemosa*	Decalactones, xestodecalactones D–F, corynesidone C	Protein kinase activity	Ebrahim et al. (2012b)
Guignardia sp.	*Scyphiphora hydrophyllacea*	Meroterpenes, guignardones F–I	Antibacterial activity	Mei et al. (2012)
Eurotium rubrum	*Hibiscus tiliaceus*	12-demethyl-12-oxoeurotechinulin B and 9-dehydroxyeurotinone	Antibacterial activity	Yan et al. (2012)
Diaporthe sp.	Mangrove trees	Dicerandrol D	Antimalarial activity	Calcul et al. (2013)
Aspergillus wentii	*Gymnogongrus flabelliformis*	Yicathin A–C	Antibacterial and antifungal activity	Sun et al. (2013)
Phompsis sp.	*Aconitum carmichaeli*	Gavodermside and Clavasterols	Inhibitory effect against *C. albicans, A. niger, Fusarium avenaceum, Hormodendrum compactin, Trichophyton gypseum*	Wu et al. (2013)
Alternaria sp. (SK11)	Mangrove plant	Atropisomer 2	MptpB inhibitor	Xia et al. (2014)
Streptomyces sp.	*Boesenbergia rotunda (L.)*	Naringenin, 3; 7-Methoxy-3, 3′,4′,6-tetrahydroxyflavone and 2′,7-Dihydroxyisoflavone, Fisetin, -Hydroxydaidzein, Xenognosin	Activity against *Bacillus cereus, Bacillus subtilis, Staphylococcus aureus*	Taechowisan et al. (2014)
Penicillium sp. FJ-1	*Avicennia marina*	4-(20,30-dihydroxy-30-methyl-butanoxy)-phenethanol, and 15-, hydroxy-6a,12-epoxy-7b,10aH,11bH-spiroax-4-ene-12-one	Antiproliferative activities	Zheng et al. (2014)

Penicillium oxalicum EN-201	*Rhizophora stylosa*	Penioxamide A, 18-hydroxydecaturin B	Brine shrimp lethality	Zhang et al. (2015)
Nigrospora sp.	Mangrove plant	Bostrycin	Anticancer activity	Gomes et al. (2015)
Unidentified	*Euphorbia hirta* and *Catharanthus roseus*	paxilline, nigricinol, Citreoisocoumarin, fatty acid, sceptrin, cladosporin	Activity against *Escherichia coli, S. aureus, Salmonella typhii, B. subtilis, Aspergillus fumigates, C. albicans*	Akpotu et al. (2015)
Xylaria sp.	*Bostrychia tenella*	Cytochalasin D	Activity against HCT-8 and SF-295 cancer cell lines	de Felício et al. (2015)
Actinoallomurus fulvus	*Capsicum frutescens*	Actinoallolides A–E	In vitro antitrypanosomal activity without cytotoxicity	Inahashi et al. (2015)
Penicillium aurantiogriseum	*Hibiscus tiliaceus*	aspermytin A, Peaurantiogriseols A–F, 1-propanone,3-hydroxy-1-(1,2,4a,5,6,7,8,8a-octahydro-2,5-dihydroxy-1,2,6-trimethyl-1-naphthalenyl)	Low inhibitory activity against human aldose reductase	Ma et al. (2015)
Botryosphaeria sp.	*Kandelia candel*	Botryoisocoumarin A	Antiinflammatory	Ju et al. (2015)
Talaromyces flavus	*Sonneratia apetala*	Talaperoxides A–D and Norsesquiterpene peroxides,	Anticancer activity	Elissawy et al. (2015)
Talaromyces amestolkiae	*Kandelia obovata*	5-hydroxy-7-methoxy-2-methylbenzofuran-3-carboxylic acid, 5-hydroxy-7-methoxy-2-methylbenzofuran-3-yl ethan-1-one aspergillumarin A, sescandelin B, 3,4-dimethyl-6,8-dihydroxyisocoumarin	Antibacterial activity α-glucosidase inhibitor	Chen et al. (2016)

(*Continued*)

TABLE 7.1 (Continued)

Endophyte	Host plant	Secondary metabolite	Activity	References
Pseudolagarobasidium acaciicola	*Bruguiera gymnorrhiza*	Terpene endoperoxides	Cytotoxic activity	Wibowo et al. (2016)
Paecilomyces variotii	Marine algae	Varioloid A and varioloid B	Anticancerous activity	Zhang et al. (2016)
Microsporum sp.	*Lomentaria catenata*	Physcion	Apoptosis in HeLa cells	Wijesekara et al. (2014)
Penicillium citrinum	*B. gymnorrhiza*	(Z)-7,40-dimethoxy-6-hydroxy-aurone-4-*O*-b-glucopyranoside	Neuroprotective activity	Huang et al. (2016)
Alternaria sp. SK6YW3L	*Sonneratia caseolaris*	Altenusin	a-glucosidase inhibitory activity	Liu et al. (2016)
Botryosphaeria sp. SCSIO	*K. candel*	Botryosphaerin B.	Antiinflammatory	Ju et al. (2016)
Fusarium sp.	Mangrove plant	Anthraquinone derivative	Anticancer activity	Cui et al. (2017)
Annulohypoxylon sp.	*Rhizophora racemosa*	Daldinone I	Cytotoxic activity	Liu et al. (2017)
Pseudomonas aeruginosa BR3	Rice	N-acylhomoserine lactone, pyochellin, phloroglucinol, lahorenoic acid, diketopiperazines, hydroxy-2-alkylquinolines, rhamnolipids	Virulence factors, biosurfactants, and antimicrobial agents	Yasmin et al. (2017)
Aeromicrobium ponti	*Vochysia divergens*	Indole-3-carbaldehyde, 1-Acetyl-β-carboline, 3-(Hydroxyacetyl)-Indole, Brevianamide F and Cyclo-(L-Pro-L-Phe)	Antibacterial activity against methicillin- resistant *Staphylococcus aureus*	Gos et al. (2017)

Lasiodiplodia sp. 318	*Excoecaria agallocha*	Secrete 2,4-dihydroxy-6-nonylbenzoate	Cytotoxic activity	Huang et al. (2017)
Pseudomonas aurantiaca	Cotton and cactus	lahorenoic acid A and 2-hydroxy-phenazine	Helps in zinc solubilization	Izzah Shahid et al. (2017)
Talaromyces sp. (HZ-YX1)	*K. obovata*	Talaramide A	Inhibition of mycobacterial PknG activity	Chen et al. (2017)
Talaromyces sp.	Mangrove plant	3-*O*-methylfunicone	Àntitumor, antifungal, and lipid-lowering activity	Nicoletti et al. (2018)
Arthrinium sp. MFLUCC16–1053	*Zingiber cassaumunar*	β-cyclocitral, 3E-cembreneA, Laurenan-2-one, Sclaresol, 2Z, 6E-farnesol, Cembrene, β-isocomene and γ-curcumene	Activity against *S. aureus and Escherichia. coli*	Pansanit and Pripdeevech, (2018)
Diaporthe spp.	*Melodorum fruticosum*	Salicylaldehyde, benzoin, benzene acetaldehyde, benzyl benzoate and benzyl cinnamate	Antioxidant	Tanapichatsakul et al. (2018)
Bacillus tequilensis GYLH001	*Angelica dahurica*	Indole-3-acetic acid and 1-aminocyclopropane-1-carboxylate deaminase.	Promote plant growth and reduce the harmful effects.	Li et al. (2018)
Bacillus mojavensis and *Bacillus atrophaeus*	*Glycyrrhiza uralensis*	Methyl ester, 1,2-bezenedicarboxyl acid, Decanodioic acid, bis(2-ethylhexyl) ester	Antifungal activity against *Fusarium fulva, Verticillum dahlia*	Mohamad et al. (2018)
Penicillium sp. LDL4.4	Huperzia serrata	Huperzine A (HupA)	Used to treat Alzheimer disease	Le et al. (2019)

inhibitor of topoisomerases I of eukaryotes (Shweta et al., 2013). Two alkaloids 7-isoprenyl indole-3-carboxylic acid and 3-acetonylidene-7-prenylindolin-2-1 isolated from endophytic actinobacteria *Streptomyces* sp. were found to have anticancer activity (Zhang et al., 2014). Endophyte *Microbiospora* sp. produced β-carbolines and 3-indole compounds with cytotoxic activities against cancer cell lines, and inhibitory activities against *Micrococcus luteus* and *Kocuria rosea* (Savi et al., 2015). *E. nigrum* isolated from *Terminalia arjuna* bark produced digoxin with anticancer activity (El-Sayed et al., 2020b).

7.4 Metabolites with antioxidant activity

Antioxidant metabolites with the ability to scavenge reactive oxygen and free radicals help in combating many diseases. Many endophytes have been reported to exhibit antioxidant activities and metabolites responsible for the activity have been extracted and purified from these endophytes (Table 7.1). Isopestacin extracted from the endophyte *Pestalotiopsis microspore* exhibited antioxidant properties by inhibiting hydroxyl and superoxide free radicals along with exhibiting antifungal activity (Strobel, 2002). In addition to anticancer activity, aryl tetralin lignans extracted from *Trametes hirsute* showed antioxidant activity (Puri et al., 2006). Likewise, the activity was shown by the metabolites produced by many endophytes: Pestacin and isopestacin isolated from an endophyte *Pestalotiopsis microspora* (Harper et al., 2003; Tejesvi et al., 2007); yangjinhualine A and 2,6-dimethoxy terephthalic acid extracted from endophyte *Streptomyces* sp. associated with *Alpinia oxyphylla* (Zhou et al., 2014); luteolin isolated from *Aspergillus fumigatus* associated with pigeon pea (Zhao et al., 2014); 1,2-(1-hydroxy-1-methyl)-2,3-dihydrobenzofuran-5-ol; 7-isopropenylbicyclo[4.2.0]octa-1,3,5-triene-2,5-diol; 2, 2-dimethylchroman-3 and 6-diol,2-(3-dihydroxy-3-methyl butyl) benzene −1,4-diol extracted from *Acremonium* sp. associated with *Cladostephus spongiosus* (Sarasan et al., 2017); S-adenosyl-N-acetylhomocysteine extracted from *Micromonospora* sp. associated with *Pueraria candollei* Graham ex Benth (Boonsnongcheep et al., 2017); Oxo-agarospiral, β-agarofuran, α-agarofuran, β-dihydro agarofuran and δ-eudesmol from endophytes *Anthrinium* sp., *Colletotrichum* sp., and *Diaporthe* sp. associated with *Aquilaria subintegra* (Monggoot et al., 2017).

7.5 Metabolites with antimicrobial activity

Majority of endophytes are reported for antagonism against pathogens, whereby they inhibit/control pathogens of animals and plants. Endophytic microorganisms isolated from various sources, including medicinal plants, have shown activity against multiple pathogens (Table 7.2). The metabolites with antimicrobial activity produced by endophytes have been identified chemically and been placed in a suitable group (Table 7.2). Few examples include Cryptocandin produced by *Cryptosporiopsis* cf. *quercina*, isolated

TABLE 7.2 Antimicrobial secondary metabolites produced by endophytes.

Nature	Metabolite	Endophyte	Plant/Host	Activity against	Action	References
Terpenes and Terpenoids	Heptelidic acid and hydroheptelidic acid	*Phyllosticta* sp.	*Abies balsamea*	*Choristoneura fumiferana*		Calhoun et al. (1992)
	Paclitaxel	*Taxomyces andreanae*	*Taxus brevifolia*	Fungi anticancer	Disruption of normal cell division	Stierle et al. (1993)
	Guanacastepene	Unidentified fungus	*Daphnopsis americana*	*Staphylococcus aureus* (Methicillin-resistant) and *Enterococcus faecium* (vancomycin-resistant)	Damages bacterial membranes	Singh et al. (2000)
	Phomadecalin C	*Phoma* sp.	Wood decay stromata	*Bacillus subtilis*		Che et al. (2002)
	Diaportheín B	*Diaporthe* sp.	Unidentified wood	*Mycobacterium tuberculosis*	Ketone group at position C-7	Dettrakul et al. (2003)
	Periconicins A and B	*Periconia* sp.	*Taxus cuspidata*	*Bacillus subtilis, Klebsiella pneumonia* and *Proteus vulgaris*		Kim et al. (2004)
	Periconicins A and B	*Periconia* sp.	-	*Candida albicans* and *Trichophyton mentagrophytes*		Oh et al. (2005)

(*Continued*)

TABLE 7.2 (Continued)

Nature	Metabolite	Endophyte	Plant/Host	Activity against	Action	References
	Scoparasin B	*Eutypella scoparia*	*G. dulcis*	*Microsporum gypseum*		Pongcharoen et al. (2006)
	3, 12-dicadalenehydroxy	*Phomopsis cassiae*	*Cassia spectabilis*	*Cladosporium sphaerospermum* and *Cladosporium cladosporioides*		Silva et al. (2006)
	Trichodermin	*Trichoderma harzianum*	*Ilex cornuta*	*Alternaria solani* and *Rhizoctonia solani*	Inhibition of peptide bond formation	Chen et al. (2007)
	Sordaricin	*Xylaria* sp.	*Garcinia dulcis*	*Candida albicans*	Inhibits fungal protein synthesis	Pongcharoen et al. (2008)
	Cycloepoxytriol B and cycloepoxylactone	*Phomopsis* sp.	*Laurus azorica*	*Chlorella fusca, Bacillus megaterium* and *Microbotryum violaceum*		Hussain et al. (2009)
	Diterpene CJ-14445	*Botryosphaeria* sp.	*Maytenus hookeri*	*Candida albicans, Penicillium avellaneum* and *Saccharomyces cerevisiae*		Yuan et al. (2009)
	8-α acetoxyphomadecalin and Phomadecalin	*Microdiplodia* sp.	*Pinus* sp.	*Pseudomonas aeruginosa*		Hatakeyama et al. (2010)

	1α-10 α-Epoxy-7α-hydroxyeremophil-11-en-12,8,β-olide	*Xylaria* sp.	*Licuala spinosa*	*Candida albicans* and *Plasmodium falciparum*	Failure to uptake nucleic acid precursors	Isaka et al. (2010)
	Phomenone	*Xylaria* sp.	*Piper aduncum*	*Cladosporium sphaerospermum* and *C. cladosporioides*	Loss of electrolyte and cell wall permeability	Silva et al. (2010)
	Helvolic acid	*Pichia guilliermondii*	*Paris polyphylla*	*Magnaporthe oryzae*	Inhibition of spore germination	Zhao et al. (2010)
	Asporyzin C and Compound JBIR 03	*Aspergillus oryzae*	*Heterosiphonia japonica*	*Artemia salina* and *Escherichia coli*		Qiao et al. (2010)
	Penicisteroid A	*Penicillium chrysogenum*	*Laurencia* sp.	*Aspergillus niger* and *Alternaria brassicae*		Gao et al. (2011)
	Botryosphaerins G and H	*Botryosphaeria* sp.	*Huperzia serrata*	Phytopathogenic fungi and nematodes		Chen et al. (2015)
	Abienol	*Diaporthe* spp.	*Melodorum fruticosum*	Antibacterial activity		Tanapichatsakul et al. (2018)
	α-terpinene, γ-terpinene, and (-)-4-terpineol	Diaporthe apiculatum	*Leucaena leucocephala*	Antifungal activity		Song et al. (2019)
Alkaloids	Peramine	*Acremoneum lolii*	*Lolium perenne* L.	Stem weevil		Rowan (1993)

(*Continued*)

TABLE 7.2 (Continued)

Nature	Metabolite	Endophyte	Plant/Host	Activity against	Action	References
	Phomopsichalasin	*Phomopsis* sp.	*Salix gracilostyla*	*B. subtilis*, *S. aureus*, *Salmonella gallinarum* and *Candida tropicalis*		Horn et al. (1995)
	Cryptocin	*Cryptosporiopsis quercina*	*Trypterigium wilfordii*	*Pyricularia oryzae*		Li et al. (2000)
	Loline alkaloid	*Neotyphodium uncinatum*	*Lolium pratense*	Insects and aphids		Blankenship et al. (2005)
	Phomoenamide	*Phomopsis* sp.	*G. dulcis*	*Mycobacterium tuberculosis*		Rukachaisirikul et al. (2008)
	Pestalachloride A	*Pestalotiopsis adusta*	Unknown Chinese tree	*Gibberella zeae*, *Fusarium culmorum*, and *Verticillium albo-atrum*		Li et al. (2008a)
	Fumigaclavine C and Pseurtotin C	*Aspergillus* sp.	*Bauhinia guianensis*	*Escherichia coli*, *Pseudomonas aeruginosa*, *Staphylococcus aureus*, *Bacillus subtilis*		Pinheiro et al. (2013)
	brevianamide F, 1-acetyl-β-carboline, indole-3-carbaldehyde,	*Aeromicrobium ponti*	*Vochysia divergens*	*Staphylococcus* spp.		Gos et al. (2017)

	tryptophol, 3-(hydroxyacetyl)-indole, cyclo-(L-Pro-L-Leu), cyclo-(L-Pro-L-Phe), cyclo-(L-Pro-L-Tyr), and cyclo-(L-Val-L-Phe)					
	Alkaloids, flavonoids, Phenol and steroids	*Methylobacterium radiotolerans* MAMP 4754	*Combretum erythrophyllum*	Inhibitory activity against *Bacillus subtilis, Bacillus cereus, Klebsiella oxytoca, Mycobacterium smegmatis*		Photolo et al. (2020)
Phenolics	2-Methoxy-4-hydroxy-6-methoxymethyl benzaldehyde	*Pezicula* sp.	Unknown tree	*Cladosporium cucumerinum*		Schulz et al. (1995)
	Colletotric acid	*Colletotrichum gloeosporioides*	*Artemesia mongolica*	*Bacillus subtilis, Staphylococcus aureus* and *Micrococcus luteus* and *helminthosporium sativum*		Zou et al. (2000)
	Cytonic acid A and B	*Cytonaema*	*Quercus* sp.	*Cytomegalovirus*	Inhibition of protease required for nucleocapsid assembly	Guo et al. (2000)

(*Continued*)

TABLE 7.2 (Continued)

Nature	Metabolite	Endophyte	Plant/Host	Activity against	Action	References
	Altenusin	*Alternaria* sp.	*Trixis vauthieri*	*Trypanosoma* and *Leishmania*	Inhibition of trypanothiome reductase	Cota et al. (2008)
	4-(2,4,7-trioxa-bicyclo [4,10]-heptan-3-yl) phenol	*Pestalotiopsis mangifera*	*Mangifera indica Linn*	*Escherichia coli, Bacillus subtilis, Pseudomonas aeruginosa Klebsiella, Candida albicans* and *Micrococcus luteus*		Subban et al. (2013)
	Phenol and steroids	*Methylobacterium radiotolerans* MAMP 4754	*Combretum erythrophyllum*	*Bacillus subtilis, Bacillus cereus, Klebsiella oxytoca, Mycobacterium smegmatis*		Photolo et al. (2020)
Aliphatic compounds	Brefeldin A	*Aspergillus clavatus* and *Paecilomyces* sp.	*Torreya grandis* and *Taxus mairei*	*Aspergillus niger, Candida albicans, Trichophyton rubrum*		Wang et al. (2002)
	Pestalofones C and E	*Pestalotiopsis fici*	Unidentified tree	*Aspergillus fumigatus*		Liu et al. (2009b)
	CR377	*Fusarium* sp.	*Selaginella pallescens*	*Candida albicans*		Brady and Clardy (2000)

Polyketides	Nodulisporins	*Nodulisporium* sp.	*Juniperus cedrus*	*Microbotryum violaceum*		Dai et al. (2006)
	6-O-Methylalaternin and Altersolanol A	*Ampelomyces* sp.	*Urospermum picroides*	*Enterococcus faecalis, Staphylococcus aureus* and *S. epidermidis*		Aly et al. (2008)
	Palmarumycin CP17 and CP18	*Edenia* sp.	*Petrea volubilis*	*Leishmania donovani*		Martínez-Luis et al. (2008)
	Pestalotheol C	*Pestaloptiopsis theae*	Unidentified tree	*HIV*	Inhibition of replication	Li et al. (2008b)
	Isofusidienol A-D	*Chalara* sp.	*Artemisia vulgaris*	*Candida albicans, Bacillus subtilis, Staphylococcus aureus* and *Escherichia coli*		Lösgen et al. (2008)
	Pestalachlorides A-C	*Pestalotiopsis adusta*	Stem of unknown tree	*Fusarium culmorum, Gibberella zeae* and *Verticillium albo-atrum*		Li et al. (2008c)
	Nodulisporins D-F	*Nodulisporium* sp.	*Erica arborea*	*Bacillus megaterium, Microbotryum violaceum* and *Chlorella fusca*		Dai et al. (2009)

(*Continued*)

TABLE 7.2 (Continued)

Nature	Metabolite	Endophyte	Plant/Host	Activity against	Action	References
	Chaetoglobosins A and C; Chaetomugilin A and D	*Chaetomium globosum*	*Ginkgo biloba*	Brine shrimp larvae and *Mucor miehei*		Qin et al. (2009)
	Pyrrocidines A and B	*Acromonium zeae*	*Zea mays*	*Fusarium verticillioides Aspergillus flavus, Eupenicillium ochrosalmoneum, Curvularia lunata* and *Clavibacter michiganense*		Wicklow et al. (2005); Wicklow and Poling (2009)
	Xanalteric acids	*Alternaria* sp.	*Sonneratia alba*	*Staphylococcus aureus*		Kjer et al. (2009)
	Maklamicin	*Micromonospora* sp.		Gram-positive bacteria		Igarashi et al. (2011)
	Peaurantiogriseols A–F, aspermytin A, 1-propanone, 3-hydroxy-1-(1, 2, 4a, 5, 6, 7, 8, 8a-octahydro-2, 5-dihydroxy-1, 2, 6-trimethyl-1-naphthalenyl)	*Penicillium aurantiogriseum*	Mangrove	Low inhibitory activity against human aldose reductase, no activity of inducing neurite outgrowth, nor antimicrobial activity.		Ma et al. (2015)

	Palitantin	*Aspergillus fumigatiaffinis*	*Tribulus terestris*	*Enterococcus faecalis* and *Streptococcus pneumoniae*		Ola et al. (2018)
	15-hydroxy-1,4,5,6-tetra-epi-koninginin G	*Trichoderma koningiopsis*	*Artemisia argyi*	*Vibrio alginolyticus*		Shi et al. (2020)
Peptides	Echinocandin A	*Cryptosporiopsis* sp. and *Pezicula* sp.	*Pinus sylvestris* and *Fagus sylvatica*	*Candida albicans* and *Saccharomyces cerevisiae*		Noble et al. (1991)
	Leucinostatin A	*Acremonium* sp.	*Taxus baccata*	*Pythium ultimum* and anticancer		Strobel et al. (1997)
	Cryptocandin	*Cryptosporiosis* cf. *quercina*	*Tripterigeum wilfordii*	*Candida albicans, Trichophyton rubrum, T. mentagrophytes, Histoplasma capsulatum* and *Sclerotinia sclerotiorum*		Strobel (2002)
	Coronamycin	*Streptomyces* sp.	*Monstera* sp.	*Cryptococcus neoformans* and *Plasmodium falciparum*		Ezra et al. (2004)
	Actinomycin X2	*Streptomyces* sp. NRRL 30562	*Kennedia nigriscans*	Plant and human pathogens		Castillo et al. (2006)

(*Continued*)

TABLE 7.2 (Continued)

Nature	Metabolite	Endophyte	Plant/Host	Activity against	Action	References
	Actinomycin D	*Streptomyces* sp.	*Alpinia galangal*	*Colletotrichum musae* and *C. albicans*		Taechowisan et al. (2007)
	Trtesin	*Fusarium tricinctum*	*Rhododendron tomentosum* Harmaja	*Candida albicans, C. utilis, Staphylococcus carnosus* and *Fusarium oxysporum*		Tejesvi et al. (2013)
	S-adenosyl-N-acetylhomocysteine	*Micromonospora* sp.	*Pueraria candollei* Graham ex Benth	An antibacterial and antioxidant compound		Boonsnongcheep et al. (2017)

from *Tripterigeum wilfordi* against *Candida albicans* (Strobel et al., 1999); Colletrotric acid extracted from *C. gloeosporioides* associated with *Artemisia mongolica* with antibacterial activity (Zou et al., 2000); Cytonic acid A and B produced by endophytic fungus *Cytonaema* sp. with inhibition activity against human cytomegalovirus protease (Guo et al., 2000); Jestoerone isolated from endophyte *Pestalotiopsis jester* with antifungal activity (Li et al., 2001); Javanicin, produced by *Chloridium* sp., associated with neem plant with activity against *Pseudomonas* sp. (Kharwar et al., 2009); alkaloids, flavonoids, phenols, and steroids produced by *Methylobacterium radiotolerans* isolated from *Combretum erythrophyllum* against *Bacillus subtilis, Bacillus cereus, Klebsiella oxytoca, Mycobacterium smegmatis* (Photolo et al., 2020).

The antimicrobial secondary metabolites produced by endophytes have been grouped chemically as:

Alkaloids

These are a class of naturally occurring basic compounds containing at least one nitrogen. They also include some neutral or weekly acidic compounds. In view of the medicinal properties, many new alkaloids have been isolated and characterized from endophytes (Table 7.2). Kakadumycin A isolated from *Streptomyces* sp. associated with fern, *Grevillea pteridifolia* showed a broad-spectrum activity against malaria parasite and Gram-positive bacteria (Castillo et al., 2003). Likewise, Pestalachloride A, isolated from the endophyte *Pestalotiopsis adusta* associated with an unknown Chinese tree, has shown activity against *Gibberella zeae*, *Verticillium albo-atrum*, and *Fusarium culmorum* (Li et al., 2008a). Four metabolites, namely Lansai A–D with antifungal and anticancer potential, have been isolated from *Streptomyces* sp.—an endophyte of *Ficus benjamina* (Tuntiwachwuttikul et al., 2008). Similarly, Fumigaclavine C and Pseurtotin C from *Aspergillus* sp. isolated from *Bauhinia guianensis* showed antibacterial activity against *Escherichia coli, P. aeruginosa, Staphylococcus aureus*, and *B. subtilis* (Pinheiro et al., 2013). Diketopiperazine belonging to this category also have substantial medicinal properties. For example, diketopiperazine gancidin W extracted from *Streptomyces* sp. associated with the *Shorea ovalis* bark showed positive activity against *Plasmodium berghei* in vivo (Zin et al., 2017).

Terpenes and terpenoids

Terpenes are naturally occurring compounds formed by the combination of isoprene units, while terpenoids are derived from terpenes with some rearrangements or containing some oxygen functionality. The two terms are used interchangeably. These are the main constituent of essential oils derived from medicinal plants and are widely used in medicines and for fragrances. Many terpenes and terpenoids have been obtained from endophytes isolated from various sources (Table 7.2). Guanacastepene extracted from an unidentified fungus isolated from *Daphnopsis americana* showed antimicrobial activity against *Staphylococcus aureus* (Methicillin-resistant) and

Enterococcus faecium (vancomycin-resistant) by damaging bacterial membranes (Singh et al., 2000). Additionally, many sesquiterpenes were reported from endophytes with antimicrobial activity: Cedarmycin A and B from *Streptomyces* sp. against *Candida glabrata* (Sasaki et al., 2001); Periconicins A and B from *Periconia* sp. isolated from *Taxus cuspidate* against *B. subtilis*, *Klebsiella pneumoniae* and *Proteus vulgaris* (Kim et al., 2004); Scoparasin B from *Eutypella scoparia* isolated from *Garcinia dulcis* against *Microsporum gypseum* (Pongcharoen et al., 2006); Trichodermin from *Trichoderma harzianum* isolated from *Ilex cornuta* against *Alternaria solani* and *Rhizoctonia solani* (Chen et al., 2007); 8-α acetoxyphomadecalin and Phomadecalin from *Microdiplodia* sp. isolated from *Pinus* sp. against *P. aeruginosa* (Hatakeyama et al., 2010); α-terpinene, γ-terpinene, and (−)-4-terpineol from *Diaporthe apiculatum* isolated from *Leucaena leucocephala* against fungi (Song et al., 2019).

Phenolics

These are the compounds with phenol as a functional group. Many plants and endophytes produce phenolics in response to environmental stress conditions. Endophytes able to produce phenolic compounds have shown antimicrobial and germicidal activity (Table 7.2). Colletotric acid isolated from endophyte *Colletotrichum gloeosporioides* associated with *Artemesia mongolica* inhibited *Bacillus subtilis*, *Staphylococcus aureus*, *Micrococcus luteus* and *Helminthosporium sativum* (Zou et al., 2000); Cytonic acid A and B isolated from *Cytonaema* associated with *Quercus* sp. inhibited *Cytomegalovirus* (Guo et al., 2000); Altenusin isolated from *Alternaria* sp. associated with *Trixis vauthieri* inhibited *Trypanosoma* and *Leishmania* (Cota et al., 2008); phenol and steroids isolated from *Methylobacterium radiotolerans* associated with *Combretum erythrophyllum* inhibited *Bacillus subtilis, B. cereus, Klebsiella oxytoca*, and *Mycobacterium smegmatis* (Photolo et al., 2020).

Polyketides

Polyketides are the secondary metabolites with varied structures, either containing alternating methylene and carbonyl groups. They are sometimes derived from the precursors containing these alternating groups, and many of them have proven antimicrobial activity. Endophytes have been proven to be a virtuous source of these compounds with antimicrobial activities (Table 7.2). Chaetoglobosins A and C; Chaetomugilin A and D from *Chaetomium globosum* associated with *Ginkgo biloba* have shown activity against brine shrimp larvae and *Mucor miehei* (Qin et al., 2009). Similarly, Nodulisporins D-F isolated from the endophyte *Nodulisporium* sp. associated with *Erica arborea* showed inhibition against *Bacillus megaterium*, *M. violaceum* and *C. fusca* (Dai et al., 2009). Likewise, Pyrrocidines A and B from *Acromonium zeae* isolated from *Zea mays* showed activity against *Fusarium verticillioides*, *Aspergillus flavus*, *Eupenicillium ochrosalmoneum*, *Curvularia lunata*, and *Clavibacter michiganense* (Wicklow et al., 2005; Wicklow and Poling, 2009). Recently, 15-hydroxy-1,4,5,6-tetra-epi-koninginin G obtained from *Trichoderma koningiopsis* associated with *Artemisia argyi* has shown activity against *Vibrio alginolyticus* (Shi et al., 2020).

Peptides

Endophytes are an important source of novel peptides with antimicrobial activity. A large number of antimicrobial peptides have been isolated from endophytes associated with medicinal and nonmedicinal plants. Leucinostatin A, Cryptocandin, Echinocandin A, Actinomycin X2, Coronamycin, and S-adenosyl-N-acetylhomocysteine are some of the examples of peptides isolated from fungal and actinomycetes endophytes exhibiting antimicrobial activity against plant and human pathogens (Table 7.2).

Flavonoids and coumarins

Flavonoids, coumarins, and tannins are also important antimicrobial natural metabolites produced by endophytes. Flavonoids, naturally found in vegetables and fruits are polyphenolics with antioxidant and antimicrobial activity. Some important flavonoids including like fisetin, naringenin, xenognosin B, 4′,6-tetrahydroxyflavone, 7-methoxy-3,3′, 2′,7-dihydroxy-4′,5′-dimethoxyisoflavone, and 3′-hydroxydaidzein isolated from *Streptomyces* sp. have shown activity against *Bacillus subtilis*, *B. cereus*, and *Staphylococcus aureus* (Taechowisan et al., 2014).

In addition to flavonoids, coumarins are also reported from endophytes. Saadamycin and 5,7-dimethoxy-4-p-methoxyl phenyl coumarin isolated from *Streptomyces* sp. associated with *Aplysina fistularis* showed inhibitory activity against clinical fungi and dermatophytes (El-Gendy and EL-Bondkly, 2010). Likewise, 5, 7-dimethoxy-4-phenylcoumarin from *S. aureofaciens* was found to prevent or delay the formation of metastases (Taechowisan et al., 2007). Decursin extracted from *Streptomyces* sp. associated with *Zingiber officilnale* inhibited the growth of pathogenic bacteria *B. subtilis*, *B. cereus*, and *S. aureus* (Taechowisan et al., 2013). Similarly, coumarins extracted from *Curvularia tsudae* associated with *Cynodon dactylon* (L.) Pers., exhibited inhibitory activity against *E. coli, Pseudomonas fluorescens, Enterococcus faecalis*, and *S. aureus* (Nischitha et al., 2020)

7.6 Organic volatiles from endophytes

Various organic volatiles or volatile organic compounds are produced by a number of microorganisms living within plant parts and have shown beneficial activities including antagonism against pathogens, plant growth promotion, the potential of being used as fuel, etc. (Table 7.3). *Myrothecium inunduatum* associated with *Acalypha indica* also produced organic volatiles, including methylcyclohexadiene, 1-methyl-, and 1-ethyl propyl cyclohexane

TABLE 7.3 Endophytes as source of volatile compounds.

Endophyte	Host plant	Compounds	Activity	References
Nodulisporium	*Myroxylon balsamum*	1, -cyclohexadiene; 1-methyl-, 1−4pentadiene and cyclohexene; 1-methyl-4-(1-methylethenyl-)-; terpenoids; alkyl alcohols, ketones, esters	Fuel and antagonistic potential against *Aspergillus fumigates, Phytopthora cinnamomi, Sclerotinia sclertiorum, Rhizoctonia solani*	Mends et al. (2012)
Stretomyces sp. *neau-D50*	*Glycine max*	3-Acetonylidene-7-prenylindolin-2-one and 7-Isoprenylindole-3-carboxylic acid	Antifungal activity against phytopathogenic fungi *Colletotrichum orbiculare, Phytophthora capsici, Corynrspora cassiicola, Fusarium oxysporum*	Zhang et al. (2014)
M. albus MOW12	*Piper nigrum L.*	Acetic acid, ethyl ester, propanoic acid, 2-methyl-methyl ester and acetic acid, 2-methyl propyl ester	Antifungal activity for *Fusarium solani, Trichoderma* sp.	Banerjee et al. (2014)
Trichoderma gamsii	*Panax notoginseng*	Dimethyl disulfide, Dibenzofuran, Methane-thiol, Ketones	Antagonistic activity against root-rot disease cause fungi *Scytalidium lignicola*	Chen et al. (2016)
Aspergillus nomius	*Eusideroxylan zwageri*	Saturated hydrocarbons, alkyl halides, alcohols and unsaturated hydrocarbons	Biofuel	Azeez et al. (2016)
Xylaria sp.	*Haematoxylon brasiletto*	3-methyl-1-butanol, thujopsene. 2-methyl-1-butanol, 2-methyl-1-propnol	Inhibits the growth of *Fusarium oxysporum*	Sánchez-Ortiz et al. (2016)
Anthrinium sp., *Colletotrichum sp., and Diaporthe sp.*	*Aquilaria subintegra*	Oxo-agarospiral, β-agarofuran, α-agarofuran, β-dihydro agarofuran and δ-eudesmol	Strong antioxidant, natural component in the fragment perfume of agarwood oil	Monggoot et al. (2017)

Annulohypoxylon sp. FPYF3050	*Neolitsea pulchella* (Meissn)Marr.	1,8-cineole	Biofuel production from biomass residues	Wang et al. (2017)
Cryptosporiopsis ericae	*Coptis chinensis*	(2Z,4Z)-cycloocta-2,4-dienol	Antagonistic against *Sclerotinia sclertiorum, S. turcica*	Zhang et al. (2018)
Diaporthe spp.	*Catharanthus roseus*	Terpenes and terpenoids: α-muurolene, α-thujene, β-phellandrene, γ-terpinene, 2-carene, caryophyllene, thujone, patchoulene, cedrene	Activity against *Botrytis cinerea, Fusarium graminearum, Phytophthora cinnamomic, Alternaria alternata, Colletotrichum gloeosporioides*	Yan et al. (2018)
Geotrichum candidum	*Solanum melongena*	3-methyl-1-butanol, ethyl 3-methylbutanol, 2-phenylethanol, Isopentyl acetate, Naphthalene and Isobutyl acetate	Mycelia growth inhibition of phytopathogen *R. solani*	Mookherjee et al. (2018)
Bacillus spp.	*Musa* spp.	DL-Proline, 5 Oxo	Antifungal activity *Fusarium oxysporum* f. sp. *cubense*	Muthulakshmi et al. (2019)
Bacillus subtilis	*Eucommia ulmoides*	2-methylbutyric acid, 2-heptane, isopentyl acetate	Inhibitory effect on the mycelia growth and conidial sporulation of *Curvularia lunata*	Xie et al. (2020)

with the potential of fuel (Banerjee et al., 2010). Organic volatiles like 1,8-cineole, 1-methyl-1,4-cyclohexadiene, and alpha-methylene-alpha-fenchocamphorone produced by fungal endophyte *Hypoxylon* sp. associated with *Persea indica* exhibited activity against *Botrytis cinerea, Sclerotinia sclerotiorum, Cercospora beticola,* and *Phytophthora cinnamomi* suggesting, thereby, that the organic volatiles may play an important role in the survival of endophyte inside its host (Tomsheck et al., 2010). 1,8-cineole, a monoterpene is an octane derivative and has the potential to be used as a fuel additive. The study further showed that organic volatiles produced by *Hypoxylon* sp. have wide applications in medicine, industry, and energy production. An endophyte *Phomopsis* sp. isolated from *Odontoglossum* sp. produced a monoterpene sabinene showing activity against pathogenic fungi (Singh et al., 2011). In addition, organic volatiles like 1-propanol, 2-methyl, 2-propanone, 1-butanol 3-methyl, and benzene ethanol were also produced by *Phomopsis* sp. Likewise, *Nodulisporium* sp. associated with *Thelypteris angustifolia* was reported to produce ketones like acetone; 2-pentanone; 2-hexanone, 4-methyl; 3-hexanone 4-methyl; 3-hexanone, 2,4- dimethyl; and 5-hepten, 2-1 (Hassan et al., 2013). Under microaerophilic conditions, *Nodulisporium* sp. produced a number of organic volatiles including alkyl alcohols including 1-hexanol, 1-pentanol, 1-propanol-2-methyl, 1-nonanol, 1-butanol-3-methyl, 1-heptanol, 1-octanol, and phenyl ethyl alcohol (Schoen et al., 2017). Additionally, the endophyte also produced secondary alcohols, alkyl ketones, terpenoids, alkyl esters, few benzene derivatives, and some hydrocarbons with the ability to be used as fuel. Two bacterial endophytes *Pseudomonas stutzeri* and *Stenotrophomonas maltophilia* showed the production of volatile organic compound dimethyl disulphide exhibiting antifungal activities against *Botrytis cinerea* and plant growth promotion (Rojas-Solís et al., 2018).

7.7 Conclusion and future prospects

Endophytes are comparatively less explored and exploited group of microbes. However, they have a vast potential of synthesizing diverse metabolites with antimicrobial, anticancer, antidiabetic and antioxidant activities, to be utilized in medicine and agriculture. Alkaloids, terpenes, terpenoids, polyketides, phenolics coumarins, tannins, and peptides are some important chemical classes of antimicrobial metabolites produced by endophytes with tremendous scope in pharmaceuticals in view of emerging multidrug resistant microorganisms. Apart from inhibiting human pathogens, these compounds are equally efficient in inhibiting plant pathogens, thus can be used in place of chemical pesticides for sustainable agriculture. However, research should also be focused on the metabolites produced in traces or small quantities as these metabolites are sometimes missed or overlooked. Moreover, culturable endophytic microorganisms constitute only a fraction of total endophytes associated with plants. What is known about endophytes is not sufficient enough to understand their diversity and functionality. There is a probability of occurrence of nonculturable endophytic microorganisms that need to be explored for the discovery of novel metabolites for use in medicine, agriculture, and industries.

References

Akpotu, M.O., Eze, P.M., Ajaghaku, D.L., Okoye, F.B.C., Esimone, C.O., 2015. Antimicrobial activity of endophytic fungi isolated from *Catharanthus roseus* and *Euphorbia hirta*. Planta Med 81 (16), PM_145.

Aly, A.H., Edrada-Ebel, R., Wray, V., Müller, W.E., Kozytska, S., Hentschel, U., et al., 2008. Bioactive metabolites from the endophytic fungus *Ampelomyces* sp. isolated from the medicinal plant *Urospermum picroides*. Phytochemistry 69 (8), 1716–1725.

Andreote, F.D., Gumiere, T., Durrer, A., 2014. Exploring interactions of plant microbiomes. Sci. Agricola 71 (6), 528–539.

Azeez, L.A., Muid, S., Hasnul, B.M., 2016. Identification of volatile secondary metabolites from an endophytic microfungus *Aspergillus nomius* KUB105. Malaysian J. Anal. Sci. 20 (4), 751–759.

Banerjee, D., Strobel, G.A., Booth, B., Sears, J., Spakowicz, D., Busse, S., 2010. An endophytic *Myrothecium inundatum* producing volatile organic compounds. Mycosphere 1 (3), 241–247.

Banerjee, D., Pandey, A., Jana, M., Strobel, G., 2014. *Muscodor albus* MOW12 an endophyte of *Pipernigrum* L. (*Piperaceae*) collected from North East India produces volatile antimicrobials. Indian. J. Microbiol 54 (1), 27–32.

Blankenship, J.D., Houseknecht, J.B., Pal, S., Bush, L.P., Grossman, R.B., Schardl, C.L., 2005. Biosynthetic precursors of fungal pyrrolizidines, the loline alkaloids. ChemBioChem 6 (6), 1016–1022.

Boonsnongcheep, P., Nakashima, T., Takahashi, Y., Prathanturarug, S., 2017. Diversity of endophytic actinomycetes isolated from roots and root nodules of *Pueraria candollei* grah. ex benth. and the analyses of their secondary metabolites. Chiang Mai. J. Sci. 44, 1–14.

Brady, S.F., Clardy, J., 2000. CR377, a new pentaketide antifungal agent isolated from an endophytic fungus. J. Nat. Products 63 (10), 1447–1448.

Calcul, L., Waterman, C., Ma, W.S., Lebar, M.D., Harter, C., Mutka, T., et al., 2013. Screening mangrove endophytic fungi for antimalarial natural products. Mar. Drugs 11 (12), 5036–5050.

Calhoun, L.A., Findlay, J.A., Miller, J.D., Whitney, N.J., 1992. Metabolites toxic to spruce budworm from balsam fir needle endophytes. Mycol Res. 96 (4), 281–286.

Castillo, U., Harper, J.K., Strobel, G.A., Sears, J., Alesi, K., Ford, E., et al., 2003. Kakadumycins, novel antibiotics from *Streptomyces* sp. NRRL 30566, an endophyte of *Grevillea pteridifolia*. FEMS Microbiol Lett. 224 (2), 183–190.

Castillo, U.F., Strobel, G.A., Mullenberg, K., Condron, M.M., Teplow, D.B., Folgiano, V., et al., 2006. Munumbicins E-4 and E-5: novel broad-spectrum antibiotics from *Streptomyces* NRRL 3052. FEMS Microbiol Lett. 255 (2), 296–300.

Che, Y., Gloer, J.B., Wicklow, D.T., 2002. Phomadecalins A – D and phomapentenone A: new bioactive metabolites from *Phoma* sp. NRRL 25697, a fungal colonist of Hypoxylon Stromata. J. Nat. Products 65 (3), 399–402.

Chen, L.Z., Chen, J.M., Zheng, X.S., Zhang, J.F., Yu, X.P., 2007. Identification and antifungal activity of the metabolite of endophytic fungi isolated from *Llex cornuta*. Chin. J. Pesticide Sci. 9 (2), 143–148.

Chen, Y.M., Yang, Y.H., Li, X.N., Zou, C., Zhao, P.J., 2015. Diterpenoids from the endophytic fungus *Botryosphaeria* sp. P483 of the Chinese herbal medicine *Huperzia serrata*. Molecules 20 (9), 16924–16932.

Chen, J.L., Sun, S.Z., Miao, C.P., Wu, K., Chen, Y.W., Xu, L.H., et al., 2016. Endophytic *Trichoderma gamsii* YIM PH30019: a promising biocontrol agent with hyperosmolar, mycoparasitism, and antagonistic activities of induced volatile organic compounds on root-rot pathogenic fungi of *Panax notoginseng*. J. Ginseng Res. 40 (4), 315–324.

Chen, S., He, L., Chen, D., Cai, R., Long, Y., Lu, Y., et al., 2017. Talaramide A, an unusual alkaloid from the mangrove endophytic fungus *Talaromyces* sp.(HZ-YX1) as an inhibitor of mycobacterial PknG. N. J. Chem. 41 (11), 4273–4276.

Cota, B.B., Rosa, L.H., Caligiorne, R.B., Rabello, A.L.T., Almeida Alves, T.M., Rosa, C.A., et al., 2008. Altenusin, a biphenyl isolated from the endophytic fungus *Alternaria* sp., inhibits trypanothione reductase from *Trypanosoma cruzi*. FEMS Microbiol Lett. 285 (2), 177–182.

Cui, C.M., Li, X.M., Meng, L., Li, C.S., Huang, C.G., Wang, B.G., 2010. 7-Nor-ergosterolide, a pentalactone-containing norsteroid and related steroids from the marine-derived endophytic *Aspergillus ochraceus* EN-31. J. Nat. Products 73 (11), 1780–1784.

Cui, H., Liu, Y., Ding, M., Zhang, Z., Liu, H., Huang, X., et al., 2017. New pyranonaphthazarin and 2-naphthoic acid derivatives from the mangrove endophytic fungus *Leptosphaerulina* sp. SKS032. Phytochem Lett. 20, 214–217.

Dai, J., Krohn, K., Flörke, U., Draeger, S., Schulz, B., Kiss-Szikszai, A., et al., 2006. Metabolites from the endophytic fungus *Nodulisporium* sp. from *Juniperus cedre*. Eur. J. Org. Chem. 2006 (15), 3498–3506.

Dai, J., Krohn, K., Draeger, S., Schulz, B., 2009. New naphthalene-chroman coupling products from the endophytic fungus, *Nodulisporium* sp. from *Erica arborea*. Eur. J. Org. Chem. 2009 (10), 1564–1569.

de Felício, R., Pavão, G.B., de Oliveira, A.L.L., Erbert, C., Conti, R., Pupo, M.T., et al., 2015. Antibacterial, antifungal and cytotoxic activities exhibited by endophytic fungi from the Brazilian marine red alga *Bostrychia tenella* (Ceramiales). Rev. Brasileira de. Farmacogn. 25 (6), 641–650.

Dettrakul, S., Kittakoop, P., Isaka, M., Nopichai, S., Suyarnsestakorn, C., Tanticharoen, M., et al., 2003. Antimycobacterial pimarane diterpenes from the fungus *Diaporthe* sp. Bioorg Med Chem. Lett. 13 (7), 1253–1255.

dos Santos, R.M.G., Rodrigues-Fo, E., 2003. Further meroterpenes produced by *Penicillium* sp., an endophyte obtained from *Melia azedarach*. Z. für Naturforschung C. 58 (9-10), 663–669.

Ebrahim, W., Kjer, J., El Amrani, M., Wray, V., Lin, W., Ebel, R., et al., 2012a. Pullularins E and F, two new peptides from the endophytic fungus *Bionectria ochroleuca* isolated from the mangrove plant *Sonneratia caseolaris*. Mar. Drugs 10 (5), 1081–1091.

Ebrahim, W., Aly, A.H., Mándi, A., Totzke, F., Kubbutat, M.H., Wray, V., et al., 2012b. Decalactone derivatives from *Corynespora cassiicola*, an endophytic fungus of the mangrove plant *Laguncularia racemosa*. Eur. J. Org. Chem. 2012 (18), 3476–3484.

El-Bialy, H.A., El-Bastawisy, H.S., 2020. Elicitors stimulate paclitaxel production by endophytic fungi isolated from ecologically altered *Taxus baccata*. J. Radiat. Res. Appl. Sci. 13 (1), 79–87.

El-Gendy, M.M., EL-Bondkly, A.M., 2010. Production and genetic improvement of a novel antimycotic agent, saadamycin, against dermatophytes and other clinical fungi from endophytic *Streptomyces* sp. Hedaya48. J. Ind. Microbiol Biotechnol. 37 (8), 831–841.

Elissawy, A.M., El-Shazly, M., Ebada, S.S., Singab, A.B., Proksch, P., 2015. Bioactive terpenes from marine-derived fungi. Mar. Drugs 13 (4), 1966–1992.

El-Sayed, E.S.R., Zaki, A.G., Ahmed, A.S., Ismaiel, A.A., 2020a. Production of the anticancer drug taxol by the endophytic fungus *Epicoccum nigrum* TXB502: enhanced production by gamma irradiation mutagenesis and immobilization technique. Appl. Microbiol Biotechnol. 104 (16), 6991–7003.

El-Sayed, E.R., Ahmed, A.S., Abdelhakim, H.K., 2020b. A novel source of the cardiac glycoside digoxin from the endophytic fungus *Epicoccum nigrum*: isolation, characterization, production enhancement by gamma irradiation mutagenesis and anticancer activity evaluation. J. Appl. Microbiol 128 (3), 747–762.

Elsebai, M.F., Saleem, M., Tejesvi, M.V., Kajula, M., Mattila, S., Mehiri, M., et al., 2014. Fungal phenalenones: chemistry, biology, biosynthesis and phylogeny. Nat. Product. Rep. 31 (5), 628–645.

Ezra, D., Castillo, U.F., Strobel, G.A., Hess, W.M., Porter, H., Jensen, J.B., et al., 2004. Coronamycins, peptide antibiotics produced by a verticillate *Streptomyces* sp.(MSU-2110) endophytic on *Monstera* sp. Microbiology 150 (4), 785–793.

Gangadevi, V., Muthumary, J., 2008. Isolation of *Colletotrichum gloeosporioides*, a novel endophytic taxol-producing fungus from the leaves of a medicinal plant, *Justicia gendarussa*. Mycol Balcanica 5 (1), 1–4.

Gao, S.S., Li, X.M., Li, C.S., Proksch, P., Wang, B.G., 2011. Penicisteroids A and B, antifungal and cytotoxic polyoxygenated steroids from the marine alga-derived endophytic fungus *Penicillium chrysogenum* QEN-24S. Bioorganic Med Chem. Lett. 21 (10), 2894–2897.

Garyali, S., Kumar, A., Reddy, M.S., 2013. Taxol production by an endophytic fungus, *Fusarium redolens*, isolated from Himalayan yew. J. Microbiol Biotechnol. 23 (10), 1372–1380.

Ge, H.M., Zhang, Q., Xu, S.H., Guo, Z.K., Song, Y.C., Huang, W.Y., et al., 2011. Chaetoglocins A–D, four new metabolites from the endophytic fungus *Chaetomium globosum*. Planta Med 77 (03), 277–280.

Gomes, N.G., Lefranc, F., Kijjoa, A., Kiss, R., 2015. Can some marine-derived fungal metabolites become actual anticancer agents? Mar. Drugs 13 (6), 3950–3991.

Gos, F.M., Savi, D.C., Shaaban, K.A., Thorson, J.S., Aluizio, R., Possiede, Y.M., et al., 2017. Antibacterial activity of endophytic actinomycetes isolated from the medicinal plant *Vochysia divergens* (Pantanal, Brazil). Front. Microbiol 8, 1642.

Guo, B., Dai, J.R., Ng, S., Huang, Y., Leong, C., Ong, W., et al., 2000. Cytonic acids A and B: novel tridepside inhibitors of hCMV protease from the endophytic fungus *Cytonaema* species. J. Nat. Products 63 (5), 602–604.

Haritakun, R., Rachtawee, P., Chanthaket, R., Boonyuen, N., Isaka, M., 2010. Butyrolactones from the fungus *Aspergillus terreus* BCC 4651. Chem. Pharm. Bull. 58 (11), 1545–1548.

Harper, J.K., Arif, A.M., Ford, E.J., Strobel, G.A., Porco Jr, J.A., Tomer, D.P., et al., 2003. Pestacin: a 1, 3-dihydro isobenzofuran from *Pestalotiopsis microspora* possessing antioxidant and antimycotic activities. Tetrahedron 59 (14), 2471–2476.

Hassan, S.R.U., Strobel, G.A., Geary, B., Sears, J., 2013. An endophytic *Nodulisporium* sp. from Central America producing volatile organic compounds with both biological and fuel potential. J. Microbiol Biotechnol. 23 (1), 29–35.

Hatakeyama, T., Koseki, T., Murayama, T., Shiono, Y., 2010. Eremophilane sesquiterpenes from the endophyte *Microdiplodia* sp. KS 75-1 and revision of the stereochemistries of phomadecalins C and D. Phytochem Lett. 3 (3), 148–151.

Horn, W.S., Simmonds, M.S.J., Schwartz, R.E., Blaney, W.M., 1995. Phomopsichalasin, a novel antimicrobial agent from an endophytic *Phomopsis* sp. Tetrahedron 51 (14), 3969–3978.

Hsiao, G., Wang, S.W., Chiang, Y.R., Chi, W.C., Kuo, Y.H., Chen, C.Y., et al., 2020. Anti-inflammatory effects of peptides from a marine algicolous fungus *Acremonium* sp. NTU492 in BV-2 microglial cells. J. Food Drug. Anal. 28 (2).

Huang, H.B., Feng, X.J., Liu, L., Chen, B., Lu, Y.J., Ma, L., et al., 2010. Three dimeric naphtho-γ-pyrones from the mangrove endophytic fungus *Aspergillus tubingensis* isolated from *Pongamia pinnata*. Planta Med 76 (16), 1888–1891.

Huang, C.H., Pan, J.H., Chen, B., Yu, M., Huang, H.B., Zhu, X., et al., 2011. Three bianthraquinone derivatives from the mangrove endophytic fungus *Alternaria* sp. ZJ9-6B from the South China Sea. Mar. Drugs 9 (5), 832–843.

Huang, G.L., Zhou, X.M., Bai, M., Liu, Y.X., Zhao, Y.L., Luo, Y.P., et al., 2016. Dihydroisocoumarins from the mangrove-derived fungus *Penicillium citrinum*. Mar. Drugs 14 (10), 177.

Huang, J., Xu, J., Wang, Z., Khan, D., Niaz, S.I., Zhu, Y., et al., 2017. New lasiodiplodins from mangrove endophytic fungus *Lasiodiplodia* sp. 318#. Nat. Product. Res. 31 (3), 326–332.

Hussain, H., Akhtar, N., Draeger, S., Schulz, B., Pescitelli, G., Salvadori, P., et al., 2009. New bioactive 2, 3-epoxycyclohexenes and isocoumarins from the endophytic fungus *Phomopsis* sp. from *Laurus azorica*. Eur. J. Org. Chem. 2009 (5), 749–756.

Igarashi, Y., Ogura, H., Furihata, K., Oku, N., Indananda, C., Thamchaipenet, A., 2011. Maklamicin, an antibacterial polyketide from an endophytic *Micromonospora* sp. J. Nat. Products 74 (4), 670–674.

Inahashi, Y., Iwatsuki, M., Ishiyama, A., Matsumoto, A., Hirose, T., Oshita, J., et al., 2015. Actinoallolides A–E, new anti-trypanosomal macrolides, produced by an endophytic actinomycete, *Actinoallomurus fulvus* MK10-036. Org. Lett. 17 (4), 864–867.

Isaka, M., Chinthanom, P., Boonruangprapa, T., Rungjindamai, N., Pinruan, U., 2010. Eremophilane-type sesquiterpenes from the fungus *Xylaria* sp. BCC 21097. J. Nat. Products 73 (4), 683–687.

Izzah Shahid, M.R., Baig, D.N., Saleem, R.S., Malik, K.A., Mehnaz, S., 2017. Secondary metabolites production and plant growth promotion by *Pseudomonas chlororaphis* and *P. aurantiaca* strains isolated from cactus, cotton, and para grass. J. Microbiol. Biotechnol. 27 (3), 480–491.

Ju, Z., Lin, X., Lu, X., Tu, Z., Wang, J., Kaliyaperumal, K., et al., 2015. Botryoisocoumarin A, a new COX-2 inhibitor from the mangrove *Kandelia candel* endophytic fungus *Botryosphaeria* sp. KcF6. J. Antibiotics 68 (10), 653–656.

Ju, Z.R., Qin, X., Lin, X.P., Wang, J.F., Kaliyaperumal, K., Tian, Y.Q., et al., 2016. New phenyl derivatives from endophytic fungus *Botryosphaeria* sp. SCSIO KcF6 derived of mangrove plant *Kandelia candel*. Nat. Product. Res. 30 (2), 192–198.

Khare, E., Mishra, J., Arora, N.K., 2018. Multifaceted interactions between endophytes and plant: developments and prospects. Front. Microbiol 9, 2732.

Kharwar, R.N., Verma, V.C., Kumar, A., Gond, S.K., Harper, J.K., Hess, W.M., et al., 2009. Javanicin, an antibacterial naphthaquinone from an endophytic fungus of neem, *Chloridium* sp. Curr. Microbiol 58 (3), 233–238.

Kim, S., Shin, D.S., Lee, T., Oh, K.B., 2004. Periconicins, two new fusicoccane diterpenes produced by an endophytic fungus *Periconia* sp. with antibacterial activity. J. Nat. Products 67 (3), 448–450.

Kjer, J., Wray, V., Edrada-Ebel, R., Ebel, R., Pretsch, A., Lin, W., et al., 2009. Xanalteric acids I and II and related phenolic compounds from an endophytic *Alternaria* sp. isolated from the mangrove plant *Sonneratia alba*. J. Nat. Products 72 (11), 2053–2057.

Kjer, J., Debbab, A., Aly, A.H., Proksch, P., 2010. Methods for isolation of marine-derived endophytic fungi and their bioactive secondary products. Nat. Protoc. 5 (3), 479–490.

Kumaran, R.S., Kim, H.J., Hur, B.K., 2010. Taxol promising fungal endophyte, *Pestalotiopsis* species isolated from *Taxus cuspidata*. J. Biosci. Bioeng. 110 (5), 541–546.

Kusari, S., Lamshöft, M., Spiteller, M., 2009. *Aspergillus fumigatus* Fresenius, an endophytic fungus from *Juniperus communis* L. Horstmann as a novel source of the anticancer pro-drug deoxypodophyllotoxin. J. Appl. Microbiol 107 (3), 1019–1030.

Le, T.T.M., Hoang, A.T.H., Le, T.T.B., Vo, T.T.B., Van Quyen, D., Chu, H.H., 2019. Isolation of endophytic fungi and screening of Huperzine A–producing fungus from *Huperzia serrata* in Vietnam. Sci. Rep. 9 (1), 1–13.

Li, J.Y., Strobel, G.A., 2001. Jesterone and hydroxy-jesterone antioomycete cyclohexenone epoxides from the endophytic fungus *Pestalotiopsis jesteri*. Phytochemistry 57 (2), 261–265.

Li, J.Y., Strobel, G., Harper, J., Lobkovsky, E., Clardy, J., 2000. Cryptocin, a potent tetramic acid antimycotic from the endophytic fungus *Cryptosporiopsis* cf. *quercina*. Org. Lett. 2 (6), 767–770.

Li, J.Y., Harper, J.K., Grant, D.M., Tombe, B.O., Bashyal, B., Hess, W.M., et al., 2001. Ambuic acid, a highly functionalized cyclohexenone with antifungal activity from *Pestalotiopsis* spp. and *Monochaetia* sp. Phytochemistry 56 (5), 463–468.

Li, E., Jiang, L., Guo, L., Zhang, H., Che, Y., 2008a. Pestalachlorides A–C, antifungal metabolites from the plant endophytic fungus *Pestalotiopsis adusta*. Bioorganic Med Chem. 16 (17), 7894–7899.

Li, E., Tian, R., Liu, S., Chen, X., Guo, L., Che, Y., 2008b. Pestalotheols A – D, bioactive metabolites from the plant endophytic fungus *Pestalotiopsis theae*. J. Nat. Products 71 (4), 664–668.

Li, E., Jiang, L., Guo, L., Zhang, H., Che, Y., 2008c. Pestalachlorides A–C, antifungal metabolites from the plant endophytic fungus *Pestalotiopsis adusta*. Bioorganic Med Chem. 16 (17), 7894–7899.

Li, J., Zhao, G.Z., Varma, A., Qin, S., Xiong, Z., Huang, H.Y., et al., 2012. An endophytic Pseudonocardia species induces the production of artemisinin in *Artemisia annua*. PLoS One 7 (12), e51410.

Li, H., Guan, Y., Dong, Y., Zhao, L., Rong, S., Chen, W., et al., 2018. Isolation and evaluation of endophytic *Bacillus tequilensis* GYLH001 with potential application for biological control of *Magnaporthe oryzae*. PLoS One 13 (10), e0203505.

Liu, J.Y., Song, Y.C., Zhang, Z., Wang, L., Guo, Z.J., Zou, W.X., et al., 2004. *Aspergillus fumigatus* CY018, an endophytic fungus in *Cynodon dactylon* as a versatile producer of new and bioactive metabolites. J. Biotechnol. 114 (3), 279–287.

Liu, K., Ding, X., Deng, B., Chen, W., 2009a. Isolation and characterization of endophytic taxol-producing fungi from *Taxus chinensis*. J. Ind. Microbiol Biotechnol. 36 (9), 1171.

Liu, L., Liu, S., Chen, X., Guo, L., Che, Y., 2009b. Pestalofones A–E, bioactive cyclohexanone derivatives from the plant endophytic fungus *Pestalotiopsis fici*. Bioorganic Med Chem. 17 (2), 606–613.

Liu, S., Yan, X., Yu, M., Chen, J., Zhang, L., 2010. A novel compound from *Penicillium* sp. (M207142). Chem. Nat. Compd. 46 (1), 116–118.

Liu, Y., Wu, Y., Zhai, R., Liu, Z., Huang, X., She, Z., 2016. Altenusin derivatives from mangrove endophytic fungus *Alternaria* sp. SK6YW3L. RSC Adv. 6 (76), 72127–72132.

Liu, Y., Stuhldreier, F., Kurtán, T., Mándi, A., Arumugam, S., Lin, W., et al., 2017. Daldinone derivatives from the mangrove-derived endophytic fungus *Annulohypoxylon* sp. RSC Adv. 7 (9), 5381–5393.

Lösgen, S., Magull, J., Schulz, B., Draeger, S., Zeeck, A., 2008. Isofusidienols: novel Chromone-3-oxepines produced by the endophytic fungus *Chalara* sp. Eur. J. Org. Chem. 2008 (4), 698–703.

Ma, Y.T., Qiao, L.R., Shi, W.Q., Zhang, A.L., Gao, J.M., 2010. Metabolites produced by an endophyte *Alternaria alternata* isolated from *Maytenus hookeri*. Chem. Nat. Compd. 46 (3), 504–506.

Ma, Y., Li, J., Huang, M., Liu, L., Wang, J., Lin, Y., 2015. Six new polyketide decalin compounds from mangrove endophytic fungus *Penicillium aurantiogriseum* 328. Mar. Drugs 13 (10), 6306–6318.

Martínez-Luis, S., Della-Togna, G., Coley, P.D., Kursar, T.A., Gerwick, W.H., Cubilla-Rios, L., 2008. Antileishmanial constituents of the Panamanian endophytic fungus *Edenia* sp. J. Nat. Products 71 (12), 2011–2014.

Mei, W.L., Zheng, B., Zhao, Y.X., Zhong, H.M., Chen, X.L.W., Zeng, Y.B., et al., 2012. Meroterpenes from endophytic fungus A1 of mangrove plant *Scyphiphora hydrophyllacea*. Mar. Drugs 10 (9), 1993–2001.

Mends, M.T., Yu, E., Strobel, G.A., Hassan, S.R.U., Booth, E., Geary, B., et al., 2012. An endophytic *Nodulisporium* sp. producing volatile organic compounds having bioactivity and fuel potential. J. Pet. Environ. Biotechnol. 3 (3).

Mitter, B., Petric, A., Shin, M.W., Chain, P.S., Hauberg-Lotte, L., Reinhold-Hurek, B., et al., 2013. Comparative genome analysis of *Burkholderia phytofirmans* PsJN reveals a wide spectrum of endophytic lifestyles based on interaction strategies with host plants. Front. Plant Sci. 4, 120.

Mohamad, O.A., Li, L., Ma, J.B., Hatab, S., Xu, L., Guo, J.W., et al., 2018. Evaluation of the antimicrobial activity of endophytic bacterial populations from Chinese traditional medicinal plant licorice and characterization of the bioactive secondary metabolites produced by *Bacillus atrophaeus* against *Verticillium dahliae*. Front. Microbiol 9, 924.

Mookherjee, A., Bera, P., Mitra, A., Maiti, M.K., 2018. Characterization and synergistic effect of antifungal volatile organic compounds emitted by the *Geotrichum candidum* PF005, an endophytic fungus from the eggplant. Microb. Ecol. 75 (3), 647–661.

Monggoot, S., Popluechai, S., Gentekaki, E., Pripdeevech, P., 2017. Fungal endophytes: an alternative source for production of volatile compounds from agarwood oil of *Aquilaria subintegra*. Microb. Ecol. 74 (1), 54–61.

Muthulakshmi, P., Thangavel, S., Seethapathy, P., 2019. Characterization of antifungal volatile organic compounds produced by bacterial endophytes against *Fusarium oxysporum* through GC-MS analysis. Microbiol Res. J. Int. 1–9.

Nicoletti, R., Salvatore, M.M., Andolfi, A., 2018. Secondary metabolites of mangrove-associated strains of Talaromyces. Mar. Drugs 16 (1), 12.

Nischitha, R., Vasanthkumari, M.M., Kumaraswamy, B.E., Shivanna, M.B., 2020. Antimicrobial and antioxidant activities and chemical profiling of *Curvularia tsudae* endophytic in *Cynodon dactylon* (L.) Pers. 3 Biotech. 10 (7), 1–12.

Noble, H.M., Langley, D., Sidebottom, P.J., Lane, S.J., Fisher, P.J., 1991. *An echinocandin from an endophytic Cryptosporiopsis sp.* and *Pezicula* sp. in *Pinus sylvestris* and *Fagus sylvatica*. Mycol Res. 95 (12), 1439–1440.

Oh, M.N., Yang, H.C., Oh, K.B., 2005. Biological characterization of periconicins, bioactive secondary metabolites, produced by *Periconia* sp. OBW-15. J. Microbiol Biotechnol. 15 (1), 216–220.

Ola, A.R., Tawo, B.D., Belli, H.L.L., Proksch, P., Tommy, D., Hakim, E.H., 2018. A new antibacterial polyketide from the endophytic fungi *Aspergillus fumigatiaffinis*. Nat. Product. Commun. 13 (12), 1934578X1801301202.

Pandi, M., Kumaran, R.S., Choi, Y.K., Kim, H.J., Muthumary, J., 2011. Isolation and detection of taxol, an anticancer drug produced from *Lasiodiplodia theobromae*, an endophytic fungus of the medicinal plant *Morinda citrifolia*. Afr. J. Biotechnol. 10 (8), 1428–1435.

Pansanit, A., Pripdeevech, P., 2018. Antibacterial secondary metabolites from an endophytic fungus, *Arthrinium* sp. MFLUCC16-1053 isolated from *Zingiber cassumunar*. Mycology 9 (4), 264–272.

Park, J.H., Choi, G.J., Lee, H.B., Kim, K.M., Jung, H.S., Lee, S.W., et al., 2005. Griseofulvin from *Xylaria* sp. strain F0010, an endophytic fungus of *Abies holophylla* and its antifungal activity against plant pathogenic fungi. J. Microbiol. Biotechnol. 15 (1), 112–117.

Photolo, M.M., Mavumengwana, V., Sitole, L., Tlou, M.G., 2020. Antimicrobial and antioxidant properties of a bacterial endophyte, methylobacterium radiotolerans MAMP 4754, isolated from combretum erythrophyllum seeds. Int. J. Microbiol. 2020, 1–11. Available from: https://doi.org/10.1155/2020/9483670.

Pinheiro, E.A.A., Carvalho, J.M., dos Santos, D.C.P., Feitosa, A.D.O., Marinho, P.S.B., Guilhon, G.M.S.P., et al., 2013. Antibacterial activity of alkaloids produced by endophytic fungus *Aspergillus* sp. EJC08 isolated from medical plant *Bauhinia guianensis*. Nat. Product Res 27 (18), 1633–1638.

Pongcharoen, W., Rukachaisirikul, V., Phongpaichit, S., Rungjindamai, N., Sakayaroj, J., 2006. Pimarane diterpene and cytochalasin derivatives from the endophytic fungus *Eutypella scoparia* PSU-D44. J. Nat. Products 69 (5), 856–858.

Pongcharoen, W., Rukachaisirikul, V., Phongpaichit, S., Kühn, T., Pelzing, M., Sakayaroj, J., et al., 2008. Metabolites from the endophytic fungus *Xylaria* sp. PSU-D14. Phytochemistry 69 (9), 1900–1902.

Puri, S.C., Nazir, A., Chawla, R., Arora, R., Riyaz-ul-Hasan, S., Amna, T., et al., 2006. The endophytic fungus *Trametes hirsuta* as a novel alternative source of podophyllotoxin and related aryl tetralin lignans. J. Biotechnol. 122 (4), 494–510.

Qiao, M.F., Ji, N.Y., Liu, X.H., Li, K., Zhu, Q.M., Xue, Q.Z., 2010. Indoloditerpenes from an algicolous isolate of *Aspergillus oryzae*. Bioorganic Med Chem. Lett 20 (19), 5677–5680.

Qin, J.C., Zhang, Y.M., Gao, J.M., Bai, M.S., Yang, S.X., Laatsch, H., et al., 2009. Bioactive metabolites produced by *Chaetomium globosum*, an endophytic fungus isolated from *Ginkgo biloba*. Bioorganic Med Chem. Lett. 19 (6), 1572–1574.

Rojas-Solís, D., Zetter-Salmón, E., Contreras-Pérez, M., del Carmen Rocha-Granados, M., Macías-Rodríguez, L., Santoyo, G., 2018. *Pseudomonas stutzeri* E25 and *Stenotrophomonas maltophilia* CR71 endophytes produce antifungal volatile organic compounds and exhibit additive plant growth-promoting effects. Biocatalysis Agric. Biotechnol. 13, 46–52.

Rowan, D.D., 1993. Lolitrems, peramine and paxilline: mycotoxins of the ryegrass/endophyte interaction. Agric, Ecosyst. Environ. 44 (1-4), 103–122.

Rukachaisirikul, V., Sommart, U., Phongpaichit, S., Sakayaroj, J., Kirtikara, K., 2008. Metabolites from the endophytic fungus *Phomopsis* sp. PSU-D15. Phytochemistry 69 (3), 783–787.

Ryan, R.P., Germaine, K., Franks, A., Ryan, D.J., Dowling, D.N., 2008. Bacterial endophytes: recent developments and applications. FEMS Microbiol Lett. 278 (1), 1–9.

Sánchez-Ortiz, B.L., Sánchez-Fernández, R.E., Duarte, G., Lappe-Oliveras, P., Macías-Rubalcava, M.L., 2016. Antifungal, anti-oomycete and phytotoxic effects of volatile organic compounds from the endophytic fungus *Xylaria* sp. strain PB 3f3 isolated from *Haematoxylon brasiletto*. J. Appl. Microbiol 120 (5), 1313–1325.

Sarasan, M., Puthumana, J., Job, N., Han, J., Lee, J.S., Philip, R., 2017. Marine algicolous endophytic fungi-a promising drug resource of the era. J. Microbiol Biotechnol. 27 (6), 1039–1052.

Sasaki, T., Igarashi, Y., Saito, N., Furumai, T., 2001. Cedarmycins A and B, new antimicrobial antibiotics from *Streptomyces* sp. TP-A0456. J. Antibiotics 54 (7), 567–572.

Savi, D.C., Haminiuk, C.W.I., Sora, G.T.S., Adamoski, D.M., Kenski, J., Winnischofer, S.M.B., et al., 2015. Antitumor, antioxidant and antibacterial activities of secondary metabolites extracted by endophytic actinomycetes isolated from *Vochysia divergens*. Int. J. Pharm, Chem. Biol. Sci. 5 (1).

Schoen, H.R., Knighton, W.B., Peyton, B.M., 2017. Endophytic fungal production rates of volatile organic compounds are highest under microaerophilic conditions. Microbiology 163 (12), 1767–1777.

Schulz, B., Sucker, J., Aust, H.J., Krohn, K., Ludewig, K., Jones, P.G., et al., 1995. Biologically active secondary metabolites of endophytic Pezicula species. Mycol Res. 99 (8), 1007–1015.

Selim, K.A., Nagia, M.M., Ghwas, D.E.E., 2017. Endophytic fungi are multifunctional biosynthesizers: ecological role and chemical diversity. Endophytic Fungi Diversity, Charact. Biocontrol 39.

Shi, X.S., Li, H.L., Li, X.M., Wang, D.J., Li, X., Meng, L.H., et al., 2020. Highly oxygenated polyketides produced by *Trichoderma koningiopsis* QA-3, an endophytic fungus obtained from the fresh roots of the medicinal plant *Artemisia argyi*. Bioorg Chem. 94, 103448.

Shweta, S., Bindu, J.H., Raghu, J., Suma, H.K., Manjunatha, B.L., Kumara, P.M., et al., 2013. Isolation of endophytic bacteria producing the anticancer alkaloid camptothecine from *Miquelia dentata* Bedd. (Icacinaceae). Phytomedicine 20 (10), 913–917.

Silva, G.H., Teles, H.L., Zanardi, L.M., Young, M.C.M., Eberlin, M.N., Hadad, R., et al., 2006. Cadinane sesquiterpenoids of *Phomopsis cassiae*, an endophytic fungus associated with *Cassia spectabilis* (Leguminosae). Phytochemistry 67 (17), 1964–1969.

Silva, G.H., de Oliveira, C.M., Teles, H.L., Pauletti, P.M., Castro-Gamboa, I., Silva, D.H., et al., 2010. Sesquiterpenes from *Xylaria* sp., an endophytic fungus associated with *Piper aduncum* (*Piperaceae*). Phytochem Lett. 3 (3), 164–167.

Singh, M.P., Janso, J.E., Luckman, S.W., Brady, S.F., Clardy, J., Greenstein, M., et al., 2000. Biological activity of guanacastepene, a novel diterpenoid antibiotic produced by an unidentified fungus CR115. J. Antibiotics 53 (3), 256–261.

Singh, S.K., Strobel, G.A., Knighton, B., Geary, B., Sears, J., Ezra, D., 2011. An endophytic *Phomopsis* sp. possessing bioactivity and fuel potential with its volatile organic compounds. Microb. Ecol. 61 (4), 729–739.

Song, Y.C., Li, H., Ye, Y.H., Shan, C.Y., Yang, Y.M., Tan, R.X., 2004. Endophytic naphthopyrone metabolites are co-inhibitors of xanthine oxidase, SW1116 cell and some microbial growths. FEMS Microbiol Lett. 241 (1), 67–72.

Song, X.Y., Wang, H., Ren, F., Wang, K., Dou, G., Lv, X., et al., 2019. An endophytic *Diaporthe apiculatum* produces monoterpenes with inhibitory activity against phytopathogenic fungi. Antibiotics 8 (4), 231.

Stierle, A., Strobel, G., Stierle, D., 1993. Taxol and taxane production by *Taxomyces andreanae*, an endophytic fungus of Pacific yew. Science 260 (5105), 214–216.

Strobel, G.A., 2002. Rainforest endophytes and bioactive products. Crit. Rev. Biotechnol. 22 (4), 315–333.

Strobel, G., Yang, X., Sears, J., Kramer, R., Sidhu, R.S., Hess, W.M., 1996. Taxol from *Pestalotiopsis microspora*, an endophytic fungus of *Taxus wallachiana*. Microbiology 142 (2), 435–440.

Strobel, G.A., Torczynski, R., Bollon, A., 1997. *Acremonium* sp.—a leucinostatin A producing endophyte of European yew (*Taxus baccata*). Plant Sci. 128 (1), 97–108.

Strobel, G.A., Miller, R.V., Martinez-Miller, C., Condron, M.M., Teplow, D.B., Hess, W.M., 1999. Cryptocandin, a potent antimycotic from the endophytic fungus *Cryptosporiopsis* cf. *quercina*. Microbiology 145 (8), 1919–1926.

Subban, K., Subramani, R., Johnpaul, M., 2013. A novel antibacterial and antifungal phenolic compound from the endophytic fungus *Pestalotiopsis mangiferae*. Nat. Prod. Res. 27, 1445–1449.

Sun, R.R., Miao, F.P., Zhang, J., Wang, G., Yin, X.L., Ji, N.Y., 2013. Three new xanthone derivatives from an algicolous isolate of *Aspergillus wentii*. Magnetic Reson. Chem. 51 (1), 65–68.

Taechowisan, T., Lu, C., Shen, Y., Lumyong, S., 2007. Antitumor activity of 4-arylcoumarins from endophytic *Streptomyces aureofaciens* CMUAc130. J. Cancer Res. Therapeutics 3 (2), 86.

Taechowisan, T., Chanaphat, S., Ruensamran, W., Phutdhawong, W.S., 2013. Antibacterial activity of Decursin from *Streptomyces* sp. GMT-8; an endophyte in *Zingiber officinale* Rosc. J. Appl. Pharm. Sci. 3 (10), 74.

Taechowisan, T., Chanaphat, S., Ruensamran, W., Phutdhawong, W.S., 2014. Antibacterial activity of new flavonoids from *Streptomyces* sp. BT01; an endophyte in *Boesenbergia rotunda* (L.) *Mansf.* J. Appl. Pharm. Sci. 4 (4), 8.

Tanapichatsakul, C., Monggoot, S., Gentekaki, E., Pripdeevech, P., 2018. Antibacterial and antioxidant metabolites of *Diaporthe* spp. isolated from flowers of *Melodorum fruticosum*. Curr. Microbiol 75 (4), 476–483.

Teiten, M.H., Mack, F., Debbab, A., Aly, A.H., Dicato, M., Proksch, P., et al., 2013. Anticancer effect of altersolanol A, a metabolite produced by the endophytic fungus *Stemphylium globuliferum*, mediated by its pro-apoptotic and anti-invasive potential via the inhibition of NF-κB activity. Bioorg Med Chem. 21 (13), 3850–3858.

Tejesvi, M.V., Kini, K.R., Prakash, H.S., Subbiah, V., Shetty, H.S., 2007. Genetic diversity and antifungal activity of species of *Pestalotiopsis* isolated as endophytes from medicinal plants. Fungal Diversity 24 (3), 1–18.

Tejesvi, M.V., Segura, D.R., Schnorr, K.M., Sandvang, D., Mattila, S., Olsen, P.B., et al., 2013. An antimicrobial peptide from endophytic *Fusarium tricinctum* of *Rhododendron tomentosum* Harmaja. Fungal Diversity 60 (1), 153–159.

Tomsheck, A.R., Strobel, G.A., Booth, E., Geary, B., Spakowicz, D., Knighton, B., et al., 2010. *Hypoxylon* sp., an endophyte of *Persea indica*, producing 1, 8-cineole and other bioactive volatiles with fuel potential. Microb. Ecol. 60 (4), 903–914.

Tuntiwachwuttikul, P., Taechowisan, T., Wanbanjob, A., Thadaniti, S., Taylor, W.C., 2008. Lansai A–D, secondary metabolites from *Streptomyces* sp. SUC1. Tetrahedron 64 (32), 7583–7586.

Varughese, T., Rios, N., Higginbotham, S., Arnold, A.E., Coley, P.D., Kursar, T.A., et al., 2012. Antifungal depsidone metabolites from *Cordyceps dipterigena*, an endophytic fungus antagonistic to the phytopathogen *Gibberella fujikuroi*. Tetrahedron Lett. 53 (13), 1624–1626.

Wang, J., Huang, Y., Fang, M., Zhang, Y., Zheng, Z., Zhao, Y., et al., 2002. Brefeldin A, a cytotoxin produced by *Paecilomyces* sp. and *Aspergillus clavatus* isolated from *Taxus mairei* and *Torreya grandis*. FEMS Immunol Med. Microbiol 34 (1), 51–57.

Wang, K.Y., Strobel, G.A., Yan, D.H., 2017. The production of 1, 8-cineole, a potential biofuel, from an endophytic strain of *Annulohypoxylon* sp. FPYF3050 when grown on agricultural residues. J. Sustain. Bioenergy Syst. 7 (2), 65–84.

Wibowo, M., Prachyawarakorn, V., Aree, T., Mahidol, C., Ruchirawat, S., Kittakoop, P., 2016. Cytotoxic sesquiterpenes from the endophytic fungus *Pseudolagarobasidium acaciicola*. Phytochemistry 122, 126–138.

Wijesekara, I., Zhang, C., Van Ta, Q., Vo, T.S., Li, Y.X., Kim, S.K., 2014. Physcion from marine-derived fungus *Microsporum* sp. induces apoptosis in human cervical carcinoma HeLa cells. Microbiol Res. 169 (4), 255–261.

Wicklow, D.T., Poling, S.M., 2009. Antimicrobial activity of pyrrocidines from *Acremonium zeae* against endophytes and pathogens of maize. Phytopathology 99 (1), 109–115.

Wicklow, D.T., Roth, S., Deyrup, S.T., Gloer, J.B., 2005. A protective endophyte of maize: *Acremonium zeae* antibiotics inhibitory to *Aspergillus flavus* and *Fusarium verticillioides*. Mycol Res. 109 (5), 610–618.

Wu, S.H., Huang, R., Miao, C.P., Chen, Y.W., 2013. Two new steroids from an endophytic fungus *Phomopsis* sp. Chem. Biodivers. 10 (7), 1276–1283.

Xia, G., Li, J., Li, H., Long, Y., Lin, S.E., Lu, Y., et al., 2014. Alterporriol-type dimers from the mangrove endophytic fungus, *Alternaria* sp.(SK11), and their MptpB inhibitions. Mar. Drugs 12 (5), 2953–2969.

Xie, S., Liu, J., Gu, S., Chen, X., Jiang, H., Ding, T., 2020. Antifungal activity of volatile compounds produced by endophytic *Bacillus subtilis* DZSY21 against *Curvularia lunata*. Ann. Microbiol 70 (1), 1–10.

Xu, T., 2008. Study of anticancer effect of winter worm and summer grass on Mcf-7 human breast cancer cells. Doctoral Dissertation, University of Missouri, Columbia.

Xu, F., Pang, J., Lu, B., Wang, J., Zhang, Y., She, Z., et al., 2009. Two metabolites with DNA-binding affinity from the mangrove fungus *Xylaria* sp.(# 2508) from the South China sea coast. Chin. J. Chem. 27 (2), 365–368.

Yan, H.J., Li, X.M., Li, C.S., Wang, B.G., 2012. Alkaloid and anthraquinone derivatives produced by the marine-derived endophytic fungus *Eurotium rubrum*. Helvetica Chim. Acta 95 (1), 163–168.

Yan, D.H., Song, X., Li, H., Luo, T., Dou, G., Strobel, G., 2018. Antifungal activities of volatile secondary metabolites of four Diaporthe strains isolated from *Catharanthus roseus*. J. Fungi 4 (2), 65.

Yasmin, S., Hafeez, F.Y., Mirza, M.S., Rasul, M., Arshad, H.M., Zubair, M., et al., 2017. Biocontrol of bacterial leaf blight of rice and profiling of secondary metabolites produced by rhizospheric *Pseudomonas aeruginosa* BRp3. Front. Microbiol 8, 1895.

Yuan, L., Zhao, P.J., Ma, J., Lu, C.H., Shen, Y.M., 2009. Labdane and tetranorlabdane diterpenoids from *Botryosphaeria* sp. MHF, an endophytic fungus of *Maytenus hookeri*. Helvetica Chim. Acta 92 (6), 1118–1125.

Zhang, G., Sun, S., Zhu, T., Lin, Z., Gu, J., Li, D., et al., 2011. Antiviral isoindolone derivatives from an endophytic fungus *Emericella* sp. associated with *Aegiceras corniculatum*. Phytochemistry 72 (11-12), 1436–1442.

Zhang, J., Wang, J.D., Liu, C.X., Yuan, J.H., Wang, X.J., Xiang, W.S., 2014. A new prenylated indole derivative from endophytic actinobacteria *Streptomyces* sp. neau-D50. Nat. Prod. Res. 28 (7), 431–437.

Zhang, P., Li, X.M., Liu, H., Li, X., Wang, B.G., 2015. Two new alkaloids from *Penicillium oxalicum* EN-201, an endophytic fungus derived from the marine mangrove plant *Rhizophora stylosa*. Phytochem Lett. 13, 160–164.

Zhang, P., Li, X.M., Mao, X.X., Mándi, A., Kurtán, T., Wang, B.G., 2016. Varioloid A, a new indolyl-6, 10b-dihydro-5aH-[1] benzofuro [2, 3-b] indole derivative from the marine alga-derived endophytic fungus *Paecilomyces variotii* EN-291. Beilstein J. Org. Chem. 12 (1), 2012–2018.

Zhang, X.H., Zhang, D.J., Liu, J.L., Pan, H.Y., Qin, J.C., Zhang, Y.H., 2018. Antifungal effects of volatile organic compounds from the endophytic fungus *Cryptosporiopsis ericae* Cc-HG-7 isolated from *Coptis chinensis* Franch. Biocontrol Sci. Technol. 28 (5), 496–508.

Zhao, P.J., Fan, L.M., Li, G.H., Zhu, N., Shen, Y.M., 2005. Antibacterial and antitumor macrolides from *Streptomyces* sp. Is9131. Arch. Pharm Res. 28 (11), 1228–1232.

Zhao, J., Mou, Y., Shan, T., Li, Y., Zhou, L., Wang, M., et al., 2010. Antimicrobial metabolites from the endophytic fungus *Pichia guilliermondii* isolated from *Paris polyphylla* var. yunnanensis. Molecules 15 (11), 7961–7970.

Zhao, J., Ma, D., Luo, M., Wang, W., Zhao, C., Zu, Y., et al., 2014. In vitro antioxidant activities and antioxidant enzyme activities in HepG2 cells and main active compounds of endophytic fungus from pigeon pea [*Cajanus cajan* (L.) *Millsp.*]. Food Res. Int. 56, 243–251.

Zheng, C., Chen, Y., Jiang, L.L., Shi, X.M., 2014. Antiproliferative metabolites from the endophytic fungus *Penicillium* sp. FJ-1 isolated from a mangrove *Avicennia marina*. Phytochem Lett. 10, 272–275.

Zhou, H., Yang, Y., Peng, T., Li, W., Zhao, L., Xu, L., et al., 2014. Metabolites of *Streptomyces* sp., an endophytic actinomycete from *Alpinia oxyphylla*. Nat. Product Res. 28 (4), 265–267.

Zin, N.M., Baba, M.S., Zainal-Abidin, A.H., Latip, J., Mazlan, N.W., Edrada-Ebel, R., 2017. Gancidin W, a potential low-toxicity antimalarial agent isolated from an endophytic *Streptomyces* SUK10. Drug Design, Dev. Ther. 11, 351.

Zou, W.X., Meng, J.C., Lu, H., Chen, G.X., Shi, G.X., Zhang, T.Y., et al., 2000. Metabolites of *Colletotrichum gloeosporioides*, an endophytic fungus in *Artemisia mongolica*. J. Nat. Products 63 (11), 1529–1530.

Chapter 8

Role and behavior of microbial volatile organic compounds in mitigating stress

Prasann Kumar[1], Khushbu Sharma[2], Lalit Saini[2] and Shipa Rani Dey[1]
[1]*Department of Agronomy, School of Agriculture, Lovely Professional University, Jalandhar, India,* [2]*Department of Agronomy, School of Agriculture, Lovely Professional University, Phagwara, India*

Chapter Outline

8.1 Introduction

A large number of secondary metabolites have been involved in microbial interactions. Some reports submitted that the three are a group of microbial volatile organic compounds (MVOCs) associated with plant and soil

Volatiles and Metabolites of Microbes. DOI: https://doi.org/10.1016/B978-0-12-824523-1.00010-9

microbes (secondary metabolites) formed by their interactions, although they are mostly unknown to date. The chemical behavior of volatile organic compounds (VOCs) is described as being compounds with low molecular mass (<300 DA), malodorous compounds (<C15) with a high vapor pressure at low boiling point and constituting lipophilic moiety. All chemical properties of VOC compounds encourage physical and chemical processes in plants and their rhizospheres such as soil evaporation, transport, and diffusion of macrosoils and micropores filled with water and gas (Insam and Seewald, 2010; Effmert et al., 2012). MVOCs are active in a variety of synthetic classes containing groups of alkenes, alcohols, ketones, benzenoids, pyrazines, sulfides, and terpenes (Schulz and Dickschat, 2007; Lemfack et al., 2014, 2017; Kanchiswamy et al., 2015; Schmidt et al., 2015). The ongoing analysis provided an extensive diagram of VOCs acquired from soil-associated microorganisms (Schenkel et al., 2015). The formation of MVOCs in the soil is affected by various elements containing the development phase of microorganisms, the availability of nutrients and oxygen, temperature, pH, and soil water content (Wheatley, 2002; Insam and Seewald, 2010).

A recent review described the creation of defined microbial VOCs maintained within the cell to be encouraged or regulated during interspecific phase microbial interactions. Product derived from primary metabolism serves as a volatile compound, although recent discoveries have shown that various MVOCs determine biological activity (Schmidt et al., 2015; Tyc et al., 2017a). In the case of bacteria, it has been observed that the creation of a specific microbial volatile compound depends on the two major regulatory components called GacS/GacA (Cheng et al., 2016; Ossowicki et al., 2017). All discoveries are noticeably dissatisfied by the judgment that all microbial volatile compounds (created by bacteria, fungi, protists, and through their interactions) simply squander items. Physiologically, due to the soluble properties of primary metabolites, they are often liable for limited interactions but are considered long-distance messengers (Tyc et al., 2017b; Westhoff et al., 2017).

8.2 Interaction of microbial volatile organic compounds

At present, there are various forms of microbial interactions occurring in subterranean, for example, microbial interactions within bacteria with bacteria, and fungi with other bacteria.

- A fungus with bacteria.
- Bacteria with protists.
- Fungi with a plant.
- Bacteria with plant.
- Bacteria with fungi with plant interactions.

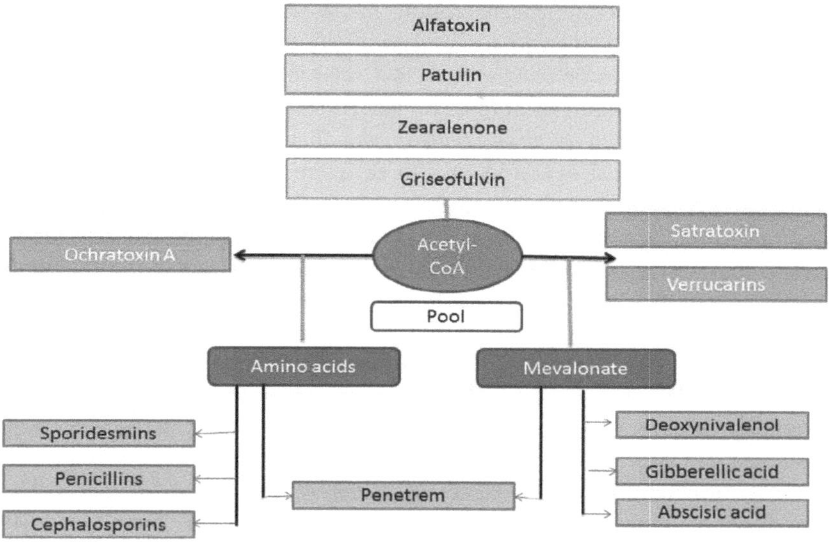

FIGURE 8.1 Showing the effects of microbial volatile organic compounds released from different sources.

However, the maximum level of subterranean-facilitated interactions between VOCs is mainly concentrated on root-emitted volatiles (recently reviewed by Delory et al., 2016). The different information collected from the previous study shows that MVOCs have both a user and a destructive impact on different living beings (Effmert et al., 2012; Schmidt et al., 2015). These metabolites help to provide living beings with quick and accurate approaches for the perception of neighborly living beings (both friends and foe) and the dispatch of appropriate reactions. The main purpose of this analysis is to encapsulate recent information on the different roles of metabolites in intra- and interkingdom interactions, including MVOCs (e.g., terpenes), associated with microbe interactions with microbe and microbe interactions with plant interactions, which leads demonstrate difficulties in studying subterranean MVOCs-mediated interactions and chances for additional studies and examination applications (Fig. 8.1).

8.2.1 Interaction of volatile organic compounds released from different bacteria

Bacterial metabolites, which show direct adverse effects on other microscopic organisms. For example, the sequester peneal baflavene produced by *Streptomyces albidoflavus* exposed to *Bacillus subtilis* (Gürtler et al., 1994) and the emission of dimethyl disulfide by two rhizospheric bacteria, *Pseudomonas fluorescens* and *Serratia plymuthica* revealed bacteriostatic effects against two plant bacterial pathogens *Agrobacterium tumefaciens* and *Agrobacterium vitis* (Danda). *Pseudomonas fluorescens* WR-1 produces volatile substances such as benzothiazole and 1-methyl naphthalene with bacteriostatic effects on *Ralstonia solanacearum* (Raza et al., 2016a). Numerous types of bacteria, such as *Pseudomonas* and *Bacillus*, can also be used as one of the biochemical operators against different plant pathogens to supply metabolite compounds with antibacterial properties (Raza et al., 2016a,b,c; Xie et al., 2016; Rajer et al., 2017; Tahir et al., 2017a,b). The ongoing report showed that bacterial species that produce metabolites of organic compounds containing benzaldehyde,1,2-benzisothiazol-3 (2 H having a strong inhibitory action against the *R. solanacearum* family, the causative agent of bacterial wilting disease (Tahir et al., 2017a). Metabolites, which alter the level of transcriptional expression of multiple genes (genetic material), are entangled in motility as well as pathogenicity and promote systemic tolerance through plants, leading to a reduction in wilting disease. Various reports indicate that bacterial virulence of VOCs occurs in bacteria due to low tolerance. The compound, 2,3butanediol and acetone have virulence in *Pectobacterium carotovorum.* Similar compounds increase the chemical formation of virulence causes in different bacteria, such as *Pseudomonas aeruginosa* (Audrain et al., 2015). On the other hand, the different volatiles produced by some bacteria has positive effects on the growth of the bacteria present in the rhizosphere. There are a few reports of the impact of VOCs on bacterial virulence. For example, 2,3butanediol and acetoin are mandatory for destruction in *P. carotovorum.* Similar compounds increase the development of virulence factors in *P. aeruginosa* (Audrain et al., 2015). Interestingly, VOCs delivered by certain microbes may also have positive effects on the development of other adjacent bacteria in the rhizosphere. This is recommended that *C. Pratensis* and *S. Plymuthica* may show stimulating and attractive properties for the development of *P. fluorescens* Pf0−1 (Garbeva et al., 2014). These VOCs stimulated the expression of genes associated with motility in *P. fluorescens* Pf0−1 and stimulated the development of secondary metabolites with antibacterial activity against Bacillus (Garbeva et al., 2014). This is suggested that *C. Pratensis* and *S. Plymuthica* may attract and stimulate the development of *P. fluorescens* in a combined attempt to increase their risk to various bacterial competitors or soil fungal pathogens.

The effects of antagonistic activity against different bacteria and the performance of different bacteria can also be adopted by the VOCs, and their protection against antimicrobials can be balanced. Various bacterial types of volatiles, such as 2-aminoacetophenone, nitric oxide, hydrogen sulfide, trimethylamine, and ammonia may affect the development of biofilm, bacterial motility, its influence, and dispersal (Raza et al., 2016a; Audrain et al., 2015). Bacterial use their utility regularly to move to different territories with more assets and fewer competitors. In *Streptomyces venezuelae*, another method of improvement, the supposed investigation, recently examined that allows nonmotile bacteria to enter more nutrients in media (Jones et al., 2017). Nonbranched types of vegetative hyphae are developing rapidly with the help of *S. venezuelae*, which can create hydrophilic settings by consumption of glucose and pH raised to escape the low-nutrition zone. The exploratory type of development promotes the various forms of species with the communication of long-distance signals that the explorer cells can deliver. Out of these signs, trimethylamine does not work as a sign to connect remotely detected streptomycin and also promotes exploratory developmental types but additionally shows antibacterial activity against *Micrococcus luteus* and *B. subtilis*, possible by increasing the pH of that medium (Jones et al., 2017).

8.2.2 Interaction bacterial MVOCs and fungal MVOCs

VOCs can play a significant part in the long-distance interaction between bacteria–fungus and leads to distinctive physiological reactions in the accomplices formed by interaction. VOCs produced by *T. viride* expanded the demonstration of a (PhlA), that is, biocontrol gene in P, 2, 4-diacetyl phloroglucinol biosynthesis encoded by the fluorescent (Lutz et al., 2004). Some latest examinations show that the development of a few species of bacteria can be through VOCs parasitic (Werner et al., 2016). For example, VOCs display the inhibitory impacts on *B. subtilis* and *B. cercus* created through *Pleurotus ostreatus* mushroom of olyster. In recent times, Schmidt et al. (2015) determined the strains of bacteria in soil responses phenotypic to volatiles released by a few strains soil infections under the distinctive condition of enhancement during various stages of development. Out of this type of attempted physiological reaction, such as motility and biofilm organization, antimicrobial activity, improvement changes, and motility of microscopic livings beings (both amassing and swimming), was on a very basic level determinedly or antagonistically affected by the contagious and oomycetal VOCs. In the examination, mutualistic bacteria attract the fungus toward itself by utilizing a potential technique that reflects consequently and by using VOCs controlling their motility to resisting the competitors (Piechulla et al., 2017). The *Fusarium culmorum* type of fungal pathogen produces VOCs presented in PRI-2 *S. plymuthica* that examined by the proteomics and transcriptomics. It shows that PRI-2C *S. plymuthica* response for the fungal VOCs that changes in expression of protein and genes, identified by the transduction signals, motility, production of secondary metabolites, biogenesis for cell enveloping, and energy metabolism (Schmidt et al., 2017). VOCs fungus presented by the metabolic investigation of PRI-2C *S. plymuthica*, the expression of the gene cluster, methyltransferase, and terpene synthesis by the coexpression of heterologous exposed the creation of sodorifen, also named as an uncommon terpene (Von Reuß et al., 2010; Kai et al., 2010), to the response of fungal type VOCs. The innovation sustains the recommended significance of VOCs as indicating the molecules in the interaction between bacteria and fungus. Different types of soil bacterial produce the VOCs by the impact of antifungal and hence the addition to it known as fungus of soil, where propagules of fungus are limited in their capacity for developing and growing (Garbeva et al., 2011). In recent times, Cordovez et al. (2015) exposed to facilitate the antifungal properties of *Streptomyces* spp. emitted volatile compounds show against *Rhizoctonia solani* and may add to suppressiveness of plant against diseases. Ossowicki et al. (2017) indicated that the chemical compound collected from the rhizosphere of tomato is *Pseudomonas* P482 might be having an antifungal response, which recommends that the antagonistic capacities of this strain against plant pathogens are because of their volatile potential (Ossowicki et al., 2017). This impact of metabolites generated from bacteria against oomycetes is not a segregated case, and various P bacteria (*Pseudomonas*) strains were described for promoting antioomycete activity (De Vrieze et al., 2015; Hunziker et al., 2015).

8.2.3 Interaction of microbial volatile organic compounds released from different fungi species

The 1-octen-3-ol, major particular parasitic VOC, well-known for the specific smell in mushroom, is delivered by a broad range of filamentous fungi and may work as a development sign among different fungus group (Miyamoto et al., 2014). Similar metabolites had been reported for showing function in *Penicillium paneum* like a self-inhibitor indication during the spore germination stage (Chitarra et al., 2004). While formative signs during populace foundation, some fungi-generated metabolites that perform in a way that focuses subordinately on specific development of mycelia and spore germination (Nemcovic et al., 2008; Stoppacher et al., 2010). The fungi-generated metabolite compounds are imposing self-inhibitory impacts and initiative interaction between different fungi metabolites, such as the *Muscodor albus* (endophytic) and *Oxyporus latemarginatus*, are able to intensely prevent the development of some plant diseases causing fungi example *Botrytis cinerea* and *R. solani* (Strobel et al., 2001). The compound VOCs generated by *Trichoderma* spp. have a solid impact in mitigating the effect of plant diseases causing fungi, for example, *Alternaria solanii*, *Sclerotium rolfsii*, and *Sclerotinia sclerotiorum* (Amin et al., 2010). In recent times, the study showed that metabolite compounds generated by fungus could affect through cata and anabolic pathways to avoid pathogenic fungus development (Fialho et al., 2016). There is a range of fungal species having the ability to inhibit the growth of antifungal compounds formed by other microbial (fungi, bacteria, and protest) competitors present in rhizosphere, such as Fusarium spp., which act as an antifungal compound against poisonous compound pentyl-alphapyrone, transmitted through *Trichoderma harzianum* species of fungus (Cooney et al., 2001). All volatile compounds generated by fungus may be preserved as significant sources of energy for fungi colonizing in carbon-constrained conditions (Cale et al., 2016). Alternatively, for fungi colonizing more carbon-rich conditions, VOCs may act as semi-or synthetic substances in a subordinate way to facilitate antagonistic and valuable interactions between fungi.

8.2.4 Interaction of protists volatile organic compounds and bacterial volatile organic compounds

A various and plentiful gathering of soil-borne microorganisms are called protists or Protozoa. The nutrient cycle remains between plant and soil, food web chain of ecosystem, and metabolism of carbon; all of these activities are carried out by protest because of their touching activities (Geisen et al., 2015, 2016). Protists are identified as major predators for bacteria and possessing antibacterial property against different bacterial species through their judicious feeding habits (Griffiths et al., 1999; Bonkowski and Brandt, 2002; Rosenberg et al., 2009; Glücksman et al., 2010). In this manner, detection of their major prey in the rhizosphere in the permeable soil medium will be a valuable characteristic of soil protists. An ongoing report submitted by Schulz-Bohm et al. (2017) discovered that secondary metabolites could assume an important part in the wider interaction of bacterial—protists. Through investigations it was found that the interactions of different protest-generated volatile compounds are genetically distinctive among soil microorganisms like protest-bacteria. Relating to those, and direct trophic interactions, they exhibited that particular bacterial generated volatiles may provide early identification of appropriate prey in environment. Stimulatingly, soil protists, for example, *Dictyostelium discoideum* produce volatiles terpenes. These volatile terpenes may be associated with barrier components, for instance, to repulse predator of nematodes. Additionally, it was indicated that soil bacteria could deliver explicit volatiles to resist predators of protists (Kai et al., 2009; Schulz-Bohm et al., 2017; Chen et al., 2016). There are some special volatile compounds that play a significant role in interorganism communications during stress, such as nitric oxide, which occurred abundantly in the atmosphere. Accumulation of NO is resultant of anthropogenic activities, for example, inadequate use of nitrogenous fertilizers, burning of fossil fuels, etc. During abiotic stress response of VOCs leads to an increase in the concentration of ozone at troposphere. In the 1990s, the properties and behavior of NO were described as unstable gases and can be diffused in liquid and aqueous media. In plants, NO is responsible for producing physiological regulators. However, NO molecule has greater stability and are easily reacted with radicals because of its lipophilic nature combined called reactive nitrogen species such as NO_2, N_2O, etc. Even these nitrogenous compounds are involved in lipid peroxidation chain reaction. NO is also considered as one of the major mediators of reactive oxygen species (ROS) signaling during stress.

8.3 The behavior of volatile organic compounds released from Trichoderma

VOCs are considered as an important agent in agriculture because they play a vital role in the interaction of abiotic and biotic factors. Recently, there are around 500 species of different bacteria and fungus have been reported for the production of a wide range of VOCs such as ketones, ester, lactones (Bitas et al., 2015; Splivallo et al., 2015; Lemfack et al., 2014; Effmert et al., 2012). The use of the Trichoderma species in agriculture is very common and found effective in many manners such as mobilization, diseases resistance, and stress tolerance, which leads to improved growth and development of the plant. It has been found that there are several VOCs secreted by trichoderma in the rhizosphere, and commonly reported are alcohols, group of ketones, and other derivatives (Ryu et al., 2003; Zhang et al., 2008a; Huang et al., 2012). The VOCs of Trichoderma have been explored intensively for their beneficial role in signaling during stress, agricultural, and antimicrobial activity (Strobel et al., 2001). The VOCs produced from Trichoderma are commonly gas-phase molecules of low and high molecular weight origin. These VOCs obtained from *Trichoderma* have been found important for agronomic, medicinal use. The VOCs of Trichoderma have a long-lasting effect in inhibiting pathogens growth in the rhizosphere and promotes a disease-free environment for plant growth (Wheatley, 2002; Huang et al., 2012).

8.3.1 Role of volatile organic compounds under abiotic stress

The major and unique behavior of VOC compounds under abiotic stress (salt stress, heavy metal stress, drought stress, etc.) provides resistance to plants to continue growth. The process responsible for providing resistance towards stress includes antioxidant production, Na accumulation, signaling and regulates the secretion of stress hormones (Sharifi and Ryu, 2016). It has been reported that the production of enzymes and the stability of membrane during stress is regulated by different organic osmolytes and glycine. Another example of VOCs secreted by *Pseudomonas* increase the tolerance of the plant toward drought stress, and salicylic acid improves the tolerance against salt stress. Furthermore, VOCs improves acceptance to drought stress by regulating hormonal balance in the plant (Vardharajula et al., 2011).

8.3.2 Volatile organic compound-mediated recruitment of beneficial insects

VOCs also improve plant health and growth by attracting natural enemies. Many reports show that VOCs also improve the interaction of microbes with plants in soil and also enhance the oil content in leaves or aerial parts

(Sharifi et al., 2018). Another biofertilizer like mycorrhizae is responsible for converting the outline of emitted microbial secondary metabolites in a French bean. Exogenous application of mycorrhizae produced two enzymes β-caryophyllene and β-ocimene, and they tend to magnetize the parasitoids of spider mites. Many researchers admit that the role of jasmonic acid in regulating essential metabolic activities of plants, and in generating secondary metabolites compounds by inducing herbivores and microbials, is very promising even under harsh conditions. The volatile metabolites generated as a byproduct of microorganisms can control the signaling pathway of jasmonic acid (Sharifi and Ryu, 2016; Sharifi et al., 2018). Endophytic fungus (*Neotyphodium uncinatum*) inoculation to aerial parts of plants due to the presence of some volatile compound leads to attract beneficial insects.

8.3.3 Volatile organic compound-mediated plant growth promotion

We all know that microbes of soil have enough potential to increase the growth of the plant without forming any relation with VOC production. At first, Ryu et al. (2003) examined that growth of plant increases by the VOCs production with *B. subtilis* GB03. In another paper, the growth of the plant was promoted two-fold by *Bacillus megaterium* XTBG34, produced by 2−pentylfuran VOC (Zou et al., 2010). Similarly, VOC (methyl ethyl ketone, 2-methyltridec-1-ene and E-11, 13-tetradecadien) produced by *Pseudomonas fluorescens* SS101 increases the *Nicotianatabacum* (Park et al., 2015). The organic compounds present as a bacterial byproduct in soil, that is, *Pseudomonas*, improves the integrity of cell division expression to maintain the protein content that increases the growth of the seedling off soybean (Vaishnav et al., 2015). Previous studies show that microbe-generated organic compounds, also responsible for the decrease in normal plant growth, is the effect of plant hormones regulation; the MVOC importance was well acknowledged in the previous decade and also determined under naturalistic conditions (Cortes-Barco et al., 2010; Raza et al., 2015). All microbes help to produce a conspicuous mixture of VOC produced by each microbe, which plays a major role in all microbial life cycles as well as their relation with different organisms. For example, some VOC microbes increase the growth of the plant (Ryu et al., 2003) while some decrease pathogen attack (Audrain et al., 2015). Few modulated the motility of bacteria, production of antibiotics, and formation of biofilm (Kim et al., 2013). Even each microbe of the VOC does not propose a single contribution. For example, VOCs suppress and kill pathogens as well as nematodes (Anyanful et al., 2005), decrease resistance to antibiotics (Audrain et al., 2015), improve the formation of biofilm in microbes (Bailly et al., 2014), and increase the growth of plants (Bansal et al., 2010).

8.3.4 Volatile organic compounds: impact on photosynthesis

Microbes of VOCs increases the health and growth of the plant through various ways and increases photosynthesis of plant and accumulation of carbohydrates. For example, Castulo-Rubio et al. (2015) and Zhang et al. (2009) examined those VOC microbes, raising the photosynthesis efficiency by increases content of chlorophyll in plants. Mostly two methods are followed for mediating this effect. One is the uptake of iron in plant cells, which is an important factor for electron transportation chain pathway, photosynthesis, and chlorophyll (Briat, 2007). Zhang et al. (2009) reported that the rhizosphere has been acidified by *subtilis* GB03 of *Bacillus* having VOCs by three-fold increase in photon release, increased iron uptake and solubility, and then converted to Fe21 and then plant uptake by Fe-regulated transporter 1 (IRTI). In this perusal, the contents of iron in VOC-unveiled plants were twice in size than the controlled treatment. In another observation, Wang et al. (2017) examined that *Bacillus amyloliquefaciens* BF06 VOCs raising Nitric oxide formation in Arabidopsis that acts as the upstream of deficiency of iron-inducing transpiration factor 1 (FITI) under conditions of iron-deficiency, mobilized, and chelated Fe. And another mechanism is to raise the unfavorable response to an accumulation of sugar on photosynthesis rate by microbes of VOCs. The observation shows that the concentration of sensing hexose sugar regulates the rate of photosynthesis negatively by Arabidopsis hexose sensor kinase-1 (HXK1) (Cho et al., 2010). *Arabidopsis* accumulates 60% less hexose sugar than VOCs of *Bacillus* than also from the controlled treatment (Zhang et al., 2008b).

8.3.5 Role of volatile organic compounds released by plant roots

It is a well-known study that plant roots of any family or genus secrete several kinds of VOCs, which may be synergistic or antagonistic with microbes present in the rhizosphere, other plants, nematodes, and in pests (Weissteiner et al., 2012). The same kind of organic volatile compounds is also released by the release by herbivore insects present in rhizosphere or hyllosphere of the plant, which further acts as signals for circumlocutory defense, and engages insect predators. Correspondingly, under abiotic or biotic stress the plant as a stress protectant secretes certain kind of VOCs. (Van Tol et al., 2001). Reports submitted that there are some plants that play a vital role in enhancing the role of beneficial microbes, improving the interaction between plant and ecosystem, and secreting 200,000 metabolites. There are about three to four classes of VOCs produced by plant roots such as alkanes, ketones, fatty acids, and sulfur compounds. It is reported that around 1% of plant metabolites are VOCs and they attract as chemostimulants, signalers, and antimicrobials. Van Dam et al. (2016) has reported that inoculation of *Janthino bacterium* and *Pseudomonas* cell in nutrient-deficient soil increases the nutrient ability to plant roots. Moreover, VOCs like limonene and borneol are the most popular VOCs excreted by plant roots, which improves bacterial quorum sensing. In the case of legumes, flavonoids are VOC compounds secreted by plant roots, which attract nitrogen-fixing bacteria and improve nitrogen's ability to crop. Correspondingly, in the case of the maize plant produced benzoxazinoids, which attract different PGPRs like *Bacillus* and *Pseudomonas*. Moreover, a sesquiterpene VOC (E)-β-caryophyllene, increase the attraction of nematodes toward rhizosphere because of root-feeding beetle larvae (Van Tol et al., 2001). The latest research estimated that the movement of microbes in soil toward plant roots could also be encouraged by VOCs released from the roots of plants. Interestingly, antifungal activity in plant root was increased by pathogen present in the rhizosphere (Schulz-Bohm et al., 2015). Application of bulk quantity of nutrient to soil influenced microbial activity (Schulz-Bohm et al., 2015). Several studies suggested that there are certain species of microbes that secrete VOCs and enhance nutrient activity under nutrient-deficient condition. Therefore, plant VOCs act as messenger chemical compounds provides attraction toward the nutrient and mineral-rich environment around the plants.

8.3.6 Volatile organic compounds improves nutrient acquisition

In soil salinity, soils are found deficient with Fe2+ mineral. Many microorganisms carried out the mechanism of mineralization of iron in soil and the compounds of VOCs, which are mutually associated with plant roots and rhizosphere (Wintermans et al., 2016). The frequent root-association of beneficial soil microbes with plant roots is strongly influenced by various biofertilizers such as *Rhizobium*, *Mycorrihza*, and *Trichoderma*. Application of these microbes to soil enhances nutrient mobilization and make them available to plant, apart from nutrient mobilization; these microbes also enhance the resistance of the plant to both biotic and abiotic stress. Stimulation of Fe uptake by the plant is strongly related to genes generated from microbial volatiles. The use of *Rhizobacterium* and *Pseudomonas* release certain types of VOCs, which promotes growth and plant development. The VOCs released from TRICHODERMA are reported for minimizing the deficiency of Fe in tomato shoots, suggesting that the phenomenon has occurred across plant species. There have been many pieces of evidence during the current year that support the suggestion that plants react strongly to MVOCs. There have been many researchers observed in which the effect of MVOCs at *Arabidopsis thaliana*. Studies have shown that microbes can cause alteration in the root system of the plant, physiology of plants, hormonal balance, and production of biomass. It has been found that a single MVOCs can perform different functions in the plant, such as production dimethyl disulfide production, which helps to improve the reduced sulfur content that improves the growth of the plant through their experiments and inferences characteristic compound of 6-pentyl-pyrone, *Trichoderm asperellum* can be raising the defense reaction of plant and sporulation of *Alternaria alternate* and *B. cinerea* decreases at the same interval of time. Plants of tobacco and maize are also protected by limiting and inducing tolerance to pathogens against *Cochliobolus heterostrophus* and *B. cinerea*. There are many studies that show that the beneficial nature of MVOCs in soil microbes can improve the growth of plants. There are only a few studies that focus on the mechanism of plant growth and development of soil-borne microbes, which are pathogenic in nature. The study indicates that soil-born *R. solani* increases the production of MVOCs for helping the growth of the plant and reduce insect resistance. It has also been approved, through a successful strategy, that pathogenic fungi improves root biomass and increases the growth of lateral root formations, which provides more surface area to help colonize fugal and its infection. Bitas et al. (2015) concluded, through his experiment, pathogenic *F. oxysporum* emitted VOCs and stimulated the *Nicotiana tobacum* and *A. thaliana* growth and effecting the signal and transportations of auxins. The pathogen of *Alternaria*-emitted VOCs has raised the early flowering, growth, and *Arabidopsis* photosynthesis rate, pepper, and maize by affecting plastic cytokine level. To analyze the impact of VOCs on plant growth, a significant factor of high-CO_2 production must be considered as contributing to inefficient plant growth. However, after so many studies, there is still limiting knowledge of the role of MVOCs in the attraction of the beneficial organism and mechanism of the various function performed by the microbes associated with plants and how they affected the quality and quantity of the plant volatile emissions. Del Giudice et al. (2008) reported that the association of bacteria with vetiver grassroots used by sesquiterpenes in the form of a carbon source. There is a variety of various experiment, which explain the chemical dialog in between plants, microbes, and organisms by the exchanging of dissolvable molecules. Most studies have reported that MVOCs mediate communication, which is unidirectional in nature, while there are very fewer studies about bidirectional interaction mediated through MVOCs. Spraker et al. (2014) observed through his experiments and stated that "VOCs decreases dialog in between fungi and bacteria where VOC present in fungal type plant pathogen, that is, *Aspergillus flavus* decreases the formation of the main EPS envy factor developed from *R. solanacearum* bacterial type plant pathogen. *Aspergillus flavus* reply to *R. solanacearum* VOCs by decreases production of conidia and raisin production of aflatoxins". For example, there is an important role played by VOCs helped in the interaction of *Verticillium longisporum*, that is, a fungal-type pathogen orbits antagonistic bacteria *P. polymyxa* in plants and experiments in vitro conditions. This interaction, which is mediated through specific VOCs, has been found to result in the disallowance of cellular metabolic activities and reduction in the growth of the fungal pathogen.

Schulz-Bohm et al. (2017), experimented using a system, that is, glass olfactometer, for the attraction of distant bacteria in soil from plant roots, and glass olfactometer equipment have been used successfully for the study of plant–herbivore or plant–nematode interactions. However, the glass olfactometer equipment was used to study the interaction in between microbes and plants. It is found that when the fungal type of infection is present than it changes the mixture of VOCs in roots and attracts the particular bacterial having antifungal properties. A strong bioinformatics technique based on the utilization of the Hidden Markov Model (HMM) and Protein Families Database (PFAM) research that gives permission to bacterial origin by terpene synthesis discovery and has shown that various phylogenetic bacteria can be a rich terpene source.

8.4 Conclusion and outlook

During the last decades, there have been many experiments performed on MVOCs, which helps reveal the functions of them and are produced by different soil microbes. From this experiment, it is evident that these small molecules, which are odorous in nature, can lead to the modification in behavior of living organism and can promote or inhibit their growth. As many researchers are focused on the unidirectional nature of MVOCs produced by a single organism and the responses of the organisms perceiving them. While there are very much fewer studies regarding the bidirectional nature of MVOCs-mediated dialog. From the experiment conducted until now regarding MVOCs, most of the interactions, which are mediated through MVOCs, take place between bacteria fungi and plants. There is less research regarding releasing of VOCs by other microbes like archaea, protista, or microbes present in rhizosphere regions like nematodes or earthworms. MVOCs could alter the composition of soil bacterial and fungal communities and significantly increased the relative abundance of proteobacteria, bacteroidetes, firmicutes, and ascomycota. Furthermore, MVOC-influenced genes are involved in important soil functions such as N-fixation (nif H), nitrification (amoA), denitrification (nirS), and antibiotic production (NRPS). Through studies, it has been observed that there are some common MVOCs released by plant roots, fungi, bacteria, and protists. For example, terpenes are the most enormous and most diverse class of metabolites, which have a great function as plant metabolites; however current studies also revealed that terpenes can be produced by all living organisms, including protista. Powerful bioinformatics method based on the use of HMMs and PFAM searches have allowed the discovery of terpene synthases of bacterial origin and showed that phylogenetically different bacteria can be a rich source of terpenes. However, while considering the implementations of MVOCs is very limited. Much research is yet to be done regarding the potential application of these compounds on large-scale agriculture and horticulture. The many experiments related to MVOCs are carried out in vitro conditions while we have to apply them under open field conditions in agriculture. There are little studies regarding the application of MVOCs under open field conditions and studies are yet to be done to know about their application in open field conditions of agriculture.

References

Amin, F., Razdan, V., Mohiddin, F., Bhat, K., Sheikh, P., 2010. Effect of volatile metabolites of *Trichoderma* species against seven fungal plant pathogens in-vitro. J. Phytol. 2, 34–37.

Anyanful, A., Dolan Livengood, J.M., Lewis, T., Sheth, S., DeZalia, M.N., Sherman, M.A., et al., 2005. Paralysis and killing of *Caenorhabditis elegans* by enteropathogenic *Escherichia coli* require the bacterial tryptophanase gene. Mol. Microbiol. 57, 988–1007.

Audrain, B., Farag, M.A., Ryu, C.M., Ghigo, J.M., 2015. Role of bacterial volatile compounds in bacterial biology. FEMS Microbiol. Rev. 39, 222–233. Available from: https://doi.org/10.1093/femsre/fuu013.

Bailly, A., Groenhagen, U., Schulz, S., Geisler, M., Eberl, L., Weisskopf, L., 2014. The inter-kingdom volatile signal indole promotes root development byinterfering with auxin signalling. Plant J. 80, 758–771. Available from: https://doi.org/10.1111/tpj.12666.

Bansal, T., Alaniz, R.C., Wood, T.K., Jayaraman, A., 2010. The bacterial signal indole increases epithelial-cell tight-junction resistance and attenuates indicators of inflammation. Proc. Natl. Acad. Sci. U.S.A. 107, 228–233.

Bitas, V., McCartney, N., Li, N., Demers, J., Kim, J.-E., Kim, H.-S., et al., 2015. Fusarium oxysporum volatiles enhances plant growth via affecting auxin transport and signaling. Front. Microbiol. 6, 1248. Available from: https://doi.org/10.3389/fmicb.2015.01248.

Bonkowski, M., Brandt, F., 2002. Do soil protozoa enhance plant growth by hormonal effects? Soil Biol. Biochem. 34, 1709–1715. Available from: https://doi.org/10.1016/S0038-0717(02)00157-8.

Briat, J.F., 2007. Iron dynamics in plants. Adv. Bot. Res. 46, 137–180.

Cale, J.A., Collignon, R.M., Klutsch, J.G., Kanekar, S.S., Hussain, A., Erbilgin, N., 2016. Fungal volatiles can act as carbon sources and semiochemicals to mediate interspecific interactions among bark beetleassociated fungal symbionts. PLoS One 11, e0162197. Available from: https://doi.org/10.1371/journal.pone.0162197.

Castulo-Rubio, D.Y., Alejandre-Ramírez, N.A., del Carmen Orozco-Mosqueda, M., Santoyo, G., Macías-Rodríguez, L.I., Valencia-Cantero, E., 2015. Volatile organic compounds produced by the rhizobacterium *Arthrobacter agilis* UMCV2 modulate *Sorghum bicolor* (strategy II plant) morphogenesis and SbFRO1 transcription in vitro. J. Plant Growth Regul. 34, 611–623.

Chen, X., Kollner, T.G., Jia, Q., Norris, A., Santhanam, B., Rabe, P., et al., 2016. Terpene synthase genes in eukaryotes beyond plants and fungi: occurrence in social amoebae. Proc. Natl. Acad. Sci. U.S.A. 113, 12132–12137. Available from: https://doi.org/10.1073/pnas.1610379113.

Cheng, X., Cordovez, V., Etalo, D.W., van der Voort, M., Raaijmakers, J.M., 2016. Role of the GacS sensor kinase in the regulation of volatile productionby plant growth-promoting *Pseudomonas fluorescens* SBW25. Front. Plant Sci. 7, 1706. Available from: https://doi.org/10.3389/fpls.2016.01706.

Chitarra, G.S., Abee, T., Rombouts, F.M., Posthumus, M.A., Dijksterhuis, J., 2004. Germination of *Penicillium paneum Conidia* is regulated by 1- octen-3-ol, a volatile self-inhibitor. Appl.

Environ. Microbiol. 70, 2823–2829. Available from: https://doi.org/10.1128/AEM.70.5.2823-2829.2004.

Cho, Y.-H., Sheen, J., Yoo, S.-D., 2010. Low glucose uncouples hexokinase1-dependent sugar signaling from stress and defense hormone abscisic acid and C_2H_4 responses in *Arabidopsis*. Plant Physiol. 152, 1180–1182.

Cooney, J.M., Lauren, D.R., di Menna, M.E., 2001. Impact of competitive fungi on trichothecene production by *Fusarium graminearum*. J. Agric. Food Chem. 49, 522–526. Available from: https://doi.org/10.1021/jf0006372.

Cordovez, V., Carrion, V.J., Etalo, D.W., Mumm, R., Zhu, H., van Wezel, G.P., et al., 2015. Diversity and functions of volatile organic compounds produced by *Streptomyces* from a disease-suppressive soil. Front. Microbiol. 6, 1081. Available from: https://doi.org/10.3389/fmicb.2015.01081.

Cortes-Barco, A.M., Goodwin, P.H., Hsiang, T., 2010. Comparison of induced resistance activated by benzothiadiazole, (2R, 3R)-butanediol and an isoparaffin mixture against anthracnose of *Nicotiana benthamiana*. Plant Pathol. 59, 643–653.

Delory, B.M., Delaplace, P., Fauconnier, M.-L., du Jardin, P., 2016. Rootemitted volatile organic compounds: can they mediate belowground plant-plant interactions? Plant Soil 402, 1–26. Available from: https://doi.org/10.1007/s11104-016-2823-3.

De Vrieze, M., Pandey, P., Bucheli, T.D., Varadarajan, A.R., Ahrens, C.H., Weisskopf, L., et al., 2015. Volatile organic compounds from native potatoassociated *Pseudomonas* as potential anti-oomycete agents. Front. Microbiol. 6, 1295. Available from: https://doi.org/10.3389/fmicb.2015.01295.

Effmert, U., Kalderas, J., Warnke, R., Piechulla, B., 2012. Volatile mediated interactions between bacteria and fungi in the soil. J. Chem. Ecol. 38, 665–703. Available from: https://doi.org/10.1007/s10886-012-0135-5.

Fialho, M.B., de Andrade, A., Bonatto, J.M.C., Salvato, F., Labate, C.A., Pascholati, S.F., 2016. Proteomic response of the phytopathogen *Phyllosticta citricarpa* to antimicrobial volatile organic compounds from *Saccharomyces cerevisiae*. Microbiol. Res. 183 (Suppl. C), 1–7. Available from: https://doi.org/10.1016/j.micres.2015.11.002.

Garbeva, P., Hol, W.H.G., Termorshuizen, A.J., Kowalchuk, G.A., de Boer, W., 2011. Fungistasis and general soil biostasis - a new synthesis. Soil Biol. Biochem. 43, 469–477. Available from: https://doi.org/10.1016/j.soilbio.2010.11.020.

Garbeva, P., Hordijk, C., Gerards, S., de Boer, W., 2014. Volatile-mediated interactions between phylogenetically different soil bacteria. Front. Microbiol. 5, 289. Available from: https://doi.org/10.3389/fmicb.2014.00289.

Geisen, S., Tveit, A.T.I., Clark, M., Richter, A., Svenning, M.M., Bonkowski, M., et al., 2015. Metatranscriptomic census of active protists in soils. ISME J. 9, 2178–2190. Available from: https://doi.org/10.1038/ismej.2015.30.

Geisen, S., Koller, R., Hunninghaus, M., Dumack, K., Urich, T., Bonkowski, M., 2016. The soil food web revisited: diverse and widespread mycophagous soil protists. Soil Biol. Biochem. 94, 10–18. Available from: https://doi.org/10.1016/j.soilbio.2015.11.010.

Glücksman, E., Bell, T., Griffiths, R.I., Bass, D., 2010. Closely related protist strains have different grazing impacts on natural bacterial communities. Environ. Microbiol. 12, 3105–3113. Available from: https://doi.org/10.1111/j.1462-2920.2010.02283.x.

Griffiths, B.S., Bonkowski, M., Dobson, G., Caul, S., 1999. Changes in soil microbial community structure in the presence of microbial-feeding nematodes and protozoa. Pedobiologia 43, 297–304.

Gürtler, H., Pedersen, R., Anthoni, U., Christophersen, C., Nielsen, P.H., Wellington, E.M., et al., 1994. Albaflavenone, a sesquiterpene ketone with a zizaene skeleton produced by a streptomycete with a new rope morphology. J. Antibiot. 47, 434–439. Available from: https://doi.org/10.7164/antibiotics.47.434.

Huang, C.-J., Tsay, J.-F., Chang, S.-Y., Yang, H.-P., Wu, W.-S., Chen, C.-Y., 2012. Dimethyl disulfide is an induced systemic resistance elicitor producedby *Bacillus cereus* C1L. Pest. Manage. Sci. 68, 1306–1310. Available from: https://doi.org/10.1002/ps.3301.

Hunziker, L., Bonisch, D., Groenhagen, U., Bailly, A., Schulz, S., Weisskopf, L., 2015. *Pseudomonas* strains naturally associated with potato plants produce volatiles with high potential for inhibition of *Phytophthora* infestans. Appl. Environ. Microbiol. 81, 821–830. Available from: https://doi.org/10.1128/AEM.02999-14.

Insam, H., Seewald, M.S.A., 2010. Volatile organic compounds (VOCs) in soils. Biol. Fertil. Soils 46, 199–213. Available from: https://doi.org/10.1007/s00374-010-0442-3.

Jones, S.E., Ho, L., Rees, C.A., Hill, J.E., Nodwell, J.R., Elliot, M.A., 2017. *Streptomyces* exploration is triggered by fungal interactions and volatile signals. Elife 6, e21738. Available from: https://doi.org/10.7554/eLife.21738.

Kai, M., Haustein, M., Molina, F., Petri, A., Scholz, B., Piechulla, B., 2009. Bacterial volatiles and their action potential. Appl. Microbiol. Biotechnol. 81, 1001–1012. Available from: https://doi.org/10.1007/s00253-008-1760-3.

Kai, M., Crespo, E., Cristescu, S.M., Harren, F.J.M., Francke, W., Piechulla, B., 2010. *Serratia odorifera*: analysis of volatile emission and biological impact of volatile compounds on *Arabidopsis thaliana*. Appl. Microbiol. Biotechnol. 88, 965–976. Available from: https://doi.org/10.1007/s00253-010-2810-1.

Kanchiswamy, C.N., Mainoy, M., Maffei, M.E., 2015. Chemical diversity of microbial volatiles and their potential for plant growth and productivity. Front. Plant Sci. 6, 151. Available from: https://doi.org/10.3389/fpls.2015.00151.

Kim, K.S., Lee, S., Ryu, C.M., 2013. Interspecific bacterial sensing through airborne signals modulates locomotion and drug resistance. Nat. Commun. 4, 1809.

Lemfack, M.C., Nickel, J., Dunkel, M., Preissner, R., Piechulla, B., 2014. mVOC: a database of microbial volatiles. Nucleic Acids Res. 42, D744–D748. Available from: https://doi.org/10.1093/nar/gkt1250.

Lemfack, M.C., Gohlke, B.O., Toguem, S.M.T., Preissner, S., Piechulla, B., Preissner, R., 2017. mVOC 2.0: a database of microbial volatiles. Nucleic Acids Res. Available from: https://doi.org/10.1093/nar/gkx1016 [Epub ahead of print].

Lutz, M.P., Wenger, S., Maurhofer, M., Défago, G., Duffy, B., 2004. Signaling between bacterial and fungal biocontrol agents in a strain mixture. FEMS Microbiol. Ecol. 48, 447–455. Available from: https://doi.org/10.1016/j.femsec.2004.03.002.

Miyamoto, K., Murakami, T., Kakumyan, P., Keller, N.P., Matsui, K., 2014. Formation of 1-octen-3-ol from *Aspergillus flavus Conidia* is accelerated after disruption of cells independently of Ppo oxygenases and is not a main cause of inhibition of germination. PeerJ 2, e395. Available from: https://doi.org/10.7717/peerj.395.

Nemcovic, M., Jakubíková, L., Víden, I., Farkaš, V., 2008. Induction of conidiation by endogenous volatile compounds in *Trichoderma* spp. FEMS Microbiol. Lett. 284, 231–236. Available from: https://doi.org/10.1111/j.1574-6968.2008.01202.x.

Ossowicki, A., Jafra, S., Garbeva, P., 2017. The antimicrobial volatile power of the rhizospheric isolate *Pseudomonas donghuensis* P482. PLoS One 12, e0174362. Available from: https://doi.org/10.1371/journal.pone.0174362.

Park, Y.S., Dutta, S., Ann, M., Raaijmakers, J.M., Park, K., 2015. Promotion of plant growth by *Pseudomonas fluorescens* strain SS101 via novel volatile organic compounds. Biochem. Biophys. Res. Commun. 461, 361–365.

Piechulla, B., Lemfack, M.C., Kai, M., 2017. Effects of discrete bioactive microbial volatiles on plants and fungi. Plant Cell Environ. 40, 2042–2067. Available from: https://doi.org/10.1111/pce.13011.

Rajer, F.U., Wu, H., Xie, Y., Xie, S., Raza, W., Tahir, H.A.S., et al., 2017. Volatile organic compounds produced by a soil-isolate, *Bacillus subtilis* FA26 induce adverse ultra-structural changes to the cells of *Clavibacter michiganensis* ssp. sepedonicus, the causal agent of bacterial ring rot of potato. Microbiology 163, 523–530. Available from: https://doi.org/10.1099/mic.0.000451.

Raza, W., Yuan, J., Ling, N., Huang, Q., Shen, Q., 2015. Production of volatile organic compounds by an antagonistic strain *Paenibacillus polymyxa* WR-2 in the presence of root exudates and organic fertilizer and their antifungal activity against *Fusarium oxysporum* f. sp. niveum. Biol. Control. 80, 89–95.

Raza, W., Ling, N., Liu, D., Wei, Z., Huang, Q., Shen, Q., 2016a. Volatile organic compounds produced by *Pseudomonas fluorescens* WR-1 restrict the growth and virulence traits of *Ralstonia solanacearum*. Microbiol. Res. 192, 103–113. Available from: https://doi.org/10.1016/j.micres.2016.05.014.

Raza, W., Ling, N., Yang, L., Huang, Q., Shen, Q., 2016b. Response of tomato wilt pathogen *Ralstonia solanacearum* to the volatile organic compounds produced by a biocontrol strain *Bacillus amyloliquefaciens* SQR-9. Sci. Rep. 6, 24856. Available from: https://doi.org/10.1038/srep24856.

Raza, W., Wang, J., Wu, Y., Ling, N., Wei, Z., Huang, Q., et al., 2016c. Effects of volatile organic compounds produced by *Bacillus amyloliquefaciens* on the growth and virulence traits of tomato bacterial wilt pathogen *Ralstonia solanacearum*. Appl. Microbiol. Biotechnol. 100, 7639–7650. Available from: https://doi.org/10.1007/s00253-016-7584-7.

Rosenberg, K., Bertaux, J., Krome, K., Hartmann, A., Scheu, S., Bonkowski, M., 2009. Soil amoebae rapidly change bacterial community composition in the rhizosphere of *Arabidopsis thaliana*. ISME J. 3, 675–684. Available from: https://doi.org/10.1038/ismej.2009.11.

Ryu, C.M., Farag, M.A., Hu, C.H., Reddy, M.S., Wei, H.X., Pare, P.W., et al., 2003. Bacterial volatiles promotes growth in *Arabidopsis*. Proc. Natl. Acad. Sci. U.S.A. 100, 4927–4932.

Schenkel, D., Lemfack, M., Piechulla, B., Splivallo, R., 2015. A meta-analysis approach for assessing the diversity and specificity of below-ground root and microbial volatiles. Front. Plant Sci. 6, 707. Available from: https://doi.org/10.3389/fpls.2015.00707.

Schmidt, R., Cordovez, V., de Boer, W., Raaijmakers, J., Garbeva, P., 2015. Volatile affairs in microbial interactions. ISME J. 9, 2329–2335. Available from: https://doi.org/10.1038/ismej.2015.42.

Schmidt, R., Jager, V., Zuhlke, D., Wolff, C., Bernhardt, J., Cankar, K., et al., 2017. Fungal volatile compounds induce production of the secondary metabolite Sodorifen in *Serratia plymuthica* PRI-2C. Sci. Rep. 7, 862. Available from: https://doi.org/10.1038/s41598-017-00893-3.

Schulz, S., Dickschat, J.S., 2007. Bacterial volatiles: the smell of small organisms. Nat. Prod. Rep. 24, 814–842. Available from: https://doi.org/10.1039/b507392h.

Schulz-Bohm, K., Zweers, H., de Boer, W., Garbeva, P., 2015. A fragrant neighbourhood: volatile mediated bacterial interactions in soil. Front. Microbiol. 6, 1212. Available from: https://doi.org/10.3389/fmicb.2015.01212.

Schulz-Bohm, K., Gerards, S., Hundscheid, M.P.J., Melenhorst, J., De Boer, W., Garbeva, P.V., 2017. Calling from distance: attraction of soil bacteria by plant root volatiles. ISME J. (accepted).

Sharifi, R., Ryu, C.-M., 2016. Are bacterial volatile compounds poisonous odors to a fungal pathogen *Botrytis cinerea*, alarm signals to *Arabidopsis* seedlings for eliciting induced resistance, or both? Front. Microbiol. 7. Available from: https://doi.org/10.3389/fmicb.2016.00196.

Sharifi, R., Lee, S.M., Ryu, C.M., 2018. Microbe-induced plant volatiles. N. Phytol. 218. Available from: https://doi.org/10.1111/nph.14955.

Splivallo, R., Deveau, A., Valdez, N., Kirchhoff, N., Frey-Klett, P., Karlovsky, P., 2015. Bacteria associated with truffle-fruiting bodies contribute to trufflaroma. Environ. Microbiol. 17, 2647–2660. Available from: https://doi.org/10.1111/1462-2920.12521.

Spraker, J.E., Jewell, K., Roze, L.V., Scherf, J., Ndagano, D., Beaudry, R., et al., 2014. A volatile relationship: profiling an inter-kingdom dialogue between two plant pathogens, *Ralstonia solanacearum* and *Aspergillus flavus*. J. Chem. Ecol. 40, 502–513. Available from: https://doi.org/10.1007/s10886-014-0432-2.

Stoppacher, N., Kluger, B., Zeilinger, S., Krska, R., Schuhmacher, R., 2010. Identification and profiling of volatile metabolites of the biocontrol fungus *Trichoderma atroviride* by HS-SPME-GC-MS. J. Microbiol. Methods 81, 187–193. Available from: https://doi.org/10.1016/j.mimet.2010.03.011.

Strobel, G.A., Dirkse, E., Sears, J., Markworth, C., 2001. Volatile antimicrobials from *Muscodor albus*, a novel endophytic fungus. Microbiology 147, 2943–2950. Available from: https://doi.org/10.1099/00221287-147-11-2943.

Tahir, H.A., Gu, Q., Wu, H., Raza, W., Hanif, A., Wu, L., et al., 2017b. Plant growth promotion by volatile organic compounds produced by *Bacillus subtilis* SYST2. Front. Microbiol. 8, 171. Available from: https://doi.org/10.3389/fmicb.2017.00171.

Tyc, O., de Jager, V.C.L., van den Berg, M., Gerards, S., Janssens, T.K.S., Zaagman, N., et al., 2017a. Exploring bacterial interspecific interactions for discovery of novel antimicrobial compounds. Microb. Biotechnol. 10, 910–925. Available from: https://doi.org/10.1111/1751-7915.12735.

Tyc, O., Song, C.X., Dickschat, J.S., Vos, M., Garbeva, P., 2017b. The ecological role of volatile and soluble secondary metabolites produced by soil bacteria. Trends Microbiol. 25, 280–292. Available from: https://doi.org/10.1016/j.tim.2016.12.002.

Vaishnav, A., Kumari, S., Jain, S., Varma, A., Choudhary, D.K., 2015. Putative bacterial volatile-mediated growth in soybean (*Glycine max* L. Merrill) and expression of induced proteins under salt stress, J. Appl. Microbiol., 119. pp. 539–551.

Van Dam, N.M., Weinhold, A., Garbeva, P., 2016. Calling in the dark: the role of volatiles for communication in the rhizosphere. In: Blande, J.D., Glinwood, R. (Eds.), Deciphering Chemical Language of Plant Communication. Springer International Publishing, Cham, p. 175210. 218 Molecular Aspects of Plant Beneficial Microbes in Agriculture.

Van Tol, R.W.H.M., Van Der Sommen, A.T.C., Boff, M.I.C., Van Bezooijen, J., Sabelis, M.W., Smits, P.H., 2001. Plants protect their roots by alerting the enemies of grubs. Ecol. Lett. 4, 292–294.

Vardharajula, S., Ali, S.Z., Grover, M., Reddy, G., Bandi, V., 2011. Drought-tolerant plant growth-promoting *Bacillus* spp., effect on growth, osmolytes, and antioxidant status of maize under drought stress. J. Plant Inter. 6.

Von Reuß, S.H., Kai, M., Piechulla, B., Francke, W., 2010. Octa methylbicyclo[3.2.1]octadienes from the rhizobacterium *Serratia odorifera*. Angew. Chem. Int. Ed. 49, 2009–2010. Available from: https://doi.org/10.1002/anie.200905680.

Wang, J., Zhou, C., Xiao, X., Xie, Y., Zhu, L., Ma, Z., 2017. Enhanced iron and selenium uptake in plants by volatile emissions of *Bacillus amyloliquefaciens* (BF06). Appl. Sci. 7, 85.

Weissteiner, S., Huetteroth, W., Kollmann, M., Weißbecker, B., Romani, R., Schachtner, J., et al., 2012. Cockchafer larvae smell host root scents in soil. PLoS One 7 (10), e45827.

Werner, S., Polle, A., Brinkmann, N., 2016. Belowground communication: impacts of volatile organic compounds (VOCs) from soil fungi on other soil-inhabiting organisms. Appl. Microbiol. Biotechnol. 100, 8651–8665. Available from: https://doi.org/10.1007/s00253-016-7792-1.

Westhoff, S., van Wezel, G.P., Rozen, D.E., 2017. Distance-dependent danger responses in bacteria. Curr. Opin. Microbiol. 36, 95–101. Available from: https://doi.org/10.1016/j.mib.2017.02.002.

Wheatley, R.E., 2002. The consequences of volatile organic compound mediated bacterial and fungal interactions. Antonie Van. Leeuwenhoek 81, 357–364. Available from: https://doi.org/10.1023/A:1020592802234.

Wintermans, P.C.A., Bakker, P.A.H.M., Pieterse, C.M.J., 2016. Natural genetic variation in *Arabidopsis* for responsiveness to plant growth-promoting rhizobacteria. Plant Mol. Biol. 90, 623–634. Available from: https://doi.org/10.1007/s11103-016-0442-2.

Xie, S., Zang, H., Wu, H., ddinRajer, F.U., Gao, X., 2016. Antibacterial effects of volatiles produced by *Bacillus* strain D13 against *Xanthomonas oryzae* pv. oryzae. Mol. Plant Pathol. Available from: https://doi.org/10.1111/mpp.12494 [Epub ahead of print].

Zhang, H., Kim, M.S., Sun, Y., Dowd, S.E., Shi, H., Pare, P.W., 2008a. Soil bacteria confer plant salt tolerance by tissue-specific regulation of the sodium transporter HKT1. Mol. Plant-Microbe Interact. 21, 737–744.

Zhang, H., Xie, X., Kim, M.S., Kornyeyev, D.A., Holaday, S., Pare, P.W., 2008b. Soil bacteria augment *Arabidopsis* photosynthesis by decreasing glucose sensing and abscisic acid levels in planta. Plant J. 56, 264273.

Zhang, H., Sun, Y., Xie, X., Kim, M.S., Dowd, S.E., Pare, P.W., 2009. A soil bacterium regulates plant acquisition of iron via deficiency-inducible mechanisms. Plant J. 58, 568–577.

Zou, C., Li, Z., Yu, D., 2010. *Bacillus megaterium* strain XTBG34 promotes plant growth by producing 2-pentylfuran. J. Microbiol. 48, 460–466.

Chapter 9

Significance of microbial volatiles in ecological health: impact on wetland systems

T.C. Prathna
Department of Irrigation and Flood Control, Government of NCT of Delhi, New Delhi, India

Chapter Outline

9.1 Introduction

The volume of wastewater generated by countries around the world has witnessed a constant increase with increasing population growth accompanied by urbanization and economic development. Conventionally, wastewater is collected through sewer networks and treated collectively in centralized wastewater treatment plants. It is critical that wastewater treatment plants (WWTPs) are constantly rehabilitated keeping in mind the national and international discharge standards. It is equally important that WWTPs are constructed with a design capacity keeping in mind the population growth projection of the entire command area it serves. For example, the WWTP at Leeuwkuil in South Africa was established in 1952 with a design capacity of 4.5 million litres per day (MLD) and was upgraded to 36 MLD in 1980. As of 2015, the plant treats ~43 MLD with plans for upgradation between the years 2020 and 2025. Increased population has led to a change in wastewater

Volatiles and Metabolites of Microbes. DOI: https://doi.org/10.1016/B978-0-12-824523-1.00020-1

characteristics and with treatment above the design capacity, has compromised on the effluent characteristics (Teklethaimanot et al., 2015). This can also affect the quality of the water bodies into which the treated water is released.

In many instances, not all parts of the city, especially in developing countries, may be connected to the sewer network. In such cases, wastewater is discharged into nearby streams, polluting the water body (Samie et al., 2009; Trang et al., 2010). The costs of bringing these localities into the existing sewer network may not be an economically feasible option. Rapid urbanization places undue stress on the limited freshwater resources available. Urban activities and agricultural usage of land around water bodies can be considered to be major contributors of pollution in water bodies (Camorani et al., 2005). Decentralized wastewater treatment is a viable option in such cases and can be implemented in both rural and urban areas.

9.1.1 Wetlands system for wastewater treatment

Among the various decentralized methods, constructed wetlands (CWs) can be considered to be a low cost, low maintenance, chemical-free, and robust technology to treat wastewater (Kivaisi, 2001). CWs also provide other landscape and social benefits such as habitat for wildlife, esthetic appeal, and places for recreation (U.S. Environmental Protection Agency U.S. EPA, 1999). CWs have been used to treat various types of wastewater, such as agricultural, municipal, industrial, aquaculture, as well as domestic (Wu et al., 2017).

9.1.1.1 Components and types of the wetland system

Water, air, wetland plants, bacteria in the biofilms, and support media (or substrate) are some of the major integral components of any wetland system. Natural wetlands, CW, and floating wetlands are some of the wetland systems used to treat water. Of these, natural and CW systems can be used for ex situ treatment of wastewater while floating wetlands can be used as a tertiary level in situ treatment system in water bodies. Depending on the influent wastewater characteristics, the wetland system can be coupled with other relevant pretreatment technologies to treat water to permissible limits. Many countries have adopted strict guidelines to minimize the use of chemicals as much as possible with regard to decentralized alternative technologies for water treatment.

9.2 Role of microbes in wastewater treatment

Among the various components of the wetland system, bacterial communities constitute the most important component for removal and transformation of hazardous contaminants (Ramond et al., 2012; Ahn et al., 2007). Bacterial communities, as indicated from experimental studies, require around 75–100 days to stabilize in CW (Truu et al., 2009). Numerical and simulation studies have also been conducted to confirm the duration that bacterial communities require to reach stability. Samso and Garcia (2013) used a numerical model, BIO_PORE to simulate the dynamics of commonly found bacteria in subsurface flow CWs and determined that time taken to achieve stability (400–700 days) was much longer than determined by experimental studies. Some of the commonly found groups of bacteria in wetlands are heterotrophs, autotrophic nitrifying, fermenting, acetotrophic methanogens, and acetotrophic sulfate reducing and sulfate oxidizing bacteria (Samso and Garcia, 2013).

Bacteria present in the wetland systems can perform various functions such as nitrification, denitrification, ammonia oxidation, phosphorus solubilization, biofilm formation, etc. Ma et al. (2018) identified the microbial communities in the roots and substrate of wetland plants with *Salicornia* sp. treated with marine aquaculture effluents. Genera of *Pseudoalteromonas, Acinetobacter, Bacillus, Pseudomonas, Vibrio, Stenotrophomonas, Comamonas, Nisaea, Nitrospina, Nitrosopumilus*, and *Planctomyces* responsible for biofilm formation, denitrification and nitrification, promotion of plant growth, nitrite and ammonia oxidation, and phosphorus solubilization were identified in the roots and substrate by sequencing studies. A difference in diversity of the bacterial community is observed with differences in the nature of the influent wastewater in the wetland system. For example, in a wetland with *Phragmites australis* and fed with oil-contaminated water, predominant bacteria in the rhizosphere soils were hydrocarbon-degrading types. *Ochrobactum* and *Pseudomonas* were the hydrocarbon-degrading bacteria sequenced from these soils (Abed et al., 2017). Bacterial communities can also vary with the location as well as with different plants from the same species.

The role of microorganisms in the transformation and removal of pollutants from CW is even more significant than plants as they can critically counteract any negative effects on plants by the production of plant growth promoters (PGP) (Li et al., 2014). Studies on a CW system with *Juncus acutus* treating contaminated water with bisphenol A revealed an increased tolerance of the plants and also ability to degrade the contaminants when the bacterial community was enriched with PGP strains (Syranidou et al., 2017). In another related study, the rhizosphere bacteria from *Phragmites australis* were isolated and characterized for their PGP potential and their ability to enhance azo dye phytodepuration with potential for Se biomethylation and volatilization from selenate and selenite (Azaizeh et al., 2003; Riva et al., 2019). Fig. 9.1 shows a representative image of a horizontal subsurface flow CW system with media supporting the plants. The extended root area with microbes surrounding the rhizosphere region is shown.

FIGURE 9.1 Representative image of a horizontal subsurface flow constructed wetland with plants supported on a suitable substrate.

9.3 Production of microbial volatiles in soil and their role in wastewater treatment

9.3.1 Microbial volatiles-their production

Volatile organic compounds are smaller odor-producing compounds with high vapor pressure, lower boiling point and molecular weight and can be either evaporated or diffused above and below ground (Schenkel et al., 2015a,b; Vespermann et al., 2007). Volatile compounds produced by microbes have different purposes and can either be pleasant or undesirable to the human nose. For example, volatiles produced during putrefaction of organic matter are unpleasant while aromas from wine, cheese fermentation, and earthy smell from soil are pleasant to smell (Lemfack et al., 2014). These microbial volatile organic compounds (MVOCs) include classes such as alcohols, terpenes, alkenes, ketones, aldehydes, ester, ethers, and sulfides (Alcega et al., 2017; Schmidt et al., 2015; Schulz-Bohm et al., 2017). Some of the factors which determine the production of MVOC in soil environment are the presence or absence of oxygen, availability of certain nutrients, pH, and moisture content of soil (Insam and Seewald, 2010). Recent studies have indicated that these MVOCs have molecular masses below 300 Da and are mostly lipophilic in nature (Kanchiswamy et al., 2015).

MVOC can be sampled by any bioaerosol sampling method as there are still no standardized protocols available. The most commonly used method to collect the bioaerosol sample is using thermal desorption tubes coated with Tenax/Carbotrap 50/50 v/v. Varied sampling times are used ranging between 30 minutes to one day (Alcega et al., 2017). Some of the commonly used detection techniques for MVOC are mass spectrometry, gas chromatography (Insam and Seewald, 2010) and other advanced techniques such as proton transfer-reaction mass spectrometry which has detection limits up to the pptv levels (Lindinger et al., 1998).

9.3.2 Microbial volatiles in wastewater treatment

WWTPs can be a site which can contribute to the majority of MVOC emissions. Different stages of wastewater treatment—primary, secondary, and tertiary—can lead to emission of different classes of MVOC, which can be measured either at the site, upwind or downwind of the site. In a study conducted by Alcega et al. (2017), some of the major MVOC detected from WWTPs were ketones, aldehydes, alcohols, and alkanes. These VOC can be produced by the microbes during different stages of wastewater treatment such as aerobic digestion, nitrification and denitrification, P solubilization and transformation, etc. Studies indicate seasonal variations in MVOC emissions from WWTPs with more percentage of emission in summer (79%) as compared to winter (43%). Aldehydes, ethers, and ketones were the common groups of MVOC emitted in summer while it was alcohols in winter. Emission of aldehydes, ketones, and ethers in summer was due to the fermentation of organic matter by the microbes due to insufficient oxygen availability. Some of the major MVOC detected in WWTPs are formic acid, decane, dodecane, and butyl ester (Alcega et al., 2018).

9.3.2.1 Microbial volatiles in wastewater treatment using wetland systems

9.3.2.1.1 Root system of wetland plants

The root system of wetland plants is significantly distinct from terrestrial plants. They are well adapted to tolerate extreme conditions such as high salinity, heavy aquatic pollution, etc. The rhizosphere region of wetland plants harbors numerous microbes, predominantly bacteria and fungi (Mendes et al., 2013). As compared to normal plants, wetland plants have an extended root system as a stress response to metal toxicity in wastewater. The rhizosphere region around the root surface provides a suitable environment for microbes to aid in degradation and assimilation of contaminants. Wetlands plants may be classified into two based on their root system-rhizomatic roots and fibrous roots. Wetland plants such as *Phragmites* and *Typha* have a rhizomatic root system while plants belonging to *Cyperus* sp. have a fibril root system. As compared to rhizomatic roots, fibrous root systems have larger biomass, increased nutrient removal due to higher root surface area and faster growth. However, the longevity of fibrous roots is shorter in comparison to rhizomatic roots. A larger root surface area may indicate a larger attachment area for the nitrifying bacteria. A higher root porosity can also indicate more oxygen transfer to the rhizosphere region by the roots and also enhanced nutrient removal. The root surface area of common wetland plants with their root porosity is shown in Table 9.1.

The root surface area can be calculated theoretically or also measured using image analysis software. This extended root surface area also ensures

TABLE 9.1 Comparison of active root surface area of wetland plants.

SNo	Wetland plants	Root surface area (m^2/plant)[a]	Root porosity (%)	
1	*Canna indica*	2.28	20.42 ± 3.20	Wenyin et al. (2007); Lai et al. (2011)
2	*Cyperus sp.*	0.20	32.16 ± 3.42	
3	*Phragmites*	0.06	40.16 ± 3.85	
4	*Typha*	0.08	16.71 ± 2.32	
5	Spider Lily	0.20	-	
6	*Scirpus validus*	-	30.00 ± 1.63	

[a]*Period = 147 days.*

increased release of oxygen to support the microbes in the rhizosphere region. The roots of wetland plants are also known to release more oxygen as compared to terrestrial plants. The oxygen release rate from wetland plants has been determined by various methods such as root respiration studies, simulation models, and radial oxygen loss (ROL). ROL is considered to be among the primary ways of oxygen input in subsurface CW. In the case of wetland plants, transport of oxygen to the roots is through the well-developed aeration tissues. Aeration tissues are well developed in wetland plants as they are adapted to survive in difficult conditions such as flooding and high pollutant loads. The input of oxygen in the CW is primarily through atmospheric reoxygenation, ROL, and influent reoxygenation (Konnerup et al., 2011). The oxygen release rate from the roots to the rhizosphere region has been documented and measured by root respiration studies. The release rate will also depend on external factors such as the surrounding rhizosphere environment. As such roots of wetland plants have barriers to prevent radial loss of oxygen, however, oxygen is released into the rhizosphere through porous regions in the root tips and the lateral roots (Armstrong et al., 2000) (Table 9.2).

Oxygen release in the rhizosphere region of wetland plants can be augmented by factors such as type of wetland plants, microbes in the rhizosphere, and also climatic conditions (Rehman et al., 2016).

9.3.2.1.2 Microbes on the rhizosphere region of wetland plants

Microbial diversity plays a critical role in the maintenance and treatment efficiency of a CW system. Colonization of the rhizosphere region by fungi and bacteria is a very dynamic and competitive process. For example, literature shows that at least a hundred species of fungi can coexist on the

TABLE 9.2 Oxygen release rate of selected wetland plants.

Plant	Plant oxygen release rate (g/m^2/d)	Based on	References
Phragmites sp.	0.014–0.015	Respiration	Ye et al. (2012)
Scirpus sp.	0.005–0.011	Radial oxygen loss	Bezbaruah and Zhang (2005)
Typha sp.	0.023	Respiration	Wu et al. (2001)
Typha sp.	0.45	Simulation	Mburu et al. (2012)
Bulrush	28.6	BOD removal	Burgoon (1993)

BOD, biological oxygen demand.

rhizosphere region of a single plant (Bahram et al., 2011). Some of the beneficial activities of microbial volatiles produced by the root colonizing fungi are to promote plant growth (Kia et al., 2017) and also inhibit the activity of other pathogenic fungi. Bacteria colonizing the rhizosphere region, on the other hand, are known to promote beneficial activities such as enhanced nutrient uptake and better stress response (Schenkel et al., 2019). At least 841 microbial volatiles that can influence plant interactions has been identified to date (Schenkel et al., 2015a,b). However, most of the studies on microbial volatiles on plant response have been carried out on only terrestrial plants, and more extensive research is needed on how they induce wetland plant response in wastewater treatment.

Nitrogen removing bacteria Nitrogen removal in CW is accomplished by a two-step process- nitrification followed by denitrification. Nitrification involves the oxidation of ammonia as well as the oxidation of nitrite. Oxidation of ammonia is regarded to be the rate-limiting step for nitrification in CW (Choi and Hu, 2008). Ammonia-oxidizing bacteria are predominantly considered to be autotrophic (Wang et al., 2010) though recent studies have also identified heterotrophic as well as mixotrophic bacteria capable of ammonia oxidation in CWs (Houda et al., 2014; Kouki et al., 2011). Nitrifying bacteria compete with other heterotrophic and sulfide oxidizing bacteria for the limited available oxygen in the CW (Wu et al., 2013). Nitrifying bacteria are known to grow abundantly only when the chemical oxygen demand (COD) concentrations are low (Samso and Garcia, 2013).

Some of the ammonia-oxidizing and nitrifying bacterial species identified in the rhizosphere of Typha wetland plants in horizontal and vertical subsurface flow CW are *Bacillus* sp., *Arthrobacter* sp., *Pseudomonas* sp., *Bordetella* sp., *Nitrosomonas* sp., and *Ochrobactrum* sp. (Houda et al., 2014; Kouki et al., 2011; Ruiz-Rueda et al., 2009). Some of the MVOCs produced by *Bacillus* strains are 3-hydroxy-2-butanone, 2,3 butanediol, other alcohols, amines and phenols including hexadecane, 2,3-dimethoxybenzamide, and O-anisaldehyde (Rudrappa et al., 2010; Zhang et al., 2013). Some of the MVOCs detected from *Pseudomonas* sp. include 2-(benzyloxy) benzonitrile, benzothiazole, and 1-methyl naphthalene (Kai et al., 2007; Raza et al., 2016). Denitrifying bacteria isolated from a salt-stressed CW with *Canna indica* showed the presence of *Pseudomonas* and *Arthrobacter* in the presence of a salt-tolerant inoculum of *Alishewanella* sp. (Wang et al., 2020). In another related study, bacteria present in the rhizosphere region of *Myriophyllum aquaticum* surface flow CW to treat swine wastewater was sequenced. Some of the denitrifying bacteria identified include *Bacillus* sp., *Acidovorax* sp., *Rhodoplanes* sp., *Hyphomicrobium* sp., *Nocardia* sp., *Bradyrhizobium* sp., *Zoogloea* sp., and *Streptomyces* sp. (Sun et al., 2017). Some of the MVOCs produced by Streptomyces sp. are ethylene, butanol, dimethyl disulfide, trimethyl disulfide, gamma butyrolactones, isoprene, cyclopentanone, geosmin, and acetone (Scholler et al., 2002; Schulz and Dickschat, 2007; Effmert et al., 2012).

Sulfate reducing and oxidizing bacteria Some of the sulfate reducing and oxidizing bacteria identified in CWs with *Typha latifolia* are *Desulfobacter* sp. and *Thiobacillus* sp. (Chen et al., 2016) in addition to commonly found *Desulfovibrio* sp. (Lloyd, 2003). Studies indicate that sulfate reducing bacteria take sufficient time to establish in newly constructed subsurface flow CWs and hence initial sulfate removal rates are not encouraging in such wetlands (Wiessner et al., 2005). Low levels of acetate in the influent wastewater can also delay the establishment of sulfate-reducing bacterial communities and can be established once enough fermenting bacterial biomass are present (Samso and Garcia, 2013). Among the different types of wetland plants, surface flow CWs with Typha have been studied to be very effective in sulfate removal as compared to other systems with Acorus and Nuphar plants with the highest number of bacteria being supported by the Typha system (Park et al., 2009). Table 9.3 shows the list of most commonly found bacteria in the root rhizosphere and their associated microbial volatiles.

Metal uptake and contaminant removal Wetland plants are especially known for their enhanced capacity to remove trace metals as well as dyes from contaminated waters. They are known to accumulate these heavy metals in their stem and roots. Wetland plants are aided by dye reducing and heavy metal accumulating microbes in the rhizosphere region of the plants. For example, the microbe *Geobacter metallireducens* can reduce U(VI) to insoluble U(IV) while microbes such as *Desulfovibrio desulfuricans* and *Desulfovibrio fructosovorans* help in the reduction of Technetium (VII) (De Luca et al., 2001; Lovely and Coates, 1997; Lloyd, 2003). The microbes are also known to release volatiles which stimulates root and shoot growth of

TABLE 9.3 Commonly found bacteria in rhizosphere region of wetland plants and their corresponding microbial volatiles.

Genus of bacteria	Function of the bacterial species	Microbial volatile produced	References
Arthrobacter sp.	Nitrogen removal	Acetamide, Benzaldehyde, Benzothiazole, 1-Decene, Phenylacetaldehyde	Schulz and Dickschat (2007); Zou et al. (2007)
Bacillus sp.	Nitrogen removal	3-hydroxy-2-butanone, 2,3 butanediol, hexadecane, 2,3-dimethoxybenzamide, O-anisaldehyde, acetoin, isoprene, acetic acid, acetone. Ethanol	Rudrappa et al. (2010); Zhang et al. (2013); Schulz and Dickschat (2007); Farag et al. (2006)
Pseudomonas sp.	Nitrogen removal	Ethyl acetate, xylene, methyl acetate, 1-undecene, ethyl benzene, 2-(benzyloxy) benzonitrile, benzothiazole; 1-methyl naphthalene	Freeman et al. (1976); Schulz and Dickschat (2007); Kai et al. (2007); Raza et al. (2016)
Zoogloea sp.	Nitrogen removal-Denitrification	Methyliodide	Schulz and Dickschat (2007)
Streptomyces sp.	Nitrogen removal-Denitrification	Ethylene, butanol, dimethyl disulphide, trimethyl disulphide, gamma butyrolactones, isoprene, cyclopentanone, geosmine, acetone	Scholler et al. (2002); Schulz and Dickschat (2007); Effmert et al. (2012)
Desulfovibrio sp.	Sulfate reduction	Dimethylsulfide, Dimethylselenium, Trimethylarsine	Schulz and Dickschat (2007); Michalke et al. (2000)

plants as well as increase their stress resistance, thus indirectly aiding in the cleaning of wastewaters.

Even though the role of microbes and plants in wastewater treatment has been extensively reported, the role played by microbial volatiles in mediating the responses of the wetland plants to respond effectively is fully not understood. There is more scope for research to identify and map all the microbial volatiles produced by the different types of microbes present in the wetland system and categorize their role in modulating the wetland plant response.

9.4 Summary

The contents of the chapter can be summarized by the following points mentioned below:

1. CW play a critical role in removing contaminants from wastewater in a sustainable manner and are one among the most commonly used alternative methods of wastewater treatment in a decentralized way.
2. Wetland plants, substrate, and microbes form the main parts of a wetland system. The substrate media form a supporting matrix for the microbes while the rhizosphere region of the wetland plants harbors the microorganisms, which aid in wastewater treatment.
3. Wetland plants have an extended root system with an increased root surface area which releases more oxygen as compared to normal plants. This supports the extensive microbial biota in the rhizosphere region.
4. Degradation of organic matter, removal of trace metals, and dyes are some of the ways by which a CW system treats wastewater. Organic matter degradation is mainly brought about by the removal of major nutrients-nitrogen and phosphorus.
5. Removal of nutrients from wastewater is accomplished with the help of microbes in the rhizosphere region. The microbes are studied to release volatiles, which affects the plant responses. The beneficial effects that these volatiles has on the wetland plants are improved tolerance to stress conditions such as high salt environment, improved resistance to pathogens and also increased root and shoot growth.
6. More research on microbial volatiles produced by specific wetland plants and their role in modulating the plant response is needed. Identification and qualitative analysis of these volatiles specific to each microbial genus in the rhizosphere region of the wetland plant will help in using the specific microbial blends to elicit desired and improved responses in wetland systems for wastewater treatment.

References

Abed, R.M.M., Al-Kharusi, S., Gkorezis, P., Prigent, S., Headley, T., 2017. Bacterial communities in the rhizosphere of *Phragmites australis* from an oil-polluted wetland. Arch. Agron. Soil. Sci. 64 (3), 360–370.

Ahn, C., Gillevet, P.M., Sikaroodi, M., 2007. Molecular characterization of microbial communities in treatment microcosm wetlands as influenced by macrophytes and phosphorus loading. Ecol. Indic. 7, 852–863.

Alcega, S.G., Tyrell, S., Coulon, S., 2017. Sampling microbial volatile organic compounds: optimization of flow rate and sampling time. WIT Trans. Ecol. Environ. WIT 211, 69–76.

Alcega, S.G., Nasir, Z.A., Ferguson, R., Noel, C., Cravo-Laureau, C., Whitby, C., et al., 2018. Can chemical and molecular biomarkers help discriminate between industrial, rural and urban environments? Sci. Total Environ. 631-632, 1059–1069.

Armstrong, W., Cousins, D., Armstrong, J., Turner, D.W., Beckett, P.M., 2000. Oxygen distribution in wetland plant roots and permeability barriers to gas-exchange with the rhizosphere: a microelectrode and modelling study with *Phragmites australis*. Ann. Bot. 86, 687–703.

Azaizeh, H.A., Salhani, N., Sebesvari, Z., Emons, H., 2003. The potential of rhizosphere microbes isolated from a constructed wetland to biomethylate selenium. J. Environ. Qual. 32, 55–62.

Bahram, M., Põlme, S., Kõljalg, U., Tedersoo, L., 2011. A single European aspen (*Populus tremula*) tree individual may potentially harbour dozens of *Cenococcum geophilum* ITS genotypes and hundreds of species of ectomycorrhizal fungi: ectomycorrhizal fungi of a single aspen tree. FEMS Microbiol. Ecol. 75, 313–320.

Bezbaruah, A.N., Zhang, T.C., 2005. Quantification of oxygen release by bulrush (*Scirpus validus*) roots in a constructed treatment wetland. Biotechnol. Bioeng. 89 (3), 308–318.

Burgoon, P.S., 1993. Oxidation of carbon and nitrogen in the root zone of emergent macrophytes grown in wetland microcosms. Ph.D. Dissertation. University of Florida, Gainesville (As reported in Kadlec and Knight, 1996).

Camorani, G., Castellarin, A., Brath, A., 2005. Effects of land-use changes on the hydrologic response of reclamation systems. Phys. Chem. Earth. 30, 561–574.

Chen, Y., Wen, Y., Zhou, Q., Huang, J., Vymazal, J., Kuschk, P., 2016. Sulfate removal and sulfur transformation in constructed wetlands: the roles of filling material and plant biomass. Water Res. 102, 572–581.

Choi, O.K., Hu, Z.Q., 2008. Size dependent and reactive oxygen species related nano silver toxicity to nitrifying bacteria. Environ. Sci. Technol. 42 (12), 4583–4588.

De Luca, G., de Philip, P., Dermoun, Z., Rousset, M., Verméglio, A., 2001. Reduction of technetium(VII) by desulfovibrio fructosovorans is mediated by the nickel-iron hydrogenase. Appl. Environ. Microbiol. 67 (10), 4583–4587.

Effmert, U., Kalderas, J., Warnke, R., Piechulla, B., 2012. Volatile mediated interactions between bacteria and fungi in the soil. J. Chem. Ecol. 38, 665–703.

Farag, M.A., Ryu, C.M., Summer, L.W., Pare, P.W., 2006. GC-MS SPME profiling of rhizobacterial volatiles reveals prospective inducers of growth promotion and induced systemic resistance in plants. Phytochemistry 67, 2262–2268.

Freeman, L.R., Silverman, G.J., Angelini, P., Merritt Jr., C., Esselen, W.B., 1976. Volatiles produced by microorganisms isolated from refrigerated chicken at spoilage. Appl. Environ. Microbiol. 32, 222–231.

Houda, N., Hanene, C., Ines, M., Myriam, B., Imen, D., Abdennaceur, H., 2014. Isolation and characterization of microbial communities from a constructed wetlands system: a case study in Tunisia. Afr. J. Microbiol. Res. 8 (6), 529–538.

Insam, H., Seewald, M.S.A., 2010. Volatile organic compounds (VOCs) in soils. Biol. Fertil. Soils 46, 199–213.

Kai, M., Effmert, U., Berg, G., Piechulla, B., 2007. Volatiles of bacterial antagonists inhibit mycelial growth of the plant pathogen *Rhizoctonia solani*. Arch. Microbiol. 187, 351–360.

Kanchiswamy, C.N., Malnoy, M., Maffei, M.E., 2015. Chemical diversity of microbial volatiles and their potential for plant growth and productivity. Front. Plant Sci. 6, 151.

Kia, S.H., Glynou, K., Nau, T., Thines, M., Piepenbring, M., Maciá-Vicente, J.G., 2017. Influence of phylogenetic conservatism and trait convergence on the interactions between fungal root endophytes and plants. ISME J. 11, 777–790.

Kivaisi, A.K., 2001. The potential for constructed wetlands for wastewater treatment and reuse in developing countries: a review. Ecol. Eng. 16 (4), 545–560.

Konnerup, D., Sorrell, B.K., Brix, H., 2011. Do tropical wetland plants possess convective gas flow mechanisms? N. Phytol. 90, 379–386.

Kouki, S., Saidi, N., Ḿhiri, F., Nasr, H., Cherif, H., Ouzari, H., et al., 2011. Isolation and characterization of facultative mixotrophic ammonia-oxidizing bacteria from constructed wetlands. J. Environ. Sci. 23 (10), 1699–1708.

Lai, W.L., Wang, S.Q., Peng, C.L., Chen, Z.H., 2011. Root features related to plant growth and nutrient removal of 35 wetland plants. Water Res. 45, 3941–3950.

Lemfack, M.C., Nickel, J., Dunkel, M., Preissner, R., Piechulla, B., 2014. mVOC: a database of microbial volatiles. Nucleic Acids Res. 42, D744–D748.

Li, Y., Zhu, G., Ng, W.J., Tan, S.K., 2014. A review on removing pharmaceutical contaminants from wastewater by constructed wetlands: design, performance and mechanism. Sci. Total Environ. 468, 908–932.

Lindinger, W., Hansel, A., Jordan, A., 1998. On-line monitoring fo volatile organic compounds at ppt levels by means of proton transfer-reaction mass spectometry (PTR-MS) medical applications, food control and environmental research. Int. J. Mass. Spectrom. 173, 191–241.

Lloyd, J.R., 2003. Microbial reduction of metals and radionuclides. FEMS Microbiol. Rev. 27, 411–425.

Lovely, D.R., Coates, J.D., 1997. Bioremediation of metal contamination. Curr. Opin. Biotechnol. 8, 285–289.

Ma, X., Song, X., Li, X., Fu, S., Li, M., Liu, Y., 2018. Characterization of microbial communities in pilot-scale constructed wetlands with salicornia for treatment of marine aquaculture effluents. Archaea 1–12Article ID 7819840. Available from: https://doi.org/10.1155/2018/7819840.

Mburu, N., Sanchez-Ramos, D., Rousseau, D.P.L., van Bruggen, J.J.A., Thumbi, G., Stein, O.R., et al., 2012. Simulation of carbon, nitrogen and Sulphur conversion in batch operated experimental wetland mesocosms. Ecol. Engi 42, 304–315.

Mendes, R., Garbeva, P., Raaijmakers, J.M., 2013. The rhizosphere microbiome: significance of plant beneficial, plant pathogenic, and human pathogenic microorganisms. FEMS Microbiol. Rev. 37, 634–663.

Michalke, K., Wickenheiser, E.B., Mehring, M., Hirner, A.V., Hensel, R., 2000. Production of volatile derivatives of metal (loid)s by microflora involved in anaerobic digestion of sewage sludge. Appl. Environ. Microbiol. 66, 2791–2796.

Park, N., Lee, J., Chon, K., Kang, H., Cho, J., 2009. Investigating microbial activities of constructed wetlands with respect to nitrate and sulfate reduction. Desalin. Water Treat. 1, 172–179.

Ramond, J.B., Welz, P.J., Cowan, D.A., Burton, S.G., 2012. Microbial community structure stability, a key parameter in monitoring the development of constructed wetland mesocosms during start-up. Res. Microbiol. 163, 28–35.

Raza, W., Ling, N., Liu, D., Wei, Z., Huang, Q., Shen, Q., 2016. Volatile organic compounds produced by *Pseudomonas fluorescens* WR-1 restrict the growth and virulence traits of *Ralstonia solanacearum*. Microbiol. Res. 192, 103–113.

Rehman, F., Pervez, A., Khattak, B.N., Ahmad, R., 2016. Constructed wetlands: perspectives of the oxygen released in the rhizosphere of macrophytes. Clean. Soil Air Water 45 (1). Available from: https://doi.org/10.1002/clen.201600054.

Riva, V., Mapelli, F., Syranidou, E., Crotti, E., Choukrallah, R., Kalogerakis, N., et al., 2019. Root bacteria recruited by phragmites australis in constructed wetlands have the potential to enhance azo-dye phytodepuration. Microorganisms 7, 384.

Rudrappa, T., Biedrzycki, M.L., Kunjeti, S.G., Donofrio, N.M., Czymmek, K.J., Pare, P.W., et al., 2010. The rhizobacterial elicitor acetoin induces systemic resistance in *Arabidopsis thaliana*. Commun. Integr. Biol. 3, 130–138.

Ruiz-Rueda, O., Hallin, S., Baneras, L., 2009. Structure and function of denitrifying and nitrifying bacterial communities in relation to the plant species in a constructed wetland. FEMS Microbiol. Ecol. 67 (2), 308–319.

Samie, A., Obi, C.L., Igumbor, J.O., Momba, M.N.B., 2009. Focus on 14 sewage treatment plants in the Mpumalanga Province, South Africa in order to gauge the efficiency of wastewater treatment. Afr. J. Biotechnol. 8 (14), 3276–3285.

Samso, R., Garcia, J., 2013. Bacteria distribution and dynamics in constructed wetlands based on modelling results. Sci. Total Environ. 461-462, 430–440.

Schenkel, D., Lemfack, M.C., Piechulla, B., Splivallo, R., 2015a. A meta-analysis approach for assessing the diversity and specificity of belowground root and microbial volatiles. Front. Plant Sci. 6, 707–718.

Schenkel, D., Lemfack, M.C., Piechulla, B., Splivallo, R., 2015b. A meta-analysis approach for assessing the diversity and specificity of belowground root and microbial volatiles. Front. Plant Sci. 6, 707.

Schenkel, D., Deveau, A., Niimi, J., Mariotte, P., Vitra, A., Meisser, M., et al., 2019. Linking soil's volatilome to microbes and plant roots highlights the importance of microbes as emitters of belowground volatile signals. Environ. Microbiol. 21 (9), 3313–3327.

Schmidt, R., Cordovez, V., de Boer, W., Raaijmakers, J., Garbeva, P., 2015. Volatile affairs in microbial interactions. ISME J. 9, 2329–2335.

Scholler, E.G., Gurtler, H., Pedersen, R., Molin, S., Wilkins, K., 2002. Volatile metabolites from actinomycetes. J. Agric. Food Chem. 50, 2615–2621.

Schulz, S., Dickschat, J.S., 2007. Bacterial volatiles: the smell of small organisms. Nat. Prod. Rep. 24, 814–842.

Schulz-Bohm, K., Martin-Sanchez, L., Garbeva, P., 2017. Microbial volatiles: small molecules with an important role in intra- and inter-kingdom interactions. Front. Microbiol. 8, 02484.

Sun, H., Liu, F., Xu, S., Wu, S., Zhuang, G., Deng, Y., et al., 2017. *Myriophyllum aquaticum* constructed wetland effectively removes nitrogen in swine wastewater. Front. Microbiol. 8, 1932.

Syranidou, E., Christofilopoulos, S., Politi, M., Weyens, N., Venieri, D., Vangronsveld, J., et al., 2017. Bisphenol-A removal by the halophyte *Juncus acutus* in a phytoremediation pilot, characterization and potential role of the endophytic community. J. Hazard. Mater. 323, 350–358.

Teklethaimanot, G., Kamika, I., Coetzee, M.A.A., Momba, M.N.B., 2015. Population growth and its impact on the design capacity and performance of the wastewater treatment plants in Sedibeng and Soshanguve, South Africa. Environ. Manage. 56, 984–997.

Trang, N.T.D., Konnerup, D., Schierup, H.H., Chiem, N.H., Tuan, L.A., Brix, H., 2010. Kinetics of pollutant removal from domestic wastewater in a tropical horizontal subsurface flow constructed wetland system: effect of hydraulic loading rate. Ecol. Eng. 36, 527–535.

Truu, M., Truu, J., Heinsoo, K., 2009. Changes in soil microbial community under willow coppice: the effect of irrigation with secondary-treated municipal wastewater. Ecol. Eng. 35, 1011–1020.

U.S. Environmental Protection Agency (U.S. EPA), 1999. U.S. EPA manual: constructed wetlands treatment of municipal wastewaters (EPA/625/R-99/010). Cincinnati, OH: National Risk Management Research Laboratory, Office of Research and Development.

Vespermann, A., Kai, M., Piechulla, B., 2007. Rhizobacterial volatiles affect the growth of fungi and *Arabidopsis thaliana*. Appl. Environ. Microbiol. 73, 5639–5641.

Wang, X., Wen, X., Criddle, C., Wells, G., Zhang, J., Zhao, Y., 2010. Community analysis of ammonia-oxidizing bacteria inactivated sludge of eight waste water treatment systems. J. Environ. Sci. 22 (4), 627–634.

Wang, X., Zhu, H., Yan, B., Shutes, B., Banuelos, G., Wen, H., 2020. Bioaugmented constructed wetlands for denitrification of saline wastewater: a boost for both microorganisms and plants. Environ. Int. 138, 105628.

Wenyin, C., Zhanghe, C., Qifan, H., Xiaoyan, W., Cairong, W., Dafeng, C., et al., 2007. Root growth of wetland plants with different root types. Acta Ecol Sin. 27 (2), 450–458.

Wiessner, A., Kappelmeyer, U., Kuschk, P., Kästner, M., 2005. Sulfate reduction and the removal of carbon and ammonia in a laboratory-scale constructed wetland. Water Res. 39, 4643–4650.

Wu, M.Y., Franz, E.H., Chen, S., 2001. Oxygen fluxes and ammonia removal efficiencies in constructed treatment wetlands. Water Environ. Res. 73 (6), 661–666.

Wu, S., Kuschk, P., Wiessner, A., Müller, J., Saad, R.A.B., Dong, R., 2013. Sulphur transformations in constructed wetlands for waste water treatment: a review. Ecol. Eng. 52, 278–289.

Wu, H.M., Zhang, J., Ngo, H.H., Guo, W.S., Liang, S., 2017. Evaluating the sustainability of free water surface flow constructed wetlands: methane and nitrous oxide emissions. J. Clean. Prod. 147, 152–156.

Ye, J., Wang, L., Li, D., Han, W., Ye, C., 2012. Vertical oxygen distribution trend and oxygen source analysis for vertical-flow constructed wetlands treating domestic wastewater. Ecol. Engi 41, 8–12.

Zhang, X., Li, B., Wang, Y., Guo, Q., Lu, X., Li, S., et al., 2013. Lipopeptides, a novel protein, and volatile compounds contribute to the antifungal activity of the bio control agent *Bacillus atrophaeus* CAB-1. Appl. Microbiol. Biotechnol. 97, 9525–9534.

Zou, C.S., Mo, M.H., Gu, Y.Q., Zhou, J.P., Zhang, K.Q., 2007. Possible contribution of volatile-producing bacteria in soil fungistasis. Soil Biol. Biochem. 39, 2371–2379.

Chapter 10

Endophytic bacteria as source of novel bioactive compounds

Navnita Srivastava
Department of Life Science, School of Science, Sardar Patel University, Balaghat, India

Chapter Outline

10.1 Endophytic bacteria

Endophytes are endosymbiotic microbes that inhabit the tissues of plants, either throughout or for a phase of their life (Wilson, 1995; Hardoim et al., 2008). Various plants, ranging from wild to domesticated perennial trees, weeds, and agronomic plants, are reported to harbor endophytes (Yuan et al., 2014; Kumar et al., 2016; Gupta et al., 2019). The entry of endophytes inside plant tissue occurs primarily through the roots (Egamberdieva et al., 2010). However, entrance through stomata of leaves, openings of flowers, stems, and cotyledons has also been documented (Ferreira et al., 2008; Girsowicz et al., 2019; Kumar et al., 2020). After entering the host, they are transported across the plant tissues by the vascular system and, thereafter, colonize within cells or intercellular spaces (Sapers et al., 2005; Compant et al., 2010). After gaining residence within the plant tissues, the endophytes synthesize a diverse range of bioactive compounds that aid in the plant growth by combating several abiotic and biotic stresses (Truyens et al., 2015; Gupta et al., 2020).

Volatiles and Metabolites of Microbes. DOI: https://doi.org/10.1016/B978-0-12-824523-1.00009-2

The symbiotic bacterial groups with the capability to invade plant tissues, without harming them, are called endophytic bacteria (EB). They could be either obligate or facultative. Most of them possess a dual staged life cycle, circulating between soil and plant environment (Reinhold-Hurek and Hurek, 1998). Being rhizospheric, EB mainly invade the root tissue and then get translocated into the other aerial parts of the plant. This pathway has been studied by confocal laser scanning microscopy, *gfp* or *gus* reporter gene labeled strains, immunomarkers, and fluorescence in situ hybridization (FISH) (Gamalero et al., 2003). However, the entrance of EB into the plant through wound, stomata, hydathodes, and insects has also been reported (Vorholt, 2012). Both plant and EB contributes to establishing symbiotic colonization. The specific compounds in root exudates are recognized by EB, while the transpiration stream aid in their translocation to different plant tissue (de Weert et al., 2002). The bacterial flagella, lipopolysaccharides, quorum sensing (QS), and capability to synthesize enzymes as cellulase and pectinase are the properties of EB that help in their colonization inside plant tissue (Böhm et al., 2007; Suárez-Moreno et al., 2010).

The EB are highly diversified. Their diversity has been investigated by culture-independent and culture-based methods. In cultivation-based techniques, the bacteria are isolated in culture media and then studied through microscopic and biochemical assays. Whereas cultivation-independent methods use DNA based molecular techniques to study endophytic communities structure dynamics in response to several factors as plant types, tissue types, different stress conditions, etc. (Menpara and Chanda, 2013). Studies revealed that the EB is mainly comprised of Proteobacteria, followed by Actinobacteria, Firmicutes, and Bacteroidetes (Fig. 10.1). The smaller fraction of EB falls under the category of Chloroflexi, Cyanobacteria, Armatimonadetes, Verrucomicrobia, Planctomycetes, and Nitrospirae (Edwards et al., 2015; Sessitsch et al., 2012). Several factors as host plant species, soil types and fluctuations in environmental CO_2 and temperature determine the community structure of EB (Liu et al., 2017). However, numerous EB still need to be explored. An understanding of the ecology of EB and bioactive compounds synthesized by them can have great potential for bioprospecting.

The EB produces an array of bioactive compounds which are directly or indirectly beneficial to plants (Singh et al., 2017). Such compounds have evidently enhanced the ability of host plants to deal with several biotic and abiotic stresses, thereby assisting their growth. These bioactive compounds are not only beneficial to plants, but are also useful to a human. Endophytes residing in medicinal plants are a significant source of valued secondary metabolites which are medically important bioactive compounds, which contribute almost 80% of the commercially available natural drugs (Singh and Dubey, 2015). The source and roles of major bioactive compounds synthesized by EB are discussed below.

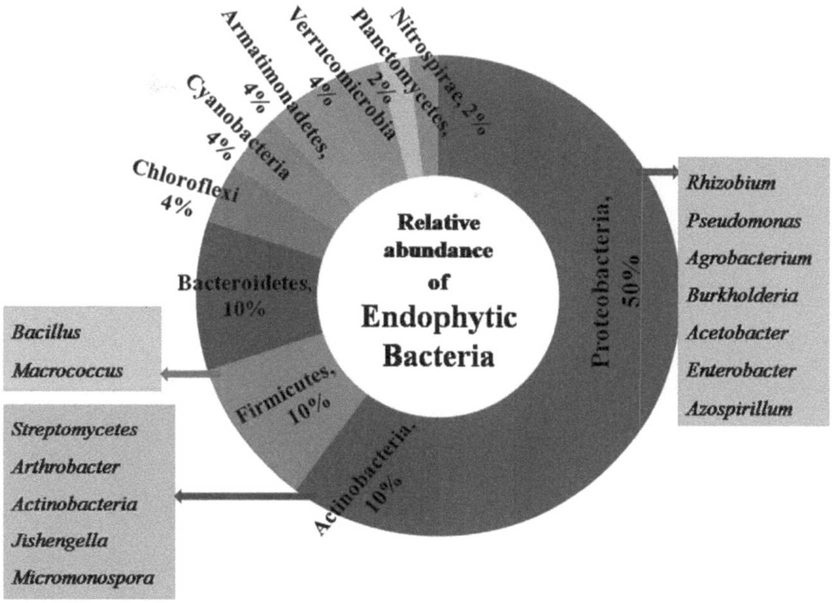

FIGURE 10.1 Relative abundance and diversity of endophytic bacteria with examples of major group. *Adapted from reference Edwards, J., Johnson, C., Santos-Medellín, C., et al., 2015. Structure, variation, and assembly of the root-associated microbiome of rice. Proc. Natl. Acad. Sci. U.S.A. 112, E911–E920; Sessitsch, A., Hardoim, P., Döring, J., et al., 2012. Functional characteristics of an endophyte community colonizing rice roots as revealed by metagenomic analysis. Mol. Plant Microbe Interact. 28–36.*

10.2 Bioactive compounds of agricultural importance

Plant growth promoting EB synthesizes various secondary metabolites, phytohormones, and novel bioactive compounds, which are capable of promoting plant growth through direct and indirect methods. The direct mechanisms include providing plants with nutrients and synthesizing certain phytohormones (Santoyo et al., 2016), while indirect aspects mainly incorporate their hostile impacts toward phytopathogens (Compant et al., 2010). Major bioactive compounds, along with their functions, produced from agriculturally significant EB have been summarized in Table 10.1.

10.2.1 Providing nutrients

Soil is a heterogeneous mixture that contains most of the macro- and micronutrients required for the growth of plants. However, some of the nutrients need to be converted into a plant's usable form. The EB perform this crucial role by converting insoluble nutrients, particularly nitrogen, phosphorus, and iron, into soluble complexes (Table 10.1A).

TABLE 10.1 Bioactive compounds from agriculturally beneficial endophytic bacteria.

Endophytic Bacteria	Source	Bioactive compound	Beneficial role	References
(A) Nutrients uptake				
Herbaspirillum seropedicae	*Oryza sativa*	Nitrogenase	Nitrogen fixation	Roncato-Maccari et al. (2003)
Paenibacillus strain P22	*Populus*	Nitrogenase	Nitrogen fixation	Scherling et al. (2009)
Gluconacetobacter diazotrophicus	*Saccharum officinarum* Pine	Nitrogenase	Nitrogen fixation	Carrell and Frank (2014)
Bradyrhizobium; Pelomonas; Bacillus sp.	*Ipomoea batatas* L.	Nitrogenase	Nitrogen fixation	Terakado-Tonooka et al. (2008)
Streptomyces sp. GMKU3100	*Oryza sativa*	Siderophore	Iron uptake	Rungin et al. (2012)
Paenibacillus polymyxa P2b-2R	*Zea mays*; Pine	Nitrogenase	Nitrogen fixation	Puri et al. (2016)
Burkholderia phytofirmans PsJN	*Arabidopsis*	Ferritin Siderophore	Iron storage	Zhao et al. (2016)
Bacillus subtilis SHHP2-1	*Triticum aestivum*	Phosphatase; Nitrogenase	Phosphate solubilization; Nitrogen fixation	Yousaf et al. (2017)
(B) Phytohormones				
Azospirillum lipoferum	*Z. mays*	Gibberellins; Abscisic acid	Combat drought stress	Cohen et al. (2009)

Pseudomonas resinovorans; *Paenibacillus polymaxa*	*Gynura procumbens*	Cytokinin	Promote plant growth	Bhore et al. (2010)
Burkholderia phytofirmans	*Arabidopsis thaliana*	Auxin	Auxin regulation	Zúñiga et al. (2013)
Bacillus amyloliquefaciens RWL-1	*Oryza sativa*	Gibberellins	Phytohormone regulation	Shahzad et al. (2016)
(C) Combating abiotic stress				
Arthrobacter sp.; *Bacillus* sp.	*Capsicum annuum* L.	Proline accumulation	Osmotic stress tolerance	Sziderics et al. (2007)
Burkholderia phytofirmans PsJN	*O. sativa*	ACC deaminase	Stress reduction	Sun et al. (2009)
Pseudomonas vancouverensis OB155; *Pseudomonas frederiksbergensis* OS261	*Solanum lycopersicum*	Genes *LeCBF1* and *LeCBF3* products	Cold stress tolerance	Subramanian et al. (2015)
B. subtilis strain B26	*Brachypodium distachyon*	Increases solubility of carbohydrates	Drought tolerance	Gagné-Bourque et al. (2015)
Mesorhizobium ciceri IC53; *B. subtilis* NUU4	*Cicer arietinum*	Proline accumulation	Salt stress tolerance	Egamberdieva et al. (2010)
Bacillus flexus KLBMP 4941	*Limonium sinense*	ACC deaminase	Stress reduction	Wang et al. (2017)
(D) Combating biotic stress				
Burkholderia sp. KJ006	*Cannabis sativa*	N-acyl-homoserine lactonase	Pathogen quorum sensing disruption	Cho et al. (2007)
Pseudomonas sp. strain G	*Vitis vinifera*	Carbamoyl-phosphate synthase	Pathogen cell–cell signaling disruption	Newman et al. (2008)

(*Continued*)

TABLE 10.1 (Continued)

Endophytic Bacteria	Source	Bioactive compound	Beneficial role	References
Actinobacteria	*Arabidopsis*	Jasmonic acid; Ethylene	Against *Erwinia carotovora*	Conn et al. (2008)
B. subtilis BSn5	*Amorphophallus konjac*	APn5 protein production	Antibacterial	Dongmei et al. (2008)
Actinobacteria	*Arabidopsis*	Salicylic acid	Against *Fusarium oxysporum*	Conn et al. (2008)
Burkholderia phytofirmans PsJN	*A. thaliana*	N-acyl-homoserine lactone synthase	Quorum Sensing	Zúñiga et al. (2013)
Bacillus methylotrophicus	*Malus domestica*	Fengycin; Surfactin; Iturin A	Against *Fusarium oxysporum, Phytophthora* sp., *Dematophora necatrix, Sclerotium rolfsii* and *Pythium aphanidermatum*	Mehta et al. (2014)
Pseudomonas sp.; *Pantoea* sp.; *Bacillus* sp.	*Cannabis sativa* L.	Quenching QS signals	Inhibit *Chromobacterium violaceum*	Kusari et al. (2014)
Enterobacter aerogenes	*Z. mays*	2,3- butanediol (2,3-BD)	Against blight disease caused by the fungus, *Setosphaeria turcica*	D'Alessandro et al. (2014)
B. methylotrophicus CKAM	*Malus domestica*	Fengycin; Iturin A; Surfactin; Chitinase, protease and pectinase	Antifungal	Mehta et al. (2014)
Bacillus amyloliquefaciens	*Gossypium hirsutum*	Iturins	Against *Verticillium dahliae*	Han et al. (2015)

Bacillus amyloliquefaciens; B. subtilis	*Z. mays*	Lipopeptides	Inhibit *F. moniliforme*	Gond et al. (2015)
Pseudomonas poae strain RE*1-1-14	*Beta vulgaris*	Poaeamide	Inhibit fungal pathogen *Rhizoctonia solani*	Zachow et al. (2015)
Bacillus mojavensis BmB 4	*Bacopa monnieri*	Fengycin; Surfactin	Antibacterial	Jasim et al. (2016)
Rhizobium radiobacter F4	*Solanum lycopersicum*	Jasmonic acid	Against *Xanthomonas translucens* pv. *translucens* and *Pseudomonas syringae* pv.	Glaeser et al. (2016)
Bacillus thuringiensis strains	*O. sativa*	Chitinase	Inhibit *F. oxysporum, F. gramineum, S. rolfsii, Pyricularia grisea* and *Physalospora piricola*	Tang et al. (2012, 2017)

Nitrogen is vital for plant growth and development. It is required for the formation of amino acids, proteins, enzymes, and nucleic acids. Although 30%–50% of required N_2 in crop fields is fixed by N_2 fixing microbes of soil (Gourion et al., 2015), the endophytic nitrogen-fixing bacteria have been found more efficient than rhizospheric microbes in supporting the plants to flourish in nitrogen-restricted soil situations (Hurek and Reinhold-Hurek, 2003). The nitrogenase enzyme present in EB fixes the atmospheric nitrogen into the plant tissue. Along with the classic example of *Rhizobium* residing in root nodule of legumes, several other EB such as *Bradyrhizobium*, *Bacillus* sp. *Pelomonas*, and *Herbaspirillum seropedicae* have been accounted to actively synthesize nitrogenase enzyme (Terakado-Tonooka et al., 2008). The nitrogenase activity in *Paenibacillus* strain P22 has been accounted to contribute in the nitrogen supply to the tissues of the poplar tree (Scherling et al., 2009). Similarly, presence of *G. diazotrophicus* has been marked in tissues of pine and sugarcane (Carrell and Frank, 2014). Furthermore, N_2 fixing *Paenibacillus polymyxa* P2b-2R was isolated from pine and maize tissue (Puri et al., 2016).

Phosphorous is a macronutrient which is required for several enzymatic activities occurring during various physiological processes. It is a component of nucleic acids, ATP, signaling molecules, phytin, etc. A major portion (approx. 70%) of the phosphorus supplied through fertilizers bind to soil and form insoluble complexes and therefore, becomes inaccessible to plant species (Ezawa et al., 2002). The EB can solubilize such complexes by chelation, ion exchange or by acidification (Nautiyal et al., 2000). Moreover, phosphatase, secreted by EB, mineralizes the organic phosphorus into usable soluble form (van der Heijden et al., 2008). The phosphate uptake by EB was established by growing cacti with EB on phosphate. The bacteria-free cacti were used as control and the growth was compared with the fertilized cacti. The growth comparable to the fertilized plant was observed in former while the latter failed to thrive (Puente et al., 2009). This provided the direct evidence of phosphate solubilizing capability of EB.

Iron is one of the most important micronutrients required for the growth and development of plants. It is a constituent of cytochromes, ferredoxins, electron transport chain, and several enzymes as peroxidase, catalase, etc. Iron is required by plants for the fundamental metabolic pathways like chlorophyll biosynthesis, photosynthesis, and nitrogen fixation. However, its bioavailability in the soil is limited. To combat this, the EB produce high-affinity siderophores which chelate insoluble Fe^{3+} bounded to the soil into soluble complexes, which is readily absorbed by the plants (Ma et al., 2016). The siderophore's importance in rice plants by endophytic *Streptomyces* sp. GMKU3100 has been deduced by the use of siderophore-deficient mutant (Rungin et al., 2012). A positive correlation between siderophore production and increment in biomass of maize was established by Marques et al. (2010). Furthermore, the growth of tomato in hydroponic culture was documented to be assisted by siderophores produced by EB (Radzki et al., 2013). The iron uptake by siderophores provides an additional advantage of depriving iron availability and thereby limiting the multiplication of phytopathogens (Verma et al., 2011; Aznar et al., 2015).

The uptake of certain micronutrients is also assisted by EB. The cadmium uptake by *B. pumilus* E2S2 enhanced the roots and shoots biomass in contrast with the untreated sorghum plants (Luo et al., 2012). Furthermore, strain NBRI-SN13 of *B. amyloliquefaciens* actuated tricalcium phosphate solubilization in rice (Nautiyal et al., 2013).

10.2.2 Phytohormones

Phytohormones production is a general attribute of EB to improve plant biomass and combat stresses. The genes encoding auxin/indole acetic acid (IAA), cytokinins, gibberellins, and abscisic acid are frequently reported in the genome of plant EB (Shahzad et al., 2016).

Auxin is involved in root initiation, synthesis of ethylene, photoperiodism, stem elongation, cell–cell signaling and induction of plant defense systems (Woodward and Bartel, 2005; Gravel et al., 2007; Glick, 2012). Studies indicate that the EB regulate the level of auxin by synthesizing it in case of low concentration, or by destroying it in case of a high amount (Table 10.1B). It has been observed that the IAA producing EB is capable of improving plant root biomass by generating lateral roots (Tsavkelova et al., 2007). Direct evidence was provided by Patten and Glick (2002) using IAA mutant GR12-2 strain of *Pseudomonas putida*, which was incapable to augment lateral root formation. On the other hand, Leveau and Lindow (2005) demonstrated IAA degrading capability of *Pseudomonas putida* 1290 that totally nullified the impacts of supplemented IAA on the roots of radish.

In another examination, the mutant PsJN strain of *Burkholderia phytofirmans*, deficient in auxin, could not restrain the impacts of externally supplied auxin in *A. thaliana* contrasted with its native strain (Zúñiga et al., 2013). Thus, the auxin by EB possesses dual characteristics to regulate optimum auxin concentration so as to create a net constructive outcome on the host plant.

Several studies have shown EB residing in plant tissues can produce cytokinins and gibberellins (Table 10.1B). Cytokinin-like compounds were extracted from *Pseudomonas resinovorans* and *Paenibacillus polymaxa* inhabiting *Gynura procumbens* plant cotyledon (Bhore et al., 2010). While *Azospirillum lipoferum* was found to synthesize gibberellins and abscisic acid in maize plants (Cohen et al., 2009). Additionally, the EB has been shown to modulate ethylene concentration in plant tissue. The synthesis of gibberellins and abscisic acid along with the regulation of ethylene by EB has found to be imperative for plant stress mitigation. This has been discussed in detail in the proceeding section.

10.2.3 Combating stresses

Phytohormones, such as abscisic acid, ethylene, and gibberellins display a crucial role in plants to combat biotic and abiotic stresses. Concentration of these hormones fluctuates during stress condition, thereby affecting the physiological processes like cell elongation, root nodulation, root initiation, abscission, leaf senescence, ripening of fruit, and transport of auxin (Sun et al., 2016). EB aid in tolerating such stress responses by regulating hormones level (Table 10.1C). For instance, it has been observed that adverse environmental factors, such as soil salinization, desertification, extreme weather, global warming, etc., resulting in accumulation of ACC (1-aminocyclopropane-1-carboxylate) in plant tissue, which is ultimately converted to ethylene by ACC-oxidase (Tudela and Primo-Millo, 1992). Upregulated ethylene production further leads to decreased growth and cell death. The EB protect the plant from stress and nullify the elevated level of ethylene by synthesizing ACC deaminase enzyme. The enzyme hydrolyzes ACC to generate ammonia and α-ketobutyrate, which is further metabolized as a source of nitrogen (Sun et al., 2009). Thus, the synthesis of ethylene, the stress hormone, is reduced, thereby aiding the plant to alleviate stress. Several EB has been studied to possess the above mentioned enzyme (Rashid et al., 2012).

Furthermore, several EB has been isolated from the stress resistance plants. The role of EB in promoting plant growth by combating stresses has been evident in several studies. The psychrotolerant EB, *P. frederiksbergensis* strain OS261, and *Pseudomonas vancouverensis* strain OB155 were observed to protect tomato from cold stress by increasing antioxidant activity and inducing the bacterial cold acclimation genes (*LeCBF1* and *LeCBF3*) (Subramanian et al., 2015). Similarly, *Burkholderia phytofirmans* strain PsJN aided *Arabidopsis* to cope up with cold stress by strengthening their cell wall (Su et al., 2015). The transcriptomics analysis revealed the up-regulation of cellular homeostasis and reactive oxygen species (ROS) detoxification in *B. phytofirmans* PsJN that minimized the effects of drought stress in potato (Sheibani-Tezerji et al., 2015). A decline in the H_2O_2 concentrations and increase in the proline contents was observed in chickpea co-inoculated with *B. subtilis* NUU4 and *M. ciceri* IC53, thereby, combating the adverse effects of salt stress. Additionally, *B. subtilis* NUU4 was capable to inhibit *F. solani*, causative agent of root rot in chickpea (Egamberdieva et al., 2010). The *Brachypodium distachyon* endophytic promotes the carbohydrate solubilization which further stimulates drought tolerance capacity in grass (Gagné-Bourque et al., 2015). The osmotic stress in pepper plants (*Capsicum annuum* L.) was managed by proline synthesis by endophytic *Arthrobacter* sp. and *Bacillus* sp. (Sziderics et al., 2007). Besides, certain enzyme activities are elevated by EB that aid to combating abiotic stresses by plants. In response to high Na^+ concentration, certain *Bacillus* sp. has been reported to induce increment of catalase, peroxidase, phenylalanine lyase, and superoxide dismutase (Damodaran et al., 2014). Recently, *Gordonia terrae* KMP456-M40 isolate were found to enhance the root length of mangrove seedlings by 65% and the biomass of salt-stressed rice under axenic conditions by 62%. The bacterium was also able to enhance barley biomass under salt stress (Soldan et al., 2019). The studies presented *G. terrae* KMP456-M40 isolate as promising candidates for sustainable agricultural production in salt-affected soils. Thus, the EB sense changes in plant's physiology and accordingly alter gene expression and hormones concentration to adjust and thrive in the modified condition. However, the specific mechanisms in EB involved in promoting plant growth during stress still need to be elucidated.

10.2.4 Protection against phytopathogens

EB are capable of inhibiting bacterial, viral and fungal phytopathogens by synthesizing an array of inhibitory compounds like siderophores, volatile organic compounds (VOCs), toxins, hydrolytic enzymes, and antimicrobial peptides (Table 10.1D). The most common EB genera possessing the antimicrobial activity belongs to *Actinobacteria*, *Enterobacter*, *Bacillus*, *Paenibacillus*, *Serratia* and *Pseudomonas* (Liu et al., 2010a,b). Overall, there are three processes by which EB confer resistance against phytopathogens: (1) by induced systemic resistance (ISR); (2) by antimicrobial compound synthesis; and (3) by disrupting QS of phytopathogens.

The ISR, initiated by EB, could effectively protect against bacterial and fungal infections (Alvin et al., 2014). The EB, particularly, *Bacillus*, *Pseudomonas* and *Serratia*, trigger the ISR by salicylic acid (SA), jasmonic acid (JA), and/or ethylene (ET) mediated pathways (Pieterse et al., 2012). The EB in *Arabidopsis*, *Bacillus cereus* AR156 and *Enterobacter radicincitans* DSM 16656 are found to trigger ISR by both, SA and JA/ET mediated pathways (Niu et al., 2011; Brock et al., 2013). While inhibition of leaf pathogens *Pseudomonas syringae* pv. and *Xanthomonas translucens* pv. *translucens* in tomato DC3000 was induced by *R. radiobacter* F4 solely *via* the JA-mediated pathway (Glaeser et al., 2016). In a further study, *Actinobacteria* of *Arabidopsis* conferred resistance against phytopathogenic fungus *Fusarium oxysporum* and bacteria *Erwinia carotovora* and by JA/ET and SA mediated pathway, respectively (Conn et al., 2008). The *B. pumilus* inhibited phytopathogen *Phaeomoniella chlamydospora*, thereby imparting immunity in grapevine (Haidar et al., 2016). These studies presented EB as an effective agent to boost plant immunity.

The bioactive compounds released by EB to suppress that growth of phytopathogen indirectly promote the growth of host plant. Numerous antagonizing compounds against pathogen have been studied to be synthesized by EB. A fungicide isolated from EDR4 strain of *B. subtilis*, endophyte of wheat, was reported to restrain growth of *R. cerealis*, *G. graminis*, *F. graminearum*, *Macrophoma kuwatsukai*, *F. oxysporum* f. sp. *vasinfectum*, and *B. cinerea*

(Liu et al., 2010a,b). The iturin A, surfactin, and fengycin, isolated from apple tree inhabiting *B. methylotrophicus*, have shown to be effective against *Sclerotium rolfsii*, *Phytophthora* sp., *Pythium aphanidermatum*, *F. oxysporum*, and *Dematophora necatrix* (Mehta et al., 2014). The isoforms of iturins were isolated from cotton endophytic *Bacillus amyloliquefaciens* that triggered immunity against infection of fungus, *Verticillium dahliae* (Han et al., 2015). The endophytic *Enterobacter aerogenes*, residing in northern corn leaf, is reported to produce volatile 2, 3- butanediol (2, 3-BD), which inhibit *Setosphaeria turcica*, causative agent of blight disease (D'Alessandro et al., 2014). Growth of fungus, *F. moniliforme*, was also restricted by lipopeptides synthesized by *B. subtilis* and *B. amyloliquefaciens* in *Zea mays* (Gond et al., 2015). Poaeamide, a novel lipopeptide, was extracted from *Pseudomonas poae* which harbored sugar beetroot tissue. The extract was capable to restrain the growth of *Rhizoctonia solani* (Zachow et al., 2015). The EB *Bacillus megaterium* BP17 and *Curtobacterium luteum* TC10 imparted resistance to black pepper plant against *Radopholus similis* Thorne, a nematode (Aravind et al., 2009). Considering the effectiveness of EB, the large-scale industrial production and usage of *Pantoea vagans* C9-1 against fire blight is in practice (Smits et al., 2011).

Some EB releases extracellular hydrolytic enzymes that cleave the membrane and disrupt the activity of plant pathogens. *B. thuringiensis* have be reported to produce chitinase which inhibit various phytopathogenic fungi, such as *S. rolfsii*, *F. oxysporum*, *Physalospora piricola*, *Pyricularia grisea*, and *F. gramineum* (Tang et al., 2012, 2017). Other hydrolytic enzymes, such as proteases, chitosanases, and glucanases inhibit the population of phytopathogenic microbes (Mehta et al., 2014). However, the application of such antimicrobial compounds from EB in field condition is still in its infancy.

QS plays a critical role in establishing biofilm formation, cell-to-cell communication, adaptation, and multiplication of phytopathogens (Miller and Bassler, 2001). Recently, some EB has been studied to utilize QS quenching technique to prevent the colonization of pathogenic microbes. The bacteria residing in tissues of *Cannabis sativa* L. is reported to interrupt cell-to-cell communication of the pathogen *Chromobacterium violaceum* by quenching its QS signals (Kusari et al., 2014). *Bacillus* species possess the capability to produce quorum quenching enzymes, in particular N−acyl-homoserine lactonase (AiiA), which hydrolyze/modify and inactivate autoinducer compounds, typically N-acyl-homoserine lactones (AHLs). The hydrolyzed nonfunctional autoinducers could no longer participate in signaling and thus the communication between pathogens gets disrupted (Chen et al., 2013; Lopes, 2018). Additionally, species of *Bacillus* are capable to disrupt fatty acid signaling molecules and hence preventing QS-mediated virulence in *Xylella fastidiosa* and *Xanthomonas* species (Newman et al., 2008). However, molecular mechanism of QS quenching capability of EB to inhibit colonization of pathogen still needs to be explored.

10.3 Bioactive compounds of pharmacological importance

Globally numerous infectious diseases cause 50% (approx.) reported fatalities annually (Menpara and Chanda, 2013). With advances of the molecular, biochemical, and computational techniques, several nano to pico drugs have been generated. However, the natural sources are still considered as the best resource for drug formulation. Plants with medicinal properties are being effectively used as an alternative and/or a complementary medicine for the treatment of various diseases. Most of the pharmacological and toxicological drugs obtain from such plants are observed to be produced from endophytes inhabiting the plant tissues. The array of novel secondary metabolites, generated by endophytes, have been found to possess therapeutic properties as antimicrobial, antidiabetic, antitumor, antiarthritic, and immunosuppressors (Godstime et al., 2014). The bioactive compounds produced from EB along with their roles in controlling diseases have been depicted in Table 10.2.

10.3.1 Antimicrobial compounds

Various antimicrobial compounds isolated from EB, belong to different chemical groups like alkaloids, peptides, steroids, quinines, terpenoids, phenols, and flavonoids (Table 10.2A). The antimicrobial peptides can additionally be used as food preservatives by controlling growth of pathogens causing food spoilage and thus, curbing food-associated diseases (Liu et al., 2008). The lipopeptides, as already discussed in Section 10.2.3, play a crucial role as antibiotic for enhancing plant immunity. As an antibiotic, they display antimicrobial, cytotoxic, and surfactant activities (Stein, 2005; Raaijmakers et al., 2010). Among the EB, lipopeptides from *Bacillus* sp. and *Paenibacillus* sp. are well characterized (Villarreal-Delgado et al., 2018). Recently, *Paenibacillus kribbensis* and *B. subtilis* isolated from *Taxus brevifolia* (Pacific yew) and *Ginkgo biloba* L., respectively, were observed to possess antibacterial activities. They were found effective against five foodborne pathogenic bacteria. Scanning electron microscopic analysis revealed ruptured and lysed cells of all the assayed foodborne pathogens, suggesting that metabolite(s) of EB could have penetrated the cell membrane and caused cell lysis, thereby leading to cell death (Islam et al., 2018, 2019).

Several studies have presented the endophytic *Streptomyces* sp. as a storehouse of numerous antimicrobial peptides. Echinomycin and Kakadumycin A are novel antibiotics possessing a notable antimalarian activity and a potential anticancer drug (Waring and Wakelin, 1974). These antibiotics were isolated from fern-leaved grevillea (*Grevillea pteridifolia*) endophytic *Streptomyces* sp. (Castillo et al., 2003). The "Coronamycin," is yet another novel peptide antibiotic obtained from *Streptomyces* sp. isolated from *Monstera* sp. It was found to be active against the human fungal pathogen *Cryptococcus neoformans*, along with the malaria parasite, *Plasmodium*

TABLE 10.2 Bioactive compounds from pharmacologically beneficial endophytic bacteria.

Endophytic Bacteria	Source	Bioactive compound	Beneficial role	References
(A) Antimicrobial peptides				
Pseudomonas syringae	*Gramineae*	Pseudomycin	Antifungal	Harrison et al. (1991)
Pseudomonas viridiflava	*Gramineae*	Ecomycins	Inhibit *Cryptococcus neoformans* and *Candida albicans*	Harrison et al. (1991)
Streptomyces sp.	*Grevillea pteridifolia*	Kakadumycin A; Echinomycin	Antimalaria	Castillo et al. (2003)
Streptomyces sp.	*Monstera* sp.	Coronamycin	Antimalaria; Antifungal	Ezra et al. (2004)
Streptomyces sp. NRRL 30562	*Kennedia nigriscans*	Munumbicins	Antimalaria	Castillo et al. (2006)
Streptomyces sp. strain GT2002/1503	*Bruguiera gymnorrhiza*	Xiamycin-A	Anti-HIV	Ding et al. (2010)
Streptomyces sp.	-	Sansanmycin A	Antituberculosis	Xie et al. (2010)
Bacillus sp.; *Pseudomonas* sp.	*Plectranthus tenuiflorus*	Extracellular proteolytic enzymes	Antibacterial	El-Deeb et al. (2013)
Jishengella endophytica	*Rhizophora mangle*	Alkaloids	Antiviral (H1N1)	Wang et al. (2014)
Bacillus sp.; *Paenibacillus* sp.	Cereal crops	Lipopetides	Antimicrobial	Villarreal-Delgado et al. (2018)
(B) Antioxidants				
Paenibacillus polymyxa	*Stemona japonica*	Exopolysaccharides	Scavenging activities on superoxide and hydroxyl radicals	Liu et al. (2009)
Macrococcus caseolyticus	*Aloe vera*	1,1-diphenyl- 2-picrylhydrazyl	Scavenging properties	Akinsanya et al. (2015)

Bacillus mycoides M31; *Citrobacter youngae* MEB5; *Raoultella ornithinolytica* MEB11	*Rubia cordifolia*	Nonribosomal peptide synthetase	Antioxidant	Nongkhlaw and Joshi (2015)
Serratia marcescens cenA; *Bacillus subtilis* cenB	*Centella asiatica*	Nonribosomal peptide synthetase	Antioxidant	Nongkhlaw and Joshi (2015)
Pseudomonas palleroniana Y1	*Acmella oleracea*	Nonribosomal peptide synthetase	Antioxidant	Nongkhlaw and Joshi (2015)
Microbispora sp. LGMB259	*Vochysi adivergens*	β carboline	Antioxidant; Antibacterial; Antitumor	Savi et al. (2015)
Streptomyces flavoviridis A3WK	*Ocimum basillicum*	1H-pyrazole-3-carboxylic acid-5-methyl	Antioxidant	Khanam and Vootla (2018)
(C) Anticancer				
Streptomyces sp.	*Grevillea pteridifolia*	Kakadumycin A; Echinomycin	Anticancer	Waring and Wakelin (1974)
Streptomyces alnumycin	*Alnus glutinosa*	Alnumycin	Inhibited the growth of K562 human leukemia cells	Bieber et al. (1998)
Streptomyces sp. strain Is9131	*Maytenus hookeri Putterlickia verrucosa; P. retrospinosa plants*	Macrolides	Inhibited human SGC7901 gastric, HL60 leukemia, BEL7402 liver, and A- 549 lung tumor cell lines growth	Lu and Shen (2003); Zhao et al. (2005); Kusari et al. (2014)
Streptomyces laceyi strain MS53	*Ricinus communis*	Salaceyins	Cytotoxic against the human breast cancer cell line SKBR3	Kim et al. (2006)
Streptomyces hygroscopicous		Pterocidin	Antitumor	Igarashi et al. (2006)
Micromonospora sp.	*Lupinus angustifolius*	Anthraquinones	Inhibited the invasion of murine colon 26-L5 carcinoma cells	Igarashi et al. (2007); Qin et al. (2011)
Bacillus amyloliquefaciens	*Ophiopogon japonicus*	Exopolysaccharides	Antitumor activity against the human gastric carcinoma cell lines MC-4 and SGC-7901	Chen et al. (2013)

(Continued)

TABLE 10.2 (Continued)

Endophytic Bacteria	Source	Bioactive compound	Beneficial role	References
Streptomyces sp. strain BO-07	*Boesenbergia rotunda*	Biphenyls	Anticancer activity against human HepG2 and Huh7 liver, and HeLa cervical tumor cell lines	Taechowisan et al. (2017)
Streptomyces flavoviridis A3WK	*Ocimum basillicum*	Aziridine, 1-methyl; 1Methyl-3-nitro-5[4-nitropyrazole-1-yl]	Antitumor	Khanam and Vootla (2018)
Streptomyces cavourensis strain YBQ59	*Cinnamomum cassia*	Bafilomycin D; Nonactic acid; 5,11-epoxy-10-cadinano	Inhibit human lung adenocarcinoma EGFR-TKI-resistant cells A549 and H1299 growth	Vu et al. (2018)
(D) Anti-inflammatory				
Streptomyces aureofaciens CMUAc130	*Zingiber officinale*	5,7,4′-trimethoxy-4-phenylcoumarin; 5,7-dimethoxy-4-phenylcoumarin	Inhibit the expression of iNOS and COX-2 protein	Taechowisan et al. (2007)
Streptomyces sp. LJK109,	*Alpinia galanga*	3-methylcarbazoles	Reduce the production of inflammatory mediators	Taechowisan et al. (2012)
Streptomyces flavoviridis A3WK	*O. basillicum*	n-Hexadecanoic acid	Antiinflammatory	Khanam and Vootla (2018)

falciparum (Ezra et al., 2004). Furthermore, *Streptomyces* sp. GT2002/1503, inhabiting mangrove (*Bruguiera gymnorrhiza*) plant was found to produce the novel pentacyclicindolosesquiterpene, named as Xiamycin-A which exhibited selective anti-HIV activity (Ding et al., 2010). Sansanmycin A isolated from *Streptomyces* sp. were observed effective against *Mycobacterium tuberculosis* H37Rv (Xie et al., 2010). Other antibiotics such as munumbicins, valinomycin, rapamycin, hypericin, and clethramycin have also been isolated from *Streptomyces* sp. which exhibited antimicrobial activities against pathogens like *Escherichia coli*, *Vibrio cholera*, Coronavirus, *Shigella* sp. etc. (Gouda et al., 2016). Among them, munumbicins display an impressive antimicrobial activity. The methicillin resistant *S. aureus* strain, vancomycin insensitive *E. faecalis* strain and multidrug resistant (MDR) strain of *M. tuberculosis* have been found sensitive to munumbicins (Castillo et al., 2002). The antimalarial action of munumbicins has been accounted for to be two-fold more than that of chloroquine (Castillo et al., 2006).

The pseudomycins represent a group of antimycotics peptides which are isolated from broth of endophytic *P. syringae* culture. These mycopeptides are effective against a variety of human and plant pathogenic fungi (Harrison et al., 1991). The endophytic bacterium, *Pseudomonas viridiflava*, residing in grass tissue is known to produce Ecomycins (Miller et al., 1998), which inhibit the human pathogens *Candida albicans* and *Cryptococcus neoformans* (Harrison et al., 1991). Significant antimicrobial activities against a group of human pathogens (*E. coli*, *Staphylococcus aureus*, *Streptococcus agalactiae*, *Klebsiella pneumoniae*, *Candida albicans* and *Proteus mirabilis*) were observed by crude extracts of the *Bacillus* sp. and *Pseudomonas* sp. isolated from roots of medicinal plant, *Plectranthus tenuiflorus* (El-Deeb et al., 2013). In recent studies, Malaysian plants-endophytic *B. subtilis* was observed to efficiently inhibit *S. aureus and Pseudomonas aeruginosa* (Fikri et al., 2018).

10.3.2 Antioxidants

The antioxidants are the free radicals scavenger which effectively nullify the damages caused by ROSs, which contribute to adverse effects like cellular degeneration, DNA damages, and carcinogenesis (Huang et al., 2007). Antioxidants have been regarded as an effective therapy for ROS-associated pathological conditions like neurodegenerative diseases (Parkinson's and Alzheimer's), cancer, aging, hypertension, etc. (Valko et al., 2007). Various antioxidants released from EB (Table 10.2B) exhibits anitimicrobial, anticarcinogenic, antiinflammatory, or antimutagenic activities (Owen et al., 2000; Sala et al., 2002; Cozma, 2004). The *Stemona japonica* endophyte *Paenibacillus polymyxa* produced "exopolysaccharides" which displayed remarkable scavenging activities on free radicals (Liu et al., 2009). The curative and therapeutic use of *Aloe vera* is attributed to free radical scavenging compounds produced by endophytic Firmicute, *Macrococcus caseolyticus*. A major portion (approx. 80%) of the EB residing *A. vera* were found to synthesize 1,1-diphenyl- 2-picrylhydrazyl, which exhibited almost 75% of the cellular scavenging activities (Akinsanya et al., 2015).

10.3.3 Anticancer

Endophyte extracts are better choice against chemotherapeutic agents due to their less toxicity on non-tumor cells and better activity against drug-resistant microbes. Owing to a better therapeutic efficacy with low side-effects, the natural endophyte-derived metabolites are being considered as the preferable human anticancer drugs (Cardoso-Filho, 2018). The endophytic novel secondary metabolites have shown promising results against the cancerous transformed cell lines in several in vitro studies (Table 10.2C). The salaceyins, obtained from *Streptomyces laceyi* MS53, an endophyte of *Ricinus communis*, possessed the cytotoxic property against SKBR3 continuous cell line derived from the human breast (Kim et al., 2006). The anthraquinones synthesized from *Lupinus angustifolius*-endophytic *Micromonospora* sp. remarkably prevented the growth of 26-L5 carcinoma cells cultured from murine colon (Qin et al., 2011). The *B. amyloliquefaciens* sp., isolated from *Ophiopogon japonicas*, produced exopolysaccharides which exhibited anticarcinogenic activity against SGC-7901 and MC-4 carcinoma cell lines obtained from gastric cells of human (Chen et al., 2013). *Streptomyces* sp. strain Is9131 from *Maytenus hookeri* were reported to inhibit several human tumor cell lines, namely, HL60 leukemia, A- 549 lung, BEL7402 liver and SGC7901 gastric (Zhao et al., 2005). Furthermore, *Streptomyces alnumycin* from *Alnus glutinosa* inhibited the propagation of human K562 leukemia cell line (Bieber et al., 1998). The anticancer activity of biphenyls obtained from *Streptomyces* sp. BO-07 of *Boesenbergia rotunda* was studied by Taechowisan et al. (2017) in human Huh7 and HepG2 liver, and HeLa cell lines. Recently, *Streptomyces cavourensis* YBQ59, endophyte of *Cinnamomum cassia*, was observed to prevent the multiplication of human lung adenocarcinoma cell lines (Vu et al., 2018). The research related to characterization and effect of endophytic bacterial antitumor metabolites in animal models is in progress. Once established, they could prove to be a potential drug for effective, low cost, and safer treatment for cancer.

10.3.4 Anti-inflammatory

The anti-inflammatory properties of endophytic *Streptomyces* sp. have been studied (Table 10.2D). The major antiinflammatory compounds were found to suppress the mediators of inflammatory responses, such as TNF-α, PGE2, NO, IL-6, IL-10, and IL-1β. The in vitro antiinflammatory activity of 5,7-dimethoxy-4-phenylcoumarin and 5,7,4′-trimethoxy-4-phenylcoumarin, obtained from *Z. officinale* endophytic *Streptomyces aureofaciens* CMUAc130, was estimated in lab condition. The results indicated the suppression of iNOS and COX-2 proteins, which further reduced the synthesis of TNF-α (Taechowisan et al., 2007). In another study, 3-methylcarbazoles, obtained from *Streptomyces* sp. LJK109, endophyte of *Alpinia galanga*, was shown to reduce the production of inflammatory mediators (Taechowisan et al., 2012). Recently, anti-inflammatory activity of *Streptomyces flavoviridis* A3WK extract was found significantly more compared to that of its host plant *Ocimum basillicum*, thereby presenting endophyte as a better anti-inflammatory agent (Khanam and Vootla, 2018).

10.4 Conclusion

EB have demonstrated to be wealthy storehouse of diverse secondary metabolites with wide range applications in agricultural and pharmaceutical arenas. The excellent novel metabolites from EB not only accelerate plant growth by employing different direct and indirect mechanisms, but also assist the plant to thrive by nullifying the effects of environmental stresses. The implementation of such beneficial bacteria in agriculture will certainly ensure the reduced usage of environmentally hazardous chemical fertilizers, pesticides, and other agrochemicals. This biopractice will pave the pathway for productive, sustainable, and environmentally friendly agriculture systems. Additionally, the EB can considerably be valuable in the biomedical field for the development of anti-infection agents, which will be vital in the treatment of human ailments. Such secondary metabolites are evidently cheaper, effective, and safer as compared to chemical drugs. Overall, the novel bioactive compounds from EB could be of huge assistance in addressing the difficulties of agriculture, medicine, and environment.

10.5 Future perspectives

Although many EB have been identified, a large group of them still remain unexplored. The possibilities of discovering unexplored distinct bacteria in the medicinal plants are enormous. The modern technologies as metagenomics and next-generation sequencing could explore the numerous pools of antimicrobials secreted by yet uncultivated EB. Moreover, a brief examination on a particular population in a host is imperative before mass production of bioactive compounds. Furthermore, a better insight of complex dynamics that define the plant-endophyte relationship is imperative to achieve consistent results under field conditions. The factors affecting relationship of plant-endophyte should be thoroughly assessed at the molecular level using genomic, transcriptomics, and metabolomics tools. Future investigations should concentrate on unraveling the metabolic pathways for the synthesis of various significant endophytic bioactive compounds.

References

Akinsanya, M.A., Goh, J.K., Lim, S.P., et al., 2015. Diversity, antimicrobial and antioxidant activities of culturable bacterial endophyte communities in *Aloe vera*. FEMS Microbiol. Lett. 362, fnv184.

Alvin, A., Miller, K.I., Neilan, B.A., 2014. Exploring the potential of endophytes from medicinal plants as sources of antimycobacterial compounds. Microbiol. Res. 169, 483–495.

Aravind, R., Kumar, A., Eapen, S., et al., 2009. Endophytic bacterial flora in root and stem tissues of black pepper (*Piper nigrum* L.) genotype: isolation, identification and evaluation against *Phytophthora capsici*. Lett. Appl. Microbiol. 48, 58–64.

Aznar, A., Chen, N.W., Thomine, S., et al., 2015. Immunity to plant pathogens and iron homeostasis. Plant Sci. 240, 90–97.

Bhore, S.J., Ravichantar, N., Loh, C.Y., 2010. Screening of endophytic bacteria isolated from leaves of Sambung Nyawa [*Gynura procumbens* (Lour.) Merr.] for cytokinin-like compounds. Bioinformation 5, 191–197.

Bieber, B., Nüske, J., Ritzau, M., et al., 1998. Alnumycin a new naphthoquinone antibiotic produced by an endophytic *Streptomyces* sp. J. Antibiot. 51, 381–382.

Böhm, M., Hurek, T., Reinhold-Hurek, B., 2007. Twitching motility is essential for endophytic rice colonization by the N_2-fixing endophyte *Azoarcus* sp. strain. Mol. Plant Microbe Interact. 20, 526–BH533.

Brock, A.K., Berger, B., Mewis, I., et al., 2013. Impact of the PGPB *Enterobacter radicincitans* DSM 16656 on growth, glucosinolate profile, and immune responses of *Arabidopsis thaliana*. Microb. Ecol. 65, 661–670.

Carrell, A.A., Frank, A.C., 2014. *Pinus flexilis* and *Picea engelmannii* share a simple and consistent needle endophyte microbiota with a potential role in nitrogen fixation. Front. Microbiol. 5, 333.

Castillo, U.F., Strobel, G.A., Ford, E.J., et al., 2002. Munumbicins, wide-spectrum antibiotics produced by *Streptomyces* NRRL 30562, endophytic on *Kennediani griscans*. Microbiology 148, 2675–2685.

Castillo, U., Harper, J.K., Strobel, G.A., 2003. Kakadumycins, novel antibiotics from *Streptomyces* sp. NRRL 30566, an endophyte of *Grevillea pteridifolia*. FEMS Microbiol. Lett. 224, 183–190.

Castillo, U.F., Strobel, G.A., Mullenberg, K., et al., 2006. Munumbicins E-4 and E-5: novel broad- spectrum antibiotics from *Streptomyces* NRRL3052. FEMS Microbiol. Lett. 255, 296–300.

Chen, F., Gao, Y., Chen, X., et al., 2013. Quorum quenching enzymes and their application in degrading signal molecules to block quorum sensing-dependent infection. Int. J. Mol. Sci. 14, 17477–17500.

Cho, H.S., Park, S.Y., Ryu, C.M., et al., 2007. Interference of quorum sensing and virulence of the rice pathogen *Burkholderia glumae* by an engineered endophytic bacterium. FEMS Microbiol. Ecol. 60, 14–23.

Cohen, A.C., Travaglia, C.N., Bottini, R., et al., 2009. Participation of abscisic acid and gibberellins produced by endophytic *Azospirillum* in the alleviation of drought effects in maize. Botany 87, 455–462.

Compant, S., Clément, C., Sessitsch, A., 2010. Plant growth-promoting bacteria in the rhizo-and endosphere of plants: their role, colonization, mechanisms involved and prospects for utilization. Soil Biol. Biochem. 42, 669–678.

Conn, V.M., Walker, A., Franco, C., 2008. Endophytic actinobacteria induce defense pathways in *Arabidopsis thaliana*. Mol. Plant Microbe Interact. 21, 208–218.

Cozma, L.S., 2004. The role of antioxidant therapy in cardiovascular disease. Curr. Opin. Lipidol. 15, 369–371.

D'Alessandro, M., Erb, M., Ton, J., et al., 2014. Volatiles produced by soil-borne endophytic bacteria increase plant pathogen resistance and affect tritrophic interactions. Plant Cell Environ. 37, 813–826.

Damodaran, T., Rai, R.B., Jha, S.K., et al., 2014. Rhizosphere and endophytic bacteria for induction of salt tolerance in gladiolus grown in sodic soils. J. Plant Interact. 9, 577–584.

de Weert, S., Vermeiren, H., Mulders, I.H., et al., 2002. Flagella-driven chemotaxis towards exudate components is an important trait for tomato root colonization by *Pseudomonas fluorescens*. Mol. Plant Microbe Interact. 15, 1173–1180.

Ding, L., Münch, J., Goerls, H., et al., 2010. Xiamycin, a pentacyclicindolosesquiterpene with selective anti-HIV activity from a bacterial mangrove endophyte. Bioorg. Med. Chem. Lett. 20, 6685–6687.

Dongmei, H., Huifang, B., Ziquan, Y., Ziniu, Y., Ming, S., 2008. Identification and characterization of novel antimicrobial protein APn5 against *Erwinia carotovora*. Wei Sheng Wu Xue Bao 48, 1192–1197.

Edwards, J., Johnson, C., Santos-Medellín, C., et al., 2015. Structure, variation, and assembly of the root-associated microbiome of rice. Proc. Natl. Acad. Sci. U.S.A. 112, E911–E920.

Egamberdieva, D., Wirth, S.J., Shurigin, V.V., et al., 2010. Endophytic bacteria improve plant growth, symbiotic performance of chickpea (*Cicer arietinum* L.) and induce suppression of root rot caused by *Fusarium solani* under salt stress. Front. Microbiol. 8, 1887.

El-Deeb, B., Fayez, K., Gherbawy, Y., 2013. Isolation and characterization of endophytic bacteria from *Plectranthus tenuiflorus* medicinal plant in Saudi Arabia desert and their antimicrobial activities. J. Plant Interact. 8, 56–64.

Ezawa, T., Smith, S.E., Smith, F.A., 2002. P metabolism and transport in AM fungi. Plant Soil 244, 221–230.

Ezra, D., Castillo, U.F., Strobel, G.A., et al., 2004. Coronamycins, peptide antibiotics produced by a verticillate *Streptomyces* sp. (MSU-2110) endophytic on *Monstera* sp. Microbiology 150, 785–793.

Ferreira, A., Quecine, M.C., Lacava, P.T., et al., 2008. Diversity of endophytic bacteria from *Eucalyptus* species seeds and colonization of seedlings by *Pantoea agglomerans*. FEMS Microbiol. Lett. 287, 8–14.

Fikri, A.S.I., Rahman, I.A., Nor, N.S.M., et al., 2018. Isolation and identification of local bacteria endophyte and screening of its antimicrobial property against pathogenic bacteria and fungi, vol. 1940. In: Proceedings of AIP Conference, AIP Publishing, Melville, NY, pp. 020072.

Gagné-Bourque, F., Mayer, B.B.F., Charron, J.B.B., et al., 2015. Accelerated growth rate and increased drought stress resilience of the model grass *Brachypodium distachyon* colonized by *Bacillus subtilis* B26. PLoS One 10, e130456.

Gamalero, E., Lingua, G., Berta, G., et al., 2003. Methods of studying root colonization by introduced beneficial bacteria. Agronomy 23, 407–418.

Girsowicz, R., Moroenyane, I., Steinberger, Y., 2019. Bacterial seed endophyte community of annual plants modulated by plant photosynthetic pathways. Microbiol. Res. 223-225, 58–62.

Glaeser, S.P., Imani, J., Alabid, I., et al., 2016. Non-pathogenic *Rhizobium radiobacter* F4 deploys plant beneficial activity independent of its host *Piriformospora indica*. ISME J. 10, 871–884.

Glick, B.R., 2012. Plant growth-promoting bacteria: mechanisms and applications. Scientifica 2012.

Godstime, O.C., Enwa, F.O., Augustina, J.O., et al., 2014. Mechanisms of antimicrobial actions of phytochemicals against enteric pathogens-a review. J. Pharm. Chem. Biol. Sci. 2, 77–85.

Gond, S.K., Bergen, M.S., Torres, M.S., 2015. Endophytic *Bacillus* spp. produce antifungal lipopeptides and induce host defence gene expression in maize. Microbiol. Res. 172, 79–87.

Gouda, S., Das, G., Sen, S.K., et al., 2016. Endophytes: a treasure house of bioactive compounds of medicinal importance. Front. Microbiol. 7, 1538.

Gourion, B., Berrabah, F., Ratet, P., et al., 2015. Rhizobium-legume symbioses: the crucial role of plant immunity. Trends Plant Sci. 20, 186–194.

Gravel, V., Antoun, H., Tweddell, R.J., 2007. Growth stimulation and fruit yield improvement of greenhouse tomato plants by inoculation with *Pseudomonas putida* or *Trichoderma atroviride*: possible role of indole acetic acid (IAA). Soil Biol. Biochem. 39, 1968–1977.

Gupta, A., Verma, H., Singh, P.P., Singh, P., Singh, M., Mishra, V., et al., 2019. Rhizome endophytes: roles and applications in sustainable agriculture. Seed Endophytes. Springer, Cham, pp. 405–421. Available from: http://doi.org/10.1007/978-3-030-10504-4_19.

Gupta, A., Singh, S.K., Singh, V.K., Singh, M.K., Modi, A., Zhimo, V.Y., et al., 2020. Endophytic microbe approaches in bioremediation of organic pollutants. In: Kumar, A., Kumar Singh, V. (Eds.), Microbial In: Microbial Endophytes: Prospects of Sustainable Agriculture. Elsevier, pp. 157–174. Available from: https://doi.org/10.1016/B978-0-12-818734-0.00007-3.

Haidar, R., Roudet, J., Bonnard, O., et al., 2016. Screening and modes of action of antagonistic bacteria to control the fungal pathogen *Phaeomoniella chlamydospora* involved in grapevine trunk diseases. Microbiol. Res. 192, 172–184.

Han, Q., Wu, F., Wang, X., et al., 2015. The bacterial lipopeptide iturins induce *Verticillium dahliae* cell death by affecting fungal signalling pathways and mediate plant defence responses involved in pathogen associated molecular pattern-triggered immunity. Environ. Microbiol. 17, 1166–1188.

Hardoim, P.R., van Overbeek, L.S., van Elsas, J.D., 2008. Properties of bacterial endophytes and their proposed role in plant growth. Trends Microbiol. 16, 463–471.

Harrison, L.H., Teplow, D.B., Rinaldi, M., et al., 1991. Pseudomycins, a family of novel peptides from *Pseudomonas syringae* possessing broad-spectrum antifungal activity. J. Gen. Microbiol. 137, 2857–2865.

Huang, W.Y., Cai, Y.Z., Xing, J., et al., 2007. A potential antioxidant resource: endophytic fungi from medicinal plants. Economic Bot 61, 14–30.

Hurek, T., Reinhold-Hurek, B., 2003. *Azoarcus* sp. strain BH72 as a model for nitrogen fixing grass endophytes. J. Biotechnol. 106, 169–178.

Igarashi, Y., Miura, S.S., Fujita, T., et al., 2006. Pterocidin, a cytotoxic compound from the endophytic *Streptomyces hygroscopicus*. J. Antibiot. 59, 193–195.

Igarashi, Y., Trujillo, M.E., Martínez-Molina, E., et al., 2007. Antitumor anthraquinones from an endophytic actinomycete *Micromonospora lupini* sp. nov. Bioorg. Med. Chem. Lett. 17, 3702–3705.

Islam, M.D., Choi, J., Baek, K.H., 2018. Antibacterial activities of endophytic bacteria isolated from *Taxus brevifolia* against foodborne pathogenic bacteria. Foodborne Pathog. Dis. 15, 269–276.

Islam, M.D., Choi, J., Baek, K.H., 2019. Control of foodborne pathogenic bacteria by endophytic bacteria isolated from *Ginkgo biloba* L. Foodborne Pathog. Dis. 16, 1–10.

Jasim, B., Sreelakshmi, S., Mathew, J., Radhakrishnan, E.K., 2016. Identification of endophytic *Bacillus mojavensis* with highly specialized broad spectrum antibacterial activity. 3 Biotech. 6, 187.

Khanam, W., Vootla, S.K., 2018. Comparative study of *Ocimum basillicum* and its endophytic actinomycetes *Streptomyces flavoviridis* A3WK: evaluation of antioxidant, anti-inflammatory and antimicrobial activity. Int. J. Pharma. Sci. Res. 9, 1023–1034.

Kim, N., Shin, J.C., Kim, W., et al., 2006. Cytotoxic 6-alkylsalicylic acids from the endophytic *Streptomyces laceyi*. J. Antibiot. 59, 797–800.

Kumar, A., Singh, R., Yadav, A., Giri, D.D., Singh, P.K., Pandey, K.D., 2016. Isolation and characterization of bacterial endophytes of *Curcuma longa* L. 3 Biotech. 1 6 (1), 60.

Kumar, A., Droby, S., Singh, V.K., Singh, S.K., White, J.F., 2020. Entry, colonization, and distribution of endophytic microorganisms in plants. Microbial Endophytes. Woodhead Publishing, pp. 1–33. Available from: https://doi.org/10.1016/B978-0-12-819654-0.00001-6.

Kusari, P., Kusari, S., Lamshöft, M., et al., 2014. Quorum quenching is an antivirulence strategy employed by endophytic bacteria. Appl. Microbiol. Biotechnol. 98, 7173–7183.

Leveau, J.H., Lindow, S.E., 2005. Utilization of the plant hormone indole-3-acetic acid for growth by *Pseudomonas putida* strain 1290. Appl. Environ. Microbiol. 71, 2365–2371.

Liu, X., Dong, M., Chen, X., et al., 2008. Antimicrobial activity of an endophytic *Xylaria* sp. YX-28 and identification of its antimicrobial compound 7-amino- 4-methylcoumarin. Appl. Microbiol. Biot. 78, 241–247.

Liu, J., Luo, J., Ye, H., et al., 2009. Production, characterization and antioxidant activities in vitro of exopolysaccharides from endophytic bacterium *Paenibacillus polymyxa* EJS-3. Carbohydr. Polym. 78, 275–281.

Liu, B., Huang, L., Buchenauer, H., et al., 2010a. Isolation and partial characterization of an antifungal protein from the endophytic *Bacillus subtilis* strain EDR4. Pesticide Biochem. Physiol. 98, 305–311.

Liu, X., Jia, J., Atkinson, S., et al., 2010b. Biocontrol potential of an endophytic *Serratia* sp. G3 and its mode of action. World J. Microbiol. Biotechnol. 26, 1465–1471.

Liu, H., Carvalhais, L.C., Crawford, M., et al., 2017. Inner plant values: diversity, colonization and benefits from endophytic bacteria. Front. Microbiol. 8, 2552.

Lopes, R., 2018. Um Conto Endofítico: Silenciamento de Fitopatógenos Por Bactérias Benéficas Do Interior de Plantas. Novas Edições Acadêmicas, Beau Bassin-Rose Hill.

Luo, S., Xu, T., Chen, L., et al., 2012. Endophyte- assisted promotion of biomass production and metal-uptake of energy crop sweet sorghum by plant-growth-promoting endophyte *Bacillus* sp. SLS18. Appl. Microbiol. Biotechnol. 93, 1745–1753.

Lu, C., Shen, Y., 2003. A new macrolide antibiotic with antitumor activity produced by *Streptomyces* sp. CS, a commensal microbe of *Maytenus hookeri*. J. Antibiot 56, 415–418.

Ma, Y., Rajkumar, M., Zhang, C., et al., 2016. Beneficial role of bacterial endophytes in heavy metal phytoremediation. J. Environ. Manage 174, 14–25.

Marques, A.P., Pires, C., Moreira, H., et al., 2010. Assessment of the plant growth promotion abilities of six bacterial isolates using *Zea mays* as indicator plant. Soil. Biol. Biochem. 42, 1229–1235.

Mehta, P., Walia, A., Kakkar, N., et al., 2014. Tricalcium phosphate solubilisation by new endophyte *Bacillus methylotrophicus* CKAM isolated from apple root endosphere and its plant growth- promoting activities. Acta. Physiol. Plant 36, 2033–2045.

Menpara, D., Chanda, S., 2013. In: Méndez-Vilas, A. (Ed.), Endophytic Bacteria-Unexplored Reservoir of Antimicrobials for Combating Microbial Pathogens in Microbial Pathogens and

Strategies for Combating Them: Science, Technology and Education. Formatex Research Center), Badajoz, pp. 1095–1103.

Miller, M.B., Bassler, B.L., 2001. Quorum sensing in bacteria. Annu. Rev. Microbiol. 55, 165–199.

Miller, C.M., Miller, R.V., Garton-Kenny, D., et al., 1998. Ecomycins, unique antimycotics from *Pseudomonas viridiflava*. J. Appl. Microbiol. 84, 937–944.

Nautiyal, C.S., Bhadauria, S., Kumar, P., et al., 2000. Stress induced phosphate solubilization in bacteria isolated from alkaline soils. FEMS Microbiol. Lett. 182, 291–296.

Nautiyal, C.S., Srivastava, S., Chauhan, P.S., et al., 2013. Plant growth-promoting bacteria *Bacillus amyloliquefaciens* NBRISN13 modulates gene expression profile of leaf and rhizosphere community in rice during salt stress. Plant Physiol. Biochem. 66, 1–9.

Newman, K.L., Chatterjee, S., Ho, K.A., et al., 2008. Virulence of plant pathogenic bacteria attenuated by degradation of fatty acid cell-to-cell signaling factors. Mol. Plant Microbe Interact. 21, 326–334.

Niu, D.D., Liu, H.X., Jiang, C.H., et al., 2011. The plant growth–promoting rhizobacterium *Bacillus cereus* AR156 induces systemic resistance in *Arabidopsis thaliana* by simultaneously activating salicylate-and jasmonate/ ethylene-dependent signaling pathways. Mol. Plant Microbe Interact. 24, 533–542.

Nongkhlaw, F.M.W., Joshi, S.R., 2015. Investigation on the bioactivity of culturable endophytic and epiphytic bacteria associated with ethnomedicinal plants. J. Infect. Dev. Ctries. 9 (9), 954–961.

Owen, R.W., Giacosa, A., Hull, W.E., et al., 2000. The antioxidant/anticancer potential of phenolic compounds isolated from olive oil. Eur. J. Cancer 36, 1235–1247.

Patten, C.L., Glick, B.R., 2002. Role of *Pseudomonas putida* indoleacetic acid in development of the host plant root system. Appl. Environ. Microbiol. 68, 3795–3801.

Pieterse, C.M., van der Does, D., Zamioudis, C., et al., 2012. Hormonal modulation of plant immunity. Annu. Rev. Cell Dev. Biol. 28.

Puente, M.E., Li, C.Y., Bashan, Y., 2009. Endophytic bacteria in cacti seeds can improve the development of cactus seedlings. Environ. Exp. Bot. 66, 402–408.

Puri, A., Padda, K.P., Chanway, C.P., 2016. Seedling growth promotion and nitrogen fixation by a bacterial endophyte *Paenibacillus polymyxa* P2b-2R and its GFP derivative in corn in a long-term trial. Symbiosis 69, 123–129.

Qin, S., Xing, K., Jiang, J.H., et al., 2011. Biodiversity, bioactive natural products and biotechnological potential of plant-associated endophytic actinobacteria. Appl. Microbiol. Biotechnol. 89, 457–473.

Raaijmakers, J.M., de Bruijn, I., Nybroe, O., et al., 2010. Natural functions of lipopeptides from *Bacillus* and *Pseudomonas*: more than surfactants and antibiotics. FEMS Microbiol. Rev. 34, 1037–1062.

Radzki, W., Mañero, F.G., Algar, E., et al., 2013. Bacterial siderophores efficiently provide iron to iron-starved tomato plants in hydroponics culture. Anton Leeuw 104, 321–330.

Rashid, S., Charles, T.C., Glick, B.R., 2012. Isolation and characterization of new plant growth-promoting bacterial endophytes. Appl. Soil Ecol. 61, 217–224.

Reinhold-Hurek, B., Hurek, T., 1998. Life in grasses: diazotrophic endophytes. Trends Microbiol. 6, 139–144.

Roncato-Maccari, L.D.B., Ramos, H.J.O., Pedrosa, F.O., et al., 2003. Endophytic *Herbaspirillum seropedicae* expresses *nif* genes in gramineous plants. FEMS Microbiol. Ecol. 45, 39–47.

Rungin, S., Indananda, C., Suttiviriya, P., et al., 2012. Plant growth enhancing effects by a siderophore producing endophytic Streptomycete isolated from a Thai jasmine rice plant (*Oryza sativa* L. cv. KDML105). Anton Leeuw 102, 463–472.

Sala, A., Recio, M.C., Giner, R.M., et al., 2002. Antiinflammatory and antioxidant properties of *Helichrysum italicum*. J. Pharm. Pharmacol. 54, 365–371.

Santoyo, G., Moreno-Hagelsieb, G., Orozco-Mosqueda, M.D.C., et al., 2016. Plant growth-promoting bacterial endophytes. Microbiol. Res. 183, 92–99.

Sapers, G.M., Gorny, J.R., Yousef, A.E., 2005. Microbiology of Fruits and Vegetables. CRC Press.

Savi, D.C., Haminiuk, C.W.I., Sora, G.T.S., et al., 2015. Antitumor, antioxidant and antibacterial activities of secondary metabolites extracted by endophytic actinomycetes isolated from *Vochysi adivergens*. Int. J. Pharm. 5, 347–356.

Scherling, C., Ulrich, K., Ewald, D., et al., 2009. A metabolic signature of the beneficial interaction of the endophyte *Paenibacillus* sp. isolate and *in vitro*-grown poplar plants revealed by metabolomics. Mol. Plant Microbe Interact. 22, 1032–1037.

Sessitsch, A., Hardoim, P., Döring, J., et al., 2012. Functional characteristics of an endophyte community colonizing rice roots as revealed by metagenomic analysis. Mol. Plant Microbe Interact. 28–36.

Shahzad, R., Waqas, M., Khan, A.L., et al., 2016. Seed-borne endophytic *Bacillus amyloliquefaciens* RWL-1 produces gibberellins and regulates endogenous phytohormones of *Oryza sativa*. Plant Physiol. Biochem. 106, 236–243.

Sheibani-Tezerji, R., Rattei, T., Sessitsch, A., et al., 2015. Transcriptome profiling of the endophyte *Burkholderia phytofirmans* PsJN indicates sensing of the plant environment and drought stress. mBio 6, e00621-15.

Singh, R., Dubey, A.K., 2015. Endophytic actinomycetes as emerging source for therapeutic compounds. Indo Glob. J. Pharm. Sci. 5, 106–116.

Singh, M., Kumar, A., Singh, R., Pandey, K.D., 2017. Endophytic bacteria: a new source of bioactive compounds. 3 Biotech. 7 (5), 315.

Smits, T.H., Rezzonico, F., Kamber, T., et al., 2011. Metabolic versatility and antibacterial metabolite biosynthesis are distinguishing genomic features of the fire blight antagonist *Pantoea vagans* C9-1. PLoS One 6, e22247.

Soldan, R., Mapelli, F., Crotti, E., et al., 2019. Bacterial endophytes of mangrove propagules elicit early establishment of the natural host and promote growth of cereal crops under salt stress. Microbiol. Res. 223-225, 33–43.

Stein, T., 2005. *Bacillus subtilis* antibiotics: structures, syntheses and specific functions. Mol. Microbiol. 56, 845–857.

Su, F., Jacquard, C., Villaume, S., et al., 2015. *Burkholderia phytofirmans* PsJN reduces impact of freezing temperatures on photosynthesis in *Arabidopsis thaliana*. Front. Plant Sci. 6, 810.

Suárez-Moreno, Z.R., Devescovi, G., Myers, M., et al., 2010. Commonalities and differences in regulation of N-acyl homoserine lactone quorum sensing in the beneficial plant-associated *Burkholderia* species cluster. Appl. Environ. Microbiol. 76, 4302–4317.

Subramanian, P., Mageswari, A., Kim, K., et al., 2015. Psychrotolerant endophytic *Pseudomonas* sp. strains OB155 and OS261 induced chilling resistance in tomato plants (*Solanum lycopersicum* Mill.) by activation of their antioxidant capacity. Mol. Plant Microbe Interact. 28, 1073–1081.

Sun, Y., Cheng, Z., Glick, B.R., 2009. The presence of a 1-aminocyclopropane-1-carboxylate (ACC) deaminase deletion mutation alters the physiology of the endophytic plant growth-promoting bacterium *Burkholderia phytofirmans* PsJN. FEMS Microbiol. Lett. 296, 131–136.

Sun, L., Wang, X., Li, Y., 2016. Increased plant growth and copper uptake of host and non-host plants by metal-resistant and plant growth-promoting endophytic bacteria. Int. J. Phytoremed. 18, 494–501.

Sziderics, A.H., Rasche, F., Trognitz, F., et al., 2007. Bacterial endophytes contribute to abiotic stress adaptation in pepper plants (*Capsicum annuum* L.). Can. J. Microbiol. 53, 1195–1202.

Taechowisan, T., Lu, C., Shen, Y., et al., 2007. *In vitro* anti-inflammatory activity of 4-arylcoumarins from endophytic *Streptomyces aureofaciens* CMUAc130 in murine macrophage RAW 264.7 cells. Nat. Prod. Res. 21, 1104–1113.

Taechowisan, T., Chanaphat, S., Ruensamran, W., et al., 2012. Anti-inflammatory effect of 3-methylcarbazoles on RAW 264.7 cells stimulated with LPS, polyinosinic-polycytidylic acid and Pam3CSK. Adv. Microbiol. 2, 98–103.

Taechowisan, T., Chaisaeng, S., Phutdhawong, W.S., 2017. Antibacterial, antioxidant and anticancer activities of biphenyls from *Streptomyces* sp. BO-07: an endophyte in *Boesenbergia rotunda* (L.) Mansf A. Food Agric. Immunol. 28, 1330–1346.

Tang, Y., Zou, J., Zhang, L., et al., 2012. Anti-fungi activities of *Bacillus thuringiensis* H3 chitinase and immobilized chitinase particles and their effects to rice seedling defensive enzymes. J. Nanosci. Nanotechnol. 12, 8081–8086.

Tang, M., Sun, X., Zhang, S., et al., 2017. Improved catalytic and antifungal activities of *Bacillus thuringiensis* cells with surface display of Chi9602DeltaSP. J. Appl. Microbiol. 122, 106–118.

Terakado-Tonooka, J., Ohwaki, Y., Yamakawa, H., et al., 2008. Expressed *nifH* genes of endophytic bacteria detected in field-grown sweet potatoes (*Ipomoea batatas* L.). Microbes. Environ. 23, 89–93.

Truyens, S., Weyens, N., Cuypers, A., et al., 2015. Bacterial seed endophytes: genera, vertical transmission and interaction with plants. Environ. Microbiol. Rep. 7, 40–50.

Tsavkelova, E.A., Cherdyntseva, T.A., Botina, S.G., et al., 2007. Bacteria associated with orchid roots and microbial production of auxin. Microbiol. Res. 162, 69–76.

Tudela, D., Primo-Millo, E., 1992. 1-Aminocyclopropane-1-carboxylic acid transported from roots to shoots promotes leaf abscission in *Cleopatra mandarin* (*Citrus reshni* Hort. ex Tan.) seedlings rehydrated after water stress. Plant Physiol. 100, 131–137.

Valko, M., Leibfritz, D., Moncol, J., et al., 2007. Free radicals and antioxidants in normal physiological functions and human disease. Int. J. Biochem. Cell Biol. 39, 44–84.

van der Heijden, M.G., Bardgett, R.D., van Straalen, N.M., 2008. The unseen majority: soil microbes as drivers of plant diversity and productivity in terrestrial ecosystems. Ecol. Lett. 11, 296–310.

Verma, V., Singh, S., Prakash, S., 2011. Bio-control and plant growth promotion potential of siderophore producing endophytic *Streptomyces* from *Azadirachta indica* A. Juss. J. Basic. Microbiol. 51, 550–556.

Villarreal-Delgado, M.F., Villa-Rodríguez, E.D., Cira-Chávez, L.A., et al., 2018. The genus *Bacillus* as a biological control agent and its implications in the agricultural biosecurity. Mex. J. Phytopathol. 36, 95–130.

Vorholt, J.A., 2012. Microbial life in the phyllosphere. Nat. Rev. Microbiol. 10, 828–840.

Vu, H.N.T., Nguyen, D.T., Nguyen, H.Q., et al., 2018. Antimicrobial and cytotoxic properties of bioactive metabolites produced by *Streptomyces cavourensis* YBQ59 isolated from

Cinnamomum cassia prels in Yen Bai province of Vietnam. Curr. Microbiol. 75, 1247–1255.

Wang, P., Kong, F., Wei, J., et al., 2014. Alkaloids from the mangrove-derived actinomycete *Jishengella endophytica* 161111. Mar. Drugs 12, 477–490.

Wang, T.T., Ding, P., Chen, P., et al., 2017. Complete genome sequence of endophyte *Bacillus flexus* KLBMP 4941 reveals its plant growth promotion mechanism and genetic basis for salt tolerance. J. Biotechnol. 260, 38–41.

Waring, M.J., Wakelin, L.P., 1974. Echinomycin: a bifunctional intercalating antibiotic. Nature 252, 653–657.

Wilson, D., 1995. Endophyte: the evolution of a term, and clarification of its use and definition. Oikos 73, 274–276.

Woodward, A.W., Bartel, B., 2005. Auxin: regulation, action, and interaction. Ann. Bot. 95, 707–735.

Xie, Y., Xu, H., Sun, C., et al., 2010. Two novel nucleosidyl-peptide antibiotics: Sansanmycin F and G produced by *Streptomyces* sp SS. J. Antibiot. 63, 143–146.

Yousaf, M., Rehman, Y., Hasnain, S., 2017. High-yielding wheat varieties harbour superior plant growth promoting-bacterial endophytes. Appl. Food Biotechnol. 4, 143–154.

Yuan, M., He, H., Xiao, L., et al., 2014. Enhancement of Cd phytoextraction by two *Amaranthus* species with endophytic *Rahnella sp.* Chemosphere 103, 99–JN104.

Zachow, C., Jahanshah, G., de Bruijn, I., et al., 2015. The novel lipopeptide poaeamide of the endophyte *Pseudomonas poae* RE*1-1-14 is involved in pathogen suppression and root colonization. Mol. Plant Microbe Interact. 28, 800–810.

Zhao, P.J., Fan, L.M., Li, G.H., et al., 2005. Antibacterial and antitumor macrolides from *Streptomyces* sp. Is9131. Arch. Pharm. Res. 28, 1228–1232.

Zhao, S., Wei, H., Lin, C.Y., et al., 2016. *Burkholderia phytofirmans* inoculation-induced changes on the shoot cell anatomy and iron accumulation reveal novel components of *Arabidopsis*-endophyte interaction that can benefit downstream biomass deconstruction. Front. Plant Sci. 7, 24.

Zúñiga, A., Poupin, M.J., Donoso, R., et al., 2013. Quorum sensing and indole-3-acetic acid degradation play a role in colonization and plant growth promotion of *Arabidopsis thaliana* by *Burkholderia phytofirmans* PsJN. Mol. Plant Microbe Interact. 26, 546–553.

Chapter 11

Bacterial metabolites: an unexplored quarry

Bishal Singh[1] and Evangeline Christina[2]
[1]School of Medical Science and Technology, Indian Institute of Technology, Kharagpur, India,
[2]Department of Molecular Biology and Genetic Engineering, School of Bioengineering and Biosciences, Lovely Professional University, Jalandhar, India

Chapter Outline

Volatiles and Metabolites of Microbes. DOI: https://doi.org/10.1016/B978-0-12-824523-1.00006-7

11.1 Introduction

Bacteria are primitive forms of life that appeared millions of year ago on Earth, which are single-cell, a few micrometers in size, and a microscopic, prokaryotic biological entity that constitutes a large domain of microorganisms. These organisms exist either alone or in groups in a particular arrangement and present in diverse environments like soil, water, hot springs, acidic condition, basic condition, radioactive waste, and deep biosphere of almost every habitat. Most of the bacteria are not completely characterized but some of them are broadly classified and studied based on various parameters like shape, size, arrangement, mode of respiration, staining property, growth condition, biochemical test, metabolic pathway, and metabolites produced. only 27%–35% of the known bacteria are understood and studied well so that it can culture in a laboratory environment.

There are various types of bacteria that can be either harmful (pathogenic), causing disease and harming the condition of health, or beneficial such as commensal and probiotics, which help to support, enhance, and protect many forms of life: animals, plants, or both. They are used in various fields like agriculture, dairy, veterinary, industrial, and medicinal purposes based on their usefulness and application. There are various organic and inorganic compounds that are used by bacteria as the source of energy and metabolism for survival, growth, reproduction, and to support basic functions of life.

Bacteria are known for various sources of diverse organic compounds and molecules having small molecular weight, which are synthesized as an intermediate product, final product, and byproduct of various metabolic pathway occurring inside the cell, which are known as metabolites. Metabolites are restricted to a small molecule having various functions as the source of energy, structural component, signaling, communication, stimulation, inhibition, catalysis, defense, and activation of various enzymes. Some of the examples are amino acids, organic acids, enzymes, peptide, alkaloids, Macrolides, pigments, vitamins, and antibiotics. Fig. 11.1 depicts the wide application of bacterial metabolites.

11.2 Physiological pathways in bacteria for metabolite generation

11.2.1 Heterotrophic metabolism in bacteria

The bacteria, which cannot synthesize its own source of energy and use the source of energy synthesized by other autotrophic organism, are generally

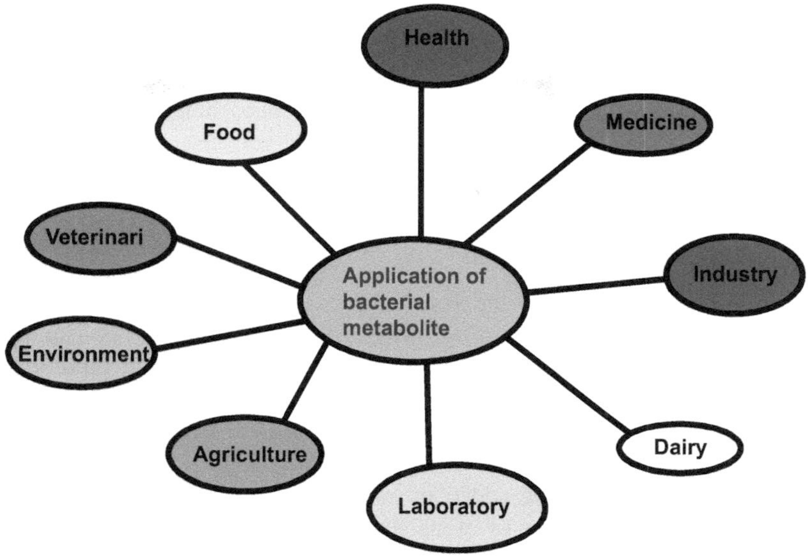

FIGURE 11.1 Wide application of bacterial metabolites.

termed as heterotrophs, and the mode of nutrition is known as heterotrophic. Most of the heterotrophic bacteria are pathogenic, which obtains energy from organic compounds like carbohydrates, proteins, peptide, amino acids, and lipids by oxidation process with the generation of ATP as the currency of energy. In this process, some precursor molecules are also formed which are required during biosynthesis or assimilatory reaction in the bacterial cell. The catabolic reaction of heterotrophic metabolism inside the bacterial cell is the major pathway that simultaneously produces energy in chemical form (ATP and NADP) and precursor molecules, which are required during biosynthesis of new cellular components.

Heterotrophs require organic compounds containing carbon and nitrogen synthesized by autotrophs, and are used for growth and development by the aerobic or anaerobic process to generate NADH + H + as a chemical energy source. These type of bacteria readily grow in enriched media containing carbohydrates, proteins, lactose, blood, or other sources of metabolite.

11.2.2 Bacterial glycolysis

It is a physiological process that occurs in almost every living organism either prokaryotes or eukaryotes, and aerobes or anaerobes. The primary site of this physiological process is cytoplasm inside the cell, where the carbohydrate molecule like glucose is modified by chemical reactions in the presence of various enzymes to produce molecules with small carbon number as pyruvic acid. In this process ATP and NADH molecules are also generated as energy currency. It is the initial or first step of respiration and consists of a total of ten steps where various intermediate and final products are formed as metabolites. Some of them include glucose 6 phosphate, fructose 6 phosphate,1,6-fructose biphosphate, dihydroxyacetone phosphate, glyceraldehyde 3 phosphate, 1,3 bisphosphoglycerate, 3 phosphoglycerates, 2 phosphoglycerates, phosphoenolpyruvate, and pyruvate.

11.2.3 Anaerobic respiration

The process of respiration in the absence of oxygen for oxidation of organic compounds to generate energy for growth, development and metabolism, and gas as a byproduct can be stated as an anaerobic mode of respiration or anaerobic oxidation. Various species of bacteria perform this method of oxidation reaction where they use certain molecules like NO^{3-}, SO_4^{2-}, fumarate, and CO_2 as terminal electron acceptors (Baron, 1996). These electron acceptor molecules must be added in media during the culture process for proper growth of these type of organisms. Most important bacteria that are anaerobes in nature are nitrate reducers, methanogens, and archaebacteria. These bacteria are a source of various metabolites like nitrate, methane, various enzymes, pigments, and antibiotics.

11.2.4 Fermentation by bacteria

Fermentation is also a heterotrophic mode of metabolism where organic compounds are used to accept electrons for oxidation phenomenon. In this process, simple organic compounds like methanol, ethanol, acetic acid, aldehydes, ketones, lactic acid, and acetic acid are produced from various carbon sources as a metabolite from bacteria. In this process, glucose molecules are partially oxidized to generate simple organic compounds, which act as terminal electron and hydrogen acceptor and is further secreted in the medium as a waste metabolite by bacteria. ATP molecules are formed by dehydrogenation reaction. This incomplete biological oxidation process generates energy, but the energy produced in this process are sufficient for bacterial growth and development.

The physiology of fermentation was discovered by Pasteur in the late 1980s which differentiate the microbial world as aerobes and anaerobes.

Based on the product formed by bacteria in the fermentation process, they can be divided into homofermentative like *Streptococcus lactis*, *Lactobacillus casei*, *Lactobacillus pentosus* and heterofermentative *Lactobacillus*, *Leuconostoc*, *Microbacterium*, *Enterobacteriaceae* group, etc. All the metabolites produced at the end of the fermentation process mostly occurs through the glycolytic pathway, where carbohydrate source like glucose dissimilates into pyruvate or further modification occurs in the presence of enzymes to generate other forms of metabolite as stated in Table 11.1.

TABLE 11.1 Summarizes the various types of bacteria and their metabolites produced during the fermentation process (Kluyver, 1956).

Name of bacteria	Product
Enterobacteriaceae (Escherichia coli, salmonella, shigella, proteus)	CO_2, H_2, formate, acetate, lactate, succinate
Aeromonas, Serratia, Erwinia, Bacillus	CO_2, H_2, ethanol, carbinol, butylene glycol
Propionibacterium, Veillonella	CO_2, propionate, acetate, succinate
Acetobacter	Acetate
Gluconobacter	Gluconate
Lactobacillus	Lactic acid

11.2.5 Kreb cycle in bacterial membrane

The Kreb cycle or TCA cycle is one of the physiological processes in which three different types of organic acids are formed in a series of chemical reaction performed by all aerobic organism in the presence of multiple enzymes. It is an oxidative reaction to oxidize acetyl-CoA to ultimately release stored form of energy as ATP and completely decarboxylated to CO_2. In the case of bacteria, the Kreb cycle occurs in the cytosol with a proton gradient across the cell surface as mitochondria are absent and synthesize various intermediate and final metabolic products like citrate, isocitrate, α-ketoglutarate, NADH, $FADH_2$, and GTP (Kornberg, 1970).

11.2.6 Glycoxalate cycle in bacteria

A modified version in the Krebs cycle is the glyoxalate cycle or shunt, which is present in certain bacteria that lack certain enzymes required for the Krebs cycle. In case of bacteria, molecular O_2 are required to oxidize L-malate through electron transport chain.

In this process, the acetic acid production by oxidation of fatty acid is further oxidized to generate acetyl-CoA by the β-oxidation pathway to produce malic acid as a metabolic product. It is an unusual pathway discovered during a study of the oxidation process of lipids in bacteria, leading to biosynthesis of carbohydrates. In this process, oxaloacetate is either converted to pyruvic acid and CO_2 or into phosphoenolpyruvic acid by bacteria like *Azotobacter vinelandii* (Hempfling, 1979).

11.2.7 Final oxidation as electron transport chain

The completing stage of respiration by multiple oxidation-reduction reactions in which transfer of electron and proton across the membrane takes places with the help of various biological compounds for the generation of ATP by oxidative phosphorylation process is known as an electron transport system; enzymes related to electron transfer in this process are present in cell membrane and are also known as mesosomes in bacteria (Jurtshuk et al., 1975).

The electron transport chain is diverse in case of bacteria where it can also oxidize lactate, malate, formate, glutamate by nonpyridine nucleotide-dependent pathway. Bacteria also contain various types of cytochromes like cytochrome p-450, p-420, c′, c′c′. In the electron transport system, energy is generated in the membrane, conserved in it, and then translocated by the coupled process for the formation of the ATP molecule. All enzymes for this process are bound to the membrane when electron transfer in various steps with the help of electron carrier ATP is formed from ADP (Jurtshuk et al., 1981).

11.2.8 Proton extrusion pump

The proton extrusion pump hypothesis is also known as the Mitchell hypothesis, which is one of the highly complex and fascinating concepts that tries to explain the conservation of energy in a biological system by a process like the chemiosmotic coupling of photosynthetic and oxidative phosphorylation. The conservation of free energy is based on osmotic potential due to the difference in proton concentration across the membrane. The various models have been proposed where this states that the coupling process is required energy production and management across an intact membrane.

11.2.9 Bacterial photosynthesis

Many bacteria have bacterial chlorophyll and possess a phototropic mode of nutrition by the process of photosynthesis. The major difference in between bacterial photosynthesis and plant or cyanobacterial photosynthesis is the compound which donates the hydrogen for reduction of CO_2 to glucose. Unlike heterotroph, the phototrophic organism can synthesize the glucose molecule, intracellularly by the process of photosynthesis in which inorganic compound like CO_2 and H_2O (H_2S) are used in presence of chlorophyll pigments and sunlight to synthesize carbohydrates by photolysis of H_2O and reduction of CO_2 through light and dark reactions, respectively. Most of the phototrophic bacteria are Gram-negative, few heliobacterium strains show the presence of endospores. *Halobacterium salinarum* an archaebacterium have purple membrane containing rhodopsin as photosynthetic pigments and a Gram-positive bacteria *Helicobacterium chlorum* phylogenetically related with family *Bacillaceae* have a new type of chlorophyll as bacteriochlorophyll G.

11.2.10 Nitrogen fixation/cycle

All the chemicals like sulfur, oxygen, nitrogen, carbon, and hydrogen must be balanced in the environment to sustain life on Earth by biochemical cycles. Generally, nitrogen is recycled from one organism to another organism as the source for growth and development. The potential of some species of bacteria for the chemical transformation of nitrogen in various compounds is highly appreciated and useful in the field of agriculture. The various process in the nitrogen cycle includes ammonification as an essential step for reuse of nitrogen from one life form to another by the decomposition process of organic compounds in an inorganic form. The other major steps in the nitrogen cycle are nitrification to produce nitrate, denitrification for nitrogen gas production, and finally nitrogen fixation to convert the inorganic form of nitrogen to the organic form. This process of the nitrogen cycle is completed with combined effects of various bacteria which can be free-living like *Azotobacter*, *Dexia*, and *Azomona* species, or can be symbiont like *Rhizobium* species.

11.3 Classification of bacteria metabolites based on their function

Bacteria produce various small molecular weight (<2500 AMU) organic compounds known as metabolites for the regulation of their metabolic process for proper growth, a development where some metabolites directly influence the growth of the organism, and some play an indirect role to support their life. Some metabolites produced by bacteria can be beneficial to other organisms, which help them in their growth and development, and some metabolites inhibit or prohibit the reproduction and development of other organisms, which are harmful to them. Bacteria produce various metabolites like antibiotics to inhibit the growth of other microbes, toxins for a defense mechanism, pigments for camouflage, antioxidants, photosynthesis, macrolides, growth-stimulating factors, organic acids, and alcohols. Bacterial metabolite plays a very important role in shaping various diversity by regulating the environmental conditions. There are diverse metabolite produced by bacteria that are synthesized naturally inside the cell during various physiological pathway; more than 25,000 types of metabolites have been reported, but only 2% from them are properly studied, and its function is well understood (Rose and Tempest, 1985)

Based on the use of and utility during the growth and development of bacterial lifespan, they are broadly classified as

1. Primary bacterial metabolite
2. Secondary bacterial metabolite

11.3.1 Primary bacterial metabolites

Primary bacterial metabolites can be said to those small molecules and bioorganic compounds which have a direct role in primary physiological pathways which sustain proper growth and reproduction. These metabolites are well used in the supply of nutrition and are also used in biotransformation of bacteria. One of the best examples of a primary metabolite produced at a large scale in the industry is C_2H_4 (ethylene).

11.3.2 Secondary bacterial metabolites

The secondary bacterial metabolites do not play a vital role in fundamental functions and physiology of life but have a very important role in ecological functions, such as defense, offense, inhibition of growth, and many more. They are generally formed at the latent stage of bacterial life with the diverse applications, such as antibiotics, pigments, growth hormones, antitumor agent, macrolides, antioxidants, and toxins. Some of them are derived from primary metabolites. These are obtained after an extraction process and are used in pharmaceutical and food industries for their promising capabilities in disease, health, and nutrition management area (Rose and Tempest, 1985).

11.4 Classification of bacterial metabolites based on their application

11.4.1 Bacterial metabolites for nutrition enhancement and food quality improvement

Bacteria are the major and key player in determination of the quality of any food product. The quality of food does not completely depend on nutritional value, but also depends on various factors like appearance, smell, flavor, tenderness, softness, shape, size, packaging, preservatives used, etc. There are so many additives that are used to increase the quality of any food product which is processed either from the kitchen or industry. There are various bacterial metabolites used as these additives.

11.4.1.1 Bacterial metabolites as food colorant

The food product processed from various industries must appear attractive to look more appealing. Various food industries use either synthetic or natural color (dye, pigment). In recent times, almost every food industry is attracted toward the utilization of natural coloring agents due to various side effects of synthetic color (Aberoumand, 2011; Venil et al., 2013).

These natural coloring agents mostly belong to various bacterial pigments or metabolites, which imparts colors to the food product, natural color from bacteria are cheap in production, easy in the extraction and have higher

TABLE 11.2 Summarizes various types of bacteria and pigment synthesize by them as metabolite (Rao et al., 2017).

Pigments	Organism
Blue pigment	*Streptomyces coelicolor*
Yellow pigments (carotenoids)	*Streptomyces* spp.
Orange pigments (canthaxanthin) and dark red (canthaxhant)	*Bradyrhizobium* sp.
Red pigment (rubrolone)	*Streptomyces echinoruber*
Yellow pigment (zeaxanthin)	*Paracoccus* spp.
Blue and green pigments (pyocyanin)	*Pseudomonas* spp.

yields (Malik et al., 2012). The metabolite that is a pigment produced by bacteria is diverse in color, which not only expands in use but also helps in various other applications. Various colored pigments like blue, yellow (carotenoids), orange (canthaxanthin) and dark red (canthaxhant), red (rubrolone), yellow (zeaxanthin), blue and green (pyocyanin) are synthesized by different bacteria, which are given in Table 11.2.

Pigments like canthaxanthin are used in food products like cheese, candy, fish, meat, beverage, wine, and beer. Riboflavins are mostly used in beverages, desserts, and ice cream, whereas carotenoids are used to prevent photoreactions (Chattopadhyay et al., 2008).

11.4.1.2 Bacterial metabolites as antioxidants

Free radicals are present inside our body or cells, which are produced during various reactions frequently happening inside the cells. An increase in the number of free radicals increases the chance of chronic diseases, such as cancer, diabetes, and a cardiovascular and autoimmune disorder (Rankovic et al., 2011). An antioxidant is used to reduce the activity of free radicals as an additive in food products. Antioxidants are the molecule which neutralizes the free radical by donating an electron and prevent cellular damages (Lobo et al., 2010). There is a various bacterial metabolite which has antioxidant property some of them are carotenoids, naphthoquinone (Tuli et al., 2015), melanin from *Streptomyces glaucescens* and *Xanthomonas*. These bacterial metabolites protect the cell from either photodamage or oxidative damage.

11.4.1.3 Bacterial metabolites as vitamins

These are organic compounds with fewer requirements but act as essential components in the regulation of normal physiology. It is not synthesized by

TABLE 11.3 Summarizes various vitamins produced by bacteria and their function in human life (Berg et al., 2002; Bolander, 2006).

Vitamins	Function	References
Vitamin K	Blood clotting, treat osteoporosis, activation of the receptor for transcription mechanism	Berg et al. (2002), Bolander (2006)
Vitamin A	Gene transcription, growth and development, and as a precursor of rhodopsin	Berg et al. (2002)
Vitamin B2	Growth and reproduction	

metabolic activity inside the human body so it is added as a dietary supplement in balanced condition inside food product (Gupta and Gupta, 2015). Several vitamins are synthesized with the help of various bacteria, which are further added as additives in food. The vitamins formed by bacteria as their metabolite have a very important role in humans, like vitamin K, helps in blood clotting, treat osteoporosis, activation of the receptor for transcription mechanism. Vitamin A helps in gene transcription, growth, and development, and as a precursor of rhodopsin and Vitamin B2 helps in growth and reproduction (Berg et al., 2002; Bolander, 2006). Industrial production of riboflavin by using two bacterial species are *Eremothecium ashbyii* and *Ashbya gossypii*, whereas Vitamin B12 are produced by using Pseudomonas *denitrifificans* and *Propionibacterium shermanii* (Kusel et al., 1984; Spalla et al., 1989; Demain et al., 2005) (Table 11.3).

11.4.1.4 Bacterial metabolites as amino acids

It is an amphoteric organic compounds containing both acidic and basic groups in a single molecule. These molecules bind with each other by peptide bond and give various protein molecules which can act as building constituent of cellular components or as enzyme for a various metabolic reaction. Some amino acids can be synthesized inside our body, hence called nonessential amino acids, but some cannot, such as essential amino acids. Amino acids are essential components for proper growth so requirements of these components are important in our balanced diet; various amino acids produced as bacterial metabolite are used as an additive in various food products.

The amino acid is used for various purposes to increase the food quality. Some of them are used as food additives like lysine, methionine, and threonine; flavor enhancers like aspartic acid, glutamate, serine, antioxidant like cysteine, tryptophan, and histidine; and sweeteners like aspartic acid and phenylalanine (Bommarius et al., 1998; Ikeda, 2003; Mueller and

Huebner, 2003; Leuchtenberger et al., 2005; Park and Lee, 2008; Ivanov et al., 2014), which can be used in various food products. These all amino acids are metabolic products of various bacteria species of genus as *Brevibacterium*, *Corynebacterium*, *Micrococcus*, and *Microbacterium* (van Doyen et al., 2012).

11.4.1.5 Bacterial metabolites as organic acids

Small molecular weight acids are produced through the fermentation process, which can be further used as an additive, preservative, and more to increase food product quality. These are generally used in beverages, pickles, and other food items. Those used in food quality improvements include acetic acid, lactic acid, glutamic acid, and citric acid (Sauer et al., 2008).

Vinyl acetate is generally used as the packaging material of food products, and citric acids are generally used in food and beverage. Seventy percent of citric acid is also used as an antioxidant to preserve or enhance the aroma of the product.

11.4.2 Role of bacterial metabolites as pigments

In this demanding era of natural dye, bacterial pigments are available as alternative sources of natural pigments since synthetic dyes can have various harmful effects. It is easy to harvest various bacterial pigments since bacterial culture is easy, rapid, and controllable. Bacterial pigments are not only helpful as a colorant but also act as antioxidants, antimicrobials, and anticancer agents. Pigment-producing bacteria are present in almost every habitat. Multiple genera of bacteria produce a variety of pigments. The most common genus to produce a high amount of pigment is *Streptomyces* (Conn and Jean, 1941). Bacteria produce a wide variety of pigments as their metabolites, which can be carotenoids, melanin, violacein, prodigiosin, pyocyanin, actinorhodin and zeaxanthin (Ahmad et al., 2012; Venil et al., 2014). Sources of various pigment metabolites along with the bacteria that produced them are given in Table 11.4.

11.4.3 Role of bacterial metabolites as a biomarker for disease diagnostic

Biomarkers are the biological molecules, which help in identification of specific organism. It is one of the most important aspects for diagnosis and differentiation of various pathological and nonpathological microbes. Every bacterial species has at least one particular organic compound in them which becomes its characteristic function, and those molecules help to identify those bacteria from the bacterial population. The early detection of bacterial infection based on biomarkers with the help of various advances tool will lead to better management of the patient (Yuan et al., 2012). Multiple

TABLE 11.4 Summarizes different type of pigments synthesized by various bacteria and their function (Tuli et al., 2015; Venil et al., 2013; Ahmad et al., 2012).

Bacteria	Pigment	Function	References
Micromonospora lupine	Anthraquinone	Antitumor agent	
Streptomyces sp.	Carotenoid	Food grade pigment	
Chromobacterium	Viocein	Antiparasitic agent, antitumor agent	
Chromobacterium sp., *NIIST*	Viocein	Antifungal agent	
Hymenobacter sp.	Carotenoid	Photosensitizer	
Streptomyces	Melanin	Antimicrobial and antioxidant	
Pseudomonas	Pyocyanin	Antimicrobial	
Hahella chejuensis	Prodiginines	Antibiotic	
Pedobacter	Carotenoid	Antioxidant	

metabolites produced during various metabolic pathways leads to the identification of various biomarkers (Nicholson and Wilson, 2003). Some of the bacterial metabolites as the signature to identify them, such as volutin granules in cytoplasm, helps to identify *Corynebacterium diphtheria*; the presence of enzyme coagulase helps to identify *Staphylococcus aureus*; the urease enzymes help to identify proteus and cytochrome oxidase enzyme to identify *Pseudomonas*, *Neisseria*, and *Vibrio*.

Sometimes bacterial pigments can also be used as their biomarker to identify them like yellow pigments (carotenoids) in *Streptomyces*, the red pigment in *Streptomyces echinoruber*. Blue and green pigments (pyocyanin) in *Pseudomonas* spp. (Tuli et al., 2015). Fig. 11.2 explains the application of bacterial metabolites in disease identification.

11.4.4 Role of bacterial metabolites for medication and therapeutics

In these long durations, humans have learned about the diversity of metabolites and find various methods to harness them for medical applications. There are diverse bacterial metabolites that are being used for the treatment of disease and to improve human health.

Some of those metabolites are as follows.

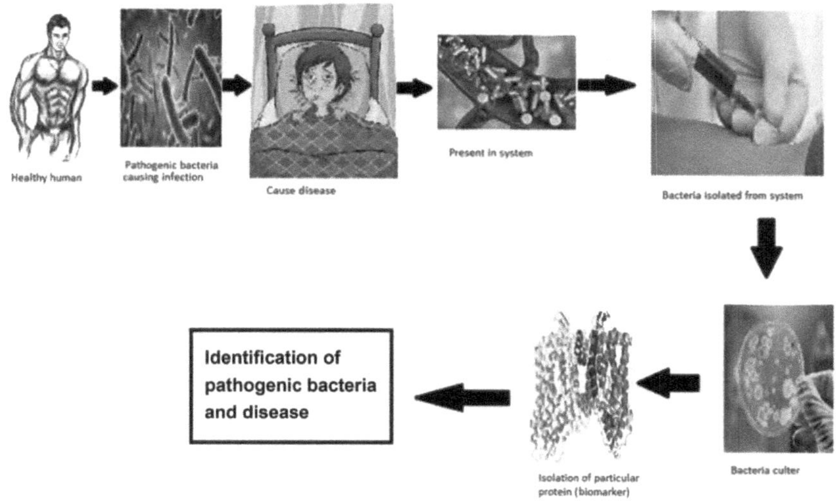

FIGURE 11.2 Application of bacterial metabolites in disease identification.

11.4.4.1 Bacterial metabolite as antibiotics

Antibiotics are those organic compounds secreted by the microorganism or synthesized in the laboratory to destroy or inhibit the growth of bacteria in its surrounding. These are also known as antibacterial agents used as a medication for the treatment of disease caused by different bacterial species. These antibiotics produced by bacteria are a form of a secondary metabolite, which are generally synthesized by the modification of primary metabolites in bacteria. There are hundreds of antibiotics used in health care for treatment of various microbial infection and disease; the new antibiotics identified or derived from bacteria are having unique characteristics and are playing a major role in the treatment of antibiotic-resistant bacteria. Some of the examples of antibiotics produced as a metabolite by bacteria are streptomycin, tetracycline, vancomycin, gentamycin, amphotericin B, chloramphenicol, daptomycin, neomycin, tetracyclin, neomycin, colistin, bacitracin, polymyxin, and others, which act on various microbes like Gram-positive bacteria, Gram-bacteria, and fungus. The most common genus of bacteria-producing antibiotics are *Streptomyces*; they produce various types of antibacterial, antifungal, antiparasitic, and a wide variety of bioactive compounds. The sources of various antibiotic-producing bacteria are given in Table 11.5.

11.4.4.2 Bacterial metabolite as macrolides

Macrolides are a class of antibiotics which was isolated first time in the 1950s from *Streptomyces erythreus* (a soil bacterium). It inhibits the process of protein synthesis in a Gram-positive organism and acts as bacteriostatic agents. It binds with the 23s rRNA of the large subunit and blocks the exit tunnel of the ribosome. Its structures consists of sugars, an amino sugar, and

TABLE 11.5 Summarizes the diversification of antibiotics produced by bacteria as their metabolite and their range of action on various organism (Omura, 1992; Miyadoh, 1993; Lazzarini et al., 2000; Berdy, 2005).

Antibiotics	Organism	Activity
Streptomycin	*Streptomyces griseus*	G − bacteria
Tetracyclin	*Streptomyces rimosus*	Broad-spectrum
Vancomycin	*Streptomyces orientalis*	G + bacteria
Gentamycin	*Micromonospora*	Broad-spectrum
Amphotericin B	*Streptomyces nodosus*	Fungus
Nystatin	*Streptomyces noursei*	Fungus
Natamycin	*Streptomyces natalensis*	Fungus
Chloramphenicol	*Streptomyces venezuelae*	G − and G + bacterium
Daptomycin	*Streptomyces roseosporus*	G + bacteria
Fosfomycin	*Streptomyces fradiae*	G + and G − bacteria
Lincomycin	*Streptomyces lincolnensis*	G +
Neomycin, streptomycin	*S. fradiae, S. griseus*	G + and G − bacteria
Tetracyclin	*Streptomyces aureofaciens*	G + and G − bacteria
Clavulanic acid	*Streptomyces clavuligerus*	G + and G − bacteria
Di-2,4-diacetylfluoroglucylmethane	*Pseudomonas aurantiaca*	G + bacteria
Colistin	*Bacillus polymixin var. colistinus*	G − bacteria
Bacitracin	*Bacillus subtilis var. tracy*	G + bacteria
Polymixin	*Paenibacillus polymyxa*	G − bacteria

aglycone ring. Some examples are clarithromycin, erythromycin, roxithromycin, azithromycin, spiramycin, tilmicosin, and tylosin, these compounds inhibit the growth of bacteria and are recommended for bacterial infection.

11.4.4.3 Bacterial metabolite as immunosupressor

There are various types of metabolites produced by bacteria and having many properties; some metabolites of bacteria also possess the property of immune suppression which help for times when the patient is immunocompromised. During various organ or bone marrow transplants, the immune system of the patient is suppressed with the help of these metabolites to reduce the chance of rejection. Some examples of immune-suppressive metabolites are rapamycin and tacrolimus. Rapamycin was first isolated from soil bacteria and have antibacterial, antifungal, and immunosuppressant properties, and it is also known as sirolimus. Tacrolimus is a macrolide produced by soil bacteria *Streptomyces tsukubaensis*, has an immunosuppressive property, and is mostly used in the allogeneic graft.

11.4.4.4 Bacterial metabolite as anticancer agent

Cancer is one of the diseases having the highest mortality rate. During cancer, the normal cell starts multiplying uncontrollably and form a bulk known as a tumor, which further spread by the process of metastasis from its site of origin to another part of the body. Still, it has a major impact on society all over the world as 8.2 million people die every year according to WHO. The rate of disease incidence is increasing every year and almost 70% of new cases have been reported in the past two decades. Cancer is caused due to various factors but the most important factor is a mutation which can be either induced due to carcinogenic compounds or naturally. Where normal physiology of a cell is hindered, cancer is treated by various methods like surgical removal, radiotherapy, and chemotherapy. Chemotherapy is considered one of the most effective methods to treat cancer like surgery and radiotherapy cannot manage cancer with metastasis (Baba and Catoi, 2007) Many bacterial metabolites have been reported having an anticancer property, and these anticancer agents of bacteria are secondary metabolites, which is mostly produced by bacteria of genus *Streptomyces*. The first anticancer metabolite isolated was Actinomycin by Waksman and Woodruff. Other anticancer metabolites are aclacinomycin, calicheamicin, doxorubicin, chromomycin A3, actinomycin D, mitomycin C, bleomycin, streptozotocin, thiocoralin, mithramycin, daunomycin, and epothilone (Singh et al., 2017). These metabolites are useful for various different types of cancer; some are under the development stage and some are present in the market for cancer treatment. The source of these anticancer metabolites is presented in Table 11.6.

11.4.4.5 Bacterial metabolite as vaccine

The vaccine consists of biological (intact organism or in part) or organic components (toxins, proteins, DNA, and adjuvant) which help to protect from various infectious diseases by providing active acquired immunity. A vaccine particularly contains complete microbes or its components like toxins or surface proteins, which are either killed, inactivated, or weakened so that it should only mimic the antigen and do not infect healthy cells. The antigen given as a vaccine helps to stimulate the body's immune system, which secrete antibodies to neutralize these antigens and further neutralize

TABLE 11.6 Summarizes various anticancer metabolite produced by diverse range of bacteria and their use in various types of cancer (Singh et al., 2017).

Anticancer agent	Source	Type of cancer
Aclacinomycin	*Streptomyces galilaeus*	Lung cancer
Calicheamicin	*Micromonospora echinospora*	Leukemia
Doxorubicin	*Streptomyces peucetius*	Uterus and ovarian cancer
Chromomycin A3	*Streptomyces griseus*	Breas and urinary bladder cancer
Actinomycin D	*Streptomyces antibioticus*	Wilm's tumor
Mitomycin C	*Streptomyces caespitosus*	Stomach, esophagus breast cancer
Bleomycin	*Streptomyces verticillus*	Hodgkin's disease, squamous cell carcinoma, testicular cancer
Streptozotocin	*Streptomyces achromogenes*	Pancreatic cancer
Thiocoraline	*Micromonospora marina*	Colon cancer
Mithramycin	*Streptmyces plicatus*	Bone and testicular cancer
Daunomycin	*S. peucetius*	Leukemia
Epothilone	*Sorangium cellulosum*	Breast cancer

the microbes associated with these antigens. The vaccine can be used both as a prophylactic and as a therapeutic agent.

There are various vaccines prepared from bacterial sources to prevent infectious disease. These vaccines contain either toxins, surface proteins, killed bacteria, or attenuated bacteria, and various disease like Tuberculosis, tetanus, meningitis, cholera, and typhoid pneumonia are prevented with vaccines produced from various bacterial metabolites. Fig. 11.3 summarizes various bacterial candidates for vaccine development.

11.4.4.6 Bacterial metabolites as antiviral agent

There are various metabolites synthesized by bacteria through the various pathways; some of them have medicinal properties and some are used for

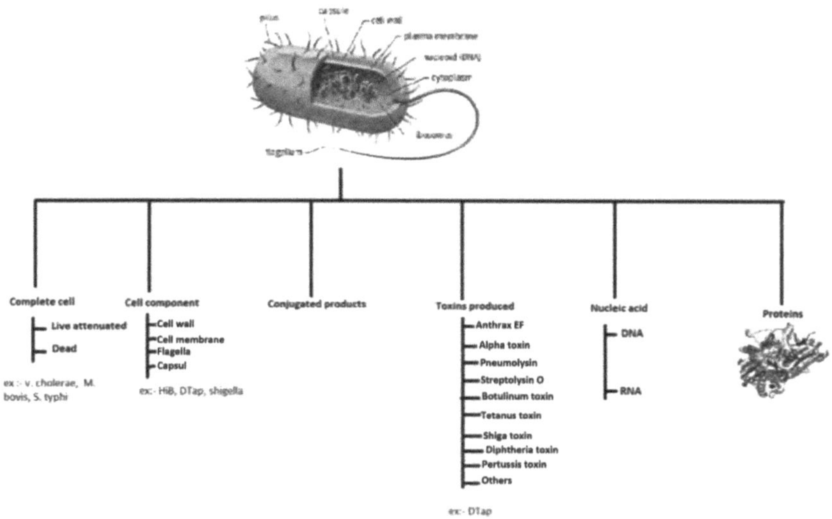

FIGURE 11.3 Various bacterial candidates for vaccine development.

other purposes. Various species of bacteria or bacterial extract shows antiviral properties like recent study suggest antiviral nature of extremophilic actinomycetes extract against influenza virus and paramyxoviruses. This is due to the presence of metabolites, which are either precursors of antiviral compounds or antiviral in nature, and several metabolites produced by bacteria act as a precursor. Some of them are formyl glycals (precursor for various C or N nucleoside) and benzimidazoles, which on modification by various chemical reaction shows an antiviral property. Many polycyclic sugars, pyrazole pyrimidines, and hydrazide derivatives give rise to the antiviral compound.

11.4.4.7 Others

Other functions of bacterial metabolites in health care and management can act as an antidiabetic effect due to lactic acid bacteria, which maintains the glucose level as well as acts as probiotics as an antiinflammatory, antifungal, and more.

11.4.5 Role of bacterial metabolites as a probe for the study of life at a molecular level

Multiple physiological pathways occur inside the cell to synthesize various component for maintenance, growth, repair, and development or to generate energy to sustain in nature. These anabolic and catabolic pathways can be easily studied and fully understood with the help of probes inside the system. Multiple researchers are using a variety of bacterial metabolites such as

BioProbes to investigate process and pathway at micro dimension in multiple ways to understand the function of the cell.

These bacterial metabolites are helping to decode the complexity of the cellular mechanism. Once it is fully decoded and understood, it can be easily manipulated for industrial and health benefits. Some important metabolites, which are used as BioProbes in cellular biology, are bafilomycin (macrolides produced by *Streptomycetes*), fostriecin (produced by *Streptomyces pulveraceus*), and geldanamycin (isolated from *Streptomyces hygroscopicus*). It inhibits the function of HSP90, herbimicin, leptomycin (secondary metabolite, antifungal), and tautomycin (phosphatase inhibitor, synthesized by *Streptomyces spiroverticillatus*).

There are various enzymes and organic compounds formed as an intermediate product during the physiological process, which is also used to track the process. There are various receptors on the cell surface which are either antagonist or agonist with bacterial metabolites reflecting the possibility to study cell and physiological pathway through multiple dimensions, these all application of metabolites helps to explore and use the knowledge in multiple ways to target any organism or any metabolic pathway and engineer at the cellular and molecular level without harming the nature.

11.4.6 Role of bacterial metabolites in health improvement

There are various species of bacteria which are beneficial for humankind, some species are pathogenic in nature and cause various illnesses, but various bacterial species coexist in nature with humankind as commensal; the bacteria which do not harm human health are also known as good bacteria, and these bacteria directly or indirectly provide benefits to human health. These good bacteria can be broadly classified into two major groups that improve human health directly or indirectly. They are

1. Probiotics
2. Commensal

11.4.6.1 Probiotics

Probiotics are those types of bacterial species that help to improve the health of the host, and their composition in host digestive tract is beneficial as they control the diversity of microbes present inside gut and control metabolic activity of existing bacteria. They are also known as good bacteria, which are generally present in fermented products. We consume these bacteria in our diet every day, and research has revealed the health benefits of probiotics.

These bacteria provide benefits through proper digestion process, to prevent diarrhea and autoimmune disease, and to decrease vaginal and urinary tract infections. In various studies it is found that these organisms produce certain enzymes or proteins which inhibit the growth of harmful bacteria. Some strains can even stimulate the immune system and it also plays an important role in hormone production, vitamins, and nutrient absorption.

These probiotics can be broadly classified into two main species.

11.4.6.1.1 Bifidobacteria

These are generally used in food supplements, which helps in supporting the immune system, maintaining gut microbiome and in proper breakdown of food particles. The most common species used of this genus are *Bifidobacterium animalis* (present in yogurt; helps in digestion and boost immunity), *Bifidobacterium breve* (present in the digestive and urinary tract, helps in digestion and inhibit the growth of harmful bacteria), *Bifidobacterium lactis* (present in raw milk of Nestle's product), and *Bifidobacterium longum* (present in GI tract, breaks down carbohydrates, and act as an antioxidant).

11.4.6.1.2 Lactobacillus

These generally contain lactase enzymes that help in lactic acid formation, which helps in regulation of pH and inhibits growth of other microbes (Abe et al., 1995). It also acts as a source of energy and helps in mineral absorption. Some common species are *Lactobacillus acidophilus* (present in GI tract, yogurt, and prevent from various infection), and *Lactobacillus reuteri* (present in mouth and intestine; prevent tooth decay and GI tract infection).

11.4.6.2 Commensals

Commensals are those type of microbes that reside on either surface of the body or at mucosa without harming human health. The microbes living in harmony with human mostly consist of bacteria, also known as commensal bacteria, which are 10 times more than the cells present in our body. These bacteria are present at a particular location of the body such as the skin surface, oral cavity, intestine, nasopharyngeal cavity, mucosal surface of the genital tract, and other anatomical places that provide a suitable environment for proper growth and multiplication (Kova et al., 2011). The microbiome present inside the human gut is highly complex and is not understood properly; 70% of the bacteria present inside our body cannot be cultivated by the present microbial technique. There are around 50 different bacterial phyla residing inside the human gut but most prominent phylum of bacteria are *Acitinobacter, Proteobactor, Firmicutes*, and *Bacteroidetes* (Eckburg et al., 2005).

These commensal bacteria have a compounding effect in human health in several ways. Some of them are protecting from various pathogenic bacteria by inhibiting their growth, by breaking various food components, and helping in absorption to prevent the colonization of other microbes by synthesizing various growth stimulatory factors. It also helps in boosting the host immune system to prevent various diseases like inflammatory bowel disease, celiac disease, rheumatic disease, allergies, and some neurological diseases (Carding et al., 2015). It also helps in shaping innate and adaptive postnatal immunity conditions.

11.4.7 Role of bacterial metabolites for dairy product enhancement

Milk and dairy products are the essential component of the daily balanced diet. The dairy product is added in our diet from prehistoric times. they are rich in protein, calcium, phosphorous, and vitamins. There are various dairy products available in the market, which may be as raw milk or fermented product of milk. Lactic acid bacteria play an important role in dairy fermentation, as these bacteria are abundantly found in nature and are also considered as "Generally Recognized as Safe" bacteria. some of the species of this group include *Lactobacillus*, *Lactococcus*, *Streptococcus*, *Leuconostoc*, and *Pediococcus* (Wouters et al., 2002). Lactic acid, which helps in precipitation of protein in milk and makes the product thicker and solid, is the product of bacterial fermentation process as a metabolite.

Some important dairy products in markets are as follows.

11.4.7.1 Curd

Curd is the coagulated product of milk, which can be made by adding an edible acidic substance or without adding any acids. In normal processes lactobacillus, a rod-shaped bacterium, converts lactose into lactic acid through the fermentation process, which imparts the sour taste of old milk. The curdling of milk can be separated into two parts: whey as a liquid part containing the protein of milk, and another as curd, the solid part containing milk protein or casein.

11.4.7.2 Yogurt

Yogurt is a fermented dairy product that is made by heating milk up to 80°C (to kill other microbes and for denaturation of the milk protein), then cooled to 45°C and inoculated with bacteria like *Lactobacillus* and *Streptococcus salivarius* subsp. Thermophilus at room temperature. Sometimes probiotic are also used which are *Streptococcus thermophilus*, *L. acidophilus*, and Bifidobacteria for yogurt production, which is also known as bioyogurt. Many studies suggest the consumption of probiotic in balance concentration results in multiple therapeutic benefits (Viljoen and Lourens-Hattingh, 2001).

11.4.7.3 Cheese

Cheese is also one type of fermented dairy product that helps in the preservation of milk for a longer duration. The process of making cheese from milk consists of three main steps. The first step includes preparation of curd by coagulating milk protein (casein) in the presence of lactobacillus species (Button and Dutton, 2012). In the next step, the solid part of the curd along with milk protein is separated and molded into various shapes and sizes. In the last step the cheese is left for aging where the flavors of cheese are produced due to amino acid catabolism by various microbes on the surface of the cheese to produce a malty, fruity, and sweet flavor, floral, chemical flavor, and buttery flavor (Yvon and Rijnen, 2001; Ardo, 2006).

11.4.7.4 Kefir

Kefir, a fermented dairy beverage containing alcohol, is made by adding milk along with kefir grain, the kefir grains have association of multiple varieties of lactic acid bacteria (*Lactobacillus fermentum, L. acidophilus, Lactobacillus helveticus, L. casei, Lactobacillus kefir, L. lactis, Lactobacillus Brevis*, and *Lactobacillus parakeferi*), acetic acid bacteria (*Acetobacter aceti, Acetobacter rasens*), and yeast, which helps in the fermentation process. The composition of bacteria may differ based method of cultivation, the origin of grains, geographical distribution, etc. (Witthuhn et al., 2005; Pintado et al., 1996).

11.4.7.5 Kumis

Kumis is also a fermented dairy product containing alcohol and used as a beverage in various countries. It is made from liquid starter culture and raw milk where the milk for kumis contains a high amount of sugar which on fermentation produce a higher amount of alcohol in compare with kefir. Based on the amount of lactic acid in kumis, they are strong kumis (acidification of milk to pH 3.6–3.3 with lactic acid from lactose conversion ratio is 80%–90% by *Lactobacillus bulgaricus* and *Lactobacillus rhamnosus*), moderate kumis (acidification up to pH 4.5–3.9 with a conversion ratio of 50% by *L. acidophilus*) light kumis (acidification to reach pH 4.5–5 by *S. thermophilus* and *Streptococcus cremoris*). (Danova et al., 2005). Another dairy products fermented by the association of various bacteria are sour cream, also known as fermented cream, a cultured buttermilk fermentation product of cow's milk by *S. lactis* and *L. bulgaricus*.

11.4.8 Role of bacterial metabolites for agriculture and crop production

11.4.8.1 Bacterial metabolite as pesticides and herbicides

Pesticides and growth regulators are the most important components used in agriculture for the cultivation of crops and to increase crop yield. Most commonly used chemical pesticides are halogenated, carbamate, and organophosphorus, but their use results in various adverse effects such as heavy toxicity

TABLE 11.7 Summarizes the function of different subspecies of *Bacillus thuringiensis* (Lambert et al., 1992; Milner, 1994).

Bacillus thuringiensis subspecies	Function
Var. tenebrionis	Control beetle larva
Var. kurstaki, entomocidus, galleriae, aizawai	Control caterpillars
Var. israeliensis	Control mosquito and blackfly larvae

in agricultural products, water and soil contamination (Canan, 2013), the mutation in various insects, and disturbances in the natural balance (Lacey and Siegel, 2000). There are so many bacterial metabolites that have pesticide-like properties, which are also known as biopesticides. These are more convenient than chemical pesticides because they are biodegradable and highly effective, with high target specificity and low environmental risk.

Biopesticides include biofungicides, bioherbicides, and bioinsecticides (*Bacillus thuringiensis, Bacillus sphaericus*) (Canan, 2013). The major role-playing bacteria as an insecticide is *B. thuringiensi* with diverse pest-controlling properties. It is used as Bt sprays on fruits and vegetables (Meadows, 1993). The roles of various species of *B. thuringiensis* is given in Table 11.7.

11.4.8.2 Bacterial metabolite as the nitrogen source

Nitrogen is an important element required for the biosynthesis of bioactive compounds like amino acids, proteins, nucleic acid, enzymes, and various others that have a proper function in the development of crops. Plants cannot absorb molecular or free form nitrogen from the atmosphere, it can only absorb as a combined form as nitrate. This form of nitrogen is made available by bacteria through a process called nitrogen fixation, in which the free forms of bacteria are *Cyanobacteria*, *Anabaena*, *Nostoc*, and *Clostridium*, and *Rhizobium* are symbiotic in root nodules of leguminous plants to increase soil fertility.

11.4.8.3 Bacterial metabolite as biofertilizer

Fertilizer consists of essential components in the organic form needed for proper development of plants and crops, and in ancient times multiple chemical fertilizers have been used to improve crop yield, but these chemical fertilizers reduce the fertility of the field and destroy the quality of agricultural soil. It also contaminates various water reservoirs (Heinrich, 2000). There has been a study suggesting the use of biofertilizers not only enhance crop yield but also increases the fertility, sustainability, and health of the soil.

Biofertilizers are ecofriendly organic compounds containing various live bacteria and microbes, which in the soil during the cultivation of crops create microbiome at the rhizosphere and promote the growth of the plant by supplying the primary nutrient. It can either be done by the process of nitrogen fixation or by making nutrients into a soluble form.

The most common form of bacteria present in biofertilizers is *Anabaena-Azolla*, *Rhizobium* spp., *Streptomyces* spp., and *Pseudomonas* spp.

11.4.9 Role of bacterial metabolites used in veterinary

Veterinary is the science which explains about the health and welfare of domestic animals, it not only deal about the health and disease associate with domesticated animals but also explains about various factor associated with them like nutrition, gut microbiome, antibiotic and chemical used and various other things. In the present era, farmers are more focused on animal husbandry to use various food products produced by animals. The most important bacterial metabolite used are antibiotics to improve poultry farming, pig and cattle husbandry. Whereas the use of probiotic which is supplemented by bacteria of genus like *Lactobacillus*, *Bifidobacterium*, *Bacillus*, *Enterococcus* helps animals to gain weight in very less time as these bacteria help in the digestion of food and protect from infectious microbes (Chang et al., 2001).

These microbes produce various enzymes like amylase and protease, which helps in digestion of carbohydrates and proteins. These bacteria increase the size of villi for better absorption of nutrition from the gastrointestinal tract (Abe et al., 1995). They also help in boosting the immune system of cattle. The fibrous feed of the cow is broken down by various bacteria present in the rumen of a cow by an enzymatic process to break cellulose and hemicellulose. The most common bacteria in the rumen of the cow are *Ruminococcus* and *Selenomonas*.

11.4.10 Role of bacterial metabolites used for industry

Bacteria are a source of various compounds used in daily life in the various fields of health, such as in medication, in agriculture as pesticides and fertilizer, in the field of dairy for nutritional enhancement, but a large-scale requirement is fulfilled only by the industrial production of bacterial metabolites. Bacteria are cultured in favorable artificial conditions inside fermenters to isolate metabolites like antibiotics, probiotics, drugs, vaccine, insecticides, enzymes, fuels, solvents, and various peptides. *Lactobacillus*, *Lactococcus*, and *Streptococcus* are used in the dairy industry. Lactic acid and acetic acid bacteria are used to produce olives, pickles, and sauerkraut through the process of pickling in food industries. Various antibiotics, vaccines, and some useful enzymes in pharmaceutical industries are also produced by various bacteria. *Actinomycetes* spp. are used to produce tetracycline, erythromycin, rifamycin, streptomycin, and ivermectin. *Bacillus* and *Paenibacillus* spp. are used for polymixin and bacitracin production. Various bacteria or their components are used for the preparation of a vaccine against various preventable bacterial disease in pharma industries.

Various other bacterial metabolite having an industrial application is as follows.

11.4.10.1 Ethyl alcohol, A bacterial metabolite

Ethanol is one of the most common chemical compound used in our daily life, which is also a primary metabolite of various bacteria. It is widely used in chemical industries as a reagent for various compounds and mixture for explosive, cosmetic products, solvents for organic compounds like dye, oil, and waxes. It is also used to prepare various hand gels, sanitizers, and disinfectants to kill various microbes. Ethanol is also produced for the application of fuel due to its combustible nature. About 90%–95% bioethanol formed from various carbon sources is due to a microbial activity known as fermentation (Sarris and Papanikolaou, 2016). Escherichia *coli*, *Klebsiella*, and *Clostridium* are commonly used bacteria in industrial production of ethyl alcohol (Moniruzzaman and Ingram, 1998).

11.4.10.2 Amino acids as bacterial metabolite

These are essential molecules needed to build diverse biological compounds. Amino acids have a wide application in the food industry as a flavoring agent (serin, aspartic acid, and glutamate), sweetener (phenylalanine, aspartic acid), food additive (lysine, methionine, threonine), and antioxidant (cysteine, threonine) (Bommarius et al., 1998). Fermentation and enzymatic process are used for the synthesis of essential amino acids in the industry as it is pure and globally accepted methods. *Corynebacterium* and *E. coli* are commonly used for the synthesis of various amino acids in the industry (van Ooyen et al., 2012).

11.4.10.3 Organic acids as bacterial metabolite

There are various applications of organic acids in day to day practice. They are important constituents of various food products, beverages, pharmaceutical products, solvents, petrochemical, detergent, textile, rubber, cosmetic products, plastics, and dye (Sun et al., 2015). It has wide application in the construction and automotive industries too. The most common organic acids produced in industries due to bacterial fermentation process are acetate, lactate, and citrate (Sauer et al., 2008). Vinyl acetate monomer is widely used as packaging material and citric acid as preservatives, flavoring, emulsifier, and buffering agent in industries.

11.5 Bacterial metabolites from genetically modified bacteria

Bacteria are well-studied forms of life whose metabolic pathway is studied more deeply at the molecular and genetic level. So, it becomes easy to manipulate with various genetic engineering and biotechnological tools and techniques. Multiple biotechnology companies produce severe compounds like enzymes, hormones, stimulating and inhibitory factors, medicine, vaccine, nucleic acids, and proteins by using bacteria for humankind. Genetic engineering requires two fundamental approaches to produce any metabolite from bacteria, which does not synthesize it. The first is to identify the gene or metabolic pathway associated with the metabolite synthesis in the natural host cell, and secondly is to manipulate the bacterial metabolic pathway by gene insertion or mutation process with the help of genetic engineering tools and methods.

Several human hormones, enzymes, and proteins are produced by the genetically modified method and are widely used as therapeutic agents for a disease like diabetes, heart attack, tuberculosis, and AIDS. Some of them are insulin (Walsh, 2005), as well as human growth hormones streptokinase, urokinase, interferon, and tumor necrosis factor. Genetically modified *Bacillus subtilis* and *Corynebacterium ammoniagenes* are used for riboflavin production (Koizumi et al., 2000). Blue pigment-producing *Streptomyces coelicolor* is used for yellow (kalafungin), orange, or yellow-red pigments. Genetic modification of any bacteria leads to change in some property of bacteria; the modified bacteria gain extra features due to genetic modification. It is not applicable in only human welfare, genetic modification of various bacteria is also done for decreasing and decontaminating various pollution and to clean the environment. Another approach of genetic engineering is to produce super strains of bacteria, which can synthesize multiple new metabolites and help in bioremediation and pollution degradation to clean the environment. Fig. 11.4 explains the genetic modification of bacteria for metabolites.

11.6 Future aspects of bacterial metabolites and their application

The bacterial metabolite is large in number and diverse in their properties and applications. Multiple bacterial metabolites have been unexplored to date, so the future of the bacterial metabolite will solely depend on exploring it, as the new metabolite will discover the new field of application and will be open. There is a wide range of application for the bacterial metabolite. There are multiple infectious and noninfectious diseases in animals, and each disease has an effect on metabolism and host condition, which leads to the generation of abnormal metabolites that can be used as a biomarker for disease diagnosis and effective management of a patient. It is not only sufficient for therapeutic application, it can only give a general idea of the disease. To start treatment a physician should know the causative agent for selection of appropriate drug.

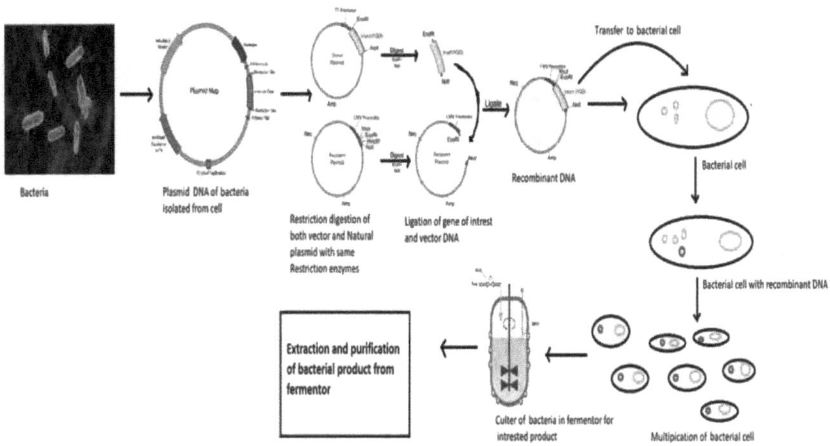

FIGURE 11.4 Genetic modification of bacteria for metabolites.

To identify the causative agent of any bacterial disease, again metabolites come to play a very important role as a biomarker for identification of bacteria. If bacterial metabolite can be used as a biomarker, the various rapid diagnostic kit can be designed and the various tests can be standardized, which not only reduce time in health care centers but also reduce the mortality rate.

Bacterial metabolites have wide application in therapeutics where they can be directly used as antibiotic, antifungal, antiviral, antiinflammatory, and anticancer compounds, or they can be used as reagents to develop various new classes of medicine having antipathogenic, anticancer, and immunosuppression properties. Bacterial metabolites are synthesized and used by bacterial cell or any other by various metabolic pathways. To understand and to study those metabolic pathways and biochemical reactions, suitable metabolites can be used as a probe to understand those pathways so that metabolic engineering can be done if required in future. A better knowledge of these metabolic pathways and their relationship at the genetic level will help to create various strains of bacteria for synthesis of the required compound in any environmental conditions. New strains of bacteria are produced either by genetic engineering, where the gene of interest is added to the bacterial genome with various tools and techniques in which bacteria gets extra genes, or it can be produced through induced mutation by suitable mutagenic compounds. Where the existing gene of bacteria is either upregulated or downregulated based on the requirement of the condition. Bacterial strains produced by these methods can be used for various purposes in industries to produce various products, in the environment for bioremediation, and bioaccumulation of toxic compounds and minerals, metals, and minerals, respectively. It helps to decrease the pollution from the environment.

Bacterial metabolites have also multiple uses in agriculture, fishery, and animal husbandry, an association of the bacteria with every life form interlinks and broadens the application of the compounds produced by it. The scientific proof behind, the mechanism of action of various metabolite produced by probiotics, and commensal are still not known, which can be explored for proper monitoring of these gut bacteria for productive results.

Bacterial metabolites have wide applications not only in the biological field, but it can be also used in the field of physics and chemistry. Various metabolites, especially pigments, can be used to harvest sunlight, but various instruments like dye-sensitized solar cells can also be designed; where metabolic bacterial pigment can be used, the piezoelectric property of various metabolite should be used to design various sensors like the cantilever. Various composites and nanomaterials can be prepared from these metabolites for mimicking an in vivo environment inside in vitro conditions, and for site-directed drug delivery. There are a lot more.

Reference

Abe, F., Shimamura, S., Ishibashi, N., 1995. Effect of administration of *Bifidobacteria* and lactic acid bacteria to newborn calves and piglets. Diary Sci. 78 (912), 2838–2846.

Aberoumand, A., 2011. A review article on edible pigments properties and sources as natural colourants in foodstuff and food industry. World J. Dairy Food Sci. 6, 71–78.

Ahmad, W.A., Ahmad, W.Y.M., Zakaria, Z.A., Yusof, N.Z., 2012. "Application of bacterial pigments as colorant. In: Ahmad, W.A., Ahmad, W.Y.M., Zakaria, Z.A., Yusof, N.Z. (Eds.), Application of Bacterial Pigments as Colorant: The Malaysian Perspective. Springer, Berlin, pp. 57–74. Available from: https://doi.org/10.1007/978-3-642-24520-6_4.

Ardo, Y., 2006. Flavour formation by amino acid catabolism. Biotechnol. Adv. 24 (2), 238–242.

Baba, A.I., Catoi, C., 2007. Comparative Oncology. The Publishing House of the Romanian Academy.

Berdy, J., 2005. Bioactive microbial metabolites a personal view. J. Antibiot. 58 (1), 1–26.

Berg, J.M., Tymoczko, J.L., Stryer, L., 2002. Biochemistry, Vitamins Are Often Precursors to Coenzymes, fifth ed. W H Freeman, New York. Available from: http://www.ncbi.nlm.nih.gov/books/NBK22549/.

Bolander, F.F., 2006. Vitamins: not just for enzymes. Curr. Opin. Investig. Drugs 7 (10), 912–915.

Bommarius, A.S., Schwarm, M., Drauz, K., 1998. Biocatalysis to amino acid-based chiral pharmaceuticals—examples and perspectives. J. Mol. Catal. B Enzym. 5 (1–4), 1–11.

Button, J.E., Dutton, R.J., 2012. Cheese microbes. Curr. Biol. 22 (15), 587–589.

Canan, U., 2013. Microorganisms in biological pest control—a review (bacterial toxin application and effect of environmental factors). In: Silva-Opps Dr., M. (Ed.), Current Progress in Biological Research. InTech. Available from: https://doi.org/10.5772/55786.

Carding, S., Verbeke, K., Vipond, D.T., Corfe, B.M., Owen, L.J., 2015. Dysbiosis of the gut microbiota in disease. Microb. Ecol. Health Dis. 26, 26191. Available from: https://doi.org/10.3402/mehd.v26.26191.

Chang, Y.-H., Kim, J.-K., Kim, H.-J., Kim, W.-Y., Kim, Y.-B., Park, Y.-H., 2001. Selection of a potential probiotic *Lactobacillus* strain and subsequent *in-vivo* studies. Antonne Van Leeuwenhoek 193–199.

Chattopadhyay, P., Chatterjee, S., Sen, S.K., 2008. Biotechnological potential of natural food-grade colourants. Afr. J. Biotechnol. 7, 2972–2985. Available from: https://doi.org/10.5897/AJB08.433.

Conn, H.J., Jean, E.C., 1941. Value of pigmentation in classifying actinomycetes. J. Bacteriol. 42, 791–799.

Danova, S., Petrov, K., Pavlov, P., Petrova, P., 2005. Isolation and characterization of *Lactobacillus* strains involved in koumiss fermentation. Intl. J. Dairy. Technol. 58 (2), 100–105.

Demain, A.L., Newcomb, M., Wu, J.H.D., 2005. Cellulase, clostridia and ethanol. Microbiol. Mol. Biol. Rev. 69, 124–154.

Eckburg, P.B., Bik, E.M., Bernstein, C.N., Purdom, E., Dethlefsen, L., Sargent, M., et al., 2005. Diversity of the human intestinal microbial flora. Science 308, 1635–1638.

Gupta, U.C., Gupta, S.C., 2015. Role of vitamins in human health and nutrition: sources and morbidity. Curr. Nutr. Food Sci. 11 (2), 105–115. Available from: https://doi.org/10.2174/1573401311666150429225548.

Heinrich, W.S., 2000. Fertilizers. Ullmann's Encyclopedia of Industrial Chemistry. Wiley-VCH, Weinheim. Available from: https://doi.org/10.1002/14356007.a10_323.pub3.

Hempfling, W.P., 1979. Microbial Respiration. Bachman Papers in Microbiology no. 13.

Ikeda, M., 2003. Amino acid production processes. Adv. Biochem. Eng. Biotechnol. 79, 1–35.

Ivanov, K., Ivanova, S., Georgieva, M., Atanasov, P., 2014. Production and regulatory analytical control of amino acids include in food additives. Pharmacia 61 (2), 48–54.

Jurtshuk Jr, P., Mueller, T.J., Acord, W.C., 1975. Bacterial terminal oxidases. Crit. Rev. Microbiol. 3, 359.

Jurtshuk Jr, P., Mueller, T.J., Wong, T.Y., 1981. Isolation and purification of the cytochrome oxidase of *Azotobacter vinelandii*. Biochim. Biophys. Acta 637, 374.

Koizumi, S., Yonetani, Y., Maruyama, A., Teshiba, S., 2000. Production of riboflavin by metabolically engineered *Corynebacterium ammoniagenes*. Appl. Microbiol. Biotechnol. 53, 674–679.

Kornberg, H.L., 1970. The role and maintenance of the tricarboxylic acid cycle in *Escherichia coli*. In: Goodwin, T.W. (Ed.), British Biochemistry Past and Present. Biochemistry Society Symposium no. 30. Academic Press, London.

Kova, H.K., Hudcovic, T., Vannucci, L., 2011. The role of gut microbiota (commensal bacteria) and the mucosal barrier in the pathogenesis of inflammatory and autoimmune diseases and cancer: contribution of germ-free and gnotobiotic animal models of human diseases. Cell Mol. Immunol. 8, 110–120. Available from: https://doi.org/10.1038/cmi.2010.67.

Kusel, J.P., Fa, Y.H., Demain, A.L., 1984. Betaine stimulation of vitamin B12 biosynthesis in *Pseudomonas denitrifificans* may be mediated by an increase in activity of d-aminolaevulinic acid synthase. J. Gen. Microbiol. 130, 835–841.

Lacey, L.A., Siegel, J.P., 2000. Safety and ecotoxicology of entomopathogenic bacteria. Entomopatgenic Bacteria: From Laboratory to Fifield Application. Springer, Dordrecht. Available from: https://doi.org/10.1007/978-94-017-1429-7_14.

Lambert, B., Hofte, H., Annys, K., Jansens, S., Soetaert, P., Peferoen, M., 1992. Novel *Bacillus thuringiensis* insecticidal crystal protein with a silent activity against coleopteran larvae. Appl. Env. Microbiol. 58 (8), 2536–2542.

Lazzarini, A., Cavaletti, L., Toppo, G., Marinelli, F., 2000. Rare genera of actinomycetes as potential producers of new antibiotics. Antonie Van Leeuwenhoek 78 (3–4), 399–405.

Leuchtenberger, W., Huthmacher, K., Drauz, K., 2005. Biotechnological production of amino acids and derivatives: current status and prospects. Appl. Microbiol. Biotechnol. 69, 1–8.

Lobo, V., Patil, A., Phatak, A., Chandra, N., 2010. Free radicals, antioxidants and functional foods: impact on human health. Pharmacogn. Rev. 4, 118–126. Available from: https://doi.org/10.4103/0973-7847.70902.

Malik, K., Tokkas, J., Goyal, S., 2012. Microbial pigments: a review. Int. J. Microb. Res. Technol. 1, 361–365.

Meadows, M.P., 1993. *Bacillus thuringiensis* in the environment—ecology and risk assessment. In: Entwistle, P.F., Cory, J.S., Bailey, M.J., Higgs, S. (Eds.), Bacillus Thuringiensis: An Environmental Biopesticide; Theory and Practice. Wiley, Chichester, pp. 193–220.

Milner, R.J., 1994. History of *Bacillus thuringiensis*. Agric. Ecosyst. Env. 49 (1), 9–13. Fertilizer.

Miyadoh, S., 1993. Research on antibiotic screening in japan over the last decade: a producing microorganisms approach. Actinomycetologica 9, 100–106.

Moniruzzaman, M., Ingram, L.O., 1998. Ethanol production from dilute acid hydrolysate of rice hulls using genetically engineered *Escherichia coli*. Biotechnol. Lett. 20, 943–947.

Mueller, U., Huebner, S., 2003. Economic aspects of amino acid production. Adv. Biochem. Eng. Biot. 79, 137–170.

Nicholson, J.K., Wilson, I.D., 2003. Understanding 'global' systems biology: metabonomics and the continuum of metabolism. Nat. Rev. Drug Discov. 2 (8), 668–676.

Omura, S., 1992. Trends in the search for bioactive microbial metabolites. J. Ind. Microbiol. 10 (3–4), 135–156.

Park, J.H., Lee, S.Y., 2008. Towards systems metabolic engineering of microorganisms for amino acid production. Curr. Opin. Biotechnol. 19, 454–460.

Pintado, M.E., Da Silva, J.A., Fernandes, P.B., Malcata, F.X., Hogg, T.A., 1996. Microbiological and rheological studies on *Portuguese kefir* grains. Intl J. Food Sci. Technol. 31 (1), 15–26.

Rankovic, B.R., Kosanic, M.M., Stanojkovic, T.P., 2011. Antioxidant, antimicrobial and anticancer activity of the lichens *Cladonia furcata*, *Lecanora atra* and *Lecanora muralis*. BMC Complement. Altern. Med. 11, 97. Available from: https://doi.org/10.1186/1472-6882-11-97.

Rao, M.P.N., Xiao, M., Li, W.-J., 2017. Fungal and bacterial pigments: secondary metabolites with wide applications. Front. Microbiol. Available from: https://doi.org/10.3389/fmicb.2017.01113.

Rose, A.H., Tempest, D.W., 1985. Advance in Microbial Physiology. vol. 25, pp. iii–vii, 1–345.

Sarris, D., Papanikolaou, S., 2016. Biotechnological production of ethanol: biochemistry, processes and technologies. Eng. Life Sci. 16, 307–329. Available from: https://doi.org/10.1002/elsc.201400199.

Sauer, M., Porro, D., Mattanovich, D., Branduardi, P., 2008. Microbial production of organic acids: expanding the markets. Trends Biotechnol. 26 (2), 100–108. Available from: https://doi.org/10.1016/j.tibtech.2007.11.006.

Singh, R., Kumar, M., Mittal, A., Mehta, P.K., 2017. Microbial metabolites in nutrition, healthcare and agriculture. 3 Biotech 7 (1), 15. Available from: https://doi.org/10.1007/s13205-016-0586-4.

Spalla, C., Grein, A., Garofano, L., Ferni, G., 1989. Microbial production of vitamin B12. In: Vandamme, E.J. (Ed.), Biotechnology of Vitamins, Pigments and Growth Factors. Elsevier Appl Science, New York, pp. 257–284. Available from: https://doi.org/10.1007/978-94-009-1111-6_15.

Sun, X., Shen, X., Jain, R., et al., 2015. Synthesis of chemicals by metabolic engineering of microbes. Chem. Soc. Rev. 44, 3760–3785. Available from: https://doi.org/10.1039/C5CS00159E.

Tuli, H.S., Chaudhary, P., Beniwal, V., Sharma, A.K., 2015. Microbial pigments as natural colour sources: current trends and future perspectives. J. Food. Sci. Technol. 52, 4669–4678. Available from: https://doi.org/10.1007/s13197-014-1601-6.

van Ooyen, J., Noack, S., Bott, M., et al., 2012. Improved L-lysine production with *Corynebacterium glutamicum* and systemic insight into citrate synthase flflux and activity. Biotechnol. Bioeng. 109, 2070–2081. Available from: https://doi.org/10.1002/bit.24486.

Venil, C.K., Zakaria, Z.A., Ahmad, W.A., 2013. Bacterial pigments and their applications. Process. Biochem. 48, 1065–1079. Available from: https://doi.org/10.1016/j.procbio.2013.06.006.

Venil, C.K., Aruldass, C.A., Dufosse, L., Zakaria, Z.A., Ahmad, W.A., 2014. Current perspective on bacterial pigments: emerging sustainable compounds with coloring and biological properties for the industry-an incisive evaluation. RSC Adv. 4, 39523–39529. Available from: https://doi.org/10.1039/c4ra06162d.

Viljoen, B.C., Lourens-Hattingh, A., 2001. Yogurt as probiotic carrier food. Intl. Dairy. J. 11 (1), 1–17.

Walsh, G., 2005. Therapeutic insulins and their large-scale manufacture. Appl. Microbiol. Biotechnol. 67, 151–159. Available from: https://doi.org/10.1007/s00253-004-1809-x.

Witthuhn, R.C., Schoeman, T., Britz, T.J., 2005. Characterisation of the microbial population at different stages of Kefir production and Kefir grain mass cultivation. Intl. Dairy J. 15 (4), 383–389.

Wouters, J.T., Ayad, E.H., Hugenholtz, J., Smit, G., 2002. Microbes from raw milk for fermented dairy products. Intl. Dairy J. 12 (2), 91–109.

Yuan, M., Breitkopf, S.B., Yang, X., Asara, J.M., 2012. A positive/negative ion-switching, targeted mass spectrometry-based metabolomics platform for bodily fluids, cells, and fresh and fixed tissue. Nat. Protoc. 7 (5), 872–881.

Yvon, M., Rijnen, L., 2001. Cheese flavour formation by amino acid catabolism. Intl. Dairy J. 11 (4), 185–201.

Chapter 12

Microbial metabolites in nutrition and healthcare

Kothandapani Sundar[1,2] and T. Ramachandira Prabu[1]
[1]*Department of Microbiology, Prist Deemed University, Puducherry Campus, Puducherry, India,*
[2]*Department of Microbiology, Sri Lakshmi Narayana Institute of Medical Sciences, Puducherry, India*

Chapter Outline

12.1 Introduction

Microbiome and their metabolites are substantially imperative in the regulation of host immunity, cellular function, overall metabolism, and human welfare. Microbial cell growth requires nearly 60–70 various metabolites (Feist and Palsson, 2010). Microbes are single-celled microorganisms such as bacteria, fungi, archaea, algae, protozoa, and parasites, although some species were

Volatiles and Metabolites of Microbes. DOI: https://doi.org/10.1016/B978-0-12-824523-1.00012-2

animal pathogens and belong to commensal groups. Intervention should produce prolonged stable microbiota with prevention of diseases at early childhood and adult human developmental stages, whereas new intervention would boost curing medical conditions like inflammatory bowel disease (IBD), influencing longer durations of development through to adulthood (Nicholson et al., 2012). Biocatalytic transformation of microbes act as programmable biofactories have potential impact on consumables and clinical biomolecule production necessitates for human race (Abubucker et al., 2012). There are scenarios of human system biology where transgenomic cross-talk among metabolic products (microbiota genome) and signaling pathways (host genome) influence dietetic machinery (Holmes et al., 2012). Discovery of novel metabolites and phytochemicals were expressed in microbes for heterologous production provides conducive condition by engineering host metabolic pathway. Bioengineering of metabolic pathway in suitable heterologous host requires sufficient data of cognate enzymes involved in biosynthesis. These unsolved pathways are still challenging the perception of the detailed structural and functional characterization of metabolic enzymes (Kotopka et al., 2018). Application of rational genetic engineering methodology for producing novel and altered metabolites through biosynthetic gene clusters. Manipulation of biosynthetic gene clusters through de nova biosynthetic pathway, protein engineering through direct evolution method (Menon and Jenner, 2018). Combinatorial biosynthesis approaches of manipulating BGCs through site-specific mutagenesis, mutasynthesis, gene insertion, gene deletion, gene alteration, biosynthetic pathway reorganization through mapping intersubunit interaction, domain swapping, module swapping, subunit swapping (Menon and Jenner, 2018). High-throughput metagenomics, metatranscriptomics, and metablomics are the foundation of "omics" platforms and are revealed as macromolecular machineries to understand the microbial-host behavior and robust gene discovery of uncultured environmental microbes for novel secondary metabolites. Pharmacometabolomics applications for personalized medicine used for drug therapy screening at the "omics" level create metabolic signatures of patients to help identify disease heterogeneity (Kaddurah-Daouk and Weinshilboum, 2013). A system-based metabolic modeling approach as genome-scale metabolic models (GEMs) are dealt with annotated genomes (genes for enzyme catalyzing reaction, metabolic reaction, and metabolites) mapping microbial metabolism from host genotype to phenotype (Jia et al., 2020), flux balance analysis (FBA) for RNAseq data (relative and absolute gene expression pattern based on reaction fluxes on active and nonactive gene), and metabolomics for understanding cellular metabolic behavior (Pinu et al., 2018). Recent developmental tools used as relative expressions and metabolomics integrations (REMI) dealt with metabolite concentration, thermodynamics, gene expression, and metabolic models based on an abundance of data from metabolites, differential gene expression validation between wild types, and test sample (mutant) conditions related to metabolic flux level predicting cellular phenotypes (Hadadi et al., 2020).

12.2 Microbial metabolites

Microbial metabolism is divided into three major categories: (1) growth requires an electron donor, (2) a carbon source is required for essential metabolism, and (3) metabolic energy is required for growth. Primary and secondary metabolites are products of microbial metabolism used for large-scale production in microbiology industries such as antibiotics, recombinant vaccine, essential amino acids, and isolated novel chemicals.

12.2.1 Primary metabolite

The evolutionarily distinct metabolism of microbes provides the fundamental aspects of conserved primary metabolite biosynthesis, which play an essential role in cellular homeostasis. Tryptophan metabolite is a precursor for actinomycin antibiotics synthesis (Laurans et al., 2018).

12.2.2 Secondary metabolite

The fundamental discovery of secondary metabolites were natural drug-like activity harnessed for pharmaceutical applications optimized during evolutionary time scale for target host-microbe interaction provide conducive environmental condition. The secondary metabolite synthesis that occurs in a stationary growth phase product were nonribosomal peptide synthase (NRPS), polyketide synthase (PKS), shikimate pathway intermediates, and terpene biosynthesis (Hug et al., 2020). Exotoxins and endotoxins were bacterial products poisonous to mammalian systems and produced at a larger scale; for example, clostridium botulinum secreted botulinum toxins used for cosmetics and muscle paralysis treatment (Fredrick et al., 2017). Diversified molecular function of secondary metabolite exhibiting clinical properties include immunosuppressive compounds, antibiotics, antifungals, and anticancer agents. Microbial indole compounds signaling interleukin-22 (IL-22) production in immune cells act as a mucosal barrier and gastrointestinal immunity (Laurans et al., 2018). Insoluble proteins and carbohydrates used as prebiotics for gut microbial targets undergo fermentation synthesis-distinct metabolites as branched-chain fatty acids (BCFA), short-chain fatty acids (SCFAs), and hydrogen sulfide (Verbeke et al., 2015).

12.3 Bioactive microbial metabolites in human welfare

The precursor-directed biosynthesis produces a new derivative of complex metabolites by manipulating biosynthetic pathways and incorporating chemical synthesis materials of nonnative origin, integrating with microbial genomes using enzymes without manipulation. Fed growing microorganism with precursor analogs in the growth medium follows substrate consumption and direct incorporation into the metabolic pathway for producing novel structural metabolites. These methodology have limitation of least substrate incorporation rate and macromolecular stability problem. This novel version of methodology is called mutasynthesis, or a combinatorial biosynthesis approach, in which mutagenized bacterium (*Streptomyces fradiae* 3535) grow in a medium containing precursors (2-epistreptamine and an aminocyclitols streptamine) resulting in four novel antibiotics named hybrimycins A and their isomer, and hybrimycins B and their isomer (Shier et al., 1969). Using heteroaromatic and aromatic aminoacids has wide potential for intermediates synthesis in *Actinosynnema pretiosum* during mutasynthesis process, which yield ansamitocin derivatives (Harmrolfs et al., 2014). Myxopyronin (α-pyrone antibiotics) produced in *Myxococcus fulvus* use advanced biosynthetic intermediates (Sahner et al., 2015). IBD patients with reduced trimethylamine and SCFA levels are indicators identified from metabolomic analysis (Marchesi et al., 2007). The higher production of SCFAs can prevent pathogen entry, infection, and epithelial integrity (Osbelt et al., 2020). In addition, SCFAs activate GPR43 and GPR41 autonomic nervous systems and are transversed to the blood-brain barrier (Park et al., 2019).

12.4 Microbial metabolites in the regulation of host immunity

Microbial metabolites involved in cellular homeostasis is highly dependent on host immune response treatment for metabolic diseases (Imai and Kurihara, 1984). Regulation of host signaling mechanism under microbe-host interaction in metabolite production connection with innate and adaptive immune response. Indoleamine 2,3-dioxygenase (IDO), an enzyme involved in tryptophan degradation in mammals and bacteria increases activity on regulatory T cell—mediated immune tolerance and promote negative feedback on cancer cell development (Laurans et al., 2018). In human, bariatric surgery have altered microbiota composition and increased in the gut such as γ-proteobacteria, reducing obesity, type-2 diabetes mellitus. Microbial composition also increases 4-cresol, propionate, acetate concentration in rat blood plasma and urine excretion (Li et al., 2014).

Bacteria lipopolysaccharides provoke innate immune response activate TLR4 (Toll-like receptor-4), an antigenic repertoire-enhanced novel symbiont acclimatization in intestinal mucosa (Fang et al., 2017). Lacking TLR5 receptor-mediated modulation of gut microbiota transferred from genetically altered mice to recipient wild-type mice (germ-free), the results in hyperphagia lead to obesity and low insulin sensitivity (Carvalho et al., 2011). NLRP6 and NLRP3 receptor-lacking mice leads to pyroptosis activation, which triggers gut microbiota immunity responses (Henao-Mejia et al., 2012). Glucagon-like peptide 2 synthesis from prebiotic treatment increases permeability in mice intestine immune functions and abolished bacterial migration (Cani et al., 2009; Lei et al., 2018). Dendritic CX3CR1 chemokine receptor for gut lumen bacteria, with clearance to prevent mesenteric lymph node migration, causes obesity (Imai and Kurihara, 1984). Intestinal bacteria translocation to kidney and liver cells in streptozotocin-treated diabetic mice leads to adipose tissue inflammation (Imai and Kurihara, 1984).

12.5 Microbial metabolites in nutrition, health, and disease

Microbial tryptophan synergistic can have effects on the indoleamine 2,3-dioxygenase activity in mammalian metabolic disease (Laurans et al., 2018). Obese mice (*ob/ob*) feces-derived stercobilin, known as gut bacterial metabolites, upregulate in type-II diabetes conditions (Sanada et al., 2020). Stercobilin in blood circulation increases reabsorption mechanisms and induces higher metabolic rates (Sanada et al., 2020). Bacteria metabolites, such as L-carnitine and phosphatidylcholine, regulate cardiometabolic diseases (Papandreou et al., 2020). Microbe-secreted bile acids and choline are linked to fatty liver disease and nonalcoholic fatty liver disease (Dumas et al., 2006). Bile acids and hydrogen sulfide are detrimental factors in pathogenic bacteria's critical role in inflammatory conditions (Hertel et al., 2019). Butyrate-mediated G-protein-coupled receptor 43 (GPR43) activation recruit regulatory T cells, histone deacetylase inhibition acts as an intestinal barrier, and GPR43 inhibition induces inflammation in mice (Zhang and Davies, 2016). Glycine under minimal levels of circulation involved in nonalcoholic fatty liver disease, obesity-related disease, and restoration of beneficial impacts using glycine supplementation or through diet regulates liver metabolism (Alves et al., 2019).

Elevated cresol levels induced depression in mice from fecal transplantation using the metabolomics platform (Casaccia et al., 2016). Choline, mannitol, GlcNAc-6-P metabolite synthesis in *Streptococcus* sp. M334, M143, and *Clostridium* sp. HGF2 are using metabolomics and metagenomics platforms (Amedei and Morbidelli, 2019). Diethylhexylphthalate (DEHP) used as a plasticizer in neurodevelopmental disorder alters gut microbe composition in mice (Lei et al., 2019). Salt-enriched diet and fruit berries linked to gut microbial modulation, predominantly *Erysipelotrichaceae* and *Proteobacteria* families of bacteria regulating polyphenol excretion, cause increased risk of cardiovascular disease in hypertensive rats (Gomes et al., 2019). Innate lymphoid cells (ILC) migration to infectious areas undergo tissue restoration, embryo development, and host defensive mechanism, which revealed organ system homeostasis controlled endogenous metabolite and microbiota (Willinger, 2019). Retinoic acid (Vitamin A), oxysterols (Cholesterol) induced ILC3 and ILC1 intestine migration control GPR183 receptor signal transduction (Willinger, 2019).

12.6 Gut microbiome: a key modulator of metabolism in health

Understanding the dynamics of gut microbiomes is crucial for digestion, enhanced immune machinery, and drug metabolism. An imbalance of healthy and unhealthy gut microbiota can be associated with metabolic disorders. Gut microbial enzyme action on bile acids metabolism is involved in steroid and cholesterol biosynthesis in gut activates, which host nuclear receptors for health and diseases (Kriaa et al., 2019). Therapeutic interventions, such as prebiotics, probiotics, diet, and fecal transplant options are metabolic axes of gut microbiome modulation (Quigley and Gajula, 2020). Gut microbiome manipulation would result novel in therapy for curing gut dysbiosis (Wong and Levy, 2019). One modality comprises target usage of specific antibiotics to eliminate or attenuate nonessential gut bacterial genera. Druggable microbiome is phenomenon for specific microbial action engineered for benefiting the host without wipeout bacteria (Lemon et al., 2012). Human gut producing secondary metabolite called lantibiotic were synthesized through posttranslationally modified peptides (RiPPs) pathways, that is, ribosomally synthesized and RiPPs. Oxazole-modified microcins, thiazole-modified microcins, bacteriocins, and ruminiococcin A were secondary metabolites producing through RiPPs Pathways (Wang et al., 2019). NRPS pathway mediated synthesis of metabolites in *Klebsiella oxytoca* produce novel tilivalline derivatives (pathogenicity factor) reconstituted with precursors anthralnilate, indole protein reducing pathogenicity, and supplemented for medicinal applications (Tesmar et al., 2018).

Probiotics are safe, live microorganisms, and the dosage would meet the criteria of US FDA standards for human consumption in terms of eatable viability during prolonged storage condition. They remain in the GI tract to provide a beneficial response, amenable for commercial production, for example, yogurt consumed by humans for thousands of years (Simone, 2019). Lactobacilli and bifidobacteria were two gut bacteria used for gut microbiome modulation importance for inhibition of harmful bacterial growth, secretion of vitamin B12, strengthening host immunity and regulation of lipid profile (Pique et al., 2019). Probiotics studies in human and mice feeding yogurt composite lactobacillus bacteria produce higher-fold gene expression for carbohydrate metabolism (McNulty et al., 2011).

Prebiotics are nondigestible dietary eatables that can be cooked in a way which will not affect their chemical nature; for example, unassimilated oligosaccharides (soyoligosaccharides, transgalactooligosaccharides, and isomaltooligosaccharides) metabolized in the large intestine, and through bifidobacteria that benefit human health (Davani-Davari et al., 2019). Medical complication such as intestinal tumor, stomach inflammation, intestine rupture, hypersensitive reactions were being treated using prebiotics and probiotics. Uptake of prebiotics and probiotics helps in gut microbiome modulation which leads to benefit host health (Kho and Lal, 2018). The advantages of prebiotics include gastrointestinal absorption, enhanced fermentation of gut microbiota, prevention of gastric acidity, hydrolysis by enzymes, colon assimilation, and growth enhancers such as xylan and cellulose (Davani-Davari et al., 2019). European volunteers have tested prebiotics using lactulose and transgalactooligosaccharides with minimal feeding duration revealing distinct gut microbial constitutions (Bruno-Barcena and Azcarate-Peril, 2015). Gut microbiota proteomics and genomics platform comprises mass spectroscopic, nuclear magnetic resonance (NMR) analysis, whole genome sequencing—454 pyrosequencing (Li et al., 2014). Gnotobiotic mice tissue sample analysis using NMR spectroscopy showed steep changes in plasma lipid profile and dynamic metabolic effects on various tissue compartments (Claus et al., 2008).

Gastrointestinal tract and aging developmental stage is associated with immune suppressive effects on gut microbiota homeostasis. Symbiotic microbial composition is associated with carbohydrates; protein metabolism in the large intestine is related to aging diseases and metabolic pathways of gut microbiota. Milk is supplemented with fructooligosaccharides as prebiotic ingredient and *Lactobacillus johnsonii* (La1) as probiotic bacteria. This supplementation is feed to 6-week-old infants as a test sample and compared to normal breastfed infants taken as control sample. Fecal sample analysis has elucidated that beneficial microbiota modulation is highly abundance in test sample only (Holscher et al., 2012). Gut microbial composition shows distinct phenomena in autistic disorder children with increased 4-cresyl sulfate, phenylacetylglutamine in urine exertion, and steep abundance of clostridia in the gut (Yap et al., 2010). A comparison of higher antibiotic user 65 years and above and younger individuals profiling gut microbiota differences in fecal samples constitutes higher clostridium species in the older groups (Tiihonen et al., 2010). *Eubacterium limosum* anaerobic bacteria present in the gut and beneficial in nature produces acetate as energy sources for all age groups, particularly centenarians, showed a 10-fold increase in gut microbiota composition than younger individuals (Zhang and Davies, 2016).

12.7 Microbial metabolites and chronic metabolic disorders with respective to molecular mechanism

Ido1$^{-/-}$ knock out mice elucidates glucose homeostasis impaired, insulin resistance, fat deposition in hepatic cell would prevent secondary metabolites indole formation from microbial conversion (Laurans et al., 2018). IBD patients with indole metabolite concentration altered level binds with host aryl hydrocarbon receptor (AhR) downregulates inflammatory protein and upregulates tight junction protein (Bansal et al., 2010; Shimada et al., 2013). Likewise, indole interaction with AHR activates colitis in mice through dextan sulfate sodium upregulation (Ji et al., 2015). Mice with CARD9 protein (caspase recruitment domain family member 9) undergo tryptophan elimination and IL-22 production to prevent microbial growth, whereas *Card9*$^{-/-}$ mice are prone to colitis. *Card9*$^{-/-}$ mice supplemented with AhR agonist and lactobacillus strain reduced intestinal inflammation (Lamas et al., 2016). Interestingly, patients with IBD extracted microbiota exemplified low production AHR ligands (Lamas et al., 2016). Berry-derived microbial metabolite urolithin A (UroA) enhances gut barrier activity mediated through Nrf2 pathway (nuclear factor erythroid 2-related factor 2) increases protein for epithelial tight junction and prevent colinic diseases (Singh et al., 2019).

Stercobilin screened from transplantation feces material obtained from murine model to GF mice as recipient expressed donor phenotypes with minimized chronic inflammation against type-II diabetes and obesity triggering IL-1β and TNF-α proinflammatory molecule expressed macrophage RAW264 cell (Sanada et al., 2020). Trimethylamine (TMA) as gut microbial metabolite derived from nutrients and enzymatic cleavage of TMAO moiety conversion in liver cells act as proinflammatory responses (Wang et al., 2020). Neurological disorders such as depression and irritable bowel syndrome are linked to gut microbial dysbiosis (Rogers et al., 2016). *Lactobacillus plantarum* is a genetically engineered strain to modify techoic acid biosynthesis pathway in mice generated immune response activating IL-10 production (Grangette et al., 2005).

12.8 Reprogramming microbe genetically in terms of agronomy and clinical products

Microbial genetic engineering and cost-effective genome sequencing leads to identification of novel metabolites. Most therapeutics phytochemicals are remain challenging to produce using microbial host tailoring plant metabolic pathway. First step is to mining plant and microbial enzymes. Secondly, to improve plant enzymes activity through cytochrome P450 and electron transfer system in the heterologous host. Finally, control multienzymes pathway through compartmentalization of heterologous host metabolism and optimization of metabolites synthesis (Kotopka et al., 2018). Rational site-specific mutation of biosynthetic gene vioprolides, an antifungal agent undergo adenylation (A) domain interaction with serine amino acid not with alanine. *N*-methyltransferase domain to site-directed mutagenesis produced diversified vioprolide conformation (Tesmar et al., 2018). Nonribosomal peptides combinatorial biosynthesis procedure on subdomain swapping adenylation (A) domain from phenylalanine to valine (Kries et al., 2015). Two bioengineered *L. plantarum* strains (ammonia-consuming) administered in mice successfully treated liver failure (hyperammonemia) (Nicaise et al., 2008). Overexpressing bile salt hydrolase in *L. plantarum* 80 containing vector pCBH1 administered as microencapsulated form to mice effectively act on taurodeoxycholic acid detoxification to lowering blood cholesterol (Jones et al., 2004).

12.9 Prospective of human health with host-microbiata-drug interaction

Novel drug exposure in the human digestive system were upregulates of drug metabolizing enzymes. This happens due to natural selection of xenobiotics, microbial toxins, and plant toxins acclimatization in gut. During the course of history our body immune system develop to detoxify new drug adverse chemical reaction. Liver and gut naturally acquired tailored detoxification system using cytochrome P450 enzymes enhancing dietary metabolism. In mice, stool transfer experiment elucidates microbiome-dependent IDO activity directly proportional to microbial metabolism and inversely proportional to growth elimination targeted action of combinatorial antibiotics as metronidazole, neomycin, ampicillin, and vancomycin (Laurans et al., 2018). Two phases of drug metabolism reactions are phase 1, which undergoes hydroxylation and redox reaction render conjugation nonnative compound with ions and metabolites catalyzed through host enzymes transfer to increase polarity for urine exertion, for example, sulfotransferases transfers sulfate to hydroxyl phenolic products (4-cresol). Phase 2 drug metabolism undergoes conjugation coupling gut metabolite or drugs, adding particular functional groups example uridine 5′-diphospo−glucuronosyltransferase in glucuronidation reaction converts D-glucuronic acid to phenolic *O*-ether glucuronides (Holmes et al., 2012). Gut producing phenylacetic acids and benzoic acids through microbial activity eliminating neuroactive amino acid production at excess levels as glutamate and glycine, which are called neuroactive deamination (Beyoglu et al., 2012).

Glucuronidated and sulfate-modified drug such as acetaminophen (Tylenol) are metabolized through cytochrome P450 reaction (Holmes et al., 2012). Drug effects were predicted based on mathematical model called pharmacometabonomics which is proposed for metabolites profiling. Modified acetaminophen conjugated with 4-cresyl sulfate and excreted in urine test samples which greatly differ among individuals (Holmes et al., 2012). Likewise, many drugs have properties to form hydroxylated metabolites, which get eliminated from excretion through sulfation, for example, 4-cresol synthesis in the distal colon could determine a drug's fate and metabolism (Holmes et al., 2012). Using novel procedure for modulation of drug action associated with gut microbes for improving safety drug metabolism and absorption in the gut for altered host phenotype. (1) utilization of metabolic substrate competition, (2) gut microbiota drugs administered orally thorough primary metabolism, (3) human metabolites from secondary metabolism, and (4) bioavailability (ionization condition) in the gut. Chemotherapeutic drug CPT-11 administered on specific dose level for

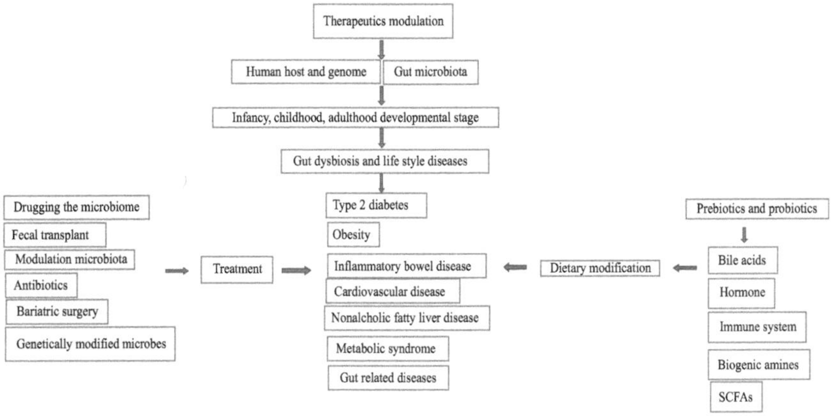

FIGURE 12.1 Dynamics of human gut modulation with perceptive to developmental stages, health, and host-microbiata-drug interaction.

treatment colon cancer (Wallace et al., 2010). Beyond CPT-11, elevated level lead to adverse side effects such as diarrhea through symbiotic bacterial enzyme action β-glucuronidases involved in phase 2 conjugation. This conversion to glucuronide conjugate results in incomplete detoxification in hepatic cells in recycled gut-secreted bile acids regenerated with CPT-11 drug action in gastrointestinal side effects (Wallace et al., 2010). Microbial β-glucuronidase inhibitors prevent enzyme action without ill effects; other microbial symbionts in mice (orally administered) and deconjugation CPT-11 drugs minimized toxicity effects (Wallace et al., 2010). *Helicobacter pylori* reduced levodopa drug action for Parkinson's disease therapy using bioavailability, whereas *H. pylori* elimination substantially increases levodopa level in serum increases during Parkinson's therapy (Disbrow et al., 2018). Microbial metabolism-enhanced β-glucuronidase activity reduces colon cancer treatment and increases gastrointestinal toxicity by blocking β-glucuronidase inhibitors, which increases irinotecan drug activity in intestinal carcinoma with minimized gastrointestinal side effects (Carmody and Turnbaugh, 2014) (Fig. 12.1).

12.10 Bioengineering biosynthetic process and heterologous production of metabolites

Perception of individual microbe and metabolites vital for microbiome engineering understanding novel functionalities. Cost-effective and efficient chemical synthesis of metabolite to improved bioengineered microbes to modify biosynthetic pathway of target metabolite production and host producing natural products three dimensional structure are known to be discovered. Heterologous expression of metabolite may offer enormous benefit without comprising chemical structure, genetic manipulation, foreign biosynthetic gene clusters (BGCs) higher expression, synthesizing optimum yield of secondary metabolites. Procedure for heterologous expression of BGCs using (1) appropriate vector selection with gene cassette, (2) heterologous host transformation, and (3) successful inducible expression of metabolite. Selection of heterologous host which is evolutionary associated with donor species for better gene expression. Example *Xenorhabdus* sp. is genetically associated with heterologous host (*Escherichia coli*) for successful BGCs expression (Huo et al., 2019; Cai et al., 2017).

Genetically distant microbial species for overexpression of protein using (1) optimization of codons biased, (2) engineering strong promoter and regulatory gene sequence, and (3) engineering of ribosome-binding site for increasing translational efficiency (Horbal and Luzhetskyy, 2016). Optimized actinomycetes BGCs expression in bioengineered *Streptomycetes albus*, *Streptomycetes lividans*, and *Streptomycetes coelicolor* (Huo et al., 2019). Antimalarial drug artemisinin overproduced in *Saccharomyces cerevisiae* in which artemisinic acid intermediate changed to artemisinin through efficient photochemical process via. singlet oxygen addition (Paddon et al., 2013). *E. coli* known for recombinant protein production as pegfilgrastim, fusion protein interleukin-2, insulin, filgrastim, endostatin (Folkman, 2006; Babaeipour et al., 2015; Molineux, 2004). Multimodular NRPS and PKS genes challenging an *E. coli* host expression to a lesser concentration of metabolic substrate, higher guanine-cytosine content, and codon biases (Cai et al., 2017). Thiolactomycin BGCs increased threefold production in commercial marine bacteria *Salinispora tropica* CNB-440 (Zhang et al., 2018). Argyrins (antibiotics A21459) extracted from actinomycetes function as potent antibacterial, anticancer, immunosuppressant activity, and BGCs expression is highly optimized in *S. coelicolor* and *Cystobacter species* (Huo et al., 2019; Sasse et al., 2002). *Myxococcus xanthus* DK 1622, genetically modified bacteria that carried the synthetic form of argyrins in BGC expressions failed due to inefficient genetic manipulation in native host NRPS pathways, which requires a substrate for megasynthetase and deficient tailoring methods. Myxobacterial synthetic DNA construct polyketide synthase-like LC-PUFA synthases BGCs were expressed in *Yarrowia lipolytica* (oleaginous yeast) consume trace level of nicotinamide adenine dinucleotide phosphate and precursors of acyl-CoA incorporation (Gemperlein et al., 2019). Docosahexaenoic acid (PUFA) de novo synthesis in *Y. lipolytica* construct produces 350 mg/L highest concentration with better quality followed by omega-3 fatty acids synthesis with same procedure (Gemperlein et al., 2019; Lenihan-Geels et al., 2013). Genomics integration of cryptic BGCs belongs to *Frankia alni* strain ACN14a were transferred to *S. albus* chassis strains J1074 encoded bacterial artificial chromosome (BAC) comprises 15 biosynthetic gene clusters. *S. albus* deletion mutant were created Del14 strain for enhancing novel fralnimycin production (Myronovskyia et al., 2018).

Cytochrome P450s from plant origin engineered in microbe for in vivo production of novel metabolites. Novel semi-biosynthetic platform for expressing high titer terpenoids in *E. coli*, which is fully functioned and cost-effective production at larger scale (Dietrich et al., 2009). Traditional complex phytochemical production of cannabinoid biosynthesis uses *S. cerevisiae* (Luo et al., 2019). The fermentation process creates better optimization in *S. cerevisiae*, which produced 300-fold thebaine and hydrocodone (opioids) than bacteria host expression (Galanien et al., 2015). Clinical taxol used as anticancer agent, which is extracted from pacific yew tree barks. This provides lower yield of metabolites extraction nearly 0.4% level. Paclitaxel precursor as taxadiene were overexpressed in *E. coli* generated 1 g/L (Kotopka et al., 2018; Ajikumar et al., 2010; Biggs et al., 2016). Optimized taxadiene-5α-ol production in *E. coli* strain bioengineered cytochrome P450 enzymes pathway favors high level oxygenation chemistry for taxol biosynthesis (Kotopka et al., 2018; Ajikumar et al., 2010; Biggs et al., 2016). Improved fermentation strategies to harvest phytochemicals of higher pharmaceutical importance uses two bacterial systems platforms due to bacterial cells lacking compartmentalization (Kotopka et al., 2018). Coculture fermentation platforms can be achieved through separating each module's metabolic pathway split and forwarded module to each dedicated microbial strain for soluble protein expression to avoid misfolded protein aggregation and membrane-bound proteins synthesis during stress conditions (Kotopka et al., 2018). Various types of *E. coli* strain with different gene constructs were growing in the same growth medium for phytochemicals production. These phytochemicals are flavonoids and resveratrol were produced through cocultivation of distinct microbes (Zhou et al., 2015; Jones et al., 2016). *S. cerevisiae* and *E. coli* co-culturing for expressing cytochrome P450 genes in order to produce scoulerine and corytuberine (Minami et al., 2008).

12.11 Microbiome functional characterization in terms of metabolites production

Several metagenomics sequencing project based on culturing novel microbes from environmental sample through environmental DNA (eDNA) and novel compound extraction for metabolite production.

12.11.1 Metagenomics

Microbiome screening using marker gene 16S ribosomal RNA sequencing platform for human and animal infectious models reveals our ideas toward diversified gut microbes. This paves the way for more accurate fingerprinting of evolutionary genomic origin and functional dynamics of microbial community even though strategies have been biased toward nonspecificity. Using high-throughput shotgun metagenomics sequencing helps gain functional insights into the gene functioning of target bacterial communities. Genome breaking and amplification of template DNA obtained short reads mapped using de novo assembly or reference genome and functional annotation done within a genome database. Whole genome mapping should be conducted using online and offline platforms such as METAPREP, MG-RAST, and HUMAnN (Abubucker et al., 2012; Meyer et al., 2008; Goll et al., 2010). Shotgun metagenomic sequencing is a technique for microbial DNA amplification without any host DNA sequence contamination. Sequencing data accuracy helps to improve mapping of human disease and strain-specific genes within the microbial community (Abubucker et al., 2012).

Malacidins A and malacidins B calcium-dependent antibiotics, discovered from metagenomics sequencing platforms, are successful in probing for novel secondary metabolite families (Hover et al., 2018). A 2000 special Earth-specimen probe for NRPS A domains for BGCs sequencing isolated a vast array of genomic collection using PCR with degenerate primers. Friulimicin and daptomycin structures distinct from malacidins, calcium-dependent antibiotics, and biosyntheses mediated through NRPS pathway is similar (Hover et al., 2018). Screening complex scaffold and novel eDNA remains challenging, but eventually functional metagenomics platforms have identified novel derivatives of chloramphenicol (Nasrin et al., 2018). One such phenomenon followed for endophytic bacterium metagenomics includes (1) sample isolated from trace level material, (2) structure determination using NMR spectroscopy, (3) collection data merged with synthetic biology, (4) eDNA sequencing provide completed functional characterization to novel discovery of metabolites, and (5) BGCs sequencing and expression in heterologous host. Exploring metagenomics-based gene discovery suitable for divamide A and anti-HIV lanthipeptide derived from tunicates, marine invertebrates carrying uncultured symbiotic bacteria (Smith et al., 2018). Fecal samples collected for potential screening of gut microbiota using illumina sequencing among 124 European volunteer datasets collected around 576.7 Gb obtained, and an annotated part reveals 3.3 million genes are nonredundant in nature from contigs assembly constitutes 99% belongs to bacteria (Luo et al., 2009). Metagenomics and metatranscriptomics datasets allow us to predict bacterial gene and transcript mapping to novel metabolites.

12.11.2 Metatranscriptomics

A snapshot metabolic pathway, cellular machinery, molecular function of specific genes, composition, and structural aspects of screening from gut microbes have impacts on human disease and health. Metatranscriptomic shotgun sequencing is one such approach for microbial gene regulation under various pathological conditions. Microbial RNA library construction using total microRNA, mRNA have been isolated and converted to complementary DNA (cDNA) using reverse transcriptase. The sequence been annotated from contigs assembly and nonredundant sequence screening novel genes and function using the reference genome database (Bashiardes et al., 2016). Pre-mRNA half-life is highly unstable, leads to poor RNA quality, however, needs to improve RNA integrity would be good for gene expression studies (Bashiardes et al., 2016). RNA-seq performed for short-period xenobiotics exposure to gut microbes revived altered protein biosynthesis, biosynthesis of vitamin, pentose phosphate pathway, host xenobiotic metabolism, and biodegradation (Haiser et al., 2013). Assessing multiomics approach to sieve best candidate microbes and metabolites.

12.11.3 Metabolomics

Lactobacillus fermentum and *Lactobacillus acidophilus* predominate gut bacteria contributions to the largest number of blood metabolites as *p*-cresol sulfate, phenylacetylglutamine, sebacate, and tartronate (Holmes et al., 2012). For example, one blood metabolite coordinated with 118 host bacterial species and 93 pathways of bacterial metabolism (Visconti et al., 2019). Fifty individuals have had feces and blood samples tested, screening for gut microbial ecosystems at functional and taxonomic levels with a perspective on human health, using whole metagenome shotgun sequencing and metabolomics studies associated with 191 microbial species, 347 metabolic pathway, and 713 metabolite retrieved from fecal sample analysis, while blood sample analysis revealed 233 microbial species, 370 metabolic pathways, and 673 blood metabolites (Visconti et al., 2019); 191 microbial species and 713 fecal metabolites showed 116,090 associations, and likewise, 713 fecal metabolites and 347 metabolic pathways showed 235,852 associations from fecal sample analyses (Visconti et al., 2019). Similarly, 233 species and 673 blood metabolites showed 150,792 associations and 673 blood metabolites, and 370 metabolic pathways showed 245,641 associations (Visconti et al., 2019).

Evaluate host metabolism and gut microbes fingerprinting using germ-free (GF) mice tissue experimentation with normal mice. Limitation factors are GF mice acknowledged with severe abnormalities in developmental stages and gastrointestinal tracts with disordered features such as a thinning of the intestinal wall, smaller villi, and enlargement of the cecum (Coates, 1975; Al-Asmakh and Zadjali, 2015; Nicklas et al., 2015). Discovering novel GF mice is a tedious process providing greater opportunities for the gut microbial associations with host metabolome, bacterial metabolites production, host phenotypes, and host pathological state. Control mice compared with GF mice revealed thousands of metabolites circulated in kidney, central nervous system, heart, liver, and peripheral blood serum (Martin et al., 2007; Wikoff et al., 2009; Claus et al., 2008; Matsumoto et al., 2013). Advances in analytical chemistry techniques such as LC-MS, and NMR for the discovery of novel metabolites such as nonribosomal peptides, glycolipids, polyketides, secondary bile acids, biogenic amines, posttranslationally modified peptides. Selected ion flow tube-mass spectrometry (SIFT-MS) used for screening markers and detecting volatile organic compounds (VOCs) for IBD.

Crohn's disease (CD) patient distinguished clearly from ulcerative colitis (UC) patients from detecting VOC, hydrogen sulfide, and dimethyl sulfide produced from microbes using SIFT-MS. Matrix-assisted laser desorption ionization time-of-flight mass-spectrometry (MALDI-TOF) application used for specific metabolite detection (Singhal et al., 2015). Evenly pure microbe culture obtained from complex fecal sample detection in spectra peaks output in high-throughput manner analysis with databases such as SetupX, Human Metabolome Database, BinBase, and MetaCyc (Kouskoumvekaki and Panagiotou, 2011). The Integrated Microbial Genomes Atlas of Biosynthetic Gene Clusters (IMP-ABC) (https://img.jgi.doe.gov/abc/) database developed for secondary metabolite discovery is based on 40,000 microbial isolates genome and 700,000 novel genes clusters (GCs) from published literature to fish out biologically active compounds using new interactive network graphs and heat map visualizations, facilitating in-depth analysis (Hadjithomas et al., 2015). Gut microbial contributions to xenobiotic metabolism and secondary metabolites production for better understanding of metabolomics research would have a substantial impact on human health. Regulation of the host drug metabolism and gut microbiota promotes ingested xenobiotic transport to hepatic cells (Haiser et al., 2013). Some drug availability to host tissues get altered due to xenobiotic metabolism, for example, cardiac glycoside (digoxin drug) for treating arrhythmias and cardiac failure (Haiser et al., 2013). In rare cases cardiac glycoside has incomplete efficiency in patients due to arginine present at higher levels, inhibiting drug action synthesis from *Eggerthella lenta* in the gut (Haiser et al., 2013).

12.12 Conclusion for future challenge

Bioengineering standards needs general understanding of microbial community function and physiological suitability, however, such intervention is fragmentary and lacking. The detoxification process of host drug toxicity and metabolism often route to microbial metabolites linked directly to drug metabolism on an enzymatic basis over thousands of years of natural selection in the gut. Drugs that target microbial genes for therapeutic purposes. Likewise, drugs that target regulated pathways of mammalian hosts that can enhance the modulation of gut microbes to improve human health. Designed drugs to help chemotherapeutics treatment with minimized side effects for patient health. Designed drugs to help prevent malicious enzyme actions eventually secreted from symbiotic microbes without harming the mammalian host. This provides a sequence of window for codevelopment age-associated gut microbiome enhancement throughout host circadian rhythms. Successful metabolic engineering is vital for the deregulation of biosynthetic and catabolic pathways. Biosynthetic machinery reprograming is a vital direction for selected metabolites and novel derivatives in heterologous host production, circumventing low yield and bottleneck associated with BGCs. Scenario of spatial metabolomics is the snapshot of small molecules with perceptive to specific location and chemical properties. Spatial metabolomics helps to understand immune modulation of the host, microbial metabolic interaction, in situ spatial relationship with host metabolism. Spatial metabolomics techniques called metaFISH to understand symbiotic relationship of host-microbes with respective to their metabolic interactions. This technique is combination of high-resolution atmospheric-pressure MALDI mass spectrometry and FISH microscopy to easy mapping of metabolite fluorescent images at micrometer scale (Geier et al., 2020). Metabolite extraction in plants is a challenging task due to difficulties in getting completed BGCs expressed in a bacterial host from cell and chromosome architectures (Azmir et al., 2013). Phytochemicals have low availability in a plant, but serve great importance in agricultural and pharmaceutical sectors. Promising alternative approaches for high-value phytochemical production that is cost-effective at an industrial-scale is achieved through bioengineered yeast and bacteria as heterologous hosts. Construction metagenomic DNA libraries of heterogeneous origin is a cumbersome process eventually for snapshotting novel BGCs in comparison with the sequence alignment of respective microorganism and logical choice of selecting suitable heterologous host. In rare cases, eDNA BGCs showing inefficient expression in heterologous hosts rather than native bacterial hosts would be a good choice of expression.

References

Abubucker, S., Segata, N., Goll, J., Schubert, A.M., Izard, J., Cantarel, B.L., Rodriguez-Mueller, B., Zucker, J., Thiagarajan, M., Henrissat, B., White, O., Kelley, S.T., Methe, B., Schloss, P. D., Gevers, D., Huttenhower, C., 2012. Metabolic reconstruction for metagenomic data and its application to the human microbiome. PLoS Comput Biol. 8 (6), e1002358.

Ajikumar, P.K., Xiao, W., Tyo, K.E.J., Wang, Y., Simeon, F., Leonard, E., Mucha, P., Phon, T.H., Pfeifer, B., Stephanopoulos, G., 2010. Isoprenoid pathway optimization for Taxol precursor overproduction in *Escherichia coli*. Science 330, 70–74.

Al-Asmakh, M., Zadjali, F., 2015. Use of germ-free animal models in microbiota-related research. J. Microbiol. Biotechnol. 25, 1583–1588.

Alves, A., Bassot, A., Bulteau, A., Pirola, L., Morio, B., 2019. Glycine metabolism and its alterations in obesity and metabolic diseases. Nutrients 11 (6), 1356.

Amedei, A., Morbidelli, L., 2019. Circulating metabolites originating from gut microbiota control endothelial cell function. Molecules 24 (21), 3992.

Azmir, J., Zaidul, I.S.M., Rahman, M.M., Sharif, K.M., Mohamed, A., Sahena, F., Jahurul, M.H. A., Ghafoor, K., Omar, A.K.M., 2013. Techniques for extraction of bioactive compounds from plant materials: a review. J. Food Eng. 117, 426–436.

Babaeipour, V., Khanchezar, S., Mofid, M.R., 2015. Efficient process development of recombinant human granulocyte colony-stimulating factor (rh-GCSF) production in *Escherichia coli*. Iran. Biomed. J. 19, 102–110.

Bansal, T., Alaniz, R.C., Wood, T.K., Jayaraman, A., 2010. The bacterial signal indole increases epithelial-cell tight-junction resistance and attenuates indicators of inflammation. Proc. Natl. Acad. Sci. U. S. A. 107, 228–233.

Bashiardes, S., Zilberman-Schapira, G., Elinav, E., 2016. Use of metatranscriptomics in microbiome research. Bioinform. Biol. Insights 10, 19–25.

Beyoglu, D., Smith, R.L., Idle, J.R., 2012. Dog bites man or man bites dog? The enigma of the amino acid conjugations. Biochem. Pharmacol. 83 (10), 1331–1339.

Biggs, B.W., Lima, C.G., Sagliania, K., Shankara, S., Stephanopoulosc, G., Meya, M.D., Ajikumara, P.K., 2016. Overcoming heterologous protein interdependency to optimize P450-mediated Taxol precursor synthesis in *Escherichia coli*. Proc. Natl. Acad. Sci. U. S. A. 113, 3209–3214.

Bruno-Barcena, J.M., Azcarate-Peril, M.A., 2015. Galacto-oligosaccharides and colorectal cancer: feeding our intestinal probiome. J Funct Foods. 12, 92–108.

Cai, X., Nowak, S., Wesche, F., Bischoff, I., Kaiser, M., Furst, R., Bode, H.B., 2017. Entomopathogenic bacteria use multiple mechanisms for bioactive peptide library design. Nat. Chem. 9, 379–386.

Carmody, R.N., Turnbaugh, P.J., 2014. Host-microbial interactions in the metabolism of therapeutic and diet-derived xenobiotics. J. Clin. Invest. 124, 4173–4181.

Cani, P.D., Possemiers, S., Van de Wiele, T., Guiot, Y., Everard, A., Rottier, O., et al., 2009. Changes in gut microbiota control inflammation in obese mice through a mechanism involving GLP-2-driven improvement of gut permeability. Gut 58 (8), 1091–1103.

Carvalho, F.A., Aitken, J.D., Gewirtz, A.T., Vijay-Kumar, M., 2011. TLR5 activation induces secretory interleukin-1 receptor antagonist (sIL-1Ra) and reduces inflammasome-associated tissue damage. Mucosal Immunol. 4 (1), 102–111.

Claus, S.P., Tsang, T.M., Wang, Y., Cloarec, O., Skordi, E., Martin, F.P., Rezzi, S., Ross, A., Kochhar, S., Holmes, E., Nicholson, J.K., 2008. Systemic multicompartmental effects of the gut microbiome on mouse metabolic phenotypes. Mol. Syst. Biol. 4, 219.

Coates, M.E., 1975. Gnotobiotic animals in research: their uses and limitations. Lab. Anim. 9, 275–282.

Davani-Davari, D., Negahdaripour, M., Karimzadeh, I., Seifan, M., Mohkam, M., Masoumi, S.J., et al., 2019. Prebiotics: definition, types, sources, mechanisms, and clinical applications. Foods. 8 (3), 92.

Dietrich, J.A., Yoshikuni, Y., Fisher, K.J., Woolard, F.X., Ockey, D., McPhee, D.J., Renninger, N.S, Chang, M.C.Y., Baker, D., Keasling, J.D., 2009. A novel semi-biosynthetic route for artemisinin production using engineered substrate-promiscuous P450BM3. ACS Chem. Biol. 4, 261–267.

Disbrow, E.A., Lu, X., McGee, D.J., 2018. Stomaching the possibility of a pathogenic role for *Helicobacter pylori* in Parkinson's disease. J. Parkinson's Dis. 8 (3), 367–374.

Dumas, M.E., Barton, R.H., Toye, A., Cloarec, O., Blancher, C., Rothwell, A., et al., 2006. Metabolic profiling reveals a contribution of gut microbiota to fatty liver phenotype in insulin-resistant mice. Proc. Natl. Acad. Sci. U. S. A. 103, 12511–12516.

Fang, W., Bi, D., Zheng, R., Cai, N., Xu, H., Zhou, R., et al., 2017. Identification and activation of TLR4-mediated signalling pathways by alginate-derived guluronate oligosaccharide in RAW264.7 macrophages. Sci. Rep. 7, 1663.

Feist A.M., Palsson B.O., 2010. The biomass objective function. Curr Opin Microbiol. 13 (3), 344–349.

Folkman, J., 2006. Antiangiogenesis in cancer therapy—endostatin and its mechanisms of action. Exp. Cell Res. 312, 594–607.

Fredrick, C.M., Lin, G., Johnson, E.A., 2017. Regulation of botulinum neurotoxin synthesis and toxin complex formation by arginine and glucose in *Clostridium botulinum* ATCC 3502. Appl. Env. Microbiol. 83 (13), e00642−17.

Galanien, S., Thodey, K., Trenchard, I.J., Smolke, C.D., 2015. Complete biosynthesis of opioids in yeast. Science 349, 1095−1100.

Geier, B., Sogin, E.M., Michellod, D., Janda, M., Kompauer, M., Spengler, B., Dubilier, N., Liebeke, M., 2020. Spatial metabolomics of in situ host−microbe interactions at the micrometre scale. Nat. Microbiol. 5, 498−510.

Gemperlein, K., Dietrich, D., Kohlstedt, M., Zipf, G., Bernauer, H.S., Wittmann, C., Wenzel, S. C., Muller, R., 2019. Polyunsaturated fatty acid production by *Yarrowia lipolytica* employing designed myxobacterial PUFA synthases. Nat. Commun. 10, 4055.

Goll, J., Rusch, D.B., Tanenbaum, D.M., Thiagarajan, M., Li, K., Methe, B.A., et al., 2010. METAREP: JCVI metagenomics reports—an open source tool for high-performance comparative metagenomics. Bioinformatics 26, 2631−2632.

Gomes, A., Oudot, C., Macia, A., Foito, A., Carregosa, D., Stewart, D., et al., 2019. Berry-enriched diet in salt-sensitive hypertensive rats: metabolic fate of (poly)phenols and the role of gut microbiota. Nutrients 11 (11), 2634.

Grangette, C., Nutten, S., Palumbo, E., Morath, S., Hermann, C., Dewulf, J., et al., 2005. Enhanced antiinflammatory capacity of a *Lactobacillus plantarum* mutant synthesizing modified teichoic acids. Proc. Natl. Acad. Sci. U. S. A. 102, 10321−10326.

Hadadi, N., Pandey, V., Chiappino-Pepe, A., Morales, M., Gallart-Ayala, H., Mehl, F., et al., 2020. Mechanistic insights into bacterial metabolic reprogramming from omics-integrated genome-scale models. NPJ Syst. Biol. Appl. 6, 1.

Hadjithomas, M., Chen, I.M., Chu, K., Ratner, A., Palaniappan, K., Szeto, E., Huang, J., Reddy, T.B, Cimermancic, P., Fischbach, M.A., Ivanova, N.N., Markowitz, V.M., Kyrpides, N.C., Pati, A., 2015. IMG-ABC: a knowledge base to fuel discovery of biosynthetic gene clusters and novel secondary metabolites. mBio 6, e00932.

Haiser, H.J., Gootenberg, D.B., Chatman, K., Sirasani, G., Balskus, E.P., Peter, J., Turnbaugh, P. J., 2013. Predicting and manipulating cardiac drug inactivation by the human gut bacterium *Eggerthella lenta*. Science 341, 295−298.

Harmrolfs, K., Mancuso, L., Drung, B., Sasse, F., Kirschning, A., 2014. Preparation of new alkyne-modified ansamitocins by mutasynthesis. Beilstein J. Org. Chem. 10, 535−543.

Hertel, J., Harms, A.C., Heinken, A., Baldini, F., Thinnes, C.C., Glaab, E., et al., 2019. Integrated analyses of microbiome and longitudinal metabolome data reveal microbial-host interactions on sulfur metabolism in Parkinson's disease. Cell Rep. 29 (7), 1767−1777.

Henao-Mejia, J., Elinav, E., Jin, C., Hao, L., Mehal, W.Z., Strowig, T., et al., 2012. Inflammasome-mediated dysbiosis regulates progression of NAFLD and obesity. Nature 482 (7384), 179−185.

Holmes, E., Kinross, J., Gibson, G.R., Burcelin, R., Jia, W., Pettersson, S., et al., 2012. Therapeutic modulation of microbiota-host metabolic interactions. Sci. Transl. Med. 4, 137rv6.

Holscher, H.D., Faust, K.L., Czerkies, L.A., Litov, R., Ziegler, E.E., Lessin, H., et al., 2012. Effects of prebiotic-containing infant formula on gastrointestinal tolerance and fecal microbiota in a randomized controlled trial. JPEN J. Parenter. Enteral. Nutr. 36 (Suppl. 1), 95S−105S.

Horbal, L., Luzhetskyy, A., 2016. Dual control system − a novel scaffolding architecture of an inducible regulatory device for the precise regulation of gene expression. Metab. Eng. 37, 11−23.

Hover, B.M., Estrela, A.B., Molina, H., Park, S., Perlin, D.S., Brady, S.F., 2018. Culture-independent discovery of the malacidins as calcium-dependent antibiotics with activity against multidrug-resistant Gram-positive pathogens. Nat. Microbiol. 3, 415–422.

Hug, J.J., Krug, D., Muller, R., 2020. Bacteria as genetically programmable producers of bioactive natural products. Nat. Rev. Chem. 4, 172–193.

Huo, L., Hug, J.J., Fu, C., Bian, X., Zhang, Y., Müller, R., 2019. Heterologous expression of bacterial natural product biosynthetic pathways. Nat. Prod. Rep. 36, 1412–1436.

Imai, A., Kurihara, Y., 1984. Endogenous infection in mice with streptozotocin-induced diabetes. A feature of bacterial translocation. Can. J. Microbiol. 30, 1344–1348.

Ji, T., Xu, C., Sun, L., Yu, M., Peng, K., Qiu, Y., et al., 2015. Aryl hydrocarbon receptor activation down-regulates IL-7 and reduces inflammation in a mouse model of DSS-induced colitis. Dig. Dis. Sci. 60, 1958–1966.

Jia, X., Osborn, L.J., Wang, Z., 2020. Trimethylamine (TMA) is a metabolite synthesis by microbes in the gut derived from enzymatic cleavage of TMA moiety. Molecules 25 (8), 1862.

Jones, M.L., Chen, H., Ouyang, W., Metz, T., Prakash, S. 2004. Microencapsulated genetically engineered Lactobacillus plantarum 80 (pCBH1) for bile acid deconjugation and its implication in lowering cholesterol. J. Biomed. Biotechnol., 61–69.

Jones, J.A., Vernacchio, V.R., Sinkoe, A.L., Collins, S.M., Ibrahim, M.H.A., Daniel, M., Lachance, D.M., Hahn, J., Koffas, M.A.G., 2016. Experimental and computational optimization of an *Escherichia coli* co-culture for the efficient production of flavonoids. Metab. Eng. 35, 55–63.

Kaddurah-Daouk, R., Weinshilboum, R.M., 2013. Pharmacometabolomics: implications for clinical pharmacology and systems pharmacology. Clin. Pharm. Ther. 95 (2), 154–167.

Kho, Z.Y., Lal, S.K., 2018. The human gut microbiome – a potential controller of wellness and disease. Front. Microbiol. 9, 1835.

Kotopka, B.J., Li, Y., Smolke, C.D., 2018. Synthetic biology strategies toward heterologous phytochemical production. Nat. Prod. Rep. 35, 902–920.

Kouskoumvekaki, I., Panagiotou, G., 2011. Navigating the human metabolome for biomarker identification and design of pharmaceutical molecules. J. Biomed. Biotechnol. 2011, 525497.

Kriaa, A., Bourgin, M., Potiron, A., Mkaouar, H., Jablaoui, A., Gerard, P., et al., 2019. Microbial impact on cholesterol and bile acid metabolism: current status and future prospects. J. Lipid Res. 60 (2), 323–332.

Kries, H., Niquille, D.L., Kries, D., 2015. A subdomain swap strategy for reengineering nonribosomal peptides. Chem. Biol. 22, 640–648.

Lamas, B., Richard, M.L., Leducq, V., Pham, H.P., Michel, M.L., Da Costa, G., et al., 2016. CARD9 impacts colitis by altering gut microbiota metabolism of tryptophan into aryl hydrocarbon receptor ligands. Nat. Med. 22, 598–605.

Laurans, L., Venteclef, N., Haddad, Y., Chajadine, M., Alzaid, F., Metghalchi, S., et al., 2018. Genetic deficiency of indoleamine 2,3-dioxygenase promotes gut microbiota-mediated metabolic health. Nat. Med. 24 (8), 1113–1120.

Lei, Q., Bi, J., Chen, H., Tian, F., Gao, X., Li, N., et al., 2018. Glucagon-like peptide-2 improves intestinal immune function and diminishes bacterial translocation in a mouse model of parenteral nutrition. Nutr. Res. 49, 56–66.

Lei, M., Menon, R., Manteiga, S., Alden, N., Hunt, C., Alaniz, R.C., et al., 2019. Environmental chemical diethylhexyl phthalate alters intestinal microbiota community structure and metabolite profile in mice. mSystems. 4 (6), e00724–19.

Lemon, K.P., Armitage, G.C., Relman, D.A., Fischbach, M.A., 2012. Microbiota-targeted therapies: an ecological perspective. Sci. Transl. Med. 4 (137), 137rv5.

Lenihan-Geels, G., Bishop, K.S., Ferguson, L.R., 2013. Alternative sources of omega-3 fats: can we find a sustainable substitute for fish? Nutrients 5, 1301–1315.

Li, J.V., Ashrafian, H., Bueter, M., Kinross, J., Sands, C., le Roux, C.W., et al., 2014. Metabolic surgery profoundly influences gut microbial–host metabolic cross-talk. Gut 60 (9), 1214–1223.

Luo, X., Reiter, M.A., Wong, J., Denby, C.M., Lechner, A., Zhang, Y., Keasling, J.D., 2019. Complete biosynthesis of cannabinoids and their unnatural analogues in yeast. Nature 567, 123–126.

McNulty, N., Yatsunenko, T., Hsiao, A., Faith, J.J., Muegge, B.D., Goodman, A.L., et al., 2011. The impact of a consortium of fermented milk strains on the gut microbiome of gnotobiotic mice and monozygotic twins. Sci. Transl. Med. 3, 106ra106.

Marchesi, J.R., Holmes, E., Khan, F., Kochhar, S., Scanlan, P., Shanahan, F., et al., 2007. Rapid and noninvasive metabonomic characterization of inflammatory bowel disease. J. Proteome Res. 6, 546–551.

Martin, F.P., Dumas, M.E., Wang, Y., Legido-Quigley, C., Yap, I.K., Tang, H., Zirah, S., Murphy, G.M., Cloarec, O., Lindon, J.C., Sprenger, N., Fay, L.B., Kochhar, S., van Bladeren, P., Holmes, E., Nicholson, J.K., 2007. A top-down systems biology view of microbiome-mammalian metabolic interactions in a mouse model. Mol. Syst. Biol. 3, 112.

Matsumoto, M., Kibe, R., Ooga, T., Aiba, Y., Sawaki, E., Koga, Y., et al., 2013. Cerebral low-molecular metabolites influenced by intestinal microbiota: a pilot study. Front. Syst. Neurosci. 7, 9.

Menon, B.R.K., Jenner, M., 2018. Biosynthesis: reprogramming assembly lines. Nat. Chem. 10 (3), 245–247.

Meyer, F., Paarmann, D., D'Souza, M., Olson, R., Glass, E.M., Kubal, M., et al., 2008. The metagenomics RAST server – a public resource for the automatic phylogenetic and functional analysis of metagenomes. BMC Bioinformatics 9, 386.

Minami, H., Kim, J., Ikezawa, N., Takemura, T., Katayama, T., Kumagai, H., et al., 2008. Microbial production of plant benzylisoquinoline alkaloids. Proc. Natl. Acad. Sci. U. S. A. 105, 7393–7398.

Molineux, G., 2004. The design and development of pegfilgrastim (PEG-rmetHuG-CSF, Neulasta®). Curr. Pharm. Des. 10, 1235–1244.

Myronovskyia, M., Rosenkränzera, B., Nadmidc, S., Pujicb, P., Normandb, P., Luzhetskyy, A., 2018. Generation of a cluster-free *Streptomyces albus* chassis strains for improved heterologous expression of secondary metabolite clusters. Metab. Eng. 49, 316–324.

Nasrin, S., Ganji, S., Kakirde, K.S., Li, X., Liles, M.R., 2018. Chloramphenicol derivatives with antibacterial activity identified by functional metagenomics. J. Nat. Prod. 81, 1321–1332.

Nicaise, C., Prozzi, D., Viaene, E., Moreno, C., Gustot, T., Quertinmont, E., Demetter, P., Suain, V., Goffin, P., Devière, J., Hols, P., 2008. Control of acute, chronic, and constitutive hyperammonemia by wild-type and genetically engineered Lactobacillus plantarum in rodents. Hepatology 48, 1184–1192.

Nicholson, J.K., Holmes, E., Kinross, J., Burcelin, R., Gibson, G., Jia, W., et al., 2012. Host gut microbiota metabolic interactions. Science 336 (6086), 1262–1267.

Nicklas, W., Keubler, L., Bleich, A., 2015. Maintaining and monitoring the defined microbiota status of gnotobiotic rodents. ILAR J. 56, 241–249.

Osbelt, L., Thiemann, S., Smit, N., Lesker, T.R., Schröter, M., Gálvez, E.J.C., et al., 2020. Variations in microbiota composition of laboratory mice influence *Citrobacter rodentium* infection via variable short-chain fatty acid production. PLoS Pathog. 16 (3), e1008448.

Paddon, C.J., Westfall, P.J., Pitera, D.J., Benjamin, K., Fisher, K., McPhee, D., et al., 2013. High-level semi-synthetic production of the potent antimalarial artemisinin. Nature 496, 528–532.

Papandreou, C., More, M., Bellamine, A., 2020. Trimethylamine N-oxide in relation to cardiometabolic health—cause or effect? Nutrients 12, 1330.

Park, J., Wang, Q., Wu, Q., Mao-Draayer, Y., Kim, C.H., 2019. Bidirectional regulatory potentials of short-chain fatty acids and their G-protein-coupled receptors in autoimmune neuroinflammation. Sci. Rep. 9, 8837.

Pinu, F.R., Granucci, N., James, D., Han, T., Carneiro, S., Rocha, I., et al., 2018. Metabolite secretion in microorganisms: the theory of metabolic overflow put to the test. Metabolomics 14 (4), 43.

Pique, N., Berlanga, M., Minana-Galbis, D., 2019. Health benefits of heat-killed (Tyndallized) probiotics: an overview. Int. J. Mol. Sci. 20, 2534.

Quigley, E.M.M., Gajula, P., 2020. Recent advances in modulating the microbiome. F1000Res 9, F1000.

Rogers, G.B., Keating, D.J., Young, R.L., Wong, M.L., Licinio, J., Wesselingh, S., 2016. From gut dysbiosis to altered brain function and mental illness: mechanisms and pathways. Mol. Psychiatry 21, 738–748.

Sahner, J.H.J., Sucipto, H., Wenzel, S.C., Groh, M., Artmann, R.W., Mìller, 2015. Advanced mutasynthesis studies on the natural α-pyrone antibiotic myxopyronin from *Myxococcus fulvus*. ChemBioChem 16, 946–953.

Sanada, S., Suzuki, T., Nagata, A., Hashidume, T., Yoshikawa, Y., Miyoshi, N., 2020. Intestinal microbial metabolite stercobilin involvement in the chronic inflammation of ob/ob mice. Sci. Rep. 10, 6479.

Sasse, F., Steinmetz, H., Schupp, T., Petersen, F., Memmert, K., Hofmann, H., et al., 2002. Argyrins, immunosuppressive cyclic peptides from myxobacteria. I. Production, isolation, physico-chemical and biological properties. J. Antibiot. 55, 543–551.

Shier, W.T., Rinehart, K.L., Gottlieb, D., 1969. Preparation of four new antibiotics from a mutant of *Streptomyces fradiae*. Proc. Natl. Acad. Sci. U. S. A. 63, 198–204.

Shimada, Y., Kinoshita, M., Harada, K., Mizutani, M., Masahata, K., Kayama, H., et al., 2013. Commensal bacteria-dependent indole production enhances epithelial barrier function in the colon. PLoS One 8 (11), e80604.

Simone, C., 2019. The unregulated probiotic market. Clin. Gastroenterol. Hepatol. 17 (5), 809–817.

Singh, R., Chandrashekharappa, S., Bodduluri, S.R., Baby, B.V., Hegde, B., Kotla, N.G., et al., 2019. Enhancement of the gut barrier integrity by a microbial metabolite through the Nrf2 pathway. Nat. Commun. 10, 89.

Singhal, N., Kumar, M., Kanaujia, P.K., Virdi, J.S., 2015. MALDITOF mass spectrometry: an emerging technology for microbial identification and diagnosis. Front. Microbiol. 6, 791.

Smith, T.E., Wyche, T.P., Matainaho, T.K., Bugni, T.S., Barrows, L.R., Ireland, C.M., Schmidt, E.W., 2018. Accessing chemical diversity from the uncultivated symbionts of small marine animals. Nat. Chem. Biol. 14, 179–185.

Tesmar, A.V., Hoffmann, M., Fayad, A.A., Huttel, S., Schmitt, V., Herrmann, J., et al., 2018. Biosynthesis of the *Klebsiella oxytoca* pathogenicity factor tilivalline: heterologous expression, in vitro biosynthesis, and inhibitor development. ACS Chem. Biol. 13 (3), 812–819.

Tiihonen, K., Ouwehand, A.C., Rautonen, N., 2010. Human intestinal microbiota and healthy ageing. Ageing Res. Rev. 9, 107–116.

Verbeke, K.A., Boobis, A.R., Chiodini, A., Edwards, C.A., Franck, A., Kleerebezem, M., et al., On behalf of the ILSI Europe prebiotics task force expert group-microbial metabolism and fermentation 2015. Towards microbial fermentation metabolites as markers for health benefits of prebiotics. Nutr. Res. Rev. 28 (1), 42–66.

Visconti, A., Roy, C.L., Rosa, F., Rossi, N., Martin, T.C., Mohney, R.P., Li, W., Rinaldis, E.D., Bell, J.T, Venter, J.C., Nelson, K.E., Spector, T.D., Falchi, M., 2019. Interplay between the human gut microbiome and host metabolism. Nat. Commun. 10 (1), 4505.

Wallace, B.D., Wang H., Venkatesh M., Jobin C., Yeh Li-An, Mani, S., Redinbo, M.R., 2010. Alleviating cancer drug toxicity by inhibiting a bacterial enzyme. Science. 330 (6005), 831–835.

Wang, L., Ravichandran, V., Yin, Y., Yin, J., Zhang, Y., 2019. Natural products from mammalian gut microbiota. Trends Biotechnol. 37 (5), 492–504.

Wikoff, W.R., Anfora, A.T., Liu, J., Schultz, P.G., Lesley, S.A., Peters, E.C., et al., 2009. Metabolomics analysis reveals large effects of gut microflora on mammalian blood metabolites. Proc. Natl. Acad. Sci. U. S. A. 106, 3698–3703.

Willinger, T., 2019. Metabolic control of innate lymphoid cell migration. Front. Immunol. 10, 2010.

Wong, A.C., Levy, M., 2019. Microbiome and microbiota manipulation would result of novel therapy for curing gut dysbiosis. Systems 4 (3), e00122–19.

Yap, I.K., Angley, M., Veselkov, K.A., Holmes, E., Lindon, J.C., Nicholson, J.K., 2010. Urinary metabolic phenotyping differentiates children with autism from their unaffected siblings and age-matched controls. J. Proteome Res. 9 (6), 2996–3004.

Zhang, L.S., Davies, S.S., 2016. Microbial metabolism of dietary components to bioactive metabolites: opportunities for new therapeutic interventions. Genome Med. 8, 46.

Zhang, J.J., Moore, B.S., Tang, X., 2018. Engineering *Salinispora tropica* for heterologous expression of natural product biosynthetic gene clusters. Appl. Microbiol. Biotechnol. 102, 8437–8446.

Zhou, K., Qiao, K., Edgar, S., Stephanopoulos, G., 2015. Distributing a metabolic pathway among a microbial consortium enhances production of natural products. Nat. Biotechnol. 33, 377–383.

Further reading

Gacias, M., Gaspari, S., Santos, P.S., Tamburini, S., Andrade, M., Zhang, F., et al., 2016. Microbiota-driven transcriptional changes in prefrontal cortex override genetic differences in social behavior. eLife 5, e13442.

Li, S., Li, Y., Smolke, C.D., 2018. Strategies for microbial synthesis of high-value phytochemicals. Nat. Chem. 10 (4), 395–404.

Maurice, C.F., Haiser, H.J., Turnbaugh, P.J., 2013. Xenobiotics shape the physiology and gene expression of the active human gut microbiome. Cell 152, 39–50.

Yan, F., Auerbach, D., Chai, Y., Keller, L., Tu, Q., Hüttel, S., et al., 2018. Biosynthesis and heterologous production of vioprolides: rational biosynthetic engineering and unprecedented 4-methylazetidinecarboxylic acid formation. Angew. Chem. Int. Ed. 57 (28), 8754–8759.

Zheng, X., Xie, G., Zhao, A., Zhao, L., Yao, C., Chiu, N.H., et al., 2011. The footprints of gut microbial-mammalian co-metabolism. J. Proteome Res. 10 (12), 5512–5522.

Chapter 13

Fungal strains as source of bioactive compounds and their potential application

Monika Singh
Centre of Advanced Study in Botany, Banaras Hindu University, Varanasi, India

Chapter Outline

13.1 Introduction

Microorganisms play a crucial role in the field of pharmaceutical and agriculture due to their nature to produce useful natural compounds. Among all microorganisms, fungi persist in many habitats, such as in terrestrial, marine, freshwater, air, and also associated with plants and in animals, although the majority of them are terrestrial, persisting independently or associated with plants. The chemical composition of fungi are highly complex because they are structurally different from plants and animals, but bioactive natural products, which have been produced as secondary metabolites, are more or less similar. Nearly all of the natural compounds hold several biotechnological applications in a different field, including pharmaceutical and agriculture. The bioactive compounds show different diverse groups of biologically active products. The diversity of these natural compounds guides them biologically active against different human diseases, so they play a key role in

Volatiles and Metabolites of Microbes. DOI: https://doi.org/10.1016/B978-0-12-824523-1.00019-5

the cure of human health. Biological natural compounds are the chief provenance of drugs in the pharmaceutical industry. According to research, nearly 61% isolated natural bioactive compounds have been developed into useful drugs in the whole world for the past 22 years (time period between 1981 and 2002) (Moore and Frazer, 2002; Cox, 2007; Zerikly and Challis, 2009).

Nearly 78% of antibacterial and 75% of anticancer compounds are bioactive natural compounds, which are useful to combat diseases. Thus, bioactive products offer a unique platform for the development and formation of natural drugs. Consequential improvement and variation in the microbial origin bioactive compounds discovery are confined to the management of nutritive supplements and environmental influenced factors, which enhanced the synthesis process of these natural products. At the same time, the little changes in environment-related factors have the ability to alter the yield, characteristics, and diversity of these compounds (Newman et al., 2003; Lopez et al., 2007).

Viral disease inhibition occurs due to the use of viral protein content; it is an important biological part in antiviral drug discovery and development. Recently, viral protease inhibitor medicines, especially HIV-1 protease inhibitor drugs, have been available for human medical use in the treatment of coronaviral diseases. Although these antiviral drugs can have unfavorable side effects, and they might be ineffective due to definitive drug resistance. Thus, the discovery of natural biologically active compounds, which were obtained from microbial sources that have inhibitory capabilities against HIV-1 protease activity. Fungi strains are a producer of natural secondary metabolites that offer therapeutic capabilities in the prevention and control of virus-induced diseases and for the modification of human immune response. Fungi, as producers of protease inhibitors and synthesized fungal bioactive compounds, have immunomodulatory activities as potential curing agents of coronaviral diseases in the near future (Suwannarach et al., 2020). Novel medicines that have a targeted mode of action are much required to treat severe diseases and other fatal health problems. However, bioactive products have proven to be potent and ecofriendly compounds to combat diseases. Plants and microorganisms are the major sources that we depend upon our need to discover bioactive compounds of pharmaceutical use (Deepika et al., 2016).

Endolichenic fungi lives within every explant of lichens, a unique type of mutualism occur between a fungal strain and supplements providing photosynthetic symbiont (Diaz et al., 2016). Endolichenic fungi are similar to the fungal endophytes found in explants. Lichens are found in at least 8% of the total terrestrial region of Earth with the capability to grow on various surfaces such as rocks, shrubs, and plants tree barks, as well as anthropogenic surfaces such as metals, plastic, glass, concrete, and gravestones. Nine endolichenic fungal strains were isolated from the long-whiskered ruffle lichen *Parmotrema rampoddense* (Nyl.) Hale. Out of these isolated fungal strains, three species, viz, *Fusarium proliferatum*, *Nemania primolutea*, and *Daldinia eschsholtzii* showed antimicrobial activities against bacterial pathogens using a disk diffusion assay method. *F. proliferatum* was the most potent fungal strain which formed maximum zone of inhibition against *Enterococcus faecalis* and *Staphylococcus aureus* bacterial strains. Further purification of the *F. proliferatum* fungal extract with the chromatographic techniques led to the extraction and characterization of bis (2-Ethylhexyl) terephthalate, acetyl tributyl citrate and fusarubin compounds. Acetyl tributyl citrate compound showed medium antimicrobial activity against *Klebsiella pneumoniae, Pseudomonas aeruginosa*, and *S. aureus* bacterial strains (Tan et al., 2020).

13.2 Marine fungi

The marine environmental condition is recently well explored as one of the most prominent sources regarding bioactive products extraction since microorganisms from the marine environment have unique biological, biochemical, and biosynthetic properties. Similarly, fungal microbes synthesized as biologically active products represent a broad area for novel drugs of human interest. Many research articles point out that fungal metabolites are main targets for discovery and development of novel therapeutic drugs, especially anticancer, antibiotics, antifungal, and antiparasitics. Marine endophytic fungi are also having different sources of novel bioactive compounds with curing potential and therapeutic applications due to their immense diversity. Fungal isolates also have the ability to synthesize secondary metabolites of pharmaceutical use (Jeewon et al., 2019). Marine fungal strains are a magnificent source of bioactive products, which are useful for the pharmaceutical industry. These bioactive natural compounds are mainly from these classes such as alkaloids, monoterpenoids, peptides, polyketides, and lactones (Youssef et al., 2019).

In the research, marine fungal strains are little worked in comparison to terrestrial fungi, a number of useful points have been obtained from the medicine discovery perspective (Molinski et al., 2009; Butler et al., 2014). A bottom sea-originated fungus *Simplicillium obclavatum* EIODSF 020 also showed medicinal values. *S. obclavatum* EIODSF 020 broth synthesized eight different novel linear peptides, simplicilliumtides A, B, C, D, E, F, G, and simplicilliumtides H. Compounds A and B are linear tetrapeptides possessing a 2-aminobenzoic acid residue. Compounds C and H are acetylated linear tri- and di-peptides. They showed antifouling, antimicrobial, cytotoxicity against cell lines, and acetylcholinesterase-inhibiting properties. Compound simplicilliumtides D showed strong antifouling activity against brown bryozoans (*Bugula neritina*) larvae and compound A, E, G, and H showed less cytotoxicity against HL-60 or K562 leukemia cell lines in humans (Liang et al., 2016). Marine-originated fungal strains are a rich and excellent source of novel carcinopreventive agents (Newman and Hill, 2006;

Bhadury et al., 2006). Higher rhizospheric, endophytic, and filamentous fungi residing in marine habitats produced bioactive natural products, such as leptospherin, leptosphaerolide, and leptosphaerodione from the lignicolous fungal strain *Leptosphaeria oraemaris* (Pallenberg and White, 1986; Guerriero et al., 1991). Four marine fungal strains *Aspergillus similanensis* KUFA 0013, *Neosartorya paulistensis* KUFC 7897, *Talaromyces trachyspermus* KUFC 0021, and *Neosartorya siamensis* KUFA 0017 were isolated, the extracts of two strains *N. paulistensis* and *N. siamensis* showed antiproliferative and apoptotic activities against A375 HepG2 (human melanoma) and HCT16 cell lines (Ramos et al., 2015).

13.3 Terrestrial fungi

Terrestrial fungal strains also play an essential role in the synthesis of bioactive products. Polypores are a group of terrestrial fungi that forms large fruiting bodies with the help of pores or tubes on the underside, and they belong to the phylum Basidiomycota (basidiomycetes). Polypores are strong producers of biologically active compounds. Near about 25,000 species of the class basidiomycetes are reported, and out of them about 500 species are present in the order Aphyllophorales; it also carries the polypores fungi. Many of the polypores have a wide range of distribution on all inhabited continents (except Antarctica) especially in North America, Europe, Africa, and Asia. The polypore fungi *Ganoderma lucidum* (Reishi or lingzhi mushroom), *Laetiporus sulphureus* (Crab-of-the-Woods or sulfur shelf), *Trametes versicolor* (Yunzhi), *Grifola umbellata* (Zhu Lin), *Inonotus obliquus* (Chaga), and *Wolfiporia cocos* (Fu-ling or Hoelen) showed strong antimicrobial activity. These polypores are also a rich source to develop novel antibiotics. Several compounds from these polypores are also having antiviral, cytotoxic, antineoplastic activities, and immunopotentiators. They also showed anticancer effects to prevent tumor metastasis (Zjawiony, 2004). A new terrestrial fungi *Aspergillus* sp. DHE 4 had broad-spectrum antibacterial activity. Fermentation on rice agar medium followed by filtration and purification got many bioactive natural compounds. Biochemical analysis of these bioactive compounds was confirmed by using some modern spectroscopic techniques. The results revealed that R (−)-mevalonolactone and poly-hydroxy steroidal compounds were produced from the *Aspergillus* sp. DHE 4 (Shokri et al., 2020).

Penicillium is one of the most frequently occurring fungal genera in the fungi kingdom. It has more than 300 identified species. It is ubiquitously found in most of the regions; terrestrial, marine and in the extreme environments (Kirk et al., 2008). A thermophilic fungal strain *Penicillium* sp. was isolated from hot spring sediments in Saudi Arabia. Fungal extract of *Penicillium* grown on rice agar medium. Five compounds were extracted, out of these two novel compounds 3-(furan 12-carboxylic acid)-6-(methoxycarbonyl)-4-hydroxy-4-methyl-4 and 5-dihydro-2H-pyran 13α-methyl-7-hydroxy-5-carboxylic acid methyl ester-1-indanone and three already known compounds austinol, emodin, and 2-methyl-penicinoline were isolated. Novel compounds of chemical structures were determined with the help of NMR spectroscopy and high-resolution mass spectrometry. All five extracted compounds were evaluated for their antimicrobial effect against some human pathogenic bacteria and assessed their toxicity against HTB-176, a human lymphoma cancer cell line. The known compound Austinol showed strong antimicrobial effect against *P. aeruginosa* bacterial strain, whereas emodin showed significant cytotoxicity against the HTB-176 cell line, while the other novel and known compounds were moderate to inactive in this antimicrobial and cancer cell line assay (Orfali and Perveen, 2019b). Endophytes are a major group of microorganisms having the capacity to bond between terrestrial and marine plants, and microbes have attracted the most attention due to their long life and helpfulness in stress conditions. The continuous supply of these bioactive compounds from fungi and especially from endophytes is severely hampered by the process of attenuation (Deepika et al., 2016).

13.4 Bioactive compounds synthesis by fungi: molecular aspects

Bioactive compounds are phytochemicals found in microorganisms independently or associated with plants, and also in edible things that are capable of enhanced metabolic processes and resulting in the promotion of good health. These compounds have a beneficial effect such as antioxidant activity, enzyme inhibition or induction, inhibition of receptor activities, and alterations in gene expressions in the organisms (Correia et al., 2012). Fungi are capable of producing metabolites with plant association. Several fungal endophytic isolates have the potential to produce bioactive compounds. Isolation of fungal strains is relatively a quite easy procedure, but an examination of the isolated fungal strains for required bioactive compound production is a sturdy process. Plant-associated fungi producing bioactive compounds may have several genes involved in the whole biosynthetic pathways. Therefore, ascertaining the presence of pivotal enzymes of a biosynthetic pathway could serve as a molecular marker for identification of these endophytes to produce the bioactive natural product (Vasundhara et al., 2016).

Bioactive compounds are derived from metabolic pathways with critical initial building blocks that are acyl-CoAs and that is fed into the synthesis of polyketide and terpenoid compounds and amino acids being used for the synthesis of nonribosomal peptide compounds such as penicillin (Keller, 2019). In contrast to molecular genes that are required for the synthesis of the bioactive compounds that are present in the entire fungal genome, the alleles coding the activities to produce any biologically active compounds are arranged in an array as a biosynthetic gene cluster. Secondary metabolites play a crucial role in fungal development and form an interaction with other organisms. Indeed, genes within a biosynthetic gene cluster are often coregulated with ecological functions. For example, the biosynthetic gene cluster that encodes pigments in *Aspergillus fumigatus* fungus is induced during spore synthesis in this strain (Lind et al., 2018). The biosynthetic gene cluster that encodes the virulence factor in *Fusarium graminearum* strain against the toxic fungal metabolites trichothecene (Lysoe et al., 2011) and the *Fusarium* sp. biosynthetic gene cluster that encodes the antibacterial compound bikaverin is expressed against with *Ralstonia solanacearum* bacterial strain (Spraker et al., 2018). Secondary metabolites produced by filamentous fungus constitute a valuable source of biologically rich compounds. In a study, authors have concluded that the molecular and biosynthetic pathways of an antibiotic yanuthone D isolated from a fungi *Aspergillus niger*. Yanuthone D is a meroterpenoid compound, and a polyketide 6-methylsalicylic acid (6-MSA) is the precursor molecule for formation of yanuthone D, its formation depends on a 10 gene cluster yan A to yan I and yan R. yan A and yan I, encodes a 6-MSA polyketide synthase and *O*-mevalon transferase, respectively. In addition, several branching spots in the synthesis pathway were also identified, revealed five yanuthone (yanuthone F, G, H, I, and R). Besides this, authors have found a different compound (yanuthone X1) that defines a different class of yanuthone antibiotic that depend on many enzymatic functions. Four novel yanuthone (1−4) antibiotics were extracted from the *A. niger* fungal strain (Holm et al., 2014; Petersen et al., 2015; Fig. 13.1).

Camptothecin (CPT) is an alkaloid compound; it is extracted from the stem wood of *Camptotheca acuminate* tree. This compound completely inhibits the DNA strand breaking enzyme DNA topoisomerase I. The bark of this tree is used on a large scale to prepare Chinese medicines (Wall et al., 1966). Later, CPT was discovered in many other species of plants. The synthesis pathway of CPT in trees is less characterized (Yamazaki et al., 2003, 2004). An initial compound of terpenoid synthesis is strictosidine synthase enzyme which forms Strictosidine from tryptamine and secologanin. Strictosidine is also considered the precursor for CPT. This CPT biosynthetic initial compound was isolated with the hairy roots of *Ophiorrhiza pumila* (Yamazaki et al., 2003), *Catharanthus roseus* (McKnight et al., 1990), and *Rauvolfia serpentina* (Kutchan et al., 1988). Sun et al. (2011) studied three different genes, namely, geraniol-10-hydroxylase, secologanin synthase and strictosidine synthase isolated and characterized from Camptotheca *acuminata* involved in CPT biosynthesis. Additionally, CPT biosynthetic gene from a endophytic fungal strain *Fusarium solani* isolated from *C. acuminata*. They have TDC gene (shikimate pathway) and G10H and SLS genes (mevalonate pathway); it concluded that the endophytic fungi may be used for the synthesis of CPT (Kusari et al., 2011; Kumara et al., 2014).

The lactam antibiotic "penicillin" synthesizing cluster genes from the filamentous fungal isolates *Aspergillus nidulans, Penicillium chrysogenum, Penicillium nalgiovense*, and *Penicillium notatum*, are more similar to genes

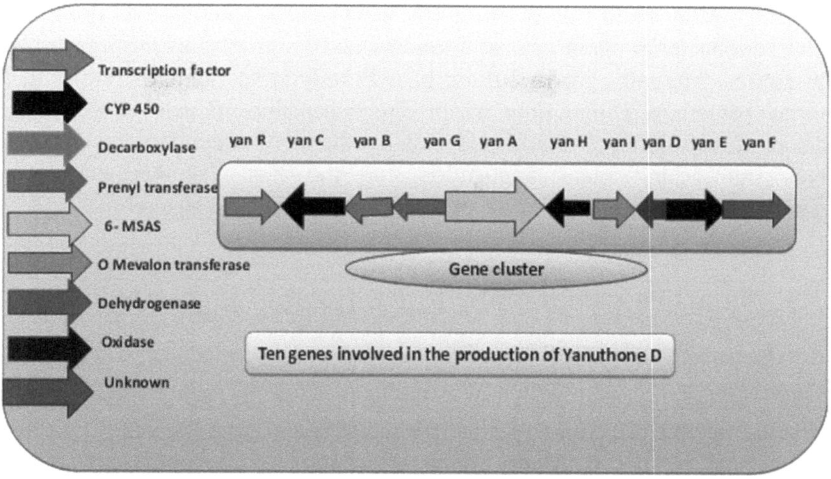

FIGURE 13.1 Pathway for the synthesis of antibiotic Yanuthone D from filamentous fungi (Holms et al., 2014).

present in *Acremonium chrysogenum* sp. for the synthesis of another lactam antibiotic cephalosporin (Gutierrez et al., 1999). In synthesis pathway of both penicillin and cephalosporin, a pcbC gene is required for the production of isopenicillin-*N*-synthase enzyme for the formation of a lactam ring. Some naturally existing strains of *P. chrysogenum* and their mutant strains were found to have the pcbC gene (lactam ring formation) but have a different ability to produce penicillin. Results concluded fungal strains synthesize penicillin at low amounts, having a single copy of the cluster gene, whereas over synthesizing fungal strains showed multiple copies of the cluster gene (Aharonowitz et al., 1992; Fierro et al., 1995, 1996). The availability of complete genome sequencing of fungi and novel bioinformatics techniques can lead to the development of many biologically active compounds (Brakhage et al., 2009; Chiang et al., 2009; Hertweck, 2009). Biosynthetic gene clusters identification of *A. nidulans* have done, there 53 well-known secondary metabolite genes were identified in their genome sequence. It is persisting in a big issue to get the work of these fungal strain gene clusters (Brakhage et al., 2008; von Dohren, 2009). Microorganisms bonding in different locations play a major role to get bioactive compounds from microbes. A study carried out with a group of 58 different species of actinomycetes simultaneously, only one strain *Streptomyces rapamycinicus* was found to activate the expression of selenic acid synthase gene in *A. nidulans* strain (Schroeckh et al., 2009; Netzker et al., 2015). The appearance of gene clusters in the synthesis of emericellamide a cyclic depsipeptides compound, and it was discovered earliest in *A. nidulans* and raised to 100 times more by the mutually grown culture of marine actinomycetes *Salinispora arenicola* and a fungal species of *Emericella* (Oh et al., 2007; Chiang et al., 2008). Pestalone is a strong antibacterial agent against *S. aureus* and *Enterococcus faecium* bacterial strains. Mutually grown culture of a *Pestalotia* sp. fungal strain with a marine bacterium *Thalassopia* sp. produced pestalone (Cueto et al., 2001; Ola et al., 2013). It is confirmed that synthesis of secondary metabolites gives out a biological application for the host (synthesizer) in a suitable environmental habitat (Firn and Jones, 2000; Brakhage et al., 2009).

13.5 Bioactive compounds from different fungal origin

13.5.1 Endophytic origin

Endophytic microorganisms reside in the plants, and they are the pool of bioactive compounds, including secondary metabolites, phenols, and flavonoids (Singh et al., 2017). Different groups of bioactive compounds such as terpenes, antibiotics, aliphatic compounds, alkaloids, polyketides, and peptides synthesize by different endophytic fungal strains from the terrestrial and marine origin. Some research deduced host plants and their associated endophytic fungi both were able to synthesize the same bioactive natural products. Modified classes of terpenes (isoprenoids or terpenoids) and polyketide antibiotics are the highly prominent bioactive compounds from endophytic strains. Fungal gene clusters coding antibacterial compounds are arranged on DNA strands. Many different fungal groups can give out the functionally identical bioactive compounds. The cluster of genes may help the sharing of antimicrobial bioactive compounds between fungal strains (Mousa and Raizada, 2013).

Fungal endophytes are also playing a major role in the synthesis of biologically active compounds (Debbab et al., 2013). Recently, several research studies on biologically active fungal compounds have rapidly increased for the use in various threatening diseases (Chen et al., 2017). In a study, nineteen endophytic fungal strains were isolated from *Lafoensia pacari* (Mangawa brava), *Guazuma ulmifolia* (bay cedar), *Campomanesia xanthocarpa* (gabiroba), and *Siparuna guianensis* (Brazilian medicinal plants). Out of these, 17 strains were identified on the basis of 16 rRNA Gene sequencing, all are from the genera *Bjerkandera, Colletotrichum, Cochliobolus, Curvularia, Diaporthe, Phaeophlebiopsis, Talaromyces*, and *Xylaraceae*. Their antioxidant activities were carried out by DPPH (2, 2-diphenyl-1-picryl-hydrazyl-hydrate) free radical scavenging method and ABTS assay. Results revealed that two *Talaromyces* and *Colletotrichium* genus species were taken to check the availability of flavonoids and antiinflammatory nature. A significant correlation was assessed between these two activities. The antiinflammatory activity of endophytic fungi was not well explored; however, the *Talaromyces* and *Colletotrichium* showed strong results for the protection of erythrocytes. This study concluded that fungal endophytes associated with host plants are excellent sources of newly synthesized bioactive compounds

(Santos et al., 2020). Fifty-six endophytic fungi were isolated from explants of *Salvia abrotanoides* plants obtained from three different geographical locations in Iran. On the basis of 16S rRNA gene sequencing, the strains were classified into 15 different genera, and some strains are *Penicillium canescens, P. chrysogenum, Penicillium charlesii, Penicillium murcianum, Penicillium* sp., *Talaromyces* sp., *Talaromyces verruculosus, Fusarium dlaminii, F. solani, Paraphoma radicina,* and *Coniolariellahispanica.* Fungal extracts of these strains were applied to the spectrometric analysis and found a broad field of useful bioactive compounds. A diterpenoid cryptotanshinone which is a major bioactive compound of *S. abrotanoides* plant was synthesized by *Penicillium* sp., and *Coniolariellahispanica* strains. Endophytic fungal isolates have a major role in the synthesis of secondary metabolites of plant, and also they give a good platform to cryptotanshinone synthesis in *S. abrotanoides* (Teimoori-Boghsani et al., 2020). Schulz et al. (2002) isolated approximately 6500 fungal isolates from shrubs, trees, and algal thalli; they were screened as endophytic fungi for their characteristics and chemical aspects. The explants of different species, that is, *Fucus vesiculosus*, *Fucus serratus*, *Fucus spiralis*, *Laminaria* sp., *Ceramium* sp., *Ascophyllum nodosum*, *Halarachnion ligulatum*, *Plumaria elegans*, *Enteromorpha* sp. were taken to isolate the endophytic fungi. A strong correlation was found between their function and ecological niche, and endophytic fungal strains were also showed as antialgal and herbicidal activities against the test organisms.

An endophytic fungal strain *Phomopsis liquidambaris* CBR-15 was isolated from wax-leaved climber plants and identified with the help of colony characteristics and 16S rRNA gene sequencing method. The antimicrobial metabolites assay was carried out using paper disks of antibiotics. PKS gene of the *P. liquidambaris* CBR-15 was discovered with the help of these degenerate primers LC1−LC2c, LC3−LC5c, and KS3−KS4c using PCR. Thin-layer chromatography method was used to identify the bioactive compounds of this strain. These results suggested that endophytic fungus *P. liquidambaris* CBR-15 harbored polyketide synthase gene, which indicates the biological ability of a fungus as a producer of antimicrobial compounds (Rao et al., 2015). Six endophytic fungal strains were isolated from Himalayan cypress or Bhutan cypress and identified with morphological and molecular methods. The fungi were further streaked on media, and grown culture was extracted using the methanol and ethyl acetate solvents. The antimicrobial activities were assessed against several human pathogenic bacterial strains. *Pestalotiopsis* sp. an ascomycete fungus from *Cupressus torulosa* or Himalayan cypress showed strong antimicrobial activity. The methanol and ethyl acetate crude extracts of *Pestalotiopsis* sp. showed minimum inhibitory concentration for *S. typhimurium* and *S. aureus*, which showed its efficacy as a potent antimicrobial agent. The phytochemical analysis revealed the existence of a diverse group of bioactive compounds in the crude fungal extracts that resembled those in the host plant extracts. On the basis of morphological characteristics and rDNA sequencing of the ITS region of the endophyte was identified as *Pestalotiopsis neglecta* strain, which discovered to be a promising source of bioactive compounds. There is obscure knowledge about endophytes isolated from *C. torulosa* D. Don. In this study, authors revealed the identification of isolated endophytic fungi *P. neglecta* from *C. torulosa* D. Don and the characterization of extracted biologically active compounds. The partially purified second fraction extracted from the fungal culture supernatant was injected and observed in gas chromatography followed by mass spectrometry, which revealed the presence of the peaks of many phytochemicals. These results indicated that endophyte fungus *P. neglecta* could be a potential source for bioactive compounds and may find potential use in pharmaceutical industries (Sharma et al., 2016). Fungal endophytes are an excellent producer of new bioactive compounds, including antibiotics, antioxidants, antiviral, anticancer, antidiabetic, and immunosuppressant compounds (Strobel et al., 2002). A fungal endophyte *Cephalosporium* sp. was isolated from the root of *Trachelospermum jasminoides* (star jasmine) plant and produced a phenolic compound (graphislactone A) having a strong antioxidant activity (Song et al., 2005). Twenty-one different endophytic fungi were isolated from the *Eugenia jambolana* (even jambul) tree, to assess the antioxidant and total phenolic content of the endophytic fungal extracts by three different antioxidant assays. Alkaloids, terpenoids, flavonoids, saponins, and phenolics were the main metabolites present in all fungal extracts. A strong correlation was recorded between total phenol and antioxidant activity in extracts. Seven endophytic fungal extracts showed a high level of phenolic content and potent antioxidant activity. Fungal strains of *Aspergillus* sp., including *A. peyronelii*, *A. niger*, and *Chaetomium* sp., exhibited higher antioxidant values between 50% and 80% among all strains. Ascorbic acid was used as a standard. The results concluded the secondary metabolites produced by endophytic fungal strains isolated from even jambul tree could be a good resource of novel natural antioxidants (Yadav et al., 2014). A fungal endophyte was isolated from green explants of *Vernonia amygdalina*, and the metabolites were extracted using ethyl acetate solvent. The fungal extract was screened for antimicrobial activity, and the bioactive compounds were present in the extract using HPLC-DAD analysis. The antimicrobial assay was performed against fungal and bacterial strains (*S. aureus, Salmonella typhi, Escherichia coli*, *P. aeruginosa, Bacillus subtilis, E. faecalis, A. niger*, and *Candida albicans*). The fungal extract inhibited all test bacteria except *S. aureus*. No antifungal activity was recorded against *A. niger and C. albicans*. Results suggested the presence of four bioactive compounds (antimicrobial agent) 4-hydroxyphenyl acetic acid, *p*-methoxycoumaric acid, indole-3-acetic acid, and acropyrone were present in the extract. 4-hydroxyphenyl acetic acid, which is a known compound, was the most abundant in the extract. Results of this entire study suggested that endophytic fungi associated with *V. amygdalina* could be a promising strain of novel compounds (Okezie et al., 2017).

13.5.2 Rhizospheric origin

Nowadays continuously increase engrossment in the research of the relation between rhizospheric microorganisms associated with the host plant. Rhizospheric microorganisms affect the physiology of plant by imparting many useful effects such as nitrogen fixation, nutritional uptake, and synthesis of active compounds in the host plants. Many fungal species are well known in the rhizospheric region of the plants that showed significant effects in secondary metabolite synthesis. There are many records present that state rhizosphere fungi not only enhanced the plant growth but also have the capability to modulate essential oil quality and quantity (Shaikh and Mokat, 2018). Soil communities have diverse characteristics and categorized taxonomically and functionally. This soil ecosystem experiences highly complex networks of bonding. Plant roots are a functional hotspot of microorganisms in the soil. Specialized bioactive compounds synthesized by host plants and their associated microbes play a critical role in various biological activities that alter the behavior of neighboring organisms. Thus, the biochemistry and metabolic activity of secondary metabolites in the rhizosphere is a key element in understanding interactions in the soil environment (Massalha et al., 2017).

The F-4 fungal strain isolated from rhizosphere soil from Kutralam hill station is a potent strain showing the highest antibacterial activity against several pathogenic bacteria. These results indicated that the F-4 strain was an excellent antimicrobial agent against *Staphylococcus* sp. Further, bioactive compounds of F-4 strain were identified by FT-IR and HPLC, GC-MS analysis. Based on the GC-MS, ten compounds (Cyclooctasiloxane, hexadecamethyl; Heptasiloxane, hexadecamethyl; 1,2-Benzenedicarboxylic acid, butyl 2-methyl propyl ester; 9,12-Octadecadienoic acid (Z, Z); 9,12-octadecadienoic acid; linolsaeure; 10,12-hexadecadien-1-ol; Docosane (CAS) n-Docosane; Hexacosane (CAS) n-Hexacosane; Tetracontane) were identified, out of these compounds, one compound Tetracontane used in antimicrobial effect (Rajalakshmi and Mahesh, 2014). Rhizospheric strain of *Phoenix dactylifera* (Date palm tree) isolated and purified on rice agar medium. Five bioactive compounds were extracted from the fungal extract of *Aspergillus* in which a novel and four already known compounds were present. The chemical structures of these extracted compounds were identified with NMR spectroscopy. The novel compound [1-(4-hydroxy-2, 6-dimethoxy-3, 5-dimethyl phenyl)-2-methyl-1-butanone] showed excellent antibacterial activity against *S. aureus* and significant growth inhibitions against *C. albicans* and *Candida parapsilosis*. It was the first report to isolate natural compounds from the *Aspergillus* sp. found in rhizospheres of Date plant (Orfali and Perveen, 2019a).

13.6 Himalayan region fungi help in the production of bioactive natural compounds

Fungal strains isolated from the Himalayan region also help in the production of bioactive natural compounds. The fungal isolates from the Western

Himalayan region plants were assessed for their bioactive potential. Seventy-two fungal endophytes were isolated and evaluated with colony morphology and molecular analysis. They represented 27 genera and belonged to class Basidiomycota and Ascomycetous. Among all plants only coniferous plants *Cedrus deodara, Pinus roxburghii*, and *Abies pindrow* harbored the various strains of diverse fungal groups. Many extracts were prepared from the broth culture of these fungal isolates. These extracts showed strong biological activity against *E. coli* and *S. aureus* strains. The study concluded that the endophytic fungal diversity from the Himalayan region has bioactive features and an antimicrobial effect. Many strains of endophytic fungi were isolated from ten different plants growing in the specified Himalayan regions, and fungal isolates were studied for their taxonomic position and bioactive potential. Endophytes isolated from *C. deodara, Pinus roxburghii*, and *A. pindrow* conifers possessed a broad range of endophytic fungal strains. These host plants (conifers) produce bioactive essential oils (Qadri et al., 2013). The fungal endophytes from the wood of *Taxus fauna* tree were used for the production and characterization of secondary metabolites. Fungal isolates have the ability to produce biologically active compounds of pharmaceutical importance. It is already known that endophytic microbes are able to synthesize unique chemotherapeutic agents (Tayung and Jha, 2010; Cragg and Pezzuto, 2015). In this study, endophytic fungal strains NFW1, NFW3, and NFW9 were isolated from a yew tree, which indicates their different ability for the production of bioactive compounds and displayed significant cancer chemopreventive activity. This study gives out the importance of these isolates to synthesize bioactive compounds for development and modulation of drugs in the near future (Fatima et al., 2016). *Ramaria* Fr. and *Clavaria* L. are coral mushrooms. They play an important role in forest ecology, and few species of these coral mushrooms are known for the rich value of bioactive compound synthesis. The biochemical analysis, antioxidant, and antimicrobial activities of twelve different coral mushroom species were done using HPLC, UPLC, and GC. Toxic metals and antioxidant activities were evaluated using fungal extracts. Antimicrobial nature was assessed against six pathogenic bacteria. Results suggested that all mushroom varieties were found to be highly rich in carbohydrates, fatty acids, protein, amino acids, phenolics, flavonoids, carotenoids, and macro- and microminerals. All mushrooms were strong antimicrobial agents. These mushrooms are reported to be free from toxic heavy metals. It will be useful for their commercial productions of antioxidant, antibacterial, and nutraceutical compounds in the pharmaceutical sector (Sharma and Gautam, 2017).

Taxol or Paclitaxel is a therapeutic medicine; it is effective against several types of cancers and other solid tumor cancers. High demands in cancer treatments and less stock of taxol have finally hiked the cost of this chemotherapeutic drug. Fungal strains, either independently or as endophytes, are novel alternative resources for taxol synthesis. However, plants resources and chemical synthesis are not able to fulfill the high demand for the chemotherapeutic agent taxol. In the present study, thirty-four different endophytic fungal isolates were obtained from yew plants (*Taxus* sp.) collected from the capital of Himachal Pradesh (India). A preliminary study of taxol-synthesizing fungal isolates was assessed based on the presence of dbat gene, which is essential for the taxol biosynthesis. A TPF-06 fungal isolate was screened, characterized and identified as *A. fumigatus* KU-837249 strain, and it is concluded that this strain belonged to *A. fumigatus* clade and it is an endophyte (Kumar et al., 2019) (Table 13.1).

13.7 Bioactive compounds and their application for human health

The incremental growth of the world population is constantly leading to a hike in the number of health diseases for flora and fauna. Among the microbes, fungi have the potential to provide health benefits to the human population by the synthesis of bioactive compounds with their use in the field of pharmaceuticals. Altomare et al. (2000) studied and reported the antifungal compounds fusapyrone and deoxyfusapyrone isolated from endophytic fungi *Fusarium semitectum* rice culture. Both the alpha-pyrone compounds showed antifungal activities against a few pathogenic fungal strains, such as *Aspergillus fumigates, Cryptococcus neoformans, C. albicans*, and *Penicillium verrucosum*. Tiny and fresh explants were collected from the medicinal plant of Thailand provinces for isolation of endophytic fungi. The results suggested that these medicinal plants provide diversity of endophytic fungi having a capacity to synthesize a biologically active compound (Wiyakrutta et al., 2004). Fungi from Ascomycetes and Basidiomycetes are the most common producers for bioactive natural compounds. These bioactive compounds, especially terpenoids, have a wide range of biological activities, which are useful to clinical and pharmaceutical industries. Taxol or paclitaxel is a diterpenoid natural compound isolated from the *Taxus brevifolia* (yew) tree. It is one of the most commonly used anticancer drugs worldwide. After the knowledge of its unique biological function and high demand in the drug industries, an extensive search was started for another source to producing this rather than the Pacific yew tree (Heinig et al., 2013). Stierle et al. (1995) reported that the biosynthesis of taxol anticancer compound in an endophytic fungal strain isolated from *T. brevifolia* (yew plant). More than 160 research articles and patents give the results in the production of Taxol and related taxanes synthesized by microorganisms, including fungi. Some common antibiotics and drugs of human uses are synthesizing with the help of many fungal strains (Fig. 13.2).

Hazalin et al. (2009) studied that microorganisms have potential to synthesize novel secondary metabolites. Approximately 300 endophytic fungi were isolated from Pahang National Park of Malaysia. Out of them,

TABLE 13.1 Fungal bioactive compounds their origin and functions.

S. No.	Strain	Source	Compound	Function	References
1.	*Alternaria alternata* AE1	*Azadirachta indica* A. Juss.	Ethyl acetate	Antibacterial	Chatterjee et al. (2019)
2.	*Fusarium oxysporum* CK F05-5	Thai orchids	Coumarins	Antifungal	Bungtongdee et al. (2018)
3.	*Curvularia* sp. T12	*Rauwolfia macrophylla*	2′-Deoxyribolactone, hexylitaconic acid ergosterol	Antimicrobial, antioxidant, acetylcholinesterase inhibition	Kaaniche et al. (2019)
4.	*Penicillium canescens, P. murcianum, Paraphoma radicina, Coniolariellahispanica*	*Salvia abrotanoides,*	Cryptotanshinone	Antibacterial, antiinflammatory and anticancer	Teimoori-Boghsani et al. (2020)
5.	*Omphalotus olearius*	–	Illudins	Antitumor, antimicrobial	Jaspers et al. (2002)
6.	*Inonotus rickii*	Fruiting bodies of mushroom *I. rickii*	3/,6b-Dihydroxycinnamolide	Anticancer	Chen et al. (2014)
7.	*Neonothopanus nambi*	Luminescent mushroom	Nambinones A–C	Anticancer	Kanokmedhakul et al. (2012)
8.	*Stereumhirsutum*	–	Hirsutane-type	Antimicrobial and antitumor	Ma et al. (2014)
9.	*Pestalotiopsis microspora*	*Terminalia morobensis*	Pestacin and isopestacin	Antimicrobial and antioxidant activity	Strobel et al. (2002)
10.	*Pestalotiopsis clavispora*	*Rhizophora harrisonii*	Pestalpolyol I	Cytotoxic against the mouse lymphoma cell line L5178Y	Perez et al. (2016)

11.	*Rhytidhysteron rufulum*	*Bruguiera gymnorrhiza*	Rhytidchromone A, B, C, and E	Cytotoxici against Kato-3 cell lines and MCF-7 cells	Chokpaiboon et al. (2016)
12.	*Penicillium* sp. ZLN29	Jiaozhou Bay of China	Polyketide compound	Cytotoxic against HepG2 cell line	Gao et al. (2013)
13.	*Halorosellinia* sp. (No. 1403), Guignardia sp. (No. 4382)	–	Anthracenedione	Anticancer	Zhang et al. (2010)
14.	*F. oxysporum*	*Catharanthus roseus*	Vincristine	Anticancer	Zhang et al. (2000)
15.	*Alternaria* sp. (ML4)	*Mussaenda luteola* L.	Saponin, Tannin, Flavaonoid, Phenol	Antioxidant and antibacterial	Gunasekaran et al. (2017)
16.	*F. oxysporum*	*C. roseus*	Vincristine	Anticancer	Wang et al. (2006)
17.	*Curvularia* sp.	*C. roseus*	Vindoline	Anticancer	Pandey et al. (2016)
18.	*Alternaria*	*C. roseus*	Vinblastine	Anticancer	Bo et al. (1998)
19.	*Fusarium solani*	*Camptotheca accuminata*	Camptothecin	Anticancer (leukemia)	Shweta et al. (2010)
20.	*Fusarium proliferatum*	*Dysoxylum binectariferum*	Rohitukine	Anticancer	Kumara et al. (2012)
21.	*P. microspora*	*T. morobensis*	Pestacin	Antioxidant and antimycotic activities	Harper et al. (2003)
22.	*P. microspora*	*T. morobensis*	Isopestacin	Antifungal and antioxidant activities	Strobel et al. (2002)
23.	*Xylaria* sp	*Piper aduncum*	Phomenone	Mycotoxic	Silva et al. (2010)

(*Continued*)

TABLE 13.1 (Continued)

S. No.	Strain	Source	Compound	Function	References
24.	*Trichoderma harzianum*	Llex cornuta	Trichodermin	Antifungal	Chen et al. (2007)
25.	*Phomopsis* sp.	*Garcinia dulcis*	Phomoenamide	Antimycobacterial	Rukachaisirikul et al. (2008)
26.	*Xylaria* sp.	*G. dulcis*	Sordaricin	Antifungal	Pongcharoen et al. (2008)
27.	*Daldinia hawksworthii*	*Xiphydria prolongata* (woodwasp)	Dalsymbiopyrone	Antimicrobial and cytotoxic effects	Pazoutova et al. (2013)
28.	*Pestalotiopsis mangiferae*	*Mangifera indica* Linn	Phenolic compound	Antibacterial and antifungal activity	Subban et al. (2012)
29.	*Pestalotiopsis neglecta* BAB-5510	*Cupressus torulosa* D. Don	Phenolic compound	Antibacterial	Sharma et al. (2016)
30.	*Cladosporium* sp. (GU214631.1), *Phoma* sp. (AY210335.1), *Preussia minima* (DQ468035.1)	*Eremophila longifolia*, *Eremophila maculata*	Ethyl extract	Human cancer cell line cytotoxicity MOLT-4	Zaferanloo et al. (2018)

Sporothrix sp. showed cytotoxicity against carcinoma cell lines, colorectal carcinoma (HCT116), and human breast adenocarcinoma (MCF7). Endophytic fungi also showed anticarcinogenic against two cell lines, that is, P388 (murine leukemia) and K562 (human chronic myeloid leukemia). Gordien et al. (2010) studied that the endophytic fungi isolated from bilberry (*Vaccinium myrtillus*) explants and their screening carried out for *Mycobacterium aurum* and *Mycobacterium tuberculosis* H37Rv bacterial strains. The fungal endophyte showed strong antibacterial activity against *M. aurum* strain. The results indicated that fungal strains isolated from Scotland are a good source of antimycobacterial agent for the future. According to Li et al. (2005), endophytic fungi were evaluated on tumor cell lines, and results indicated that about 9% and 30% of fungal isolates described antitumor and antifungal activity, respectively. For the discovery of biologically natural compounds of human interest, fungal strains in association with medicinal plant or independently are a good source to synthesize natural compounds.

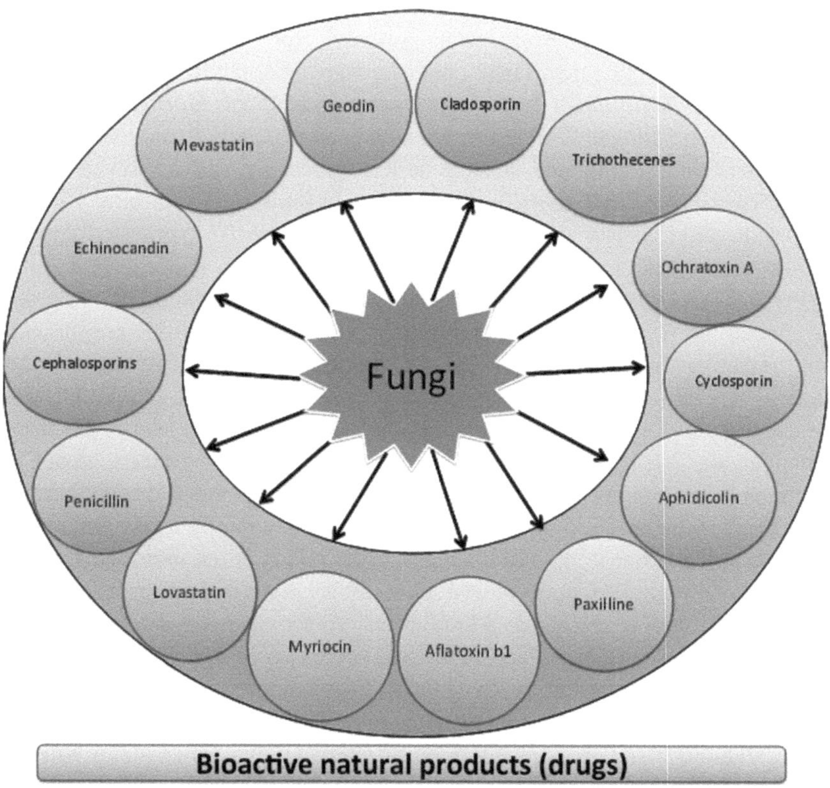

FIGURE 13.2 Bioactive natural products of fungi.

In the last thirty years, taxol is a natural compound having the potential to cure against various type of cancers, such as lung, ovarian, head, neck, bladder, prostate, melanoma, esophagal, and other solid tumor cancers. It is also used to treat Kaposi's sarcoma. The bark of the Taxus family tree is a real source of taxol. The quantity of taxol compound found in yew trees is very much less with high effort, which leads to higher demand in the pharmaceutical industry with high cost. The discovery of taxol producing fungi lead to the possibility and availability in a copious amount at a very cheaper rate. An Ascomycota fungi *Bartalinia robillardoides* was isolated from the explants of medicinal plant *Aegle marmelos* from Chennai (India). By the use of different techniques, the quantity of Taxol synthesized by fungal endophyte was quantified. Results suggested that bioactive compounds of fungal endophyte showed strong cytotoxicity against human cancer cell lines (Gangadevi and Muthumary, 2008). CPT is the most promising drug of this century. Both under in vivo and in vitro conditions, CPT and Secalonic acid exhibit antitumor and anticancer activity, respectively (Uma et al., 2008; Qi et al., 2009).

Vennila et al. (2010) studied on the evaluation of the anticancer activity of Taxol compound on mammary tumor cell lines. Taxol extracted from endophyte *Pestalotiopsis pauciseta* strain isolated from *Tabebuia pentaphylla* (Trumpet Tree), a medicinal plant. The endophytic fungal strain *Lasiodiplodia theobromae* is isolated from explants of *Morinda citrifolia*. The endophytic fungus *L. theobromae* serve as a potential microbe for the synthesis of taxol compound using genetic engineering, and it showed potential anticancer activity against breast cancers in humans (Pandi et al., 2011). The crude ethyl acetate extract of *Phomopsis* sp. GJJM07 was tested against the test pathogens, and in vitro antioxidant activity was also checked using DPPH radical scavenging assay. The antibacterial activity showed the highest against the *B. subtilis* strain (Jayanthi et al., 2011). Alkaloids are the natural bioactive compounds from plants origin. The chemical structure of alkaloid has minimum one nitrogen heterocyclic ring, having unique physiological activities in humans. Alkaloids have so many therapeutic functions (Khan et al., 2014). These are a large group of plant-extracted bioactive compounds, and many alkaloids play a crucial role in the field of pharmaceutical as a therapeutic agent. The secondary metabolite class originated from the strictosidine compound is a glucoside which is the intermediate biosynthetic compound of some biosynthetically related alkaloid groups and all monoterpene indole alkaloid. Modulation of biosynthetic pathways is a promising strategy for rapid and low-cost production of complex and desired molecules that are present in the plants. Strictosidine intermediate can be synthesized in a yeast host (*Saccharomyces cerevisiae*) in monoterpene indole alkaloid pathway using fourteen different known genes along with an additional seven genes and three deleted genes that enhance secondary metabolism. This method provides a novel source for developing the synthesis of highly complex plant originated monoterpene indole alkaloid so that this method may be used as alkaloid synthesis in copious amounts for human uses (Brown et al., 2015).

13.8 Future prospects

In the present scenario in the midst of a worldwide pandemic, there is a large-scale requirement for the innovation and synthesis of bioactive products from natural resources, which can be utilized in the pharmaceutical industry to treat various diseases. Recently, researchers are focused on the biosynthesis of bioactive compounds from the microbial origin such as fungi. These microbes are a potent producer of bioactive compounds. Several research articles reported that fungal strains isolated from plants of marine and terrestrial origin synthesize a variety of bioactive compounds. They provide the opportunity for new researchers to deal with secondary metabolites of pharmaceutical applications such as alkaloids, terpenoids, peptides, flavonoids, phenolics, antioxidants, etc. Plants are an excellent source of interesting novel drugs and therapeutic compounds. Endophytic and rhizospheric fungi isolated from plants and adhered soil provided the attention to many researchers in basic and applied science research-related fields due to their ability to synthesize various bioactive compounds, which are beneficial for us. The most attentive challenge is the lower yield of the desired bioactive compounds obtained from different fungal strains. However, to fulfill the demand of pharmaceutical industries for increasing large-scale production of useful drugs, several molecular techniques such as genetic engineering technology, proteomic, drug designing, and microbial fermentation should be emphasized for the synthesis of novel, useful drugs in the near future.

Acknowledgments

The author thanks the Department of Science and Technology-Inspire, DST, New Delhi, for fellowship in the form of JRF/SRF.

References

Aharonowitz, Y., Cohen, G., Martin, J.F., 1992. Penicillin and cephalosporin biosynthetic genes: structure, organization, regulation and evolution. Annu. Rev. Microbiol. 46, 461–495.

Altomare, C., Perrone, G., Zonno, M.C., Evidente, A., Pengue, R., Fanti, F., et al., 2000. Biological characterization of fusapyrone and deoxyfusapyrone, two bioactive secondary metabolites of *Fusarium semitectum*. J. Nat. Prod. 63, 1131–1135.

Bhadury, P., Mohammad, B.T., Wright, P.C., 2006. The current status of natural products from marine fungi and their potential as anti-infective agents. J. Ind. Microbiol. Biotechnol. 33, 325.

Bo, G., Haiyan, L., Lingqi, Z., 1998. Isolation of an fungus producing vinbrastine. J. Yunnan Univ. (Nat. Sci.) 20 (3), 214–215.

Brakhage, A.A., Schuemann, J., Bergmann, S., Scherlach, K., Schroeckh, V., Hertweck, C., 2008. Activation of fungal silent gene clusters: a new avenue to drug discovery. Prog. Drug Res. 66, 1–12.

Brakhage, A.A., Thon, M., Sprote, P., Scharf, D.H., Al-Abdallah, Q., Wolke, S.M., et al., 2009. Aspects on evolution of fungal -lactam biosynthesis gene clusters and recruitment of trans-acting factors. Phytochemistry 70, 1801–1811.

Brown, S., Clastre, M., Courdavault, V., O'Connor, S.E., 2015. De novo production of the plant-derived alkaloid strictosidine in yeast. Proc. Natl. Acad. Sci. U. S. A. 112 (11), 3205–3210.

Bungtongdee, N., Sopalun, K., Laosripaiboon, W., Iamtham, S., 2018. The chemical composition, antifungal, antioxidant and antimutagenicity properties of bioactive compounds from fungal endophytes associated with Thai orchids. J. Phytopathol. 167, 56–64.

Butler, M.S., Robertson, A.A., Cooper, M.A., 2014. Natural product and natural product derived drugs in clinical trials. Nat. Prod. Rep. 31, 1612–1661.

Chatterjee, S., Ghosh, R., Mandal, N.C., 2019. Production of bioactive compounds withbactericidal and antioxidant potential by endophyticfungus *Alternaria alternate* AE1 isolated from *Azadirachta indica* A. Juss. PLoS One 14 (4), e0214744.

Chen, L., Chen, J., Zheng, X., Zhang, J., Yu, X., 2007. Identification and antifungal activity of the metabolite of endophytic fungi isolated from *Llex cornuta*. Chin. J. Pestic. Sci. 9, 143–150.

Chen, H.P., et al., 2014. Four new sesquiterpenoids from fruiting bodies of the fungus *Inonotus rickii*. J. Asian Nat. Prod. Res. 16, 581–586.

Chen, S., He, L., Chen, D., Cai, R., Long, Y., Lu, Y., et al., 2017. Talaramide A, anunusual alkaloid from the mangrove endophytic fungus *Talaromyces*sp. (HZYX1) as an inhibitor of mycobacterial PknG. New. J. Chem. 41, 4273–4276.

Chiang, Y.M., Szewczyk, E., Nayak, T., Sanchez, J.F., Lo, H., Simityan, H., et al., 2008. Discovery of the emericellamide gene cluster: an efficient gene targeting system for the genomic mining of natural products in *Aspergillus nidulans*. Chem. Biol. 15, 527–532.

Chiang, Y.M., Lee, K.H., Sanchez, J.F., Keller, N.P., Wang, C.C.C., 2009. Unlocking fungal cryptic natural products. Nat. Prod. Commun. 4, 1505–1510.

Chokpaiboon, S., Choodej, S., Boonyuen, N., Teerawatananond, T., Pudhom, K., 2016. Highly oxygenated chromones from mangrove-derivedendophytic fungus *Rhytidhysteron rufulum*. Phytochemistry 122, 172–177.

Correia, R.T., Borges, K.C., Medeiros, M.F., Genovese, M.I., 2012. Bioactive compounds and phenolic-linked functionality of powdered tropical fruit residues. Food Sci. Tech. Int. 18, 539–547.

Cox, R.J., 2007. Polyketides, proteins and genes in fungi: programmed nano-machines begin to reveal their secrets. Org. Biomol. Chem. 5, 2010–2026.

Cragg, G.M., Pezzuto, J.M., 2015. Natural products as a vital source for the discovery of cancer chemotherapeutic and chemopreventive agents. Med. Princ. Pract. Available from: https://doi.org/10.1159/000443404.

Cueto, M., Jensen, P.R., Kauffman, C., Fenical, W., Lobkovsky, E., Clardy, J., 2001. Pestalone, a new antibiotic produced by a marine fungus in response to bacterial challenge. J. Nat. Prod. 64, 1444–1446.

Debbab, A., Aly, A.H., Proksch, P., 2013. Mangrove derived fungal endophytesea chemical and biological perception. Fungal Divers. 61, 1–27.

Deepika, V.B., Murali, T.S., Satyamoorthy, K., 2016. Modulation of genetic clusters for synthesis of bioactive molecules in fungal endophytes: a review. Microbiol. Res. 182, 125–140.

Diaz, E., Sanchez-Elordi, E., Santiago, R., 2016. Algal-fungal mutualism: cell recognition and maintenance of the symbiotic status of lichens. J. Vet. Med. Res. 3, 1052–1057.

Fatima, N., Kondratyuk, T.P., Park, E., Marler, L.E., MunibaJadoon, M., Qazi, M.A., et al., 2016. Endophytic fungi associated with *Taxus fuana* (West Himalayan Yew) of Pakistan: potential bio-resources for cancer chemopreventive agents. Pharm. Biol. 54, 2547–2554.

Fierro, F., Barredo, J.L., Diez, B., Gutierrez, S., Fernandez, F.J., Martin, J.F., 1995. The penicillin gene cluster is amplified in tandem repeats linked by conserved hexanucleotide sequences. Proc. Natl. Acad. Sci. U. S. A. 92, 6200–6204.

Fierro, F., Montenegro, E., Gutierrez, S., Martin, J.F., 1996. Mutants blocked inpenicillin biosynthesis show a deletion of the entire penicillin gene cluster at aspecific site within a conserved hexanucleotide sequence. Appl. Microbiol. Biotechnol. 44, 597–604.

Firn, R.D., Jones, C.G., 2000. The evolution of secondary metabolism—a unifying model. Mol. Microbiol. 37, 989–994.

Gangadevi, V., Muthumary, J., 2008. Taxol, an anticancer drug produced by an endophytic fungus *Bartalinia robillardoides* Tassi, isolated from a medicinal plant *Aegle marmelos* Correa ex Roxb. World J. Microbiol. Biotechnol. 24, 717.

Gao, H., Zhou, L., Li, D., Gu, Q., Zhu, T.J., 2013. New cytotoxic metabolites from the marine-derived fungus *Penicillium* sp. ZLN29. Helv. Chim. Acta. 96, 514–519.

Gordien, A.Y., Gray, A.I., Ingleby, K., Franzblau, S.G., Seidel, V., 2010. Activity of Scottish plant, lichen and fungal endophyte extracts against *Mycobacterium aurum* and *Mycobacterium tuberculosis*. Phytother. Res. 24, 692–698.

Guerriero, A., D'Ambrosio, M., Cuomo, V., Pietra, F., 1991. A novel, degraded polyketidic lactone, leptosphaerolide, and its likely diketone precursor, leptosphaerodione. Isolation from cultures of the marine ascomycete *Leptosphaeria oraemaris* (Linder). Helv. Chim. Acta 74, 1445–1450.

Gunasekaran, S., Sathiavelu, M., Arunachalam, S., 2017. In vitro antioxidant and antibacterial activity of endophytic fungi isolated from *Mussaenda luteola*. J. Appl. Pharm. Sci. 7, 234–238.

Gutierrez, S., Fierro, F., Casqueiro, J., Martin, J.F., 1999. Gene organization and plasticity of the β-lactam genes in different filamentous fungi. Antonie Van Leeuwen 75, 81–94.

Harper, J.K., Arif, A.M., Ford, E.J., Strobel, G.A., Porco, J.A., Tomer, D.P., et al., 2003. Pestacin: a 1 3-dihydro isobenzofuran from *Pestalotiopsis microspora* possessing antioxidant and antimycotic activities. Tetrahedron 59, 2471–2476.

Hazalin, N.A., Ramasamy, K., Lim, S.S.M., Wahab, I.A., Cole, A.L., Majeed, A.B.A., 2009. Cytotoxic and antibacterial activities of endophytic fungi isolated from plants at the National Park, Pahang, Malaysia. BMC Complem. Altern. M 9, 46.

Heinig, U., Scholz, S., Jennewein, S., 2013. Getting to the bottom of Taxol biosynthesis by fungi. Fungal Divers. 60, 161–170.

Hertweck, C., 2009. Hidden biosynthetic treasures brought to light. Nat. Chem. Biol. 5, 450–452.

Holm, D.K., Petersen, L.M., Klitgaard, A., Knudsen, P.B., Jarczynska, Z.D., Nielsen, K.F., et al., 2014. Molecular and chemical characterizationof the biosynthesis of the 6-MSA-derived meroterpenoid Yanuthone D in *Aspergillus niger*. Chem. Biol. 21, 519.

Jaspers, N.G., et al., 2002. Anti-tumour compounds Illudin S and Irofulven induce DNA lesions ignored by global repair and exclusively processed by transcription- and replication-coupled repair pathways. DNA Repair 1, 1027–1038.

Jayanthi, G., Kamalraj, S., Karthikeyan, K., Muthumary, J., 2011. Antimicrobial and antioxidant activity of the endophytic fungus *Phomopsis* sp. GJJM07 isolated from *Mesua ferrea*. Int. J. Curr. Sci. 1, 85–90.

Jeewon, R., Luckhun, A.B., Bhoyroo, V., et al., 2019. Pharmaceutical potential of marine fungal endophytes. In: Jha, S. (Ed.), Endophytes and Secondary Metabolites. Springer, Cham, pp. 283–305. , Reference Series in Phytochemistry.

Kaaniche, F., Hamed, A., Abdel-Razek, A.S., Wibberg, D., Abdissa, N., El Euch, I.Z., 2019. Bioactive secondary metabolites from newendophytic fungus *Curvularia*. sp isolated from *Rauwolfia macrophylla*. PLoS One 14 (6), 0217627.

Kanokmedhakul, S., et al., 2012. Cytotoxic sesquiterpenes from luminescent mushroom *Neonothopanus nambi*. Tetrahedron 68, 8261–8266.

Keller, N.P., 2019. Fungal secondary metabolism: regulation, function and drug discovery. Nat. Rev. Microbiol. 17, 167–180.

Khan, A.A., Bacha, N., Ahmad, B., Lutfullah, G., Farooq, U., Cox, R.J., 2014. Fungi as chemical industries and genetic engineering for the production of biologically active secondary metabolites. Asian Pac. J. Trop. Biomed. 4 (11), 859–870.

Kirk, P., Cannon, P., Minter, D., Stalpers, J., 2008. Dictionary of the Fungi, tenth ed. CABI, Wallingford.

Kumar, P., Singh, B., Thakur, V., 2019. Hyper-production of taxol from *Aspergillus fumigatus*, an endophytic fungus isolated from Taxus sp. of the Northern Himalayan region. Biotechnol. Rep. (Amst.) 24, e00395.

Kumara, P.M., Zuehlke, S., Priti, V., Ramesha, B.T., Shweta, S., Ravikanth, G., et al., 2012. *Fusarium proliferatum* an endophytic fungus from *Dysoxylum binectariferum* Hook. f, produces Rohitukine, a chromane alkaloid possessinganti-cancer activity. Antonie Van Leeuwenhoek 101, 323–329.

Kumara, P.M., Shweta, S., Vasanthakumari, M.M., Sachin, N., Manjunatha, B.L., Jadhav, S.S., et al., 2014. Endophytes and plant secondary metabolite synthesis: molecular and evolutionary perspective. In: Verma, V., Gange, A. (Eds.), Advances in Endophytic Research. Springer, New Delhi, pp. 177–190.

Kusari, S., Zuhlke, S., Spiteller, M., 2011. Effect of artificial reconstitution of the interaction between the plant *Camptotheca acuminata* and the fungal endophyte *Fusarium solani* on Camptothecin biosynthesis. J. Nat. Prod. 74, 764–775.

Kutchan, T.M., Hampp, N., Lottspeich, F., Beyreuther, K., Zenk, M.H., 1988. The cDNA clone for strictosidine synthase from *Rauvolfia serpentina* DNA sequence determination and expression in *Escherichia coli*. FEBS Lett. 237, 40–44.

Li, H., Qing, C., Zhang, Y., Zhao, Z., 2005. Screening for endophytic fungi with antitumour and antifungal activities from Chinese medicinal plants. World J. Microbiol. Biotechnol. 21, 1515–1519.

Liang, X., Zhang, X.Y., Nong, X.H., Wang, J., Huang, Z.H., Qi, S.H., 2016. Eight linear peptides from the deep-sea-derived fungus *Simplicillium clavatum* EIODSF 020. Tetrahedron 72, 3092–3097.

Lind, A.L., Lim, F.Y., Soukup, A.A., Keller, N.P., Rokas, A., 2018. An LaeA- and BrlA- dependent cellular network governs tissue- specific secondary metabolism in the human pathogen *Aspergillus fumigatus*. mSphere 3, 18–50.

Lopez, S.N., Ramallo, I.A., Sierra, M.G., Zacchino, S.A., Furlan, R.L.E., 2007. Chemically engineered extracts as an alternative source of bioactive natural product-like compounds. Proc. Natl. Acad. Sci. U. S. A. 104 (2), 441–444.

Lysoe, E., Seong, K.Y., Kistler, H.C., 2011. The transcriptome of *Fusarium graminearum* during the infection of wheat. Mol. Plant Microbe Interact. 24, 995–1000.

Ma, K., et al., 2014. New benzoate derivatives and hirsutane type sesquiterpenoids with antimicrobial activity and cytotoxicity from the solid-state fermented rice by the medicinal mushroom Ster-eumhirsutum. Food Chem. 143, 239–245. 79.

Massalha, H., Korenblum, E., Tholl, D., Aharoni, A., 2017. Small molecules below-ground: the role of specialized metabolites in the rhizosphere. Plant J. 90, 788–807.

McKnight, T.D., Roessner, C.A., Devagupta, R., Scott, A.I., Nessler, C.L., 1990. Nucleotide sequence of a cDNA encoding the vacuolar protein strictosidine synthase from *Catharanthus roseus*. Nucleic Acids Res. 18, 4939.

Molinski, T.F., Dalisay, D.S., Lievens, S.L., Saludes, J.P., 2009. Drug development from marine natural products. Nat. Rev. Drug Disc. 8, 69.

Moore, D., Frazer, L.N., 2002. Essential Fungal Genetics. Springer, New York.

Mousa, W.K., Raizada, M.N., 2013. The diversity of antimicrobial secondary metabolites produced by fungal endophytes: an interdisciplinary perspective. Front. Microbiol. 4, 65.

Netzker, T., Fischer, J., Weber, J., 2015. Microbial communication leading to the activation of silent fungal secondary metabolite gene clusters. Front. Microbiol. 6, 299.

Newman, D.J., Hill, R.T., 2006. New drugs from marine microbes: the tide is turning. J. Ind. Microbiol. Biotechnol. 33, 539–544.

Newman, D.J., Cragg, G.M., Snader, K.M., 2003. Natural products as sources of new drugs over the period. J. Nat. Prod. 66 (7), 1022–1037.

Oh, D.C., Kauffman, C.A., Jensen, P.R., Fenical, W., 2007. Induced production of emericellamides A and B from themarine-derived fungus *Emericella* sp. in competingco-culture. J. Nat. Prod. 70, 515–520.

Okezie, U.M., Eze, P.M., Okoye, F.B., Ikegbunam, M.N., Ugwu, M.C., Esimone, C.O., 2017. Biologically active metabolites of an endophytic fungus isolated from *Vernonia amygdalina*. Afr. J. Pharm. Res. Dev. 9, 24–26.

Ola, A.R.B., Thomy, D., Lai, D., Brotz-Oesterhelt, H., Proksch, P., 2013. Inducing secondary metabolite production by the endophytic fungus *Fusarium tricinctum* through coculture with *Bacillus subtilis*. J. Nat. Prod. 76, 2094–2099.

Orfali, R., Perveen, S., 2019a. Secondary metabolites from the *Aspergillus* sp. in the rhizosphere soil of *Phoenix dactylifera* (Palm tree). BMC Chem. 13, 103.

Orfali, R., Perveen, S., 2019b. New bioactive metabolites from the thermophilic fungus *Penicillium* sp. isolated from Ghamiqa Hot Spring in Saudi Arabia. Hindawi J. Chem. 7. Article ID 7162948.

Pallenberg, A.J., White, J.D., 1986. The synthesis and absolute configuration of (+)-leptosphaerin. Tetrahedron Lett. 27, 5591–5594.

Pandey, S.S., Singh, S., Babu, C.V., Shanker, K., Srivastava, N., Shukla, A.K., et al., 2016. Fungal endophytes of *Catharanthus roseus* enhance vindoline content by modulating structural and regulatory genes related to terpenoid indole alkaloid biosynthesis. Sci. Rep. 6, 26583.

Pandi, M., Kumaran, R.S., Choi, Y.K., Kim, H.J., Muthumary, J., 2011. Isolation and detection of taxol, an anticancer drug produced from *Lasiodiplodia theobromae*, an endophytic fungus of the medicinal plant *Morinda citrifolia*. Afr. J. Biotechnol. 10, 1428–1435.

Pazoutova, S., Follert, S., Bitzer, J., Keck, M., Surup, F., Srutka, P., et al., 2013. A new endophytic insect-associated *Daldinia* species, recognized from a comparison of secondary metabolite profiles and molecular phylogeny. Fungal Divers. 60, 107–123.

Perez, H.C.F., Daletos, G., Liu, Z., Lin, W., Proksch, P., 2016. Polyketides from the mangrove-derived fungal endophyte *Pestalotiopsis clavispora*. Tetrahedron Lett. 57, 2078–2083.

Petersen, L.M., Holm, D.K., Knudsen, P.B., Nielsen, K.F., Gotfredsen, C.H., Mortensen, U.H., et al., 2015. Characterization of four new antifungal yanuthones from *Aspergillus niger*. J. Antibiot. 68, 201–205.

Pongcharoen, W., Rukachaisirikul, V., Phongpaichit, S., Kuhn, T., Pelzing, M., Sakayaroj, J., et al., 2008. Metabolites from the endophytic fungus *Xylaria* sp. PSU-D14. Phytochemistry 69, 1900–1902.

Qadri, M., Johri, S., Shah, B.A., Khajuria, A., Sidiq, T., Lattoo, S.K., et al., 2013. Identification and bioactive potential of endophytic fungi isolated from selected plants of the Western Himalayas. Springer Plus 2, 8.

Qi, S.H., Xu, Y., Xiong, H.R., Qian, P.Y., Zhang, S., 2009. Antifouling and antibacterial compounds from a marine fungus *Cladosporium* sp. F14. World J. Microbiol. Biotechnol. 25, 399.

Rajalakshmi, S., Mahesh, N., 2014. Production and characterization of bioactive metabolites isolated from *Aspergillus terreus*in rhizosphere soil of medicinal plants. Int. J. Curr. Microbiol. App. Sci. 3, 784–798.

Ramos, A.A., Prata-Sena, M., Castro-Carvalho, B., Dethoup, T., Buttachon, S., Kijjoa, A., et al., 2015. Potential of four marine-derived fungi extracts as anti-proliferative and cell death-inducing agents in seven human cancer cell lines. Asian Pac. J. Tropic. Med. 8 (10), 798–806.

Rao, H.C.Y., Santosh, P., Rakshith, D., Satish, S., 2015. Molecular characterization of an endophytic *Phomopsis liquidambaris* CBR-15 from *Cryptolepis buchanani* Roem. and impact of culture media on biosynthesis of antimicrobial metabolites. 3 Biotech. 5, 165–173.

Rukachaisirikul, V., Sommart, U., Phongpaichit, S., Sakayaroj, J., Kirtikara, K., 2008. Metabolites from the endophytic fungus *Phomopsis* sp. PSU-D15. Phytochemistry 69, 783–787.

Santos, T.F.B., dos Santos Carvalho, C., de Almeida, M.A., 2020. Endophytic fungi isolated from Brazilian medicinal plants as potential producers of antioxidants and their relations with anti-inflammatory activity. 3 Biotech. 10, 223.

Schroeckh, V., Scherlach, K., Nutzmann, H.W., 2009. Intimate bacterial–fungal interaction triggers biosynthesis of archetypal polyketides in *Aspergillus nidulans*. Proc. Natl. Acad. Sci. U. S. A. 106, 14558–14563.

Schulz, B., Boyle, C., Draeger, S., Rommert, A., Krohn, K., 2002. Endophytic fungi: a source of novel biologically active secondary metabolites. Mycol. Res. 106, 996–1004.

Shaikh, M.N., Mokat, D.N., 2018. Role of rhizosphere fungi associated with commercially explored medicinal and aromatic plants: a review. Curr. Agric. Res. J. 6, 72–77.

Sharma, S.K., Gautam, N., 2017. Chemical and bioactive profiling, and biological activities of coral fungi from Northwestern Himalayas. Sci. Rep. 7, 46570.

Sharma, D., Pramanik, A., Agrawal, P.K., 2016. Evaluation of bioactive secondary metabolites from endophytic fungus *Pestalotiopsis neglecta* BAB-5510 isolated from leaves of *Cupressus torulosa* D. Don. 3 Biotech. 6, 210.

Shokri, A., El-Ghonemy, D.H., Shaaban, M., 2020. Production and purification of bioactive compounds with potent antimicrobial activity from a novel terrestrial fungus *Aspergillus* sp. DHE4. Biocatalysis Agric. Biotechnol. Available from: https://doi.org/10.1016/j.bcab.2020.101726.

Shweta, S., Zuehlke, S., Ramesha, B., Priti, V., Kumar, P.M., Ravikanth, G., et al., 2010. Endophytic fungal strains of *Fusarium solani*, from *Apodytes dimidiata* E. Mey. ex Arn (Icacinaceae) produce camptothecin 10-hydroxycamptothecin and 9-methoxycamptothecin. Phytochemistry 71, 117–122.

Silva, G.H., de Oliveira, C.M., Teles, H.L., Pauletti, P.M., Castro-Gamboa, I., Silva, D.H., et al., 2010. Sesquiterpenes from *Xylaria* sp., an endophytic fungus associated with *Piper aduncum* (Piperaceae). Phytochem. Lett. 3, 164–167.

Singh, M., Kumar, A., Singh, R., Pandey, K.D., 2017. Endophytic bacteria: a new source of bioactive compounds. 3 Biotech. 7 (5), 315.

Song, Y.C., Huang, W.Y., Sun, C., Wang, F.W., Tan, R.X., 2005. Characterization of graphislactone A as the antioxidant and free radical scavenging substance from the culture of *Cephalosporium* sp. IFB-E001, an endophytic fungus in *Trachelospermum jasminoides*. Biol. Pharm. Bull. 28, 506–509.

Spraker, J.E., Wiemann, P., Baccile, J.A., Venkatesh, N., Schumacher, J., Schroeder, F.C., et al., 2018. Conserved responses in a war of small molecules between a plant-pathogenic bacterium and fungi. mBio 9, e00820-18.

Stierle, A., Stierle, D., Strobel, G., Grothaus, P., Bignami, G., 1995. The search for a taxol-producing microorganism among the endophytic fungi of the Pacific yew, *Taxus brevifolia*. J. Nat. Prod. 58, 1315–1324.

Strobel, G., Ford, E., Worapong, J., Harper, J.K., Arif, A.M., Grant, D.M., 2002. Isopestacin, a unique isobenzofuranone from *Pestalotiopsis microspora* possessing antifungal and antioxidant properties. Phytochemistry 60, 179–183.

Subban, K., Subramani, R., Johnpaul, M., 2012. A novel antibacterial and antifungal phenolic compound from the endophytic fungus *Pestalotiopsis mangiferae*. Nat. Prod. Res. 1, 5.

Sun, Y., Lu, H., Li, Y., Sun, C., Song, J., Niu, Y., et al., 2011. Pyrosequencing of the *Camptotheca acuminata* transcriptome reveals putative genes involved in camptothecin biosynthesis and transport. BMC Genomics 533 (12), 1–11.

Suwannarach, N., Kumla, J., Sujarit, K., Pattananandecha, T., Saenjum, C., Lumyong, S., 2020. Natural bioactive compounds from fungi as potential candidates for protease inhibitors and immunomodulators to apply for Coronaviruses. Molecules 25, 1–21.

Tan, M.A., Castro, S.G., Oliva, P.M.P., Yap, P.R.J., Nakayama, A., Magpantay, H.D., et al., 2020. Biodiscovery of antibacterial constituents from the endolichenic fungi isolated from *Parmotrema rampoddense*. 3 Biotech. 10, 212.

Tayung, K., Jha, D., 2010. Antimicrobial endophytic fungal assemblages inhabiting bark of *Taxus baccata* L. of Indo-Burma mega biodiversity hotspot. Indian J. Microbiol. 50, 74–81.

Teimoori-Boghsani, Y., Ganjeali, A., Cernava, T., Muller, H., Asili, J., Berg, G., 2020. Endophytic fungi of native *Salvia abrotanoides* plants reveal high taxonomic diversity and unique profiles of secondary metabolites. Front. Microbiol. 10, 3013.

Uma, S.R., Ramesha, B.T., Ravikanth, G., Rajesh, P.G., Vasudeva, R., Ganeshaiah, K.N., 2008. Chemical profiling of *Nothapodytes nimmoniana* for camptothecin, an important anticancer alkaloid: towards the development of a sustainable production system in bioactive molecules and medicinal plants. In: Ramawat, K.G., Merillon, J.M. (Eds.), Bioactive Molecules and Medicinal Plants. Springer, Berlin, pp. 197–213.

Vasundhara, M., Kumar, A., Reddy, M.S., 2016. Molecular approaches to screen bioactive compounds from endophytic fungi. Front. Microbiol. 7, 1774.

Vennila, R., Thirunavukkarasu, S., Muthumary, J., 2010. Evaluation of fungal taxol isolated from an endophytic fungus *Pestalotiopsis pauciset* AVM1 against experimentally induced breast cancer in sprague dawley rats. Res. J. Pharmacol. 4, 38–44.

von Dohren, H., 2009. A survey of nonribosomal peptide synthetase (NRPS) genes in *Aspergillus nidulans*. Fungal Genet. Biol. 46, S45–S52.

Wall, M.E., Wani, M.C., Cook, C.E., 1966. Plant antitumor agents-I. The isolation and structure of camptothecin, a novel alkaloidal leukemia and tumor inhibitor from *Camptotheca acuminata*. J. Am. Chem. Soc. 88, 3888–3890.

Wang, W., Guo, Q., You, Q., Zhang, K., Yang, Y., Yu, J., et al., 2006. Involvement of bax/bcl-2 in wogonin-induced apoptosis of human hepatoma cell line SMMC-7721. Anticancer Drugs 17, 797–805.

Wiyakrutta, S., Sriubolmas, N., Panphut, W., Thongon, N., Danwisetkanjana, K., Ruangrungsi, N., et al., 2004. Endophytic fungi with antimicrobial, anticancer and anti-malarial activities isolated from Thai medicinal plants. World J. Microbiol. Biotechnol. 20, 265–272.

Yadav, M., Yadav, A., Yadav, J.P., 2014. In vitro antioxidant activity and total phenolic content of endophytic fungi isolated from *Eugenia jambolana* Lam. Asian Pac. J. Trop. Med. 7, 256–261.

Yamazaki, Y., Sudo, H., Yamazaki, M., Aimi, N., Saito, K., 2003. Camptothecin biosynthetic genes in hairy roots of *Ophiorrhiza pumila*: cloning, characterization and differential expression in tissues and by stress compounds. Plant Cell Physiol. 44, 395–403.

Yamazaki, Y., Kitajima, M., Arita, M., Takayama, H., Sudo, H., Yamazaki, M., et al., 2004. Biosynthesis of camptothecin. *In silico* and *in vivo* tracer study from [1-13 C] glucose. Plant Physiol. 134, 161–170.

Youssef, F.S., Ashour, M.L., Singab, A.N.B., Wink, M., 2019. A comprehensive review of bioactive peptides from marine fungi and their biological significance. Mar. Drugs 17, 559.

Zaferanloo, B., Pepper, S.A., Coulthard, S.A., Redfern, C.P.F., Palombo, E.A., 2018. Metabolites of endophytic fungi from Australian native plants as potential anticancer agents. FEMS Microbiol. Lett. 365 (9), 1–8.

Zerikly, M., Challis, G.L., 2009. Strategies for the discovery of new natural products by genome mining. Chembiochem 10 (4), 625–633.

Zhang, L., Guo, B., Li, H., Zeng, S., Shao, H., Gu, S., et al., 2000. Preliminary study on the isolation of endophytic fungus of *Catharanthus roseus* and its fermentation to produce products of therapeutic value. J. Tradit. Chin. Med. 31, 805–807.

Zhang, J., Tao, L., Liang, Y., Chen, L., Mi, Y., Zheng, L., et al., 2010. Anthracenedione derivatives as anticancer agents isolated from secondary metabolites of the mangrove endophytic fungi. Mar. Drugs 8, 1469–1481.

Zjawiony, J.K., 2004. Biologically active compounds from Aphyllophorales (polypore) fungi. J. Nat. Prod. 67, 300–310.

Chapter 14

Cyanobacteria-derived small molecules: a new class of drugs

Atif Khurshid Wani[1], Nahid Akhtar[1], Banhishikha Datta[1], Janmejay Pandey[2] and M. Amin-ul Mannan[1]

[1]*Department of Molecular Biology and Genetic Engineering, School of Bioengineering and Biosciences, Lovely Professional University, Jalandhar, India,* [2]*Department of Biotechnology, Central University of Rajasthan, Ajmer, India*

Chapter Outline

14.1 Introduction

The drug discovery from bacterial and fungal sources has been the focal point in the last couple of decades. The discovery of pharmaceuticals originating from microorganisms started with the exploration of penicillin discovered in 1928, from *Pencillium notatum*. However, it is only in 1941 that the first clinical trials and commercial production of penicillin were undertaken (Singh et al., 2011). Presently, the derived drugs are generally from terrestrial sources. However, a good number of drugs have also been identified from marine sources as well. Both the terrestrial and marine sources are known for their

Volatiles and Metabolites of Microbes. DOI: https://doi.org/10.1016/B978-0-12-824523-1.00003-1

efficacy, diversity, and novelty (Lindequist, 2016). Recent trends have revealed that Cyanobacteria are source of broad-spectrum secondary metabolites and antimicrobials. These are great source of nutraceuticals and find its applications in pharmaceuticals. Cyanobacteria is part of the kingdom Monera; division Eubacteria. Unlike other prokaryotes Cyanobacteria are Gram-negative bacterium and is capable of photosynthesis. Cyanobacteria are good source of natural bioactive compounds. Many of them are known for their ability to produce toxins in the form of exotoxins, endotoxins, and lipopolysaccharide like *Anabaena flosaque, Microcystis aeruginosa*, etc. (Percival and Williams, 2014). Several evidence suggests that it can produce various biologically active compounds. The derived bioactive compounds possess antimicrobial, anticancer, anti-HIV, antiinflammatory, and antimalarial activities (Falaise et al., 2016; Vijayakumar and Menakha, 2015). These bioactive metabolites are usually low molecular weight compounds, however with great degree of structural complexity and chemical diversity (Singh, 2014). Cyanobacteria has enormous potential to produce these small bioactive metabolites through various pathways including nonribosomal peptide synthetase (NRPS) and polyketide synthase (PKS), and/or of both NRPS-PKS pathway (Micallef et al., 2015b; Ochi et al., 2004; Nielsen et al., 2016). Small molecules derived from these pathways include microviridin (Han et al., 2006), Belamide-A (MacMillan and Molinski, 2002), Cryptophycin-1, Largamides D–G, and Wewakazole derived from organisms like *Microcystis aeruginosa, Symploca* spp., *Nostoc linckia, Oscillatoria* spp., and *Lyngbya sordida*, respectively. These noncanonical pathways provide a fascinating tool for diverse metabolite entity production with great therapeutic potential (Vijayakumar and Menakha, 2015). In the present chapter, we have comprehended the genomics of Cyanobacteria involved in production of the various bioactive compounds. The focus is to get familiar with the different pathways involved in the bioactive compound production. Thus, revealing Cyanobacteria as a promising bioactive metabolite production factory with great selectivity and efficacy taking its therapeutic value into consideration.

14.2 Bioactive compounds

Bioactive compounds, regardless of their source, are defined as the components of the food that show an immediate response in a particular tissue or a cell (Aisbl, 2016). They generally include essential oils, carotenoids, and phytochemicals with a specific therapeutic property (Mahfoudhi et al., 2016). These metabolites are found in different natural reservoirs like fruits, vegetables, and some grains. They are an extremely heterogeneous and ubiquitous compounds with predominantly hydrophilic chemical nature (Galanakis, 2017). Besides their abundance in vegetative sources, these bioactive compounds are also produced by a wide range of bacterial and fungal yeast strains

TABLE 14.1 Occurrence of some bioactive producing Cyanobacteria.

Bioactive compounds	Organism	Occurrence	References
Aeruginosin	*Microcystis aeruginosa*	Fresh water bloom	Elkobi-Peer and Carmeli (2015)
Aurilide	*Lyngbya majuscule*	Marine	Han et al. (2006)
Cyanopeptolin	*Nostoc edaphicum*	Baltic sea	Mazur-Marzec et al. (2018)
Barbaleucamide	*L. majuscule*	Marine	Harrigan et al. (2001)
Dolastatin	*Cyanobacterium Symploca*	Marine	Luesch et al. (2001)
Grassystatin	*Lyngbya confervoides*	Marine	Kwan et al. (2009)
Kulolide	*Rivularia* sp.	Shallow surfaces	Boudreau et al. (2012)
Lyngbyaloside	*Lyngbya bouillonii*	Marine	Klein et al. (1997)
Microcystin	*Microcystis and Anabaena*	Fresh water bloom	Turner et al. (2018)
Microcyclamide	*M. aeruginosa*	Fresh water	Ziemert et al. (2008)
Saxitoxin	*Anabaena, Aphanizomenon*	Fresh water	Hackett et al. (2013)
Microginin	*Microcystis* sp.	Fresh water bloom	Lodin-Friedman and Carmeli (2018)
Trungapeptin	*L. majuscule*	Marine	Bunyajetpong et al. (2006)

(Mazzoli et al., n.d.). Cyanobacteria is prominent in producing variety of compounds with a wide range of bioactivities. The well-studied compounds of Cyanobacteria origin include: microcystins, anatoxins, micropeptins, and microviridin (Namikoshi and Rinehart, 1996). These small bioactive compounds are prolific in the activity against cellular enzymes and thus interfering in signaling pathways. One of the studies carried out in 1996 shows death of about 60 people who were consuming water having microcystinins, a bioactive metabolite (Pouria et al., 1998). In year 2016 the United States Environmental Protection Agency placed several bioactive compounds of Cyanobacteria origin like anatoxin, saxitoxin, microcystin, and cylindrospermopsin on its fourth contamination list (Huang and Zimba, 2019) (Table 14.1).

14.2.1 Synthesis of bioactive compounds

Microorganisms have a tremendously complex system for biosynthesizing its bioactive compounds. These compounds are diverse and complex with regard to their structure and occurrence. One of the research has analyzed 279 and 281 bioactive compounds from Cyanobacteria of fresh water and marine origin, respectively. These compounds are well studied for their therapeutic and ecological implications (González-Medina and Medina-Franco, 2019). The different compounds derived from various Cyanobacteria genera are classified into alkaloids, terpenes, peptides, polyketides, and fatty acids. The biosynthesis generally occurs via ribosomal pathways or nonribosomal pathways. The polyketides and nonribosomal peptides are produced by an active and functional module of enzymes known as PKS and NRPS (Fig. 14.1).

14.2.2 Nonribosomal peptide synthetases: NRPS's biosynthetic pathway

NRPSs are modular enzymes that catalyze synthesis of peptides or peptide products from a various standard and nonproteinogenic amino acid substrate. The proteins are synthesized on ribosomes but this is not always true. The small peptides of 20–50 amino acids are arranged by the action of peptide synthetases. The assembly is done similarly as fatty acid assembly is done by different synthases. The modular enzyme mechanism is known to assemble amino acids sequentially and thus elongating polypeptide chains. The amino acid combination along with different enzyme modules and functional domains reveal the activity and structure of the synthesized peptide product (Pearson et al., 2008). The NRP synthesis in prokaryotes and very rarely in lower eukaryotes is achieved by these functional enzymes of modular

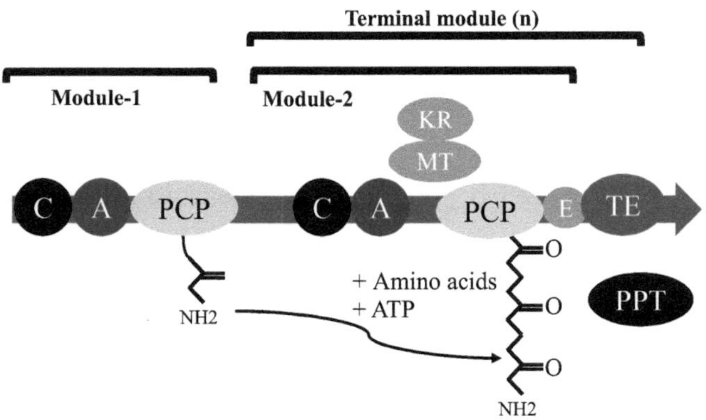

FIGURE 14.1 Nonribosomal peptide synthetase (NRPS) domain structures. *A*, Adenylation domain; *C*, condensation domain; *E*, epimerase; *KR*, kito-reductase domain; *MT*, methyltransferase; *PCP*, peptidyl carrier protein; *PPT*, phosphopantetheine transferase domain.

complexes called peptide synthetases. The main source of these NRPS are Actinobacteria, Probacteria, and Ascomycota. These along with polyketide synthases are important for the biosynthesis of bioactive compounds. The NRPs from various organisms are regarded as bioactive metabolites with specific roles in defense and the least role in growth, metabolism, or reproduction (Dittmann et al., 2013). Based on the molecular phylogeny Cyanobacteria belongs to a very coherent class of prokaryotes that mostly dwells in water blooms. Its potential as a rich source of biologically active metabolites is relatively unexplored (Neilan et al., 1999). The composition of modified and unconventional amino acids with complex cyclic structures in peptides symbolizes NR synthesis of bioactive compounds in Cyanobacteria. The NRPS enzymes are generally found in modular complexes in hybrids of 100–1500 kDa. Initially, this family of NRPS enzymes was studied from the formation of cyclosporin, vancomycin and penicillin (Fidor et al., 2019). Each module of NRP is known to perform condensation, modification, thiolation and activation. The peptide product depends on the order, organization and sequence of amino acids. The analysis of 150 NRPs peptide products along with gramicidin synthetase has led to the study of NRPs specificity (Liu et al., 2017). The enzyme modules are categorized into domains that are specific for a unique amino acid incorporation in a multistep process of peptide chain (Fig. 14.1). Initially, from NRPs the domain specific for adenylation activates an amino acid residue after being selected based on its specificity toward NRPs domain. After the activation the activated residue is joined through a phosphopantothiene containing a thiol group with a carrier protein. Finally, the domain of condensation releases peptides from NRPs and thus terminating the elongation of the peptide chain. The modification in the structure with the help of some tailoring enzymes is done by the process of glycosylation, methylation, oxygenation, and epimerization reactions (Fidor et al., 2019).

14.2.3 Polyketide synthase biosynthesis pathway

Polyketide are formed by the process of condensation between thioester units like methyl-melonyl-CoA and malonyl-CoA. These polyketides have been found to be important medicinal component in numerous drugs. Different types of PKS biosynthetic pathways have been studied in bacteria which includes the modular PKSs (type I) with multiple domains and PKSs (type III) in which elongating chain is not attached directly to a protein. The Cyanobacteria are well studied for type I PKS pathway in which the polyketide chain elongates sequentially in its active sites (Fig. 14.2) (Ridley et al., 2008). Though there are numerous differences in the chemistry of NRPS and PKS pathways but similarities are also known. Both are organized into modular domains with functional synthases. Like NRPS, the type I module also consists of three functional and active domains: ACP, AT, and KS. The basic

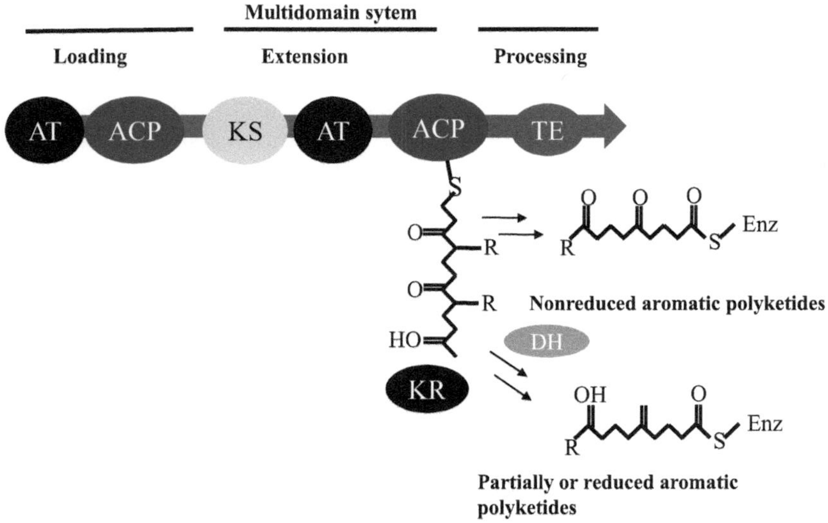

FIGURE 14.2 **Polyketide synthetase (PKS) domain structures.** *AT*, Acyltransferase domain; *ACP*, acyl carrier protein; *DH*, dehyratase domain; *KR*, kito-reductase domain; *KS*, ketosynthase domain; *TE*, thioesterase domain.

C–C bond formation is catalyzed by KS domain thus mediating the chain transfer of the polyketide to the cysteine active site of another domain, that is, KS domain of the module. This is followed by malonyl-coA or (CH3)-malonyl-ACP condensation. Hence (CH3)-malonyl-CoA is formed by a transacylation at AT domain. TE domain is cleaved finally and the product is macrocyclized (Zhou et al., 2019; Khosla et al., 2014) (Table 14.2).

14.3 Bioactive compounds of Cyanobacteria as antibiotics and nutraceuticals

The diversity of bioactive compounds in Cyanobacteria and other potential microorganisms also account for their enormous applications in pharmaceuticals and nutraceuticals. The discovery of new bioactive compounds is the need of the hour because of the increasing resistance in pathogens against conventional antibiotics (de Morais et al., 2015). The biological active compounds can originate from both the primary metabolic processes of proteins, vitamins, and fatty acids as well as the secondary metabolism (Singh et al., 2011). On the aspect of nutraceuticals, the bioactive compounds originating from Cyanobacteria are well known as a natural source of valuable nutraceuticals. These precious compounds can be carotenoids, long-chain unsaturated fatty acids, vitamins, and some antioxidants (Subudhi, 2017).

TABLE 14.2 Biosynthetic pathways used by Cyanobacteria for bioactive compound production along with their gene cluster size.

Bioactive compound	Cyanobacteria	Pathway	Gene cluster size	References
Aeruginosin	*Microcystis aeruginosa*	NRPS/PKS (aer)	34 kb	Latifi et al. (2013)
Ambiguine	*Fischerella ambigua*	Alkaloid (amb)	40–50 kb	Micallef et al. (2015a)
Anatoxin-a	*Anabaena* sp.	NRPS/PKS (ana)	28–30 kb	Calteau et al. (2014)
Apratoxin	*L. bouillonii*	PKS/NRPS (apr)	55 kb	Grindberg et al. (2011)
Barbamide	*Lyngbya majuscula*	NRPS/PKS (bar)	26 kb	Frangeul et al. (2008)
Cyanopeptolin	*M. aeruginosa*	NRPS (anb)	32 kb	Latifi et al. (2013)
Cylindrospermopsins	*Cylindrospermopsis raciborskii*	PKS (cyr)	40–45 kb	Sinha et al. (2014)
Hassallidin	*Anabaena* sp.	NRPS	61 kb	Wang et al. (2014)
Lyngbyatoxin	*L. majuscule*	Modified NRPS	10–12 kb	Edwards and Gerwick (2004)
Microginins	*Planktothrix rubescensi*	PKS/NRRPS (mic)	22–25 kb	Rounge et al. (2009)
Nostophycins	*Nostoc* sp.	NRPS/PKS (npn)	46 kb	Fewer et al. (2013)
Saxitoxin	*Anabaena circinalis*	PKS-sxt (modified aminoacids)	35 kb	Kellmann et al. (2008)
Welwitindolinone	*Westiella intricata*	Alkaloid (wel)	59 kb	Micallef et al. (2015a)

14.3.1 Bactericidal and fungicidal compounds

Multidrug resistance is a major concern for the emerging pathogens. The drug resistance in some bacteria like *Staphylococcus aureus* against methicillin, *Enterococci* against vancomycin, and *Enterobacteriaceae* against β-lactamase is commonly known. Scientists have studied various bioactive compounds isolated from Cyanobacteria as the promising antibacterial agents (Singh et al., 2011). Nosocomin produced from *Nostoc commune* has shown good antibacterial effects against *Staphylococcus epidermidis*, *Bacillus cereus*, and *Escherichia coli* at the concentration of 8, 32, and 128 parts per million respectively. The bioactive compounds like saxitoxin, microcystin, and hassallin produced by *Anabaena* spp. has also been reported to show antibacterial activity at 30–64 μg/mL against *S. aureus* a vancomycin-resistant bacterium (Mundt et al., 2003). The Nostophycins and Cyanopeptolin obtained from *Nostoc* spp. have been also studied for their antibacterial activity. The study showed bactericidal effect of these bioactive compounds against three-gram negative bacteria like *E. coli*, *Pseudomonas aeruginosa*, and *Proteus vulgaris* at the concentrations of 100, 200, and 250 μg/mL, respectively (Niveshika et al., 2016). The bioactive compounds of Cyanobacteria produced by the genera *Tolypothrix* and *Scytonema* are said to be congeneric chemically. The scytophycins, macrolides and tolytoxin are potentially good antifungal agents and also show cytotoxic effect toward cancerous cells (Bui et al., 2014). One of the several studies suggested that out of 194 strains of Cyanobacteria isolated from terrestrial, freshwater, and brackish water habitats were reported to produce antifungal compounds among which ten inhibited the *Candida albicans* were nine of them inactivated *Aspergillus* (Shishido et al., 2015).

14.3.2 Virucidal activity

The spread of deadly viruses across the globe like HIV, Dengue and Influenza are raising serious concerns among the scientists about the availability of some efficient and potent antiviral compounds. Some researchers have reported bioactive compounds derived from Cyanobacteria as the efficient antiviral agents (Mukherjee et al., 2013). One such antiviral agent is the spirulan, which is effective against Influenza, HIV-1, HIV-2, and some other coated viruses. They are known to cease the binding of uninfected and HIV-infected CD4 + cells that increase the infectivity of the virus. Another polysaccharide with acidic nature known as nostoflan is a very effective virucidal agent. It is derived from the genera *Nostoc flagelliforme* and is effective against HSV-1 (Encarnação et al., 2015). There are some glycan binding proteins in the form of scytovirin and cyanovirin that show virucidal properties by preventing viral fusion (Dixit et al., 2010). Scytovirin is a

9.7 kDa peptide isolated from *Scytonema varium*. It is known to inactivate the HIV virus by binding to the capsid (glycoprotein) of the HIV. Cyanovirin, isolated from *Nostoc ellipsosporum*, is a 11 kDa peptide that is effective against lentivirus and HIV. It mediates its antiviral activity by inhibiting the fusion of HIV envelope with CD4 membrane (Vijayakumar and Menakha, 2015; Lau et al., 2015). The Cyanobacteria-derived sulfoglycolipids show inhibitory effects towards DNA polymerase and reverse transcriptase and thus showing anti-HIV Properties (Silva et al., 2018).

14.3.3 Protozocidal activity

The protozoans are part of eukaryotic microorganism with some of its members behaving as parasites. The studies have reported their role as the causative agents in the human diseases like malaria, Chagas disease, trypanosomiasis and leishmaniosis (Capela et al., 2019). Among the bioactive compounds of Cyanobacteria, the depsipeptide, lactam, and dolastatins produced from *Neoscytalidium hyalidum*, *Oscillatoria* sp., and *Lyngbya majuscula*, respectively; showed protozocidal effect against some drug-resistant strains. Companeramides (depsipeptide) is reported to have shown antimalarial effects in chloroquine-resistant strains of *Plasmodium falciparum*. Another well-studied compound of Cyanobacteria origin is the dolastatins (Singh et al., 2011). These are produced from Cyanobacteria genera such as; *L. majuscula, Symplocahydnoides* etc. Besides showing protozocidal effect with IC_{50} of 0.1 nm they also stop tubulin treadmilling, thus inhibiting its polymerization of α-β tubulins at the plus end of the growing chain. But dolastatin has shown less antiprotozoal effect compared to the lactam and depsipeptide (Salvador-Reyes and Luesch, 2015) (Table 14.3).

14.4 Cyanobacteria bioactive compounds as promising nutraceuticals

The term "nutraceutical" is the blend of two root words, that is, nutrition and pharmaceutical. It is defined it as food/food supplement that has health benefits and may help in the prevention of any disease as well (Kalra, 2003). In addition to the therapeutic activities of bioactive compounds of Cyanobacteria, they also represent a class of metabolites with nutraceutical action (de Morais et al., 2015). Owing to production of nutraceuticals like pigments, fatty acids, vitamins and other bioactive compounds by Cyanobacteria, they are thus regarded as the nutritional source for humans and some other animals as well (Parsaeimehr et al., 2015). These nutraceutical compounds represent a diverse structural class of PKS, NRPS, and PKS-NRPS hybrids. However, the presence of very low quantity of these compounds in most of the genera of Cyanobacteria continues to be a hurdle in their large-scale production (Tan,

TABLE 14.3 Antimicrobial activity of bioactive compounds of the different genera of Cyanobacteria with their diverse chemical classes.

Bioactive compound	Organism	Chemical class	Activity	References
Abietic acid	*Plectonema radiosum, Nostoc* sp.	Terpene	Bactericidal, Algicidal	Costa et al. (2016)
Acutiphycin	*Oscillatoria acutissima*	Recombinant protein	Anticancer	Barchi et al. (1984), Moslin and Jamison (2006)
Calothrixin	*Calothrix* sp.	Cyclodepsipetide	Antimalarial	Xu et al. (2016)
Diarrhetic toxin	*Oscillatoria* sp.	Lippopolysacchride	Cytotoxic	Kubickova et al. (2019)
Debromoaplysia	*Lyngbya gracilis*	Protein	Cytotoxic	Rastogi et al. (2015a)
Didemnin	*Synechocystis tridemni*	Lipopeptide	Virucidal	Mukherjee et al. (2013)
Cryptophycin	*Nostoc* sp.	Glycoprotein	Antifungal	Mukherjee et al. (2013)
Eucapsitrione	*Eucapsis* sp.	Quinone derivative	Antibacteria, cytotoxic	Kosalec et al. (2013)
Cyanopeptolin	*Microcystis aeruginosa, Nostoc minutum, Symploca* sp.	Depsipeptide	Protease inhibitor, antifungal, Protozocidal	Lifshits and Carmeli (2012)
Hoshinolactam	*Oscillatoria* sp.	Lactam	Antiprotozoal	Ogawa et al. (2017)
Spirulan	*Arthrospira platensis*	Glycan	Virucidal	Mader et al. (2016)
Cyanovirin-N	*Nostoc ellipsosporum*	Protein	Virucidal	Matei et al. (2016)
Kulolide	*Lyngbya majuscula*	Depsipeptide	Antibacterial	Boudreau et al. (2012)
Balticidins	*Anabaena cylindrica*	Lipopeptide	Antifungal	Bui et al. (2014)

Carbamidocyclophanes	*Nostoc* sp.	Alkylresorcinols	Bactericidal (*Staphylococcus aureus*)	Luo et al. (2014)
Coriolic acid	*Oscillatoria redekei syn.*	unsaturated fatty acid	Antibacterial	Gómez-García et al. (2019)
Kawaguchipeptin	*M. aeruginosa*	Cyclic peptide	Antibacterial	Leikoski et al. (2012)
Tubercidin	*P. radiosum, Tolypothrix byssoidea*	Nucleoside	Cytotoxic	Mooberry et al. (1995)
Oscillatoxin	*Oscillatoria nigroviridis*	Protein	Anticancer	Nagai et al. (2019)
Largazole	*Symploca* sp.	Depsipeptide	Anticancer (Propoptotic)	Liu et al. (2010)
Phycocyanin	*Nostoc punctiforme, Anabaena ambigua, Calothrix* sp.	Peptide	Antioxidant, antiinflammatory	Kuddus et al. (2013)
Scytonemin	*Scytonema hoffmani, Nostoc microscopium*	Alkaloid	Enzyme inhibition	Rastogi et al. (2015b)
Fischerellins	*Fischerella musicola, Fischerella ambigua*	Polyketide	Antialgal, antifungal	Singh et al. (2001), Hagmann and Jüttner (1996)
Cyanobacterin	*Scytonema hofmanni*	Modified lactone	Antialga, growth inhibition	Ishibashi et al. (2005)
Hapalindole	*Hapalosiphon fontinalis, Hapalosiphon delicatulus*	Alkaloid	Antifungal, antifungal, antialgal, cytotoxic	Acuna et al. (2015)

2011). With enzyme activity in consideration the bioactive compounds such as micropeptins and cyanopeptolins showed much higher activity in alkaline and natural ranges than the proteases of microorganisms and some other plants (Homaei et al., 2016). Cyanobacteria biomass in association with a metallic ion can be also utilized in the manufacture of silver, platinum as well as gold nanoparticles (da Silva Ferreira et al., 2017).

14.4.1 Carotenoids

Carotenoids are ubiquitous pigments which is tremendously beneficial for food processing, health care and cosmetics. The major biological function performed by bioactive compounds of carotenoid nature is the antioxidant property. The major carotenoids produced from Cyanobacteria include: β-carotene, Astaxanthin, Zeaxanthin, Myxol, etc. each useful for a different set of biological functions. Cyanobacteria-produced carotenoids offer multiple applications with regard to the increase in its production, that is, the contamination risk is reduced because of the higher salinity (Markou and Nerantzis, 2013). Cyanobacteria is sometimes classified on the basis of the composition of the carotenoids. *Nostoc punctioforme, Anabaena*, and *Globacterviolacen* are rich in β-carotene but are devoid of zeaxanthin, on the other hand, *Synechocystis* spp. is known to contain them both. Cyanobacteria genera, *Synechococcus* are also rich in zeaxanthin, β-carotene, and nostoxanthin (Takaichi et al., 2005).

Carotenoid	*Cyanobacteria* sp.	Function(s)	References
β-Carotene	*Anabaena variabili, Nostoc punctiforme*	Vision, cell growth, differentiation	Takaichi et al. (2005), Souyoul et al. (2018)
Zeaxanthin	*Gloeobacter violaceus, Synechocystis*	Prevention against atherosclerosis, ARC, AMD	Takaichi et al. (2005), Jia et al. (2017)
Nostoxanthin	*N. punctiforme, Thermosynechococcus*	Maintain GSH level, neuroprotective	Takaichi et al. (2005), Davinelli et al. (2018)
Caloxanthin	*Synechococcus* sp., *Prochlorococcus marinus*	Probiotic, hydrolysis of bile salt	Takaichi et al. (2005), Kechagia et al. (2013)
Ketomyxol G	*N. punctiforme, A. variabili*	Reduced cancer risk, Improvement in memory power	Takaichi et al. (2005), Kechagia et al. (2013)
Hydroxymyxol	*Thermosynechococcus*	Antifungal, food supplement	Takaichi et al. (2005), Kechagia et al. (2013)

14.4.2 Fatty acids

Fatty acids, as a class of biomolecules, are a vital component in the membranes of the organelles and also act as regulatory component. PUFA's are important dietary lipids that act as precursors in the various metabolic pathways. The aquatic Cyanobacteria are a good source of fatty acids including, MUFA and PUFA with a single or multiple double bonds in the backbone. The benefit of exploring Cyanobacteria for fatty acid production is because of their photosynthetic nature transforming CO_2 and ultimately converting it into fatty acids. These derived fatty acids are generally divided into two classes, (1) omega-3 fatty acid, and (2) omega-6-fatty acids (Yu et al., 2014; Santos-Merino et al., 2018).

The fatty acids (omega-3) are vital dietary components with tremendous nutraceutical applications. Humans generally cannot synthesize the PUFA with double bond at cis-position (n-3) or (n-6). Though fish oil is also a good source of PUFAs but recent studies have revealed some of the fatty acids (omega-3) being synthesized from *Synechococcus, Gloeobacter violaceus, Syncheocystis*, etc. (Santos-Merino et al., 2018; Takahama et al., 2004). They produce the bioactive fatty acids like lyngbic acid, eicosapentanoic acid, ecosahexanoic acid, etc. They act as a nutraceutical agent by helping in the treatment of cardiovascular treatments. Besides being active metabolites, they are well known for reducing the heart attacks by decreasing the possibility of clot formation, decreasing arthymias and controlling the growth of arteries' PUFA (omega-3) decrease the levels of LDL and also inhibit the aggregation of platelets (Cassier-Chauvat et al., 2016). Some species of Cyanobacteria, such as, *Anabaena cylindrica, Nostoc canina, Nostoc muscorum* are known to be very good source of certain lipids and thus fatty acids (omega-6), these include digalactosyldiacylglycero, monogalactosyldiacylglycerol, and sulphoquinovosyldiacylglycerol. Fatty acid (omega-6) are fatty acids with its double bond present at C-6. They are beneficial for human health but cannot be synthesized in human bod. The Cyanobacteria-derived bioactive fatty acids help in maintaining healthy bones, regulate metabolism and also help in the growth of hair (Los and Mironov, 2015; Vargas et al., 1998).

14.5 Conclusion

Cyanobacteria, a unique photosynthetic bacterium, constitutes a diverse and widespread habitat around the world. Their ability to act as a potent antimicrobial, therapeutic and nutraceutical agent has been widely studied from last couple of decades. We have given our attention to the two common pathways, that is, PKS and NRPS for biosynthesis of various bioactive compounds. These compounds with variety of functions may be either PKS, or NRPS or NRPS-PKS products like cylindrospermopsins, Hassallidin, and aeruginosin, respectively. The pathways provide an excellent opportunity to explore the new bioactive compounds with functional diversification. Here we have elaborated antibacterial, antifungal, antiviral, and nutraceutical properties of the bioactive compounds of the Cyanobacteria origin. However, these bioactive compounds of microbial origin require a more robust and interdisciplinary approach to explore its further biological aspects with molecular mechanisms.

Acknowledgments

The lab funding from the Scientific and Engineering Research (SERB), Core Research Grant, file no. EMR/2017/002299, India is duly acknowledged.

Conflict of the interest

The authors declare that they have no conflict of interest.

References

Acuna, U.M., Zi, J., Orjala, J., Carcache de Blanco, E.J., 2015. Ambiguine I isonitrile from *Fischerella ambigua* induces caspase-independent cell death in MCF-7 hormone dependent breast cancer cells. Int. J. Cancer Res. 49 (1), 1655–1662.

Aisbl, E., 2016. Nutrition and Health. pp. 1–6. https://doi.org/10.1016/B978-0-08-100596-5.03425-9.

Barchi, J.J., Moore, R.E., Patterson, G.M.L., 1984. Acutiphycin and 20,21-didehydroacutiphycin, new antineoplastic agents from the cyanophyte *Oscillatoria acutissima*. J. Am. Chem. Soc. 106 (26), 8193–8197. Available from: https://doi.org/10.1021/ja00338a031.

Boudreau, P.D., Byrum, T., Liu, W.-T., Dorrestein, P.C., Gerwick, W.H., 2012. Viequeamide A, a cytotoxic member of the kulolide superfamily of cyclic depsipeptides from a marine button cyanobacterium. J. Nat. Prod. 75 (9), 1560–1570. Available from: https://doi.org/10.1021/np300321b.

Bui, T.-H., Wray, V., Nimtz, M., Fossen, T., Preisitsch, M., Schroder, G., et al., 2014. Balticidins A-D, antifungal hassallidin-like lipopeptides from the Baltic Sea cyanobacterium *Anabaena cylindrica* Bio33. J. Nat. Prod. 77 (6), 1287–1296. Available from: https://doi.org/10.1021/np401020a.

Bunyajetpong, S., Yoshida, W.Y., Sitachitta, N., Kaya, K., 2006. Trungapeptins A-C, cyclodepsipeptides from the marine cyanobacterium *Lyngbya majuscula*. J. Nat. Prod. 69 (11), 1539–1542. Available from: https://doi.org/10.1021/np050485a.

Calteau, A., Fewer, D.P., Latifi, A., Coursin, T., Laurent, T., Jokela, J., et al., 2014. Phylum-wide comparative genomics unravel the diversity of secondary metabolism in Cyanobacteria. BMC Genomics 15 (1), 977. Available from: https://doi.org/10.1186/1471-2164-15-977.

Capela, R., Moreira, R., Lopes, F., 2019. An overview of drug resistance in protozoal diseases. Int. J. Mol. Sci. 20 (22), 5748. Available from: https://doi.org/10.3390/ijms20225748.

Cassier-Chauvat, C., Veaudor, T., Chauvat, F., 2016. Comparative genomics of DNA recombination and repair in cyanobacteria: biotechnological implications. Front. Microbiol. 7, 1809. Available from: https://doi.org/10.3389/fmicb.2016.01809.

Costa, M.S., Rego, A., Ramos, V., Afonso, T.B., Freitas, S., Preto, M., et al., 2016. The conifer biomarkers dehydroabietic and abietic acids are widespread in Cyanobacteria. Sci. Rep. 6, 23436. Available from: https://doi.org/10.1038/srep23436.

da Silva Ferreira, V., ConzFerreira, M.E., Lima, L.M.T.R., Frasés, S., de Souza, W., Sant'Anna, C., 2017. Green production of microalgae-based silver chloride nanoparticles with antimicrobial activity against pathogenic bacteria. Enzyme Microb. Technol. 97, 114–121. Available from: https://doi.org/10.1016/j.enzmictec.2016.10.018.

Davinelli, S., Nielsen, M.E., Scapagnini, G., 2018. Astaxanthin in skin health, repair, and disease: a comprehensive review. Nutrients 10 (4), 522. Available from: https://doi.org/10.3390/nu10040522.

de Morais, M.G., Vaz, B., da, S., de Morais, E.G., Costa, J.A.V., 2015. Biologically active metabolites synthesized by microalgae. BioMed. Res. Int. 2015, 835761. Available from: https://doi.org/10.1155/2015/835761.

Dittmann, E., Fewer, D.P., Neilan, B.A., 2013. Cyanobacterial toxins: biosynthetic routes and evolutionary roots. FEMS Microbiol. Rev. 37 (1), 23–43. Available from: https://doi.org/10.1111/j.1574-6976.2012.12000.x.

Dixit, R., Sharma, N., Rai, A., 2010. Anticancerous/antiviral compounds from cyanobacteria. Algas 43, 4–6.

Edwards, D., Gerwick, W., 2004. Lyngbyatoxin biosynthesis: sequence of biosynthetic gene cluster and identification of a novel aromatic prenyltransferase. J. Am. Chem. Soc. 126, 11432–11433. Available from: https://doi.org/10.1021/ja047876g.

Elkobi-Peer, S., Carmeli, S., 2015. New prenylated aeruginosin, microphycin, anabaenopeptin and micropeptin analogues from a Microcystis bloom material collected in Kibbutz Kfar Blum, Israel. Mar. Drugs 13 (4), 2347–2375. Available from: https://doi.org/10.3390/md13042347.

Encarnação, T., Pais, A.A.C.C., Campos, M.G., Burrows, H.D., 2015. Cyanobacteria and microalgae: a renewable source of bioactive compounds and other chemicals. Sci. Prog. 98 (2), 145–168. Available from: https://doi.org/10.3184/003685015X14298590596266.

Falaise, C., François, C., Travers, M.-A., Morga, B., Haure, J., Tremblay, R., et al., 2016. Antimicrobial compounds from eukaryotic microalgae against human pathogens and diseases in aquaculture. Mar. Drugs 14 (9), 159. Available from: https://doi.org/10.3390/md14090159.

Fewer, D.P., Wahlsten, M., Österholm, J., Jokela, J., Rouhiainen, L., Kaasalainen, U., et al., 2013. The genetic basis for O-acetylation of the microcystin toxin in cyanobacteria. Chem. Biol. 20 (7), 861–869. Available from: https://doi.org/10.1016/j.chembiol.2013.04.020.

Fidor, A., Konkel, R., Mazur-Marzec, H., 2019. Bioactive peptides produced by cyanobacteria of the genus *Nostoc*: a review. Mar. Drugs 17 (10). Available from: https://doi.org/10.3390/md17100561.

Frangeul, L., Quillardet, P., Castets, A.-M., Humbert, J.-F., Matthijs, H.C.P., Cortez, D., et al., 2008. Highly plastic genome of *Microcystis aeruginosa* PCC 7806, a ubiquitous toxic freshwater cyanobacterium. BMC Genomics 9 (1), 274. Available from: https://doi.org/10.1186/1471-2164-9-274.

Galanakis, C.M., 2017. Chapter 1. Introduction. Nutraceutical and Functional Food Components. Elsevier Inc. Available from: https://doi.org/10.1016/B978-0-12-805257-0.00001-6.

Gómez-García, M., Sol, C., de Nova, P.J.G., Puyalto, M., Mesas, L., Puente, H., et al., 2019. Antimicrobial activity of a selection of organic acids, their salts and essential oils against swine enteropathogenic bacteria. Porcine Health Manag. 5, 32. Available from: https://doi.org/10.1186/s40813-019-0139-4.

González-Medina, M., Medina-Franco, J.L., 2019. Chemical diversity of cyanobacterial compounds: a chemoinformatics analysis. ACS Omega 4 (4), 6229–6237. Available from: https://doi.org/10.1021/acsomega.9b00532.

Grindberg, R.V., Ishoey, T., Brinza, D., Esquenazi, E., Coates, R.C., Liu, W., et al., 2011. Single cell genome amplification accelerates identification of the apratoxin biosynthetic pathway from a complex microbial assemblage. PLoS One 6 (4), e18565. Available from: https://doi.org/10.1371/journal.pone.0018565.

Hackett, J.D., Wisecaver, J.H., Brosnahan, M.L., Kulis, D.M., Anderson, D.M., Bhattacharya, D., et al., 2013. Evolution of saxitoxin synthesis in cyanobacteria and dinoflagellates. Mol. Biol. Evol. 30 (1), 70–78. Available from: https://doi.org/10.1093/molbev/mss142.

Hagmann, L., Jüttner, F., 1996. Fischerellin A, a novel photosystem-II-inhibiting allelochemical of the cyanobacterium Fischerella muscicola with antifungal and herbicidal activity. Tetrahedron Lett. 37 (36), 6539–6542. Available from: https://doi.org/10.1016/0040-4039(96)01445-1.

Han, B., Gross, H., Goeger, D.E., Mooberry, S.L., Gerwick, W.H., 2006. Aurilides B and C, cancer cell toxins from a Papua New Guinea collection of the marine cyanobacterium *Lyngbya majuscula*. J. Nat. Prod. 69 (4), 572–575. Available from: https://doi.org/10.1021/np0503911.

Harrigan, G.G., Goetz, G.H., Luesch, H., Yang, S., Likos, J., 2001. Dysideaprolines A-F and barbaleucamides A-B, novel polychlorinated compounds from a Dysidea species. J. Nat. Prod. 64 (9), 1133–1138. Available from: https://doi.org/10.1021/np0101999.

Homaei, A., Lavajoo, F., Sariri, R., 2016. Development of marine biotechnology as a resource for novel proteases and their role in modern biotechnology. Int. J. Biol. Macromol. 88, 542–552. Available from: https://doi.org/10.1016/j.ijbiomac.2016.04.023.

Huang, I.S., Zimba, P.V., 2019. Cyanobacterial bioactive metabolites—a review of their chemistry and biology. Harmful Algae 83, 42–94. Available from: https://doi.org/10.1016/j.hal.2018.11.008.

Ishibashi, F., Park, S., Kusano, T., Kuwano, K., 2005. Synthesis and algicidal activity of (+)-cyanobacterin and its stereoisomer. Biosci. Biotechnol. Biochem. 69 (2), 391–396. Available from: https://doi.org/10.1271/bbb.69.391.

Jia, Y.-P., Sun, L., Yu, H.-S., Liang, L.-P., Li, W., Ding, H., et al., 2017. The pharmacological effects of lutein and zeaxanthin on visual disorders and cognition diseases. Molecules (Basel, Switz.) 22 (4), 610. Available from: https://doi.org/10.3390/molecules22040610.

Kalra, E.K., 2003. Nutraceutical—definition and introduction. AAPS PharmSci. 5 (3), E25. Available from: https://doi.org/10.1208/ps050325.

Kechagia, M., Basoulis, D., Konstantopoulou, S., Dimitriadi, D., Gyftopoulou, K., Skarmoutsou, N., et al., 2013. Health benefits of probiotics: a review. ISRN Nutr. 2013, 481651. Available from: https://doi.org/10.5402/2013/481651.

Kellmann, R., Mihali, T.K., Jeon, Y.J., Pickford, R., Pomati, F., Neilan, B.A., 2008. Biosynthetic intermediate analysis and functional homology reveal a saxitoxin gene cluster in cyanobacteria. Appl. Environ. Microbiol. 74 (13), 4044–4053. Available from: https://doi.org/10.1128/AEM.00353-08.

Khosla, C., Herschlag, D., Cane, D.E., Walsh, C.T., 2014. Assembly line polyketide synthases: mechanistic insights and unsolved problems. Biochemistry 53 (18), 2875–2883. Available from: https://doi.org/10.1021/bi500290t.

Klein, D., Braekman, J.C., Daloze, D., Hoffmann, L., Demoulin, V., 1997. Lyngbyaloside, a novel 2,3,4-tri-O-methyl-6-deoxy-α-mannopyranoside macrolide from *Lyngbya bouillonii* (Cyanobacteria). J. Nat. Prod. 60 (10), 1057–1059. Available from: https://doi.org/10.1021/np9702751.

Kosalec, I., Kremer, D., Locatelli, M., Epifano, F., Genovese, S., Carlucci, G., et al., 2013. Anthraquinone profile, antioxidant and antimicrobial activity of bark extracts of Rhamnus alaternus, R. fallax, R. intermedia and R. pumila. Food Chem. 136 (2), 335–341. Available from: https://doi.org/10.1016/j.foodchem.2012.08.026.

Kubickova, B., Babica, P., Hilscherová, K., Šindlerová, L., 2019. Effects of cyanobacterial toxins on the human gastrointestinal tract and the mucosal innate immune system. Environ. Sci. Europe 31 (1), 31. Available from: https://doi.org/10.1186/s12302-019-0212-2.

Kuddus, M., Singh, P., Thomas, G., Al-Hazimi, A., 2013. Recent developments in production and biotechnological applications of C-phycocyanin. BioMed. Res. Int. 2013, 742859. Available from: https://doi.org/10.1155/2013/742859.

Kwan, J.C., Eksioglu, E.A., Liu, C., Paul, V.J., Luesch, H., 2009. Grassystatins A-C from marine cyanobacteria, potent cathepsin E inhibitors that reduce antigen presentation. J. Med. Chem. 52 (18), 5732–5747. Available from: https://doi.org/10.1021/jm9009394.

Latifi, A., Gugger, M., Calteau, A., Castelli, V., Oztas, S., Longin, C., et al., 2013. A tribute to disorder in the genome of the bloom-forming freshwater cyanobacterium *Microcystis aeruginosa*. 8 (8). https://doi.org/10.1371/journal.pone.0070747.

Lau, N.-S., Matsui, M., Abdullah, A.A.-A., 2015. Cyanobacteria: photoautotrophic microbial factories for the sustainable synthesis of industrial products. BioMed. Res. Int. 2015, 754934. Available from: https://doi.org/10.1155/2015/754934.

Leikoski, N., Fewer, D.P., Jokela, J., Alakoski, P., Wahlsten, M., Sivonen, K., 2012. Analysis of an inactive cyanobactin biosynthetic gene cluster leads to discovery of new natural products from strains of the genus microcystis. 7 (8). https://doi.org/10.1371/journal.pone.0043002.

Lifshits, M., Carmeli, S., 2012. Metabolites of *Microcystis aeruginosa* bloom material from Lake Kinneret, Israel. J. Nat. Prod. 75 (2), 209–219. Available from: https://doi.org/10.1021/np200909x.

Lindequist, U., 2016. Marine-derived pharmaceuticals – challenges and opportunities. Biomol. Ther. 24 (6), 561–571. Available from: https://doi.org/10.4062/biomolther.2016.181.

Liu, Y., Salvador, L.A., Byeon, S., Ying, Y., Kwan, J.C., Law, B.K., et al., 2010. Anticolon cancer activity of largazole, a marine-derived tunable histone deacetylase inhibitor. J. Pharmacol. Exp. Ther. 335 (2), 351–361. Available from: https://doi.org/10.1124/jpet.110.172387.

Liu, T., Mazmouz, R., Ongley, S.E., Chau, R., Pickford, R., Woodhouse, J.N., et al., 2017. Directing the heterologous production of specific cyanobacterial toxin variants. ACS Chem. Biol. 12 (8), 2021–2029. Available from: https://doi.org/10.1021/acschembio.7b00181.

Lodin-Friedman, A., Carmeli, S., 2018. Microginins from a *Microcystis* sp. bloom material collected from the Kishon Reservoir, Israel. Mar. Drugs 16 (3), 78. Available from: https://doi.org/10.3390/md16030078.

Los, D.A., Mironov, K.S., 2015. Modes of fatty acid desaturation in cyanobacteria: an update. Life (Basel, Switz.) 5 (1), 554–567. Available from: https://doi.org/10.3390/life5010554.

Luesch, H., Moore, R.E., Paul, V.J., Mooberry, S.L., Corbett, T.H., 2001. Isolation of dolastatin 10 from the marine cyanobacterium Symploca species VP642 and total stereochemistry and biological evaluation of its analogue symplostatin 1. J. Nat. Prod. 64 (7), 907–910. Available from: https://doi.org/10.1021/np010049y.

Luo, S., Kang, H.-S., Krunic, A., Chlipala, G.E., Cai, G., Chen, W.-L., et al., 2014. Carbamidocyclophanes F and G with anti-*Mycobacterium tuberculosis* activity from the cultured freshwater cyanobacterium *Nostoc* sp. Tetrahedron Lett. 55 (3), 686–689. Available from: https://doi.org/10.1016/j.tetlet.2013.11.112.

MacMillan, J.B., Molinski, T.F., 2002. Caylobolide A, a unique 36-membered macrolactone from a Bahamian *Lyngbya majuscula*. Org. Lett. 4 (9), 1535–1538. Available from: https://doi.org/10.1021/ol025759p.

Mader, J., Gallo, A., Schommartz, T., Handke, W., Nagel, C.-H., Gunther, P., et al., 2016. Calcium spirulan derived from *Spirulina platensis* inhibits herpes simplex virus 1 attachment to human keratinocytes and protects against herpes labialis. J. Allergy Clin. Immunol. 137 (1), 197–203.e3. Available from: https://doi.org/10.1016/j.jaci.2015.07.027.

Mahfoudhi, N., Ksouri, R., Hamdi, S., 2016. 11 - Nanoemulsions as potential delivery systems for bioactive compounds in food systems: preparation, characterization, and applications in food industry. In: Grumezescu, A.M.B.T.-E. (Ed.), Nanotechnology in the Agri-Food Industry. Academic Press, pp. 365–403. Available from: https://doi.org/10.1016/B978-0-12-804306-6.00011-8.

Markou, G., Nerantzis, E., 2013. Microalgae for high-value compounds and biofuels production: a review with focus on cultivation under stress conditions. Biotechnol. Adv. 31 (8), 1532–1542. Available from: https://doi.org/10.1016/j.biotechadv.2013.07.011.

Matei, E., Basu, R., Furey, W., Shi, J., Calnan, C., Aiken, C., et al., 2016. Structure and glycan binding of a new cyanovirin-N homolog. J. Biol. Chem. 291 (36), 18967–18976. Available from: https://doi.org/10.1074/jbc.M116.740415.

Mazur-Marzec, H., Fidor, A., Cegłowska, M., Wieczerzak, E., Kropidłowska, M., Goua, M., et al., 2018. Cyanopeptolins with trypsin and chymotrypsin inhibitory activity from the cyanobacterium *Nostoc edaphicum* CCNP1411. Mar. Drugs 16 (7), 220. Available from: https://doi.org/10.3390/md16070220.

Mazzoli, R., Riedel, K., Pessione, E., n.d. Bioactive compounds from microbes. https://doi.org/10.3389/978-2-88945-185-2.

Micallef, M.L., D'Agostino, P.M., Al-Sinawi, B., Neilan, B.A., Moffitt, M.C., 2015a. Exploring cyanobacterial genomes for natural product biosynthesis pathways. Mar. Genomics 21, 1–12. Available from: https://doi.org/10.1016/j.margen.2014.11.009.

Micallef, M.L., D'Agostino, P.M., Sharma, D., Viswanathan, R., Moffitt, M.C., 2015b. Genome mining for natural product biosynthetic gene clusters in the Subsection V cyanobacteria. BMC Genomics 16 (1), 669. Available from: https://doi.org/10.1186/s12864-015-1855-z.

Mooberry, S.L., Stratman, K., Moore, R.E., 1995. Tubercidin stabilizes microtubules against vinblastine-induced depolymerization, a taxol-like effect. Cancer Lett. 96 (2), 261–266. Available from: https://doi.org/10.1016/0304-3835(95)03940-X.

Moslin, R.M., Jamison, T.F., 2006. Highly convergent total synthesis of (+)-acutiphycin. J. Am. Chem. Soc. 128 (47), 15106–15107. Available from: https://doi.org/10.1021/ja0670660.

Mukherjee, G., Nayak, B.K., Das, D., 2013. Cyanobacteria as a valuable source of antiviral, antibacterial and antifungal compounds? An overview. Algological Stud. 143 (1), 3–25. Available from: https://doi.org/10.1127/1864-1318/2013/0143.

Mundt, S., Kreitlow, S., Jansen, R., 2003. Fatty acids with antibacterial activity from the cyanobacterium *Oscillatoria redekei* HUB 051. J. Appl. Phycol. 15 (2), 263–267. Available from: https://doi.org/10.1023/A:1023889813697.

Nagai, H., Sato, S., Iida, K., Hayashi, K., Kawaguchi, M., Uchida, H., et al., 2019. Oscillatoxin I: a new aplysiatoxin derivative, from a marine cyanobacterium. Toxins 11 (6). Available from: https://doi.org/10.3390/toxins11060366.

Namikoshi, M., Rinehart, K.L., 1996. Bioactive compounds produced by cyanobacteria. J. Ind. Microbiol. 17 (5), 373–384. Available from: https://doi.org/10.1007/BF01574768.

Neilan, B.A., Dittmann, E., Rouhiainen, L., Bass, R.A., Schaub, V., Sivonen, K., et al., 1999. Nonribosomal peptide synthesis and toxigenicity of cyanobacteria. J. Bacteriol. 181 (13), 4089–4097. Available from: https://pubmed.ncbi.nlm.nih.gov/10383979.

Nielsen, A.Z., Mellor, S.B., Vavitsas, K., Wlodarczyk, A.J., Gnanasekaran, T., Perestrello Ramos H de Jesus, M., et al., 2016. Extending the biosynthetic repertoires of cyanobacteria and chloroplasts. Plant J. 87 (1), 87–102. Available from: https://doi.org/10.1111/tpj.13173.

Niveshika, Verma, E., Mishra, A.K., Singh, A.K., Singh, V.K., 2016. Structural elucidation and molecular docking of a novel antibiotic compound from cyanobacterium *Nostoc* sp. MGL001. Front. Microbiol. 7, 1899. Available from: https://doi.org/10.3389/fmicb.2016.01899.

Ochi, K., Okamoto, S., Tozawa, Y., Inaoka, T., Hosaka, T., Xu, J., et al., 2004. Ribosome engineering and secondary metabolite production. Adv. Appl. Microbiol. 56, 155–184. Available from: https://doi.org/10.1016/S0065-2164(04)56005-7.

Ogawa, H., Iwasaki, A., Sumimoto, S., Iwatsuki, M., Ishiyama, A., Hokari, R., et al., 2017. Isolation and total synthesis of hoshinolactam, an antitrypanosomal lactam from a marine cyanobacterium. Org. Lett. 19 (4), 890–893. Available from: https://doi.org/10.1021/acs.orglett.7b00047.

Parsaeimehr, A., Sun, Z., Dou, X., Chen, Y.-F., 2015. Simultaneous improvement in production of microalgal biodiesel and high-value alpha-linolenic acid by a single regulator acetylcholine. Biotechnol. Biofuels 8, 11. Available from: https://doi.org/10.1186/s13068-015-0196-0.

Pearson, L.A., Moffitt, M.C., Ginn, H.P., B A, N., 2008. The molecular genetics and regulation of cyanobacterial peptide hepatotoxin biosynthesis. Crit. Rev. Toxicol. 38 (10), 847–856. Available from: https://doi.org/10.1080/10408440802291513.

Percival, S.L., Williams, D.W., 2014. Chapter Five – Cyanobacteria. In: Percival, S.L., Yates, M.V., Williams, D.W., Chalmers, R.M., Gray, N.F. (eds.), Microbiology of Waterborne Diseases, second ed. Academic Press, pp. 79–88. https://doi.org/10.1016/B978-0-12-415846-7.00005-6.

Pouria, S., de Andrade, A., Barbosa, J., Cavalcanti, R.L., Barreto, V.T.S., Ward, C.J., et al., 1998. Fatal microcystin intoxication in haemodialysis unit in Caruaru, Brazil. Lancet 352 (9121), 21–26. Available from: https://doi.org/10.1016/S0140-6736(97)12285-1.

Rastogi, R.P., Madamwar, D., Incharoensakdi, A., 2015a. Bloom dynamics of cyanobacteria and their toxins: environmental health impacts and mitigation strategies. Front. Microbiol. 6, 1254. Available from: https://doi.org/10.3389/fmicb.2015.01254.

Rastogi, R.P., Sonani, R.R., Madamwar, D., 2015b. Cyanobacterial sunscreen scytonemin: role in photoprotection and biomedical research. Appl. Biochem. Biotechnol. 176 (6), 1551–1563. Available from: https://doi.org/10.1007/s12010-015-1676-1.

Ridley, C.P., Lee, H.Y., Khosla, C., 2008. Evolution of polyketide synthases in bacteria. Proc. Natl. Acad. Sci. 105 (12), 4595 LP – 4600. Available from: https://doi.org/10.1073/pnas.0710107105.

Rounge, T.B., Rohrlack, T., Nederbragt, A.J., Kristensen, T., Jakobsen, K.S., 2009. A genome-wide analysis of nonribosomal peptide synthetase gene clusters and their peptides in a Planktothrix rubescens strain. BMC Genomics 10 (1), 396. Available from: https://doi.org/10.1186/1471-2164-10-396.

Salvador-Reyes, L.A., Luesch, H., 2015. Biological targets and mechanisms of action of natural products from marine cyanobacteria. Nat. Prod. Rep. 32 (3), 478–503. Available from: https://doi.org/10.1039/c4np00104d.

Santos-Merino, M., Garcillán-Barcia, M.P., de la Cruz, F., 2018. Engineering the fatty acid synthesis pathway in *Synechococcus elongatus* PCC 7942 improves omega-3 fatty acid production. Biotechnol. Biofuels 11 (1), 239. Available from: https://doi.org/10.1186/s13068-018-1243-4.

Shishido, T.K., Humisto, A., Jokela, J., Liu, L., Wahlsten, M., Tamrakar, A., et al., 2015. Antifungal compounds from cyanobacteria. Mar. Drugs 13 (4), 2124–2140. Available from: https://doi.org/10.3390/md13042124.

Silva, T., S Salomon, P., Hamerski, L., Walter, J., Menezes, R.B., Siqueira, J.E., et al., 2018. Inhibitory effect of microalgae and cyanobacteria extracts on influenza virus replication and neuraminidase activity, PeerJ, 6. p. e5716. Available from: https://doi.org/10.7717/peerj.5716.

Singh, S.B., 2014. Confronting the challenges of discovery of novel antibacterial agents. Bioorg. Med. Chem. Lett. 24 (16), 3683–3689. Available from: https://doi.org/10.1016/j.bmcl.2014.06.053.

Singh, D.P., Tyagi, M.B., Kumar, A., Thakur, J.K., Kumar, A., 2001. Antialgal activity of a hepatotoxin-producing cyanobacterium, *Microcystis aeruginosa*. World J. Microbiol. Biotechnol. 17 (1), 15–22. Available from: https://doi.org/10.1023/A:1016622414140.

Singh, R., Tiwari, S., Rai, A., Mohapatra, T., 2011. Cyanobacteria: an emerging source for drug discovery. J. Antibiotics 64, 401–412. Available from: https://doi.org/10.1038/ja.2011.21.

Sinha, R., Pearson, L.A., Davis, T.W., Muenchhoff, J., Pratama, R., Jex, A., et al., 2014. Comparative genomics of *Cylindrospermopsis raciborskii* strains with differential toxicities. BMC Genomics 15 (1), 83. Available from: https://doi.org/10.1186/1471-2164-15-83.

Souyoul, S.A., Saussy, K.P., Lupo, M.P., 2018. Nutraceuticals: a review. Dermatol. Ther. 8 (1), 5–16. Available from: https://doi.org/10.1007/s13555-018-0221-x.

Subudhi, S., 2017. Bioprospecting for algal based nutraceuticals and high value added compounds. 4 (2), 145–150. https://doi.org/10.15436/2377-1313.17.1651.

Takahama, K., Matsuoka, M., Nagahama, K., Ogawa, T., 2004. High-frequency gene replacement in cyanobacteria using a heterologous rps12 gene. Plant Cell Physiol. 45 (3), 333–339. Available from: https://doi.org/10.1093/pcp/pch041.

Takaichi, S., Mochimaru, M., Maoka, T., Katoh, H., 2005. Myxol and 4-ketomyxol 2'-fucosides, not rhamnosides, from *Anabaena* sp. PCC 7120 and *Nostoc punctiforme* PCC 73102, and proposal for the biosynthetic pathway of carotenoids. Plant Cell Physiol. 46 (3), 497–504. Available from: https://doi.org/10.1093/pcp/pci049.

Tan, L.T., 2011. Biomedical potential of marine cyanobacteria. J. Coast. Dev. 9 (3), 2006. Available from: https://ejournal.undip.ac.id/index.php/coastdev/article/view/1184.

Turner, A.D., Dhanji-Rapkova, M., O'Neill, A., Coates, L., Lewis, A., Lewis, K., 2018. Analysis of microcystins in cyanobacterial blooms from freshwater bodies in England. Toxins 10 (1), 39. Available from: https://doi.org/10.3390/toxins10010039.

Vargas, M.A., Rodrı, H., Moreno, J., Olivares, H., Campo, J.A., Rivas, J., et al., 1998. Biochemical composition and fatty acid content of filamentous nitrogen-fixing cyanobacteria. J. Psychol. 34 (5), 812–817.

Vijayakumar, S., Menakha, M., 2015. Pharmaceutical applications of cyanobacteria—a review. J. Acute Med. 5 (1), 15–23. Available from: https://doi.org/10.1016/j.jacme.2015.02.004.

Wang, H., Fewer, D.P., Holm, L., Rouhiainen, L., Sivonen, K., 2014. Atlas of nonribosomal peptide and polyketide biosynthetic pathways reveals common occurrence of nonmodular enzymes. Proc. Natl. Acad. Sci. U. S. A. 111 (25), 9259–9264. Available from: https://doi.org/10.1073/pnas.1401734111.

Xu, S., Nijampatnam, B., Dutta, S., Velu, S.E., 2016. Cyanobacterial metabolite calothrixins: recent advances in synthesis and biological evaluation. Mar. Drugs 14 (1), 17. Available from: https://doi.org/10.3390/md14010017.

Yu, A.-Q., Pratomo Juwono, N.K., Leong, S.S.J., Chang, M.W., 2014. Production of fatty acid-derived valuable chemicals in synthetic microbes. In: Frontiers in Bioengineering and Biotechnology, vol. 2, p. 78. https://www.frontiersin.org/article/10.3389/fbioe.2014.00078.

Zhou, Q., Bräuer, A., Adihou, H., Schmalhofer, M., Saura, P., Grammbitter, G.L.C., et al., 2019. Molecular mechanism of polyketide shortening in anthraquinone biosynthesis of *Photorhabdus luminescens*. Chem. Sci. 10 (25), 6341–6349. Available from: https://doi.org/10.1039/C9SC00749K.

Ziemert, N., Ishida, K., Quillardet, P., Bouchier, C., Hertweck, C., de Marsac, N.T., et al., 2008. Microcyclamide biosynthesis in two strains of *Microcystis aeruginosa*: from structure to genes and vice versa. Appl. Environ. Microbiol. 74 (6), 1791–1797. Available from: https://doi.org/10.1128/AEM.02392-07.

Chapter 15

Endophytic fungi as a potential source of cytotoxic drugs: a fungal solution to cancer

H.C. Yashavantha Rao, D. Sruthi, Subban Kamalraj, Ramalingam Parthasarathy and Chelliah Jayabaskaran
Department of Biochemistry, Indian Institute of Science, Bengaluru, India

Chapter Outline

Volatiles and Metabolites of Microbes. DOI: https://doi.org/10.1016/B978-0-12-824523-1.00015-8

15.1 Introduction

The natural products are structurally unique agents from natural sources with their novel mechanism of action, plays an important role to treat human diseases, especially for cancer treatment. With the enormous diversity of nature, it is possible to discover underexplored natural products that can interact with several therapeutic targets (Rao et al., 2017a). Hence, natural product research is a good source to discover potential cancer chemotherapeutic agents (Cragg and Pezzuto, 2016). Natural products can be considered as an unrivaled source for cancer drugs as more than 60% of the available drugs for cancer were derived from natural sources in one way or another (Newman and Cragg, 2012). However, the isolated natural products that are developed into clinically effective drugs are relatively few. Terrestrial plants, marine microbes and slime molds have given different chemotherapeutic natural agents (Cragg and Pezzuto, 2016; Rao et al., 2015a).

15.2 Natural products as drugs in cancer treatment

Natural products are used by humans from the 1000 years as a source of medicine. Despite the extensive use of medicinal plants and marine algae from ancient times, the active principles that justify their medicinal properties largely remained unexplored until the 18th and 19th centuries. In 1960s, National Cancer Institute (NCI) introduced an anticancer drug screening program intending to find natural products with an anticancer activity that led to the discovery of taxanes (taxol, baccatin III, and 10-deacetylbaccatin III) (1964) and camptothecins (1966) (Chabner and Roberts, 2005). Since then, several types of anticancer drugs like apigenin, colchicine, resveratrol, and most of them in clinical use are belongs to the class of small molecule natural products. Natural products have a particular functional group to make them enable to inhibit the growth and development of cancer cells.

Taxol originally isolated from *Taxus* species plants is the most effective drug against breast and ovarian cancer, lead mitotic arrest by stabilizing the tubulin polymerization (Horwitz, 1994). Vinca alkaloids are one of the important anticancer agents isolated from *Catharanthus roseus*, bind to tubulin and avert the polymerization of tubulin, lead the disturbance in mitotic spindle assembly (Moreno et al., 1995). Several other compounds such as epothilones, discodermolide, eleutherobin, laulimalide, and isolaulimalide were found antimitotic to cells by inhibiting tubulin depolymerization taxanes. Plant-derived anticancer drug camptothecins (irinotecan, topotecan) is found to be an inhibitor of topoisomerase-I, whereas plant-derived epipodophyllotoxins (etoposide and teniposide) and anthracyclines isolated from microbial source (e.g., doxorubicin, epirubicin) are an inhibitor of topoisomerase II (Salerno et al., 2010).

15.3 Recent advances in the treatment of cancer

Cancer is a heterogeneous disease, develops from a diverse type of tissues, and shows great genetic diversity. There are more than 100 different types of cancer. Despite the advancements made in the diagnosis and treatment of cancer, the disease remains a major health issue around the globe due to its high rate of morbidity and mortality. According to GLOBOCAN database 2008, 12 million new cancer cases were diagnosed in the globe and 7.6 million people died of cancer (around 14% of all deaths worldwide) in 2008. According to a report, the global cancer incidence is projected to increase by 75% (12.7 million in 2008 to 22.2 million) by 2030 (Bray et al., 2012). More than 70% of all cancer deaths worldwide occur in developing countries and underdeveloped countries due to poor living conditions and inadequate or limited medical facilities. In India, the prevalence of cancer is estimated at around 2.5 million with about 8 lakhs new cases and 5.5 lakhs death every year (Nandhakumar et al., 2001). Most frequently in Indian populations are lungs, breast, colon, rectum, stomach, and liver cancers (Rao and Ganesh, 1998).

The conventional cancer treatment involves surgery, radiation therapy, chemotherapy, or combinations of these modules (Urruticoechea et al., 2010). The mode of treatment is decided upon based on the type, stage, and location of cancer.

- *Surgery*: This is one of the oldest methods used in the treatment of several types of cancer. In this method of treatment, a solid tumor is removed especially when it is not spread in another part of the body.
- *Radiotherapy*: In this method, ionizing radiation is used for cancer treatment. It is commonly implemented in addition to surgery and/or chemotherapy, but for some particular types of cancer like in the early stage of head and neck used alone.
- *Chemotherapy*: This is a frequently used method for the treatment of cancer. In this method, a different group of drugs is used which works indifferently to inhibit the cancer cells. Although there are several side effects, chemotherapy remains the modules of cancer treatment. Chemotherapy is an established module for treating many different cancers and its efficiency is often challenged by toxicity to other tissues in the body. Chemotherapeutic agents are generally classified based on their mode of action. Major classes of chemotherapeutic agents are: antimetabolites, alkylating agents, mitotic inhibitors, antitumor antibiotics, topoisomerase inhibitors, hormone therapy, immunotherapy, and targeted therapy.

15.4 Cancer chemotherapeutic drugs from marine microbes

In the process of discovery of anticancer metabolites, an approach like combinatorial chemistry contributes only to minor innovation in drug discovery and effective therapy, while the investigation of unexplored sources is logically postulated to leads us to the discovery of potent bioactive natural products. In the marine ecosystem, which is home to an immense diversity of life, the knowledge about marine microbes is limited. Even though many of the marine-derived natural agents have been isolated and identified with potent anticancer efficacy, only four of them received approval for use in humans. These agents are cytarabine, Yondelis, eribulin, and ADC Adcetris (Newman and Cragg, 2014). The following are the cancer chemotherapeutic agents isolated from marine organisms.

15.5 Cytarabine and nucleoside analogues

The discovery of the bioactive nucleosides spongothymidine and spongouridine flashed to the marine environment as a natural source for novel high-value anticancer drugs, and this further led to the development of analogues like cytarabine as a potent antileukemic agent (Newman and Cragg, 2014). Clofarabine is a second-generation purine nucleoside analogue with potency to damage the DNA of leukemia cells (McGregor et al., 2009). Sapacitabine is another nucleoside analogue prodrug which is bioavailable orally and effective against myelogenous leukemia. It causes single-strand breaks after incorporation into DNA and further converted to double-strand breaks when the cell enters the second S-phase (Kantarjian et al., 2012; Lim and Jamieson, 2014).

15.5.1 Trabectedin (Yondelis)

Trabectedin is the most active ecteinascidins and is approved in Europe for its use against advanced soft-tissue sarcoma. It is the first unmodified natural product from the marine environment which is approved for cancer treatment and is still under the clinical studies for the treatment of different cancers including pediatric sarcomas, prostate and breast cancers (Cuevas et al., 2012).

15.5.2 Halichondrin B and eribulin

Halichondrin B is reported for the first time from the marine sponge *Halichondria okadai* along with several other halichondrin derivatives and is shown to act as a tubulin-destabilizing agent. Eribulin, a halichondrin B analogue is shown to be more potent than pure halichondrin B and thus subjected to advanced clinical studies and after extended clinical trials, it is approved for the treatment of refractory breast cancer (Yu and

Kishi, 2012; Newman and Cragg, 2014). These examples highlight the increasing importance given to and attention gained by the marine organisms in the search for novel lead bioactive molecules. Therefore, we can conclude that the search for compounds from the marine ecosystem will provide us with promising secondary metabolites having potent biological activity.

15.6 Cancer chemotherapeutic drugs from bacteria and fungi

Antitumor agents are one of the most crucial categories among cancer chemotherapeutic drugs. These include members of the anthracycline, ansamycin, actinomycin, epothilone, staurosporine, and bleomycin classes. Except for epothilones, which are metabolites from the myxobacterium *Sorangium cellulosum*, other metabolites were isolated from various *Streptomyces* species (Cragg et al., 2012). Some recent strategies for the development of microbial-associated anticancer agents are discussed later in the chapter.

15.6.1 Rapamycins

Rapamycin is a 31 membered macrocyclic antibiotic that is derived from the fermentation of *Streptomyces hygroscopicus* strain. It is approved as an immunosuppressive agent and led to the synthesis of a wide range of cytotoxic and other pharmacologic drugs. Chemical modifications have given two clinically approved anticancer agents—everolimus (for the treatment of kidney, pancreatic, breast, and brain cancers) and temsirolimus (for the treatment for renal carcinoma). Another rapamycin derivative with anticancer efficacy is ridaforolimus for the treatment of bone cancer and soft-tissue carcinoma (Cragg et al., 2014).

15.6.1.1 Carfilzomib (Kyprolis)

Carfilzomib is a proteasome inhibitor that is approved for the treatment of multiple myeloma patients who has received prior treatment with bortezomib, thalidomide, or lenalidomide (Redic, 2013; Steele, 2013; Thompson, 2013).

15.6.1.2 Midostaurin

Midostaurin is a semisynthetic derivative of staurosporine, an indolocarbazole alkaloid derived from *Streptomyces staurosporeus*. It is a Flt3 (FLK2/STK1) and PKC inhibitor which has finished the clinical trial for acute myeloid leukemia and high-risk myelodysplastic syndrome with either mutated or Flt3 wild-types (Fischer et al., 2010; Prudhomme, 2012).

15.7 Antitumor efficacy of secondary metabolites of endophytic fungi associated with medicinal plants

Plant-associated fungal endophytes have shown diverse applications in the field of agriculture and medicine (Rao and Satish, 2016; Rao et al., 2017b) The enormous potential and scope of natural products are markedly expanded by the studies of plant endophytic fungi, which is shown to be an important source of bioactive metabolites, especially anti-cancer agents. The following sections discuss the anticancer metabolites discovered from plant endophytic fungi. The compounds have been listed based on their chemical nature and biosynthetic origin (Chandra, 2012; Bedi et al., 2017).

15.7.1 Anticancer compounds from Ascomycetes

A large number of novel anticancer compounds belonging to various classes of metabolites are discovered from plant endophytic fungi belonging to the class Ascomycetes.

15.7.2 Polyketides

Chaetoglobosin X (from endophytic fungus *Chaetomium globosum* L18 isolated from the leaves of *Curcuma wenyujin*), Dothiorelone F (from endophytic fungus *Dothiorella* sp. isolated from *Aegiceras corniculatum* bark), Epicocconigrone A (from *Epicoccum nigrum* residing in *Mentha suaveolens* leaves), and Periconiasins A and B (isolated from the endophytic fungus *Periconia* sp. F-31 associated with *Annona muricata*) are the major polyketides with potent anticancer efficacy (Wang et al., 2012; El Amrani et al., 2013; Zhang et al., 2013; Du and Su, 2014; Rao et al., 2015b).

15.7.3 Alkaloids and nitrogen-containing anticancer compounds

Chaetomugilide A, B, and C are the natural Azaphilones alkaloids with anti-cancer activity and were isolated from endophytic fungus *C. globosum* TY1 inhabiting *Ginkgo biloba* bark. Mycoleptodiscin B is another anticancer alkaloid extracted from *Mycoleptodiscus* sp. Associated with the leaves of *Desmotes incomparabilis*. PM181110 is a novel depsipeptide isolated from an endophytic fungus *Phomopsis glabrae* associated with *Pongamia pinnata* leaves (Li et al., 2013; Ortega et al., 2013; Verekar et al., 2014).

15.7.4 Lactones

Photipyrone B is a novel anticancer Y' lactone associated with endophytic fungus *Pestalotiopsis photiniae* residing in *Roystonearegia* whereas (4S,8S) Foedanolide and (+)-(4R,8R)-Foedanolide are two spiro-Y'-lactone enantiomers isolated from *Pestalotiopsis foedan* an endophytic fungus residing in *Bruguiera sexangula* branch. Another novel anticancer bicyclic lactone Myrotheciumone A was isolated from the endophytic fungus *Myrothecium roridum* obtained from the stem *Ajuga decumbens* medicinal herb plant. Further, 3-epi-WaolA was isolated from *Libertellable pharis* associated with the mature leaves of *Oryza latifolia* (Ding et al., 2012; Yang and Li, 2013; Lin et al., 2014; Adames et al., 2015).

15.7.5 Xanthones

Phomopsis, another important genus which exists as a plant endophyte is also highly biochemically diverse. 4,5-dihydroxy-3-(2-hydroxyethyl)-1-methoxy-8-methoxycarbonylxanthone and 1,8-dihydroxy-4-(2-hydroxyethyl)-3-methoxy xanthone were the cytotoxic xanthones isolated from *Phomopsis* species associated with the rhizomes of *Paris axialis* whereas 1,5-dihydroxy-3-(2-oxopropyl)-6-methoxycarbonylxanthone and 1-hydroxy-3-(2-oxopropyl)-8-methoxycarbonyl xanthone were extracted from *Phomopsis* sp. isolated from rhizome of *Paris polyphylla* var. yunnanensis (Yang et al., 2013; Yuan et al., 2015).

15.7.6 Peroxides and quinones

Talaperoxides B and D were derived from *Talaromyces flavus* associated with the leaves of a mangrove plant *Sonneratia apetala.* A novel cytotoxic quinine compound 2,3-didehydro-19α-hydroxy-14-epicochlioquinone B extracted from *Nigrospora* sp. MA75 isolated from the stem of a marine mangrove plant *P. pinnata* (Li et al., 2011; Shang et al., 2012).

15.7.7 Terpenoids

Phomoarcherins A–C are novel anticancer sesquiterpenes obtained from the plant endophyte *Phomopsis archeri* isolated from the stem of *Vanilla albidia.* Cercosporene F is another novel Guanacastane diterpene extracted from *Cercospora* sp. isolated *Fallopia japonica* (Hemtasin et al., 2011; Feng et al., 2014).

The genus *Pestalotiopsis* exists as an endophyte in rainforests and diverse in biochemical point of view. Some of the anticancer compounds isolated from this group include Pestalols B and C, Chloropestolides B, Chloropupukeanolides C-E, Pestaloquinols A and B (Ding et al., 2011; Liu et al., 2013; Sun et al., 2014). *Diaporthe* is another genus producing diverse anticancer compounds including Isochromophilone X and Diaporine A (Zang et al., 2012; Song et al., 2014). The studies on the endophytic fungus *Emericella* sp. AST0036 residing in the leaves of *Astragalus lentiginosus,* afforded a new anticancer compound, Secoemestrin D (Xu et al., 2013).

15.7.8 Anticancer compounds from Basidiomycetes

Cytospolide B and Cytospolide E are the anticancer lactones isolated from the endophytic fungus *Cytospora* sp. strain residing in *Ilex canariensis* (Lu et al., 2011). 3-Epi-steperoxide A is a peroxide compound extracted from *Pseudolagarobasidiumac aciicola* associated with *Bruguiera gymnorrhiza* (Wibowo et al., 2014). Perenniporin A is the anticancer terpenoid extracted from *Perenniporia tephropora* Z41, an endophytic fungus obtained from *Taxus chinensis* var. mairei bark (Wu et al., 2013).

15.7.9 Compounds from Hyphomycetes

Hyphomycete class of fungi represents endophytic fungi that produce the asexual spores (*Aspergillus, Penicillium, Fusarium*, and *Alternaria*) and are well known to produce diverse anticancer metabolites.

15.7.10 Polyketides

Penicillum is an important genus to produce cytotoxic polyketides including duclauxamide A and penicitide A from *Penicillium manginii* (roots of plant *Panax notoginseng* associated with the roots of plant *P. notoginseng*), and *Penicillium chrysogenum*, respectively. Another novel hetero dimeric polyketide, Acremoxanthone E was isolated from the endophytic fungus *Acremonium camptosporum* residing in *Bursera simaruba* leaves (Gao et al., 2010; Cao et al., 2015; Meléndez-González et al., 2015).

15.7.11 Alkaloids and nitrogen-containing compounds

Aspergillines A − C, oxygenated cyclopiazonic acid-derived alkaloids with anticancer efficacy were isolated from *Aspergillus vesicolor* a plant endophyte isolated from rhizome of *P. polyphylla*. Four disulfide-bridged diketopiperazine derivatives viz., Penicibrocazine A B, E, and F were extracted from the endophytic fungus, *Penicillium brocae* isolated from fresh tissue of *Avicennia marina* (Zhou et al., 2014; Meng et al., 2014).

15.7.12 Quinones and terpenoids

Alterporriol K and L are the cytotoxic quinines extracted from endophyte *Alternaria* sp. ZJ9-6B isolated from *A. corniculatum* fruit whereas 2,14-dihydrox-7-drimen-12,11-olide is the cytotoxic sesquiterpene isolated from *Aspergillus glaucus* associated with *Ipomoea batatase* leaves (Huang et al., 2011; Mohammad, 2014).

15.7.13 Pyrans and pyrones

5-butyl-6-(hydroxymethyl)-4-methoxy-2H-pyran-2-one and 4-methoxy-6-methyl-5-(3-oxobutyl)-2H-pyran-2-one are two novel pyrans obtained from *Alternaria phragmospora*, associated with the leaves of *phragmospora* whereas Nigerapyrone B is an α-Pyrone derivative from *Aspergillus niger* obtained from the inner tissue of *A. marina* (Liu et al., 2011; Mohammad, 2014).

15.7.14 Coumarins and phenolic compounds

A new chlorine-containing anticancer isocoumarin Dichlorodiaportinol A was isolated from *Trichoderma* sp., the endophytic fungus residing in *Myoporum bontioides* roots. The phenolic compounds dihydronaphthalenone, 5-hydroxydihydrofusarubins A and B were obtained from the endophyte *Fusarium* sp. of bamboo leaves (Kornsakulkarn et al., 2011; Li et al., 2014).

A novel anticancer compound 6-methyl-1,2,3-trihydroxy-7,8-cyclohepta-9,12-diene-11-one-5,6,7,8-tetralene-7-acetamide was isolated from endophytic *Aspergillus* sp., living in the seeds of medicinal plant *Gloriosa superba*. New furandiones Asperterone B and C were isolated from *Aspergillus terreus* associated with *Malus halliana* leaves. Ginsenocin have been isolated from the endophytic *Penicillium melinii* and *Penicillium janthinellum* from the *Panax ginseng* root. Embellicines A and B were isolated from *Embellisia eureka* found in *Cladanthus arabicus*. Cladosporone A were extracted from *Cladosporium* sp. associated with the flower of the mangrove plant *Kandelia candel*. Further, a new diphenyl ether, 4-dihydroxy-2′, 6-diacetoxy-3′-methoxy-5′-methyl-diphenyl was isolated from *Verticillium* sp. from *Rehmannia glutinosa* root (Gu and Qiao, 2012; Budhiraja et al., 2013; Zheng et al., 2013; Wei et al., 2013; Ebrahim et al., 2013; Ai et al., 2014).

15.8 Taxol-producing endophytic fungi

Some of the plant-derived endophytic fungi producing anticancer drugs. It is a natural product with antitumor activity such as paclitaxel. It is obtained from *Taxus baccata*. Applications of taxol is to treat cancer of breast, ovary, and nonsmall cell cancer of the lung. Taxol can be used in combination or alone with other anticancer drugs. Stierle et al. (1993) documented a taxol-producing endophytic *Taxomyces andreanea*. Following this finding, several other taxol-producing endophytic fungi, such as *Alternaria* spp., *Fusarium* spp., *Tubercularia* spp. *Pestalotiopsis guepinii*, and *Pestalotiopsis microspora* (Strobel et al., 1997; Li et al., 1998). Further, several studies have reported taxol production from endophytic fungi isolated from non-Taxus host plants such as *Bartalinia robillardoides* from *Aegle marmelos*,

Chaetomella raphigera from *Terminalia arjuna*, *Colletotrichum gloeosporioides* from the leaves of *Justicia gendarussa*, *Pestalotiopsis pauciseta* from *Cardiospermum helicacabum*, a leaf spot fungus *Phyllosticta citricarpa* from the angiosperm *Citrus medica*, *Pestalotiopsis terminaliae* from *T. arjuna* (Gangadevi and Muthumary, 2008, 2009; Kumaran and Hur, 2009) produces taxol in liquid cultures. These findings suggest that plants and fungi are independently capable of producing important taxol. As evidence to support the independent production of taxol, several biosynthetic pathway genes such as TDS, DBAT, BAPT, 13αH, TS, 10βH, and TAT, have been isolated from endophytic fungi *Cladosporium cladosporioides* MD2, *Fusarium rodlens*, *Fusarium solani*, and *Ozonium* spp. BT2, (Kusari et al., 2014). It is clear that an endophytic fungi source for taxol production could be better if it could be simply grown, would produce taxol, which applies the biotechnology industry fermentation abilities. Most of these studies, however, have used protoplast mutagenesis and genome shuffling to find the high Taxol yielding strains (Xu et al., 2006; Zhou et al., 2005; Zhao et al., 2008, 2011). For example, *Fusarium mairei* protoplasts subjected to UV plus and diethyl sulfate mutagenesis resulted in a strain reporting 20–225.2 μg paclitaxel per liter of culture (Xu et al., 2006). Endophytic fungi such as *F. solani* isolated from *Taxus celebica* stem cuttings and others such as *P. microspora* and *Lasiodiplodia theobromae* have been reported to produce paclitaxel as well as its precursors baccatin III, deacetylbaccatin III and 7-epi-10-deacetyltaxol (Chakravarthi et al., 2008; Kamalraj et al., 2017, 2019; Sah et al., 2017). The fungal paclitaxel and baccatin III exhibited cytotoxicity toward several human cancer cell lines. Also, recently shown that the plant hormone salicylic acid is an effective elicitor of paclitaxel production in *P. microspora* (Kamalraj et al., 2019). had reported the TDS overexpression by protoplast transformation in a paclitaxel-producing endophytic fungus while Bian et al. (2017) had shown the same through *Agrobacterium tumefcaiens* in efforts towards taxadiene metabolic engineering in fungus. Some of these were partially successful. However, the poor understanding of the paclitaxel biosynthetic pathway in most of the fungi remains the biggest challenge in the way of a thorough success.

15.9 Endophytic fungi producing vinblastine/vincristine

The isolation of vinca alkaloids from *C. roseus* G. which is commonly called *Vinca rosea* for the first time. The phytochemical study of this *C. roseus* extract led to the discovery of anticancer alkaloids such as vinrosidine, vincristine, vinblastine, and vinleunosine (Johnson et al., 1963; Noble, 2016). It is generally a type of chemotherapy drug called vinca alkaloid. It is involved by interfering with cancer cells growth and into two new cells. This ultimately leads to cell death in cancer cells. Since cancer cells divide fast than normal cells, cancer cells more likely than normal cells affected by vinca alkaloids. Among vinca alkaloids, vinblastine, and vincristine are more efficient drugs. In order to find out substitute way vinblastine and vincristine, the microbial sources are a very easy and needless procedure to purify vinblastine and vincristine. Ever since several endophytic fungal diversity has been investigated (Kharwar et al., 2008; Geethanjali et al., 2019) from the *C. roseus*, but there are so far few reports of vincristine and vinblastine producing endophytic fungi (Zhang et al., 2000; Kumar et al., 2013; Padmini et al., 2015; Parthasarathy et al., 2020).

15.10 Endophytic fungi producing camptothecin (CPT) and its analogues

Camptothecin, an anticancer drug was obtained from *Camptotheca acuminata*. It is a kind of alkaloid and irinotecan which are parent compounds for which used as anticancer drugs. The camptothecin is obtained by *Nothapodytes foetida* and *C. acuminata* plants. The camptothecin purified and characterized from the wood extract of the *C. acuminata* for the first time (Takimoto, 2002). In the chloroform extract, camptothecin yield was found to be 18 μg/mg. After four decades of documentation of antitumor potential of *C. acuminata* extract, two camptothecins, topotecan and irinotecan were FDA approved to treat ovarian cancer, colorectal cancer, and small-cell lung cancer (Blagosklonny, 2004). Indeed, camptothecin was recognized to target topoisomerase-I in cancer cells to impart its activity. The mechanism of action of specifically inhibits DNA topoisomerase-I and DNA topoisomerase II (topoisomerase). The study indicated that fungus produced camptothecin (4.96 mg/100 g of dry mass) (Amna et al., 2006).

Two endophytic fungi of *F. solani* isolated from *Apodytes dimidiate* in the Western Ghats of India produces camptothecine 37 and 53 μg/100 g biomass. Endophytic fungi *A. alternate*, *Fomitopsis* sp., and *Phomopsis* sp., inhabiting *Miquelia dentata* produces camptothecin (73.9 μg/g), 9-methoxycamptothecin (55.49 μg/g), and 10-hydroxycamptothecin (42.06 μg/g). Endophytic fungi *Trichoderma atroviride* LY357 *Aspergillus* sp. LY341 and *Aspergillus* sp. LY355, inhabiting *C. acuminate*, produces camptothecin. According to Ding et al. (2013), among 161 endophytic fungi isolated from *C. acuminate*, *Botryosphaeria dothidea* X4 fungus produced 9-methoxycamptothecin. Endophytic *Neurospora* sp. inhabiting *N. foetida* produces camptothecin which showed anticancer activity against OVCAR-5 and A549 cancer cells (Rehman et al., 2008).

15.11 Endophytic fungi producing podophyllotoxin (PDT)

Podophyllotoxin, also been examined as an anticancer drug and an aryl tetralin lignan derived, which obtained from both angiosperms and gymnosperms plant. It extensively occurs in the genera of *Podophyllum*, *Dysosma*,

Juniperus, and *Diphylleia*. Recently, podophyllotoxin was obtained from *Sinopodophyllum* plant (Kusari et al., 2009). It is a podophyllotoxin is crystalline polycyclic compound with an empirical formula $C_{22}H_{22}O_8$ and a molecular weight of 414.41 m/z. It is mainly obtained from roots and rhizomes of Podophyllum species. The source of podophyllotoxin basically from traditional medicinal plants, which remains limited due to less abundance in plants. To increase the yield of podophyllotoxin is of prime importance. Various alternative strategies have been carried out for the production of podophyllotoxin, like tissue culture (Ochoa-Villarreal et al., 2016). It is an efficient inhibitor of microtubule assemblage, which binds at the site of colchicine tubulin and plays a role in DNA topoisomerase II inhibitors. The strains *Phialocephala fortinii* were isolated was from *Podophyllum peltatum* rhizomes which produced podophyllotoxin (0.5 and 189 μg/L) (Eyberger et al., 2006). Correspondingly, endophytic *Fusarium oxysporum* inhabiting *Juniperus recurve* was found to produce podophyllotoxin 28 μg/g of dry mass (Kour et al., 2008). Endophytic fungus *Trametes hirsute* recorded to produce dimethoxy-podophyllotoxin, podophyllotoxin and podophyllotoxin glycoside (Puri et al., 2006). Similarly, *Aspergillus fumigatus* was found to produce deoxypodophyllotoxin 100 μg/g of dry weight of mycelia (Kusari et al., 2009) and podophyllotoxin was produced from endophyte fungus *Alternaria tenuissima* (Liang et al., 2016).

15.12 Closing opinion and a path forward

Natural product research is a strong tool for the discovery of biologically active high-value constituents with unique structures and mechanisms of action. The plant endophytic fungi represent an underexplored group of microorganisms for the discovery of novel anticancer metabolites. With the immeasurable diversity of nature, it is possible to generate the chemical leads that are capable of interacting with all therapeutic anticancer targets. It is clear from the preceding sections that, multiple opportunities are still open for the development of novel analogs and prodrugs from well-established natural drug classes and these can further lead to the development of drugs with high clinical efficiency and decreased toxicity. The conjugation of potent cytotoxic natural products to monoclonal antibodies, which specifically target epitopes on tumors of interest offers another promising approach in the field of drug discovery. Further, new high-value drugs can be discovered from natural sources and these are providing new path for the development of novel and effective chemotherapeutic drugs.

The ability of endophytic fungi to produce phytochemicals originally produced by host plants adds to the benefits of using these microorganisms. In the future, improved cultivation, fermentation techniques, and genetic engineering will permit researchers to isolate and identify new endophytic fungal strains producing antitumor compounds. Research focuses on the molecular characterization of endophytes and optimization of their culture conditions will also improve the chances of success in new drug discovery. Thus, endophytic fungi will continue to be a promising natural source of underexplored secondary high-value metabolites and might be exploited further for the discovery of new drugs for cancer treatment.

Acknowledgments

Dr. H.C. Yashavantha Rao acknowledges the award from the University Grants Commission, New Delhi, Government of India (DSKPDF Scheme-No.F.4-2/2006 (BSR)/BL/17-18/0234). This work was supported by Grants (EMR/2015/001071 and SERB/EEQ/2017/000068) from the Department of Science and Technology (DST), and a grant (BT/PR/14569) from the Department of Biotechnology (DBT), Government of India, New Delhi to CJB, DHR (YSS/2019/000035/PRCYSS), DST-SERB (PDF/2017/000339) to Dr. D. Sruthi and DST-SERB (PDF/2017/001184) to Dr. R. Parthasarathy.

Conflicts of interest

All authors declare no conflict of interest.

References

Adames, I., Ortega, H.E., Asai, Y., et al., 2015. 3-epi-Waol A and Waol C: polyketide-derived γ-lactones isolated from the endophytic fungus *Libertella blepharis* F2644. Tetrahedron Lett. 56 (1), 252–255.

Ai, W., Lin, X., Wang, Z., Lu, X., Mangaladoss, F., Yang, X., et al., 2014. A new dimeric tetralone from fungus *Cladosporium* sp. KcFL6 derived of mangrove plant *Kandelia candel*. J. Antibiot. 68 (3), 213–215.

Amna, T., Puri, S.C., Verma, V., Sharma, J.P., Khajuria, R.K., Musarrat, J., et al., 2006. Bioreactor studies on the endophytic fungus *Entrophospora infrequens* for the production of an anticancer alkaloid camptothecin. Can. J. Microbiol. 52, 189–196.

Bedi, A., Adholeya, A., Deshmukh, S.K., 2017. Novel anticancer compounds from endophytic fungi. Curr. Biotechnol. 6, 1–18.

Bian, G., Yuan, Y., Tao, H., Shi, X., Zhong, X., Han, Y., et al., 2017. Production of taxadiene by engineering of mevalonate pathway in *Escherichia coli* and endophytic fungus *Alternaria alternata* TPF6. Biotechnol. J. 12, 1600697.

Blagosklonny, M.V., 2004. Analysis of FDA approved anticancer drugs reveals the future of cancer therapy. Cell Cycle 3, 1033–1040.

Bray, F., Jemal, A., Grey, N., Ferlay, J., Forman, D., 2012. Global cancer transitions according to the human development index (2008-2030): a population-based study. Lancet Oncol. 13 (8), 790–801.

Budhiraja, A., Nepali, K., Sapra, S., Gupta, S., Kumar, S., Dhar, K.L., 2013. Bioactive metabolites from an endophytic fungus of *Aspergillus* species isolated from seeds of *Gloriosa superba* Linn. Med. Chem. Res. 22 (1), 323–329.

Cao, P., Yang, J., Miao, C.P., et al., 2015. New duclauxamide from *Penicillium manginii* YIM PH30375 and structure revision of the duclauxin family. Org. Lett. 17 (5), 1146–1149.

Chabner, B.A., Roberts, T.G., 2005. Chemotherapy and the war on cancer. Nat. Rev. Cancer 5 (1), 65–72.

Chakravarthi, B.V., Das, P., Surendranath, K., Karande, A.A., Jayabaskaran, C., 2008. Production of paclitaxel by *Fusarium solani* isolated from *Taxus celebica*. J. Biosci. 33, 259–267.

Chandra, S., 2012. Endophytic fungi: novel sources of anticancer lead molecules. Appl. Microbiol. Biotechnol. 95, 4–49.

Cragg, G.M., Pezzuto, J.M., 2016. Natural products as a vital source for the discovery of cancer chemotherapeutic and chemopreventive agents. Med. Princ. Pract. 25, 41–59.

Cragg, G.M., Kingston, D.G.I., Newman, D.J., 2012. Anticancer Agents From Natural Products, second ed. CRC/Taylor & Francis, Boca Raton, FL.

Cragg, G.M., Grothaus, P.G., Newman, D.J., 2014. New horizons for old drugs and drug leads. J. Nat. Prod. 77, 703–723.

Cuevas, C., Francesch, A., Galmarini, C.M., et al., 2012. Ecteinascidin-743 (Yondelis®), Aplidin®, and Irvalec®. In: Cragg, G.M., Kingston, D.G.I., Newman, D.J. (Eds.), Anticancer Agents From Natural Products, second ed. CRC/ Taylor & Francis, Boca Raton, FL, pp. 291–316.

Ding, G., Zhang, F., Chen, H., Guo, L., Zou, Z., Che, Y., 2011. Pestaloquinols A and B, isoprenylated epoxyquinols from *Pestalotiopsis* sp. J. Nat. Prod. 74 (2), 286–291.

Ding, G., Qi, Y., Liu, S., Guo, L., Chen, X., 2012. Photipyrones A and B, new pyrone derivatives from the plant endophytic fungus *Pestalotiopsis photiniae*. J. Antibiot. 65 (5), 271–273.

Ding, X., Liu, K., Deng, B., Chen, W., Li, W., Liu, F., 2013. Isolation and characterization of endophytic fungi from *Camptotheca acuminata*. World J. Microbiol. Biotechnol. 29, 1831–1838.

Du, X.P., Su, W.J., 2014. Two new polyketides from mangrove endophytic fungus sp. Chem. Nat. Comp. 2 (50), 214–216.

Ebrahim, W., Aly, A.H., Wray, V., Mándi, A., Teiten, M.-H., Gaascht, F., 2013. Embellicines A and B: absolute configuration and NF-κB transcriptional inhibitory activity. J. Med. Chem. 56 (7), 2991–2999.

El Amrani, M., Lai, D., Debbab, A., et al., 2013. Protein kinase and HDAC inhibitors from the endophytic fungus *Epicoccum nigrum*. J. Nat. Prod. 77 (1), 49–56.

Eyberger, A.L., Dondapati, R., Porter, J.R., 2006. Endophyte fungal isolates from *Podophyllum peltatum* produce podophyllotoxin. J. Nat. Prod. 69, 1121–1124.

Feng, Y., Ren, F., Niu, S., et al., 2014. Guanacastane diterpenoids from the plant endophytic fungus *Cercospora* sp. J. Nat. Prod. 77 (4), 87381.

Fischer, T., Stone, R.M., Deangelo, D.J., et al., 2010. Phase IIB trial of oral midostaurin (PKC412), the FMS-like tyrosine kinase 3 receptor (FLT3) and multi-targeted kinase inhibitor, in patients with acute myeloid leukemia and highrisk myelodysplastic syndrome with either wild-type or mutated FLT3. J. Clin. Oncol. 28, 4339–4345.

Gangadevi, V., Muthumary, J., 2008. Taxol, an anticancer drug produced by an endophytic fungus *Bartalinia robillardoides Tassi*, isolated from a medicinal plant, *Aegle marmelos* Correa ex Roxb. World J. Microb. Biot. 24, 717–724.

Gangadevi, V., Muthumary, J., 2009. Taxol production by *Pestalotiopsis terminaliae*, an endophytic fungus of *Terminalia arjuna* (arjun tree). Biotechol. App. Biochem. 52, 9–15.

Gao, S.S., Li, X.M., Du, F.Y., Li, C.S., Proksch, P., Wang, B.G., 2010. Secondary metabolites from a marine-derived endophytic fungus *Penicillium chrysogenum* QEN-24S. Mar. Drugs 9 (1), 5970.

Geethanjali, D., Subban, K., Chelliah, J., 2019. Diversity and biological activities of endophytic fungi associated with *Catharanthus roseus*. BMC Microbiol. 19, 22.

Gu, W., Qiao, C., 2012. Furandiones from an endophytic *Aspergillus terreus* residing in *Malus halliana*. Chem. Pharm. Bull. 60 (11), 1474–1477.

Hemtasin, C., Kanokmedhakul, S., Kanokmedhakul, K., et al., 2011. Cytotoxic pentacyclic and tetracyclic aromatic sesquiterpenes from *Phomopsis archeri*. J. Nat. Prod. 74 (4), 609–613.

Horwitz, S.B., 1994. Taxol (paclitaxel): mechanisms of action. Ann. Oncol. 5, 3–6.

Huang, C.H., Pan, J.H., Chen, B., Yu, M., Huang, H.B., Zhu, X., 2011. Three bianthraquinone derivatives from the mangrove endophytic fungus *Alternaria* sp. ZJ9-6B from the South China Sea. Mar. Drugs 9 (5), 832–843.

Johnson, I.S., Armstrong, J.G., Gorman, M., Burnett Jr., J.P., 1963. The vinca alkaloids: a new class of oncolytic agents. Cancer Res. 23, 1390–1427.

Kamalraj, S., Singh, S., Subramani, R., Johnpaul, M., Chelliah, J., 2017. Fungal 7-epi-10-deacetyltaxol produced by an endophytic *Pestalotiopsis microspora* induces apoptosis in human hepatocellular carcinoma cell line (HepG2). BMC Compl. Altern. Med. 17, 504.

Kamalraj, S., Subramani, R., Madambakkam Srinivasan, V.P., Johnpaul, M., Chelliah, J., 2019. Salicylic acid as an effective elicitor for improved taxol production in endophytic fungus *Pestalotiopsis microspora*. PLoS One 14, 1–17.

Kantarjian, H., Faderl, S., Garcia Manero, G., et al., 2012. Oral sapacitabine for the treatment of acute myeloid leukaemia in elderly patients: a randomised phase 2 study. Lancet Oncol. 13, 1096–1104.

Kharwar, R.N., Verma, V.C., Strobel, G., Ezra, D., 2008. The endophytic fungal complex of *Catharanthus roseus* (L.) G. Don. Res. Commun. 95, 228–233.

Kornsakulkarn, J., Dolsophon, K., Boonyuen, N., Boonruangprapa, T., Rachtawee, P., Prabpai, S., 2011. Dihydrona phthalenones from endophytic fungus *Fusarium* sp. BCC14842. Tetrahedron 67 (39), 7540–7547.

Kour, A., Shawl, A.S., Rehman, S., Sultan, P., Qazi, P.H., Suden, P., et al., 2008. Isolation and identification of an endophytic strain of *Fusarium oxysporum* producing podophyllotoxin from *Juniperus recurva*. World J. Microbiol. Biotechnol. 24, 1115–1121.

Kumar, A., Patil, D., Rajamohanan, P.R., Ahmad, A., 2013. Isolation, purification and characterization of from endophytic fungus *Fusarium oxysporum* isolated from *Catharanthus roseus*. PLoS One 8, e71805.

Kumaran, R.S., Hur, B.K., 2009. Screening of species of the endophytic fungus *Phomopsis* for the production of the anticancer drug taxol. Biotechol. App. Bioc 54, 21–30.

Kusari, S., Lamshoft, M., Spiteller, M., 2009. *Aspergillus fumigatus Fresenius*, an endophytic fungus from *Juniperus communis* L. Horstmann as a novel source of the anticancer pro-drug deoxypodophyllotoxin. J. Appl. Microbiol. 107, 1019–1030.

Kusari, S., Singh, S., Jayabaskaran, C., 2014. Rethinking production of Taxol (paclitaxel) using endophyte biotechnology. Trends Biot. 32, 304–311.

Li, J.Y., Sidhu, R.S., Ford, E.J., Long, D.M., Hess, W.M., Strobel, G.A., 1998. The induction of taxol production in the endophytic fungus - *Periconia* sp from *Torreya grandifolia*. J. Ind. Microbiol. Biot. 20, 259–264.

Li, H., Huang, H., Shao, C., et al., 2011. Cytotoxic norsesquiterpene peroxides from the endophytic fungus *Talaromyces flavus* isolated from the mangrove plant *Sonneratia apetala*. J. Nat. Prod. 74 (5), 1230–1235.

Li, X., Tian, Y., Yang, S.X., Zhang, Y.M., Qin, J.C., 2013. Cytotoxic azaphilone alkaloids from *Chaetomium globosum* TY1. Bioorg. Med. Chem. Lett. 23 (10), 2945–2947.

Li, C., Gong, B., Cox, D.G., Li, C., Wang, J., Ding, W., 2014. Dichloro diaportinol A–A new chlorine-containing isocoumarin from an endophytic fungus *Trichoderma* sp. 09 from *Myoporum bontioides* A. Gray and its cytotoxic activity. Phcog. Mag. 10 (Suppl. S1), 153–158.

Liang, Z., Zhang, J., Zhang, X., Li, J., Zhang, X., Zhao, C., 2016. Endophytic fungus from *Sinopodophyllum emodi* (Wall.) ying that produces podophyllotoxin. J. Chromatogr. Sci. 54, 175–178.

Lim, M.Y., Jamieson, K., 2014. Profile of sapacitabine: potential for the treatment of newly diagnosed acute myeloid leukemia in elderly patients. Clin. Interv. Aging 9, 753–762.

Lin, T., Wang, G., Shan, W., et al., 2014. Myrotheciumones: bicyclic cytotoxic lactones isolated from an endophytic fungus of *Ajuga decumbens*. Bioorg. Med. Chem. Lett. 24 (11), 2504–2507.

Liu, L., Bruhn, T., Guo, L., et al., 2011. Chloropupukeanolides C–E: cytotoxic pupukeanane chlorides with a spiroketal skeleton from *Pestalotiopsis fici*. Chem-A Eur. J. 17 (9), 2604–2613.

Liu, L., Li, Y., Li, L., et al., 2013. Spiroketals of *Pestalotiopsis fici* provide evidence for a biosynthetic hypothesis involving diversified diels–Alder reaction cascades. J. Org. Chem. 78 (7), 2992–3000.

Lu, S., Kurtán, T., Yang, G., et al., 2011. Cytospolides A–E, new nonanolides from an endophytic fungus, *Cytospora* sp. Eur. J. Org. Chem. 28, 5452–5459.

McGregor, B.A., Brown, A.W., Osswald, M.B., et al., 2009. The use of higher dose clofarabine in adults with relapsed acute lymphoblastic leukemia. Am. J. Hematol. 84, 228–230.

Meléndez-González, C., Murià-González, M.J., Anaya, A.L., Hernández-Bautista, B.E., Hernández-Ortega, S., González, M.C., 2015. Acremoxanthone E, a novel member of heterodimeric polyketides with a bicyclo [3.2. 2] nonene ring, produced by *Acremonium camptosporum* W. Gams (Clavicipitaceae) endophytic fungus. Chem. Biodivers. 12 (1), 133–147.

Meng, L.H., Li, X.M., Lv, C.T., Huang, C.G., Wang, B.G., 2014. Brocazines A–F, cytotoxic bisthiodi ketopiperazine derivatives from *Penicillium brocae* MA-231, an endophytic fungus derived from the marine mangrove plant *Avicennia marina*. J. Nat. Prod. 77 (8), 1921–1927.

Mohammad, A.E.I., 2014. Antileukemic α-pyrone derivatives from the endophytic fungus *Alternaria phragmospora*. Tetrahedron Lett. 55 (24), 3478–3481.

Moreno, P.R., van der Heijden, R., Verpoorte, R., 1995. Cell and tissue cultures of *Catharanthus roseus*: a literature survey. Plant Cell Tiss. Org. 42, 125.

Nandhakumar, A., Prabhakaran, P.S., Dinshaw, K.A., et al., 2001. Consolidated report of the population-based cancer registries 1990–1996. Co-ordinating unit NCRP (Indian Council of Medical Research), Bangalore, pp. 116–117.

Newman, D.J., Cragg, G.M., 2012. Natural products as sources of new drugs over the 30 years from 1981 to 2010. J. Nat. Prod. 75, 311–335.

Newman, D.J., Cragg, G.M., 2014. Marine-sourced anti-cancer and cancer pain control agents in clinical and late preclinical development. Mar. Drugs 12, 255–278.

Noble, R., 2016. Anti-cancer alkaloids of *Vinca rosea*. Pharmacol. Orient. Plants 7, 61–78.

Ochoa-Villarreal, M., Howat, S., Hong, S., Jang, M.O., Jin, Y.W., Lee, E.K., et al., 2016. Plant cell culture strategies for the production of natural products. BMB Rep. 49, 149–158.

Ortega, H.E., Graupner, P.R., Asai, Y., TenDyke, K., Qiu, D., Shen, Y.Y., et al., 2013. Mycoleptodiscins A and B, cytotoxic alkaloids from the endophytic fungus *Mycoleptodiscus* sp. F0194. J. Nat. Prod. 76 (4), 741–744.

Padmini, P.C., Kurikose, G.C., Jayabaskaran, C., 2015. An endophytic fungus, *Talaromycesr adicus* isolated from *Catharanthus roseus* produces vincristine and vinblastine, which induce apoptotic cell death. PLoS One 10, 12.

Parthasarathy, R., Shanmuganathan, R., Pugazhendhi, A., 2020. Vinblastine production by the endophytic fungus *Curvularia verruculosa* from the leaves of *Catharanthus roseus* and its in vitro cytotoxicity against HeLa cell line. Anal. Biochem. 593, 113530.

Prudhomme, M., 2012. Staurosporines and structurally related indocarbazoles as antitumor agents. In: Cragg, G.M., Kingston, D.G.I., Newman, D.J. (Eds.), Anticancer Agents From Natural Products, second ed. CRC/Taylor & Francis, Boca Raton, FL, pp. 647–670.

Puri, S.C., Nazir, A., Chawla, R., Arora, R., Riyaz-Ul-Hasan, S., Amna, T., et al., 2006. The endophytic fungus *Trametes hirsuta* as a novel alternative source of podophyllotoxin and related aryl tetralin lignans. J. Biotechnol. 122, 494–510.

Rao, D.N., Ganesh, B., 1998. Estimate of cancer incidence in India in 1991. Indian J. Cancer 35 (1), 10.

Rao, H.C.Y., Satish, S., 2016. Intra-specific differentiation of fungal endosymbiont *Alternaria longissima* CLB44 using RNA secondary structure analysis and their anti-infective potential. Sci. Nat. 103, 69.

Rao, H.C.Y., Baker, S., Rakshith, D., Satish, S., 2015a. Molecular profiling and antimicrobial potential of endophytic *Gliomastix polychroma* CLB32 inhabiting *Combretum latifolium* Blume. Mycology 6, 176–181.

Rao, H.C.Y., Santosh, P., Rakshith, D., Satish, S., 2015b. Molecular characterization of an endophytic *Phomopsis liquidambaris* CBR-15 from *Cryptolepis buchanani* Roem. and impact of culture media on biosynthesis of antimicrobial metabolites. 3Biotech 5, 165–173.

Rao, H.C.Y., Rakshith, D., Harini, B.P., Satish, S., 2017a. Antimicrobial profiling and molecular identification of *Alternaria arborescens* CLB12, a myco-endosymbiont inhabiting *Combretum latifolium* Blume. J. Biol. Act. Prod. Nat. 7 (1), 1–9.

Rao, H.C.Y., Rakshith, D., Harini, B.P., Gurudatt, D.M., Satish, S., 2017b. Chemogenomics driven discovery of endogenous polyketide anti-infective compounds from endosymbiotic *Emericella variecolor* CLB38 and their RNA secondary structure analysis. PLoS One 12 (2), e0172848.

Redic, K., 2013. Carfilzomib: a novel agent for multiple myeloma. J. Pharm. Pharmacol. 65, 1095–1106.

Rehman, S., Shawl, A.S., Kour, A., Andrabi, R., Sudan, P., Sultan, P., et al., 2008. An endophytic *Neurospora* sp. from *Nothapodytes foetida* producing camptothecin. Appl. Biochem. Microbiol. 44, 203–209.

Sah, B., Subban, K., Chelliah, J., 2017. Cloning and sequence analysis of 10-deacetylbaccatin III-10-O-acetyl transferase gene and WRKY1 transcription factor from taxol producing endophytic fungus *Lasiodiplodia theobromae*. FEMS Microbiol. Lett. 364 (24), fnx253.

Salerno, S., Settimo, D.F., Taliani, S., Simorini, F., La-Motta, C., Fornaciari, G., et al., 2010. Recent advances in the development of dual topoisomerase I and II inhibitors as anticancer drugs. Curr. Med. Chem. 17, 4270–4290.

Shang, Z., Li, X.M., Li, C.S., Wang, B.G., 2012. Diverse secondary metabolites produced by marine-derived fungus *Nigrospora* sp. MA75 on various culture media. Chem. Biodivers. 9 (7), 1338–1348.

Song, Y., Dou, H., Wang, P., et al., 2014. A novel small-molecule compound diaporine A inhibits non-small cell lung cancer growth by regulating miR-99a/mTOR signaling. Cancer Biol. Ther. 15 (10), 1423–1430.

Steele, J.M., 2013. Carfilzomib: a new proteasome inhibitor for relapsed or refractory multiple myeloma. J. Oncol. Pharm. Pract. 19, 348–354.

Stierle, A., Strobel, G., Stierle, D., 1993. Taxol and taxane production by *Taxomyces andreanae*, an endophytic fungus of pacific yew. Science 260, 214–216.

Strobel, G.A., Hess, W.M., Li, J.Y., Ford, E., Sears, J., Sidhu, R.S., et al., 1997. *Pestalotiopsis guepinii*, a taxol-producing endophyte of the Wollemi pine, *Wollemia nobilis*. Aust. J. Bot. 45, 1073–1082.

Sun, J.F., Lin, X., Zhou, X.F., et al., 2014. Pestalols A–E, new alkenyl phenol and benzaldehyde derivatives from endophytic fungus Pestalotiopsis sp. AcBC2 isolated from the Chinese mangrove plant *Aegiceras corniculatum*. J. Antibiot. 67 (6), 451–457.

Takimoto, C.H., 2002. Camptothecins A2 - Bertino, Joseph R. In: Bertino, J. (Ed.), Encyclopedia of Cancer, second ed. Academic Press, New York, pp. 301–309.

Thompson, J.L., 2013. Carfilzomib: a second-generation proteasome inhibitor for the treatment of relapsed and refractory multiple myeloma. Ann. Pharmacother. 47, 56–62.

Urruticoechea, A., Alemany, R., Balart, J., Villanueva, A., Vinals, F., Capella, G., 2010. Recent advances in cancer therapy: an overview. Curr. Pharm. Des. 16 (1), 3–10.

Verekar, S.A., Mishra, P.D., Sreekumar, E.S., et al., 2014. Anticancer activity of new depsipeptide compound isolated from an endophytic fungus. J. Antibiot. 67 (10), 697–701.

Wang, Y., Xu, L., Ren, W., Zhao, D., Zhu, Y., Wu, X., 2012. Bioactive metabolites from *Chaetomium globosum* L18, an endophytic fungus in the medicinal plant *Curcuma wenyujin*. Phytomedicine 19 (3), 364–368.

Wei, P., Fei, Y., Xiao Li, L., Min, J., Zheng, C.J., et al., 2013. A new diphenyl ether from the endophytic fungus Verticillium sp. isolated from *Rehmannia glutinosa*. Chin. J. Nat. Med. 11 (6), 673–675.

Wibowo, M., Prachyawarakorn, V., Aree, T., et al., 2014. Tricyclic and spirobicyclic norsesquiterpenes from the endophytic fungus *Pseudolagarobasidium acaciicola*. Eur. J. Org. Chem. 19, 3976–3980.

Wu, L.S., Hu, C.L., Han, T., et al., 2013. Cytotoxic metabolites from *Perenniporia tephropora*, an endophytic fungus from *Taxus chinensis* var. mairei. App. Microbiol. Biotechnol. 97 (1), 30515.

Xu, F., Tao, W., Cheng, L., Guo, L., 2006. Strain improvement and optimization of the media of taxol-producing fungus *Fusarium maire*. Biochem. Eng. J. 31, 67–73.

Xu, Y.M., Espniosa Artiles, P., Liu, M.X., Arnold, A.E., Gunatilaka, A.L., 2013. Secoemestrin D, a cytotoxic epitetrathiodiox opiperizine, and emericellenes A–E, five sesterterpenoids from *Emericella* sp. AST0036, a fungal endophyte of *Astragalus lentiginosus*. J. Nat. Prod. 76 (12), 2330–2336.

Yang, X.L., Li, Z.Z., 2013. New spiral γ-lactone enantiomers from the plant endophytic fungus *Pestalotiopsis foedan*. Molecules 18 (2), 2236–2242.

Yang, H.Y., Gao, Y.H., Niu, D.Y., et al., 2013. Xanthone derivatives from the fermentation products of an endophytic fungus *Phomopsis* sp. Fitoterapia 91, 189–193.

Yu, M.J., Kishi, Y., 2012. Littlefield BA: discovery of E7389, a fully synthetic macrocyclic ketone analog of halichondrin B. In: Cragg, G.M., Kingston, D.G.I., Newman, D.J. (Eds.), Anticancer Agents From Natural Products, second ed. CRC/Taylor & Francis, Boca Raton, FL, pp. 317–345.

Yuan, L., Huang, W., Du, G., et al., 2015. Isolation of xanthones from the fermentation products of the endophytic fungus of *Phomopsis amygdali*. Chem. Nat. Comp. 51 (3), 460–463.

Zang, L.Y., Wei, W., Wang, T., Guo, Y., Tan, R.X., Ge, H.M., 2012. Isochromophilones from an endophytic fungus *Diaporthe* sp. Nat. Prod. Bioprospect. 2 (3), 117–120.

Zhang, L.B., Gou, L.H., Zeng, S.V., 2000. Preliminary study on the isolation of endophytic fungus of *Catharanthus roseus* and its fermentation to produce product of therapeutic value. Chin. Trad. Herb. Drugs 11, 805–807.

Zhang, D., Ge, H., Xie, D., et al., 2013. Periconiasins A–C, new cytotoxic cytochalasans with an unprecedented 9/6/5 tricyclic ring system from endophytic fungus *Periconia* sp. Org. Lett. 15 (7), 1674–1677.

Zhao, K., Ping, W., Zhou, D., 2008. Recent advance and prospect on taxol production by endophytic fungus fermentation—a review. Acta Microbiol. Sin. 48, 403–407.

Zhao, K., Li, Z., Ge, N., Li, X., Wang, X., Zhou, D., 2011. Investigation of fermentation conditions and optimization of medium for taxol production from taxol-producing fungi. J. Med. Plant Res. 5, 6528–6535.

Zheng, C.J., Xu, L.L., Li, Y.Y., Han, T., Zhang, Q.Y., Ming, Q.L., 2013. Cytotoxic metabolites from the cultures of endophytic fungi from *Panax ginseng*. App. Microbiol. Biotechnol. 97 (17), 7617–7625.

Zhou, D., Zhao, K., Ping, W., Ge, J., Ma, X., Jun, L., 2005. Study on the mutagensis of protoplasts from taxol-producing fungus *Nodulisporium sylviforme*. Am. J. Sci. 1, 55–62.

Zhou, M., Miao, M.M., Du, G., Li, X.N., Shang, S.Z., Zhao, W., 2014. Aspergillines A-E, highly oxygenated hexacyclic indoletetra hydrofuran-tetramic acid derivatives from *Aspergillus versicolor*. Org. Lett. 16 (19), 5016–5019.

Chapter 16

Volatile organic compounds for enhancement of plant growth through plant growth promoting rhizobacteria

Mrunal S. Wagh[1], W. Jabez Osborne[2] and Saravanan Sivarajan[3]

[1]*School of BioSciences and Technology, VIT, Vellore, India,* [2]*Biomolecules Lab, Department of Bio-Sciences, School of Bio Sciences and Technology, Vellore Institute of Technology, Vellore, India,* [3]*VIT School of Agricultural Innovations & Advanced Learning (VAIAL)*

Chapter Outline

Volatiles and Metabolites of Microbes. DOI: https://doi.org/10.1016/B978-0-12-824523-1.00008-0

16.1 Introduction

Plant-microbe interactions have been a thrust area of research for several decades. Bacteria associated with the rhizosphere termed as plant growth-promoting rhizobacteria (PGPR) have been shown to enhance the growth of the plants in various ways. Due to the nutrient-rich rhizospheric environment, plants are naturally associated with many microorganisms and also interact with each other (Liu and Zhang, 2015). These microorganisms increment in the seedling rise, crop yield, plant weight, and malady opposition (Kai et al., 2009), which has come about because some PGPR applied in horticulture. Most plant-related rhizobacteria are commensals in which the bacteria create a harmless relationship with the host plants with a little noticeable impact on the host's growth and overall physiology (Ojuederie et al., 2019).

Many substances are considered as the thin, low-molecular-weight of approximately <300 Da, and even sometimes stinking agents with transitional boiling points and higher vapor pressure resulting in dispersion of air and water. Due to this nature, thousands of unstable microbial compounds are characterized as alcohols, ketones, aromatic compounds, organic acids, aldehydes, sulfur, alkanes, and alkene (Audrain et al., 2015). Research has shown that bacteria release and yield many unstable substances that are extremely variable within the organisms as well as in multiple species that are influenced by a variety of considerations (Vespermann et al., 2007). Bacterial volatile compounds (BVCs) may ultimately lead to physiological reactions in other bacteria and fungi and may influence the development and health of higher-order organisms such as plants. Lipolysis, glycolysis, and proteolysis like catabolic pathways produce bacterial volatile organic compounds (VOCs) and also are distinguished into various classes of compounds of chemical (Penuelas et al., 2014). Due to the aroma obtained (Tahir et al., 2017), the monitoring and evaluation of such BVCs have been of considerable interest in the food and cosmetic industry.

In the rhizospheric region of many species of plants, PGPR are considered as a root-colonizing bacterial group that promote plant productivity and mostly induce plant immunity against distinct plant pathogens (Lee et al., 2012). Due to its high versatility in a wide range of conditions and metabolic stability to metabolize a broad spectrum of organic and xenobiotic substances, PGPR happens to be an efficient rhizobacterium in creating for soil habitats. Studies show VOCs synthesized by rhizobacterium-promoting plant development (PGPR) can monitor plant pathogens, stimulate plant growth, and generate systemic resistance to disease (Groenhagen et al., 2013). Growth of the plant stimulating rhizobacteria has been stated by administering plant hormones, such as auxins, gibberellins, cytokinins, and ethylene, to induce plant growth and development. Therefore more than 350 bacterial species were recorded developing around 846 < multiple volatile compounds (Effmert et al., 2012).

Several PGPRs with prospective growth-promoting activity tend to involve *Pseudomonas*, *Bacillus*, *Stenotrophomonas*, *Serratia*, and *Arthrobacter* species have been identified to aggravate growth promotion and two VOCs such as 2,3-butanediol and acetoin also have been documented (Létoffé et al., 2014). Also noted were findings of the interface between PGPR-produced VOCs and plants that demonstrate the implications of such VOCs on the factors associated with plant growth.

Throughout this chapter, we summarize the current developments on how biofilms aligned with PGPR and plant root impair the association between plant and bacteria, as well as how many pathways help improve the ecological importance and conserve plant health (Gupta et al., 2014). This study looks at recent case studies assessing the impact of BVC on plant development and immune regulation.

16.2 Bacterial volatile compounds (BVCs) and its biosynthesis

Microbes create and discharge profoundly assorted inorganic and naturally unstable compounds. Such compounds are further beneficial for plant growth in the form of biologically active BVCs (Dickschat et al., 2004). There are several organic and inorganic compounds that are produced by the bacteria.

16.2.1 Inorganic compounds

Several Bacteria emits inorganic volatile compounds which include ammonia or hydrogen cyanide (HCN), nitric oxide (NO), hydrogen sulfide (H_2S). By degrading the cysteine, most of the H_2S producing bacteria generate the gas. Nitric oxide (NO) is produced by NO synthases from L-arginine, which has a similar structure that of chemicals produced by mammalians (Mattila and Thomas, 2014). Similarly, HCN has been identified as transmitted from a couple of bacterial species categories, and its creation is until further notice limited to certain types of *Pseudomonas*, *Chromobacterium*, and *Rhizobium*. HCN biosynthesis is set off by HCN synthase, transmitted by glycine-forming hcnABC genes which create HCN and CO_2. HCN advancement mainly happens toward the end of the exponential phase under the low focus of oxygen (contribution of the anaerobic controller ANR); additionally, cyanogenesis agreements via quorum sensing (QS) appears for being established strain because, in *Pseudomonas aeruginosa* PAO1 or *Chromobacterium violaceum* CV0 but not in *Pseudomonas fluorescens* 2P24, QS moderators are considered reasonable (Blom et al., 2011).

16.2.2 Organic substances

BVCs that are mixes of the natural framework incorporate a few synthetic classes, for example, unsaturated fat subordinates such as hydrocarbons, alcohols, and ketones also acids, sulfur-nitrogen containing mixes, and terpenes (Aziz and Kim, 2017).

16.2.2.1 Hydrocarbons

Hydrocarbon is an organic compound made up both of hydrogen and carbon. Linear-chain hydrocarbons are derived either by the elongation-decarboxylation mechanism or head-to-head condensation from the biosynthetic fatty acid pathways. Even though short-chain alkanes are contained periodically in lifeforms, longer hydrocarbons, for example, are abundantly present in cyanobacteria, which are sometimes known to exhibit to consolidate extended hydrocarbons (Ladyniga et al., 2006; Tellez et al., 2001).

16.2.2.2 Ketones/alcohols

The ketones are formulated by fatty acid disintegration. Acetoin, with its oxidized form 2,3-butanedione, are extracted via anaerobic pyruvate fermentation (Ryu et al., 2003). *Proteobacteria and firmicutes* retain short-chain alcohols, such as 2,3-butanediol in low-oxygen conditions (i.e., sputum in patients with cyst fibrosis or rhizosphere-growing bacteria) (Farag et al., 2013; Whiteson et al., 2014).

16.2.2.3 Acids

In bacteria organic acids are less concentrated than in ketones and alcohols. For example, acidic, propionic, or butyric acids would be depicted as released bacteria in the chain unsaturated fats (Dickschat et al., 2004). They are infested aliphatic organic acids expressing byproducts of anaerobic digestion, accumulated under carbohydrate bacterial fermentation, and therefore are incredibly active in human intestines. Glyoxylic acid is yet another vital metabolite that might be distinguished from certain metabolic reactions which include digestion of serine, glycine, arginine proline like amino acid metabolism and ethyl glycol (Muckschel et al., 2012).

16.2.2.4 Sulfur

Besides its function in the scent of fermented products, such as cheese and wine, sulfur compounds have been identified. Bacteria are most often modulated by the biogenesis of volatiles derived from methionine, for example, dimethyl sulfide and 1(methyl thio)-3-pentanone. The function for composing such volatile compounds of sulfur introduces 3-dimethylsulfoniopropionate cleavage, inferred from L-methionine by greater plants and marine algae (Stefels, 2000).

16.2.2.4.1 Nitrogen-containing compounds

Trimethylamine (TMA) is known as a tertiary volatile amine derived by the biogenic lowering of trimethylamine oxide (TMAO), thereby leading to the spoiling fish stench. In all anaerobic conditions, TMAO can also be used as a substitution electron acceptor but the bacterial shift from TMAO to TMA persisted in aerobic and anaerobic settings. Both substances (TMA and TMAO) are particularly common in fish as well as in the intestines of both humans and animals (Audrain et al., 2015)

16.2.2.5 Terpenes

Terpenes are extracted from the structural systems of dimethylallyl pyrophosphate and isopentenyl pyrophosphate terpene, which can originate from the pathways of mevalonate or deoxy xylulose phosphate. Bacterial volatile blends registered only in specific, most terpenoid compounds produced by the bacteria is the earthy odorous geosmine and the antibiotic albaflevone; In so many bacterial species, although geosmin, depleted sesquiterpene, albaflavenone was first isolated from *Streptomyces albidoflavus* and is discovered specifically in *Streptomyces* (Schulz and Dickschat, 2007). (C10), sesquiterpene (C15), and its derivatives or products for degradation (Schulz and Dickschat, 2007). Following are some of the BVCs with their practical assignments mentioned in Table 16.1.

16.3 Effect of BVCs on plant growth and its functions

Volatile compounds which enhance the growth of plants can be an eco-friendly alternative to chemical fertilizers. These compounds has the ability to produce from the free-living bacteria or the endophytic bacteria which benefits the plants in several forms (Janssens et al., 2019). It enhances nutrient availability, stimulates defense response, stress, and also metabolic activities. These benefits are due to the mechanism promoting the plant growth by BVCs obtained using proteome or transcriptome tools which are known as molecular probes (Audrain et al., 2015). Studies conducted by Blom et al. (2011) explain how BVCs affect plant growth significantly. BVCs improve over-the-ground plant cell growth, leaf size, and several leaves enhance natural product productivity and seed development and optimize side long root and root hair formation, as well as supplement take-up, photosynthetic activity, and sugar accumulation. The BVCs control hormone signaling to boost plant growth and health (Tahir et al., 2017). These compounds thus can be used in fields for better productivity (Ryu, 2015). Some of the BVCs that affects the plant physiology and growth are given below in Table 16.2.

16.4 Benefits of BVCs for plant growth and system

BVCs are the compounds that are even beneficial to plants in many aspects including plant growth, nutrient uptake, biotic and abiotic stress, photosynthetic modulation, etc.

TABLE 16.1 List of BVCs with its functional effects.

Sr No.	Compound	Chemical class	Organism	Functional effect	References
1.	Acetaldehyde	Alcohol	*Escherichia coli*	Antibacterial activity	Létoffé et al. (2014)
2.	Glyoxylic acid	Aldehyde	*Bacillus subtilis*	Bacterial gene expression alteration	Kim et al. (2013)
3.	Formamide	Amide	*Pseudomonas fluorescens*	Enhancement of plant growth	Zhou et al. (2016)
4.	*N, N*-Dimethylformamide	Amide	*P. fluorescens*	Enhancement of plant growth	
5.	Trimethylamine	Amide	*Streptomyces venezuelae*, *E. coli*	Raises pH, antibacterial activity, communication molecule	Jones and Elliot (2018), Jones et al. (2019)
6.	2,5-Bis(1-methylethyl) pyrazine	Aromatic compound	*Paenibacillus* spp.	Antibacterial activity	Janssens et al. (2019)
7.	2,3-Butanedione	Di-Ketone	*B. subtilis*	Bacterial gene expression alteration	Kim et al. (2013)
8.	Ethyl-isovalerate	Ester	*Pseudomonas* sp.	Enhancement of plant growth	Camarena-Pozos et al. (2019)
9.	2-Heptanone	Ketone	*Pseudomonas* sp.	Antibacterial activity	Plyuta et al. (2016)
10.	2-Undecanone	Ketone	*Pseudomonas* sp.	Antibacterial activity	
11.	*O*-aminoacetophenone	Ketone	*E. coli*	Alters—antibiotic resistance profile	Groenhagen et al. (2013)
12.	Schleiferon A, B	Ketone	*Staphylococcus schleiferi*	Antibacterial activity	Lemfack et al. (2016)
13.	3-Pentadecenenitrile	Nitrile	*Pseudomonas* sp., *Micromonospora* sp.	Antibacterial activity	Montes Vidal et al. (2017)
14.	Nitric oxide	Nitrogenous compound	*B. subtilis*	Bacterial gene expression alteration	Gusarov et al. (2009)
15.	Dimethyl sulfide	Sulfur	*Pseudomonas aeruginosa*	Bacterial growth promoter	Cordovez et al. (2018)

Source: Partly adopted from Netzker, T., Shepherdson, E.M.F., Zambri, M.P., Elliot, M.A., 2020. Bacterial volatile compounds: functions in communication, cooperation, and competition. Annu. Rev. Microbiol. 74 (1). doi:10.1146/annurev-micro-011320-015542(Netzker et al., 2020)

TABLE 16.2 List of bacterial VOCs and its effects on plant growth.

Sr No.	Bacterial volatile compound (BVCs)	Plant	Target part of plant	Experimental condition	References
1.	Dimethyl hexadecylamin	Sorghum	Cholorophyll content	Petri dish	Castulo-Rubio et al. (2015)
2.	Dimethylhexadecylamine	Alfalfa	Shoot weight	Petri dish	Velázquez-Becerra et al. (2011)
3.	Albuterol and 1,3-propanediol	Tomato	Shoot weight	Pot study	Tahir et al. (2017)
4.	Dimethyl disulfide	Nicotiana	Leaf surface	Petri dish	Meldau et al. (2013)
5.	3-Pentanol	Cucumber	Fruit production	Field Study	Song and Ryu (2013)
6.	Indole	Arabidopsis	Root proliferation	Vertical plate	Bailly et al. (2014)
7.	Dimethylhexadecylamine	Sorghum	Photosynthesis	Glass flask assay	Castulo-Rubio et al. (2015)
8.	Dimethylhexadecylamine	Sorghum	Iron acquisition		
9.	Indole	Arabidopsis	Auxin	I-plate	Bhattacharyya et al. (2015)
10.	*Bacillus subtilis* SYST2	Tomato	Auxin	Pot study	Tahir et al. (2017)
11.	Indole	Arabidopsis	Cytokinin	I-plate	Bhattacharyya et al. (2015)
12.	*B. subtilis* SYST2	Tomato	Ethylene	Pot assay	Tahir et al. (2017)

Source: Partly adopted from Sharifi, R., Lee, S.M., Ryu, C.M., 2018. Microbe-induced plant volatiles. New Phytol. 218, doi:10.1111/nph.14955(Sharifi et al., 2018)

16.4.1 Bacterial volatiles promoting the growth of plants

Ryu et al. (2003) first discovered the influence of BVCs on growing plants and noticed that treatment with *Bacillus subtilis* GB03 volatiles significantly enhanced the growth of plants under arabidopsis. Evaluation of unstable profiles of compounds reported that 2,3-butanediol, and its counterpart acetoin are compounds that stimulate the production of plants (Farag et al., 2006). Roots secure plants in soil, and with water and minerals, maintain plant production. Roots provide the microorganisms with a highly nutritious environment. Plant growth-enhancing extreme volatility rhizobacteria (PGPR), dimethyl hexadecyl amine (DMHDA), and hair density (Bailly et al., 2014; Castulo-Rubio et al., 2015). Therefore, these modifications raise the quantity of root and the surface area. Some BVCs, however, may constrain principal root development while empowering proper development of the lateral root and the root hair.

16.4.2 Increasing the acquisition of minerals

PGPR embraces plant secretion of macroelements and microelements (Aziz and Kim, 2017; Meldau et al., 2013; Zhang et al., 2009). BVCs from many other forms of PGPR facilitates the absorption of iron, copper, selenium, and sulfur. The volatile UMCV2 *Nitrobacter agilis* strengthen the acquisition of iron for monocotype and dicotype plants (Castulo-Rubio et al., 2015). Medication *B. subtilis* GB03 risen the absorption of iron to twofold in arabidopsis, even in alkaline conditions (Zhang et al., 2008). *Bacillus amyloliquefaciens* BF06 volatiles exceeds the ingestion of selenium by enhancing the induction of sulfate in this plant as a transporter genes: selenium composition in volatile treated plants was 23% greater as in plants left untreated (Wang et al., 2017). These findings together imply that BVCs can enhance the uptake and transportation of nutrients in plants. These processes should be considered in developing approaches for increasing the efficiency of fertilizer uptake using a holistic approach to nutrient management.

16.4.3 BVCs modulate photosynthesis in plants

By growing the chlorophyll content and photosynthetic ability BVCs can help boost major steps in plant physiology, such as photosynthesis and carbohydrate intensification. In *Arabidopsis*, chlorophyll concentrations increase by BVCs (Zhang et al., 2009). There have been two mechanisms to exhibit the overall impact of BVCs on chlorophyll quality and photosynthesis. The first is iron, considered necessary for the biosynthesis of plant chlorophyll, electron transport system activity, and photosystem exercise (Briat, 2007). Rhizosphere acidification enhances iron solubility and embraces iron absorption.

16.4.4 BVCs enhances crop yield and quality

BVC treatment can greatly improve seed, plant, herb, biomass, mineral oil, secondary metabolite, and sugar stability and benefits. In determining seed content such as potato and beet sugar, the aggregation of sugars such as glucose, sucrose, and starch is key. The clustering of these sugars can rise from many other bacteria and fungi with volatiles (Sánchez-López et al., 2016). These BVCs do not substantially enhance plant biomass but stimulate the blooming and development of fruits. For reference, 3-pentanol and 2-pentanone had no direct effect on cucumber biomass but between six and four times normal field fruit production (Song and Ryu, 2013). Unstable *B. subtilis* GB03 and benzaldehyde, collectively *Codonopsis pilosula* and *Atractylodes lancea*, enhance the biomass and important oil content of herbal medicines (Wu et al., 2016; Zhou et al., 2016).

16.4.5 Biotic and abiotic stress relievers

Softening the biotic and abiotic stress, the BVCs improve plant growth. Also, some BVCs can cause extreme defense responses for salinization of soil and drought stress, which pose dangerous challenges to crop implementation. Most of the BVCs are highly toxic to plant pathogens, such as dimethyl disulfide (DMDS) and 2-methyl pentanoate (Cordovez et al., 2018; Groenhagen et al., 2013; Raza et al., 2016). Rhizobacteria treatment can help to mitigate these problems by changing the design of the root system that makes water take-up more effective. Rhizobacteria establish systemic resistance to abiotic stress by upregulating proline, antioxidants, decreasing accumulation of Na^+ in plants and the development of hormones (Liu and Zhang, 2015; Sharifi et al., 2018).

16.4.6 Effect of BVCs on bacterial growth

Several researchers have reported groups investigated the effect on bacterial differentiation and the evolution of BVCs formed by soil-associated bacteria. The development of headspace geosmin, an usually observed terpenoid compound in this bacterial genus, is affiliated with the sporulation in *Streptomyces* spp., *Albidoflavus* AMI 246 (Scholler et al., 2002) possesses antimicrobial properties against *B. subtilis* all with albaflavenone, a transitory antibiotic sesquiterpene ketone (Gurtler et al., 1994). Likewise, the two rhizospheric bacteria; *P. fluorescens* and *Serratia plymuthica*, have bacteriostatic effects on *Agrobacterium tumefaciens* and *Agrobacterium vitis*, two bacterial pathogens in plants (Dandurishvilin et al., 2011).

BVC has also been documented as regulating the bacterial response to various stresses, along with antibiotic exposure (Heal and Parsons, 2002). For both Gram-negative and Gram-positive bacteria, unstable ammonia inferred from a high-density bacterial population rises tetracycline and ampicillin resistance at a radius and decreasing aminoglycoside susceptibility (Bernier et al., 2011). TMA, again an adaptable compound developed by many *Enterobacteriaceae*, also transmits considerate tetracycline and aminoglycosides in all bacterial strains assessed, but greatly increases immunity of chloramphenicol and seems to lower stress resistance (Létoffé et al., 2014). The volatile compounds released from B, as specified for ammonia. Also, *B. subtilis* creates ampicillin and E-resistance to *E. coli* tetracycline (Kim et al., 2013). Although H_2S is set to release mostly by bacteria, little is recognized about its significant role in nonsulfuric microorganisms. Indole, a heteroaromatic agent, can stimulate divergent gene expressions including the multidrug encoding export industries. Indole also progresses to several phenotypes, which include drug adherence in nonindole (*P. aeruginosa* and *Salmonella enterica*) bacteria, and indole-producing bacteria.

16.5 VOCs by plant growth-promoting rhizobacteria: its secretion, biocontrol and role in plant growth

BVCs act as biostimulants well as bioprotectants (Chung et al., 2015). Several mechanisms are there which are used by PGPR to improve development. The VOCs that are the production of gaseous organic molecules is yet another pathway used by microbes in soil (Schulz and Dickschat, 2007). These VOCs emitted by the microbes in the soil serve as ISR elicitors, that is, triggered systemic resistance response that stimulates further plant defense mechanisms without any physical contact or damage to the plants (Kanchiswamy et al., 2015). Except for the molecular interfacial reaction between plants and soil microbes involving several PGPR, signaling also plays an important role in the maintenance of plant health in the form of VOCs generated (Lemfack et al., 2013). Similarly, VOCs are ecofriendly and also may be used for sustainably handling crops, also the MVOCs are beneficial for agricultural production and environmental management. Bacterial VOCs include alcohol, HCN, ammonia, and carboxylic acids, and sulfur which can biocontrol due to the antifungal activities (Choudhary et al., 2008). Volatile compound indole released due to indole-producing *Escherichia coli* and *Proteus vulgaris* has been documented as to improvement of plant biomass and lateral root development in *Arabidopsis thaliana* (Fincheira and Quiroz, 2018). These VOCs also act as a signaling molecule for facilitating interactions for diffusion. Similarly, volatiles like indole and DMHDA produced for the PGPR help improve the root length, root hair density in the *A. thaliana* plant (Kai et al., 2016). Several VOCs emitted from some PGPR acts to activate the FIT1, FRO2 gene expressions which improves the uptake of iron and also several plant growth parameters including the sulfur deficiency and also assimilation of dimethyl disulfide (Liu and Zhang, 2015). Some of the BVCs released by PGPR are given later in the chapter.

16.5.1 Hydrogen cyanide (HCN)

HCN has been reported to have originated from these few bacterial species and is now confined to certain species of *Chromobacterium*, *Pseudomonas*, and *Rhizobium*. It induces HCN biosynthesis by HCN synthase, regulated by the glycine-forming HCN and CO_2 genes. HCN development also occurs at the end of the exponential step under the low concentration of oxygen [including anaerobic regulator (ANR)], Due to the necessity for QS regulators in *P. aeruginosa* PAO1 or *C. violaceum* CV0 *butnotin*, *P. fluorescens* 2P24, the regulation of cyanogenesis is thus concerning (Blom et al., 2011).

16.5.2 Indole

Indole is an omnipresent chemical created across both Gram-positive bacteria and Gram-negative ones. Indole play vital role in spore production, signaling, plasmid stabilization, drug tolerance, biofilm formation, and virulence (Lee and Lee, 2010). Indole facilitates Arabidopsis progress and plays a significant role in lateral root development that can be modulated by hindering the machinery for auxin signaling (Bailly et al., 2014). The latest research has been carried out with the auxin reporter DR5::GUS lines and classic physiological and transport assays to illustrate that plants can transform bacterial indole to auxin (indole acetic acid), and the polar auxin transport system is crucial for indole-induced lateral root development. These findings indicate that bacterial indole satisfies as a carrier and manages plant growth and yield (Lee and Lee, 2010). Indole is a BVC that modulates root growth through interfering with auxin signals governed by interkingdom communication between plant-bacteria. For example, *P. vulgaris* yields indole as its dominant volatile, which helps to improve the vigor index of Chinese cabbage at an advantageous formation of just 0.63 ng per 44.18-cm^3 I-plate by up to 40% (Yu et al., 2000).

16.5.3 Dimethyl disulfide

Bacillus cereus C1L induces ISR and secures plants from the fungal necrotrophic pathogen B. Films. The analysis by solid-phase microextraction (SPME), GC/MS identified *cereus* C1L ISR elicitor as dimethyl disulfide, an erratic component that serves to protect tobacco and maize against *B. cinerea* and heterotrophic *Cochliobolus* in tobacco and maize. The dimethyl disulfide feature has shown it can be a possible ISR elicitor in *Bacillus cereus* C1L (Huang et al., 2012). DMDS is among the most bioactive substances that the bacteria produce. Cultured on LB agar, *Bacillus ambifaria* improved considerably lateral root structure and biomass in *A. thaliana* plants 21 days after their release.

16.5.4 Dimethylhexadecylamine

BVCs emitted by *Arthrobacter agilis* UMCV2, a maize-insulated PGPR strain, fostered the growth of *Medicago sativa* seedlings as assessed by an increment at root and stem length, and biomass of plants (Velázquez-Becerra et al., 2010). The unstable substances *N,N*-dimethyl hexadecyl amine (dimethyl hexadecyl amine), and lipoamino acids functionally attributed to QS signals through bacteria have emerged as main substances modulating enriched plant development.

16.5.5 Tridecane

Paenibacillus polymyxa E681 one of the PGPR has been segregated from the roots of winter barley; it stimulates the development of plant biomass and control functions for phytopathogens (Lee et al., 2012). Tridecane is a hydrocarbon released by the E681 alkane C13. In *Arabidopsis*, Tridecane inhibits systemic tolerance to *Pseudomonas syringae pv.* ES4326 *maculicola* by priming expression of genes within the signaling pathways for salicylic acid (SA), JA, and ethylene (ET).

16.6 Role of bacterial volatiles in biofilm production

Several bacteria begin to form biofilms, and this multicellular growth model is highly probable to predominate in nature as a technique for protecting against hostile environmental conditions, for example, *P. aeruginosa* (Harper et al., 2014). Biofilms usually have specific developmental features that are different from freely planktonic or nonbiofilm forming cells. The presence of biofilm architecture in plant-microbe interactions cannot be negligible, and the detection of plant growth enhancements by evolved biofilm inoculum will have great scope to promote plant growth (Son et al., 2010). Root exudates are commonly thought to play important roles in the interaction between plants and microbes on the rhizosphere (Bais et al., 2006). Recent studies have also shown the impacts of volatile compounds on the varying stages of bacterial biofilm advancement, from bacterial motility to biofilm dispersal. Formation of bacterial biofilm involves several steps, where it first attaches to a substrate and then forms microcolonies before forming matured three-dimensional structures and detaches from the substrate.

Formation of biofilm has five main steps as given below (Jacques et al., 2010):

1. Initial attachment: Planktonic cells assembly.
2. Irreversible attachment: Irreversible attachment of planktonic cells.
3. Microcolony formation: Cell aggregation and accumulation in layers, the formation of a matrix, and surface adhesion.
4. Maturation: Biofilm growth and cells structural, metabolic heterogeneity.
5. Dispersion: Cells dispensation into planktonic cells released into the environment. The BVCs involved in the process are given below in Fig. 16.1.

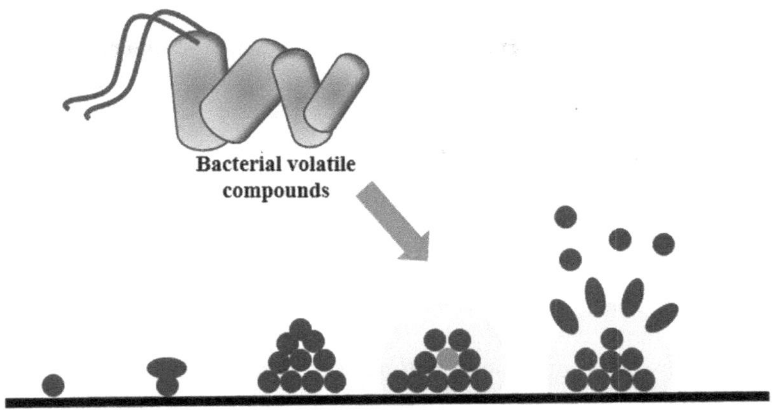

Stimulation	**Inhibition**
Ammonia	Indole
Indole	Nitric oxide (NO)
Nitric Acid	Glyoxylic acid
Trimethylamine	2,3-Butanedione
1-Butanol	Acetoin
2-Butanone	Ethanol
Acetoin	1-Butanol
Hexadecane	2-Butanone

FIGURE 16.1 Bacterial volatile compounds (BVCs) involved during biofilm formation. *Partly adapted from Audrain, B., Farag, M.A., Ryu, C.M., Ghigo, J.M., 2015. Role of bacterial volatile compounds in bacterial biology. FEMS Microbiol. Rev. 39, 222–233. doi:10.1093/femsre/fuu013.*

Of the 12 nontoxic volatile compounds assessed, *E. coli* was minimized by aerial exposure to 1-butanol or indole. Adhesion of *P. aeruginosa*, respectively, while adhesion of *P. aeruginosa* reduced by 2-butanone and acetoin (Létoffé et al., 2014). Many other BVCs can impact the ability of the bacteria to form biofilms. For example, ammonia inhibited the accumulation of biofilms in *Bacillus licheniformis*, *B. subtilis*, and *Staphylococcus aureus* (Nijland and Burgess, 2010; Létoffé et al., 2014). While soluble indole in *Vibrio cholerae*, *P. fluorescens*, and *P. aeruginosa* have formerly been shown to enhance the production of biofilms, other research reported conflicting effects on the shape of *E. coli* biofilms formation (Mueller et al., 2009; Lee and Lee, 2010). Thus it prevents the development of biofilm in both when used as an airborne volatile signal. *E. coli* and *P. aeruginosa*, thus continues to promote *S. aureus* biofilm development (Létoffé et al., 2014). Plenty other BVCs like 1-butanol, 2-butanone, acetoin, ammonia, ethanol, hexadecane, glyoxylic acid, and TMA have a significantly beneficial or negative effect on bacterial growth in one or more of the bacterial species described including *P. aeruginosa*, *E. coli*, *S. aureus*, and *B. subtilis*. Finally, given that sublethal antibiotic concentrations increase or decrease biofilm development in most other bacterial species such as *P. aeruginosa* and *E. coli*, we presume that BVC responsiveness, which stimulates antibiotic resistance thresholds, during antibiotic treatment may adversely affect the biofilm formation (Bernier and Surette, 2013).

16.7 Root biofilm formation

The root-system gives harbor, play a significant role in water and supplement take up from the soil, and is a site of synthesis of numerous metabolites, for example, cytokinins, and auxins, which assume a significant job in the development and formative procedures (Ortíz-Castro et al., 2009). VOCs are effective chemical communication mediators, functioning exclusively as a favorable, repellent, or cautioning signals in all realms of life. Microbial species can emanate volatiles for various purposes, for example, communication and protection (Kai et al., 2009). Because of the unpredictability properties of this sort of compounds, it is enticing to hypothesize that the plant, explicitly the root framework may detect rapidly and viably the VOCs discharged by its related microorganisms, for example, PGPR. Studies conducted by Gutiérrez-Luna et al. (2010) can see just how to use unstable natural compound outflow to PGPR balance root-framework engineering in *A. thaliana*. This examination indicated how the plants can see VOCs adjust the morphological and genetic procedures that are significant in the numerous capacities played by the root framework, for example, water and supplement osmosis. Another study performed by Das et al. (2017) explained that for the lead extraction process, *Pennisetum purpureum* is a good plant source for the bio removal of lead. *P. purpureum* is also been considered as a root accumulator for the lead as it translocates Pb for the aerial part of roots.

16.8 Microbial volatile compounds in quorum sensing

QS is a type of population density-dependent cell to cell signaling that triggers changes in behavior when the population reaches a critical density (Chernin et al., 2013). These QS molecules are found in both Gram-positive and negative bacteria, including pathogenic as well as beneficial plant-related strains (Faure et al., 2009). Several volatile compounds are identified bearing the ability to communicate between cells. 2-amino-acetophenone (2-AA) is a known aromatic compound showing similar odor that of *P. aeruginosa* and is used to detect the infection of *P. aeruginosa* which is the QS small volatile compound that promotes the antibiotic tolerance level in several bacteria (Que et al., 2013). The first volatile regulated QS molecule identified is 2-AA alkylquinoline formulation of 2-AA, a non-4-hydroxy-2 alkylquinoline volatile molecule driven by multiple virulence factor regulator (MvfR) by all pqs ABCD operon control and including pqsA and pqsD genes, too (Audrain et al., 2015; Kesarwani et al., 2011).

16.9 Analysis of bacterial volatile compounds (BVCs)

BVC analysis is an arduous task because of its large variety of volatile ample supply and the nature of mixtures and/or matrixes in which they are typically found and are extracted from. The analysis is done in three main steps which include BVCs extraction, analysis, and identification.

16.9.1 Extraction of BVCs

Different techniques have been proposed for obtaining volatile expelled of all bacteria, and the amount of observable Instability gradually expanded with the complexities and affect ability of the detachment methods used (Wenke et al., 2012). Although the closed-loop stripping technique (CLSA) can be commonly used to preconcentrate volatiles from aqueous samples (Meruva et al., 2004), SPME is a rapid sample preparation approach that involves the collection, isolation, concentration, and infusion into a single solvent-free analytical instrument (Goupry et al., 2000; Marilley and Casey, 2004). Finally, bacterial headspace direct and real-time gas sequencing may be obtained utilizing secondary electrospray ionization-mass spectrometry (SESI-MS), it also has many advantages, such as an adaptable sensing cap and high-throughput sample processing possibilities (Zhu and Hill, 2013). SESI-MS was used for human breath vapor monitoring and clinically relevant pathogens (Zhu et al., 2010).

16.9.2 BVCs analysis

Standard technique for examining BVC profiles tends to focus on gas chromatography coupled with mass spectrometry (GC-MS), differentiated by separating capability and particularly vulnerable output detection; in fact, modified GC-MS software boosts the detection range because of the corresponding peak demodulation and background subtraction (Farag, 2014). For real-time analysis complex, electronic noses (eNoses), ion flow-tube mass spectrometry (SIFT-MS), and ion-mobility spectrometer are often used (Dolch et al., 2012; Lirk et al., 2003). Nevertheless, precarious extracting for clinical or tissue samples was evidenced only occasionally and was carried out on a selected fragile or volatile population (Whiteson et al., 2014).

16.9.3 BVCs identification

Compounds found in a specific material may be analyzed by contrasting mass spectra to spectra from different methods, such as libraries Wiley or NIST. Nevertheless, such libraries incline to confuse the naive user, because the next link in the database can be treated uncritically as a true recognition. A transient microbial database, known as MVOC, is now available online in the form of a

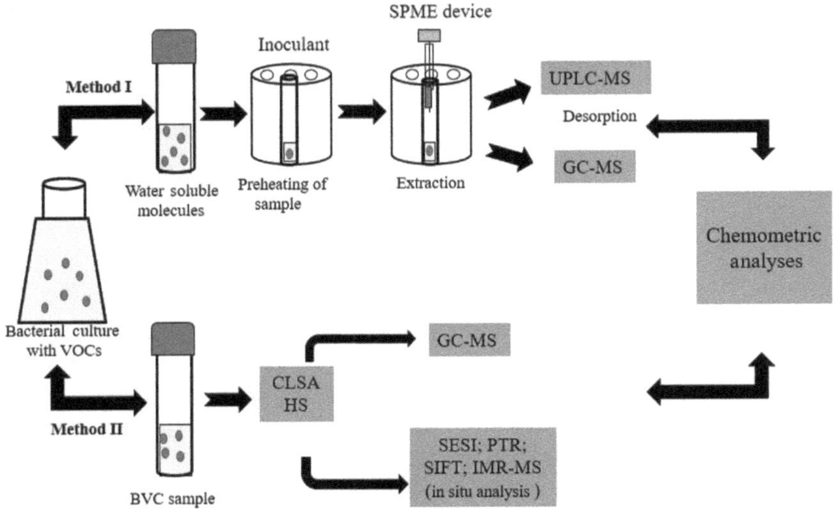

FIGURE 16.2 The systematic workflow of analysis of bacterial VOCs. Method I: Water-soluble molecules in the sample, volatile analysis was done using GC-MS, SPME, and UPLC-MS. Method II: Bacteria inoculant in broth or agar medium further analyses using SIFT-MS, SESI-MS, and proton-transfer-reaction MS (PTR-MS) multicapillary column, ion-molecule reaction (IMR-MS) in situ analysis. *Partly adapted from Tang, X., Misztal, P.K., Nazaroff, W.W., et al., 2016. Volatile organic compound emissions from humans indoors. Environ. Sci. Technol. 50 (23), 12686–12694.*

bioinformatics probe. Around 1000 reactive microbial compounds and comprehensive information around their species emission, biosynthetic processes, and biological effects are described on this website (Lemfack et al., 2014). Multivariate statistical techniques, generally split into two unmonitored and regulated categories, are mandated to estimate distinct change observed in BVC released from different species, particularly when analyzing multiple strains. Using supervised methods such as differential lowest possible squares (PLS) or discriminant analysis of PLS requires comparing the samples with separate ones. A differentiation–multiblock hierarchical main component evaluation (MB-PCA) and MB-PLS strengthen, that is, the comparative analysis of data sets derived from separate theoretical frameworks: NMR versus MS. Fig. 16.2 offers a comprehensive flowchart for the study of bacterial VOCs.

16.10 Benefits and drawbacks of volatile molecules

- Two of the biggest advantages bacteria receive from being used volatile compounds is their inability to evaporate and their capacity to spread through water, air, and dirt. This makes the dispersion of these molecules quickly, entering locations and covering distances that would be unavailable to larger secondary metabolites. This increases the variety of effect, allowing the creator to interact and engage distantly with other species.

- At the similar time, this ensures that the influence of volatile compounds would be more reactive, with higher rates of production required to produce these molecules' high local concentrations relative to slightly diffusible secondary metabolites.
- The compact size and fairly basic chemical structure of volatile compounds imply that the genetic and energy expense of developing them is significantly smaller than that of secondary metabolites, which are more complex. For example, antibiotics need vast clusters of biosynthetic genes to be obtained; several enzymes are required to assemble precursors and alter biosynthetic intermediates before yielding the resulting bioactive molecule.
- By comparison, the synthesis of most volatiles needs much fewer enzymes and modifications, and in certain cases during the synthesis of more complex molecules, they are released as byproducts.
- The nature of their dispersal means that volatile compounds are common goods—all these molecules can be reached or detected by neighboring species in principle. If that's gain or an annoyance to the producer organism is likely to rely on the molecule impacts on both the producer and his neighbor and the energy is taken to generate it.
- Although these are basic materials, it can be difficult to study volatile molecules. In comparison to secondary metabolites, their genes for biosynthesis are often not easy to observe, and can also be extracted via several biochemical pathways. Resultantly, pure substances are frequently used to determine the value or role of individual compounds in place of genetic tests.
- The intricate bouquet of volatile compounds formed by one organism often poses difficulties when it comes to dissecting how functional results arise through one or more volatile compounds.
- These VOCs emitted from several bacteria are useful in cosmetics, food, textile as well as agricultural products like fertilizers, pesticides, etc.

16.11 Conclusion

Bacterial volatiles has led to the function and production of various major physiological characters and also have been helpful in plant-microbe interaction. BVCs are nothing more than the chemical dialect often used bacteria to share information with the plants. These compositions manipulate the potential of hormonal pathways to calibrate and activate, plant physiology leading to greater biomass, higher yield. Several MVOCs are used for treating the plants on a large scale. These plants exhibit good features like higher root volume, increased leaf number and size, flowering ability, and seed fruit production. Due to these properties microbes used as BVCs can be used as a good source of fertilizers for farming. Biological volatiles has outstanding transitional potential. Research in this field has provided the information on over 100 MVOCs, however, are identified to date, but very few of them are characterized and are used as an alternative to fertilizers. VOCs, which are identified, have shown their capabilities as antibacterial and antifungal hence these molecules can be used for controlling the growth, development, and also toxicity of microbial pathogens in agricultural, industrial, and clinical sectors. In any case, we should think about the symptoms of these volatiles, which are profoundly dynamic and possibly hazardous. Many volatiles that is suitable for plant use including effects on living beings that are not a target, such as nematodes, bugs, and humans. Further examinations are expected to distinguish more successful volatiles and to decide their viable focuses, just as to explore the impacts of manufactured unpredictable blends on plant development under field conditions. This chapter focused on the BVCs and its recent advances for plant growth. Several technical challenges in the identification of novel BVCs can be analyzed using synthetic biology, genomics, and bioengineering-like inputs.

References

Audrain, B., Farag, M.A., Ryu, C.M., Ghigo, J.M., 2015. Role of bacterial volatile compounds in bacterial biology. FEMS Microbiol. Rev. 39, 222–233. Available from: https://doi.org/10.1093/femsre/fuu013.

Aziz, A., Kim, K.S., 2017. Adsorptive volatile organic removal from air onto NaZSM-5 and HZSM-5: kinetic and equilibrium studies. Water Air Soil Pollut. 228, 319.

Bailly, A., Groenhagen, U., Schulz, S., Geisler, M., Eberl, L., Weisskopf, L., 2014. The inter-kingdom volatile signal indole promotes root development by interfering with auxin signaling. Plant J. 80, 758–771. Available from: https://doi.org/10.1111/tpj.12666.

Bernier, S.P., Surette, M.G., 2013. Concentration-dependent activity of antibiotics in natural environments. Front. Microbiol. 4, 20.

Bais, H.P., Weir, T.L., Perry, L.G., et al., 2006. The role of root exudates in rhizosphere interactions with plants and other organisms. Annu. Rev. Plant Biol. 57.

Bernier, S.P., Letoffe, S., Delepierre, M., et al., 2011. Bio genic ammonia modifies antibiotic resistance at a distance in physically separated bacteria. Mol. Microbiol. 81, 705–716.

Bhattacharyya, D., Garladinne, M., Lee, Y.H., 2015. Volatile indole produced by rhizobacterium *Proteus vulgaris* JBLS202 stimulates the growth of *Arabidopsis thaliana* through auxin, cytokinin, and brass in osteroid pathways. J. Plant Growth Regul. 34, 158–168.

Blom, D., Fabbri, C., Connor, E.C., et al., 2011. Production of plant growth modulating volatiles is widespread among rhizosphere bacteria and strongly depends on culture conditions. Environ. Microbiol. 13, 3047–3058.

Briat, J.F., 2007. Iron dynamics in plants. Adv. Bot. Res. 46, 137–180.

Camarena-Pozos, D.A., Flores-Núñez, V.M., López, M.G., López-Bucio, J., Partida-Martínez, L. P., 2019. Smells from the desert: microbial volatiles that affects plant growth and development of native and non-native plant species. Plant Cell Env. 42, 1368–1380.

Castulo-Rubio, D.Y., Alejandre-Ramírez, N.A., del Carmen Orozco-Mosqueda, M., Santoyo, G., Macías-Rodríguez, L.I., Valencia-Cantero, E., 2015. Volatile organic compounds produced by the rhizobacterium *Arthrobacter agilis* UMCV2 modulate Sorghum bicolor (strategy II plant) morphogenesis and SbFRO1 transcription in vitro. J. Plant Growth Regul. 34, 611–623.

Chernin, L., Toklikishvili, N., Ovadis, M., Khmel, I., 2013. Quorum-sensing quenching by volatile organic compounds emitted by Rhizosphere bacteria. Mol. Microb. Ecol. Rhizosphere 791–800. Available from: https://doi.org/10.1002/9781118297674.ch75.

Choudhary, D.K., Johri, B.N., Prakash, A., 2008. Volatiles as priming agents that initiate plant growth and defense responses. Curr. Sci. 595–604.

Chung, J., Song, G.C., Ryu, C.-M., 2015. Sweet scents from good bacteria: case studies on bacterial volatile compounds for plant growth and immunity. Plant Mol. Biol. 90 (6), 677–687. Available from: https://doi.org/10.1007/s11103-015-0344-8.

Cordovez, V., Schop, S., Hordijk, K., de Boulois, H.D., Coppens, F., et al., 2018. Priming of plant growth promotion by volatiles of root-associated *Microbacterium* spp. Appl. Environ. Microbiol. 84 (22), e01865-18.

Dandurishvilin, N., Toklikishvili, N., Ovadis, M., et al., 2011. Broad-range antagonistic rhizobacteria *Pseudomonas fluorescens* and *Serratia plymuthica* suppress Agrobacterium crown gall tumors on tomato plants. J. Appl. Microbiol. 110, 341–352.

Das, A., Belgaonkar, P., Raman, A.S., Banu, S., Osborne, J.W., 2017. Bioremoval of lead using *Pennisetum purpureum* augmented with *Enterobacter cloacae-VITPASJ1:* a pot culture approach. Environ. Sci. Pollut. Res. 24 (18), 15444–15453.

Dickschat, J.S., Wenzel, S.C., Bode, H.B., et al., 2004. Biosynthesis of volatiles by the *myxobacterium Myxococcus Xanthus*. Chem. Biochem. 5, 778–787.

Dolch, M.E., Hornuss, C., Klocke, C., et al., 2012. Volatile compound profiling for the identification of Gram-negative bacteria by ion-molecule reaction–mass spectrometry. J. Appl. Microbiol. 113 (5), 1097–1105.

Effmert, U., Kalderas, J., Warnke, R., et al., 2012. Volatile mediated interactions between bacteria and fungi in the soil. J. Chem. Ecol. 38, 665–703.

Farag, M.A., 2014. Comparative mass spectrometry and nuclear magnetic resonance metabolomic approaches for nutraceuticals quality control analysis: a brief review. Recent. Pat. Biotechnol. 8 (1), 17–24.

Farag, M.A., Ryu, C.-M., Sumner, L.W., Paré, P.W., 2006. GC–MS SPME profiling of rhizobacterial volatiles reveals prospective inducers of growth promotion and induced systemic resistance in plants. Phytochemistry 67, 2262–2268.

Farag, M.A., Zhang, H., Ryu, C.M., 2013. Dynamic chemical communication between plants and bacteria through airborne signals: induced resistance by bacterial volatiles. J. Chem. Ecol. 39, 100718.

Faure, D., Vereecke, D., Leveau, J.H., 2009. Molecular communication in the rhizosphere. Plant Soil 321, 279–303.

Fincheira, P., Quiroz, A., 2018. Microbial volatiles as plant growth inducers. Microbiol. Res. 208, 63–75.

Goupry, S., Rochut, N., Robins, R.J., et al., 2000. Evaluation of solid-phase microextraction for the isotopic analysis of volatile compounds produced during fermentation by lactic acid bacteria. J. Agr. Food Chem. 48, 2222–2227.

Groenhagen, U., Baumgartner, R., Bailly, A., Gardiner, A., Eberl, L., et al., 2013. Production of bioactive volatiles by different *Burkholderia ambifaria* strains. J. Chem. Ecol. 39, 892–906.

Gupta, G., Parihar, S.S., Ahirwar, N.K., et al., 2014. Plant growth promoting rhizobacteria (PGPR): current and future prospects for development of sustainable agriculture. J. Biotech. Pharm. Res. 5, 001–006.

Gurtler, H., Pedersen, R., Anthoni, U., et al., 1994. Albaflavenone, a sesquiterpene ketone with a zizaene skeleton produced by a streptomycete with a new rope morphology. J. Antibiot. (Tokyo) 47, 434–439.

Gusarov, I., Shatalin, K., Starodubtseva, M., Nudler, E., 2009. Endogenous nitric oxide protects bacteria against a wide spectrum of antibiotics. Science 325, 1380–1384.

Gutiérrez-Luna, F.M., López-Bucio, J., Altamirano-Hernández, J., Valencia-Cantero, E., de la Cruz, H.R., Macías-Rodríguez, L., 2010. Plant growth-promoting rhizobacteria modulate root-system architecture in *Arabidopsis thaliana* through volatile organic compound emission. Symbiosis 51 (1), 7583. Available from: https://doi.org/10.1007/s13199-010-0066-2.

Harper, D., Parracho, H., Walker, J., Sharp, R., Hughes, G., Werthén, M., et al., 2014. Bacteriophages and biofilms. Antibiotics 3 (3), 270–284. Available from: https://doi.org/10.3390/antibiotics3030270.

Heal, R.D., Parsons, A.T., 2002. A novel intercellular communication system in *Escherichia coli* that confers antibiotic resistance between physically separated populations. J. Appl. Microbiol. 92, 1116–1122.

Huang, C.J., Tsay, J.F., Chang, S.Y., Yang, H.P., Wu, W.S., Chen, C.Y., 2012. Dimethyl disulfide is an induced systemic resistance elicitor produced by *Bacillus cereus* C1L. Pest. Manag. Sci. 68, 1306–1310. Available from: https://doi.org/10.1002/ps.3301.

Jacques, M., Virginia, A., Tremblay, Y.D.N., 2010. Biofilm formation in bacterial pathogens of veterinary importance. Anim. Health Res. Rev. 11 (2), 97–121.

Janssens, T.K.S., Tyc, O., Besselink, H., de Boer, W., Garbeva, P., 2019. Biological activities associated with the volatile compound 2,5-bis(1-methyl ethyl)-pyrazine. FEMS Microbiol. Lett. 366, fnz023.

Jones, S.E., Elliot, M.A., 2018. 'Exploring' the regulation of *Streptomyces* growth and development. Curr. Opin. Microbiol. 42, 25–30.

Jones, S.E., Pham, C.A., Zambri, M.P., McKillip, J., Carlson, E.E., Elliot, M.A., 2019. Streptomyces volatile compounds influence exploration and microbial community dynamics by altering iron availability. mBio 10, e00171–19.

Kai, M., Haustein, M., Molina, F., Petri, A., Scholz, B., Piechulla, B., 2009. Bacterial volatiles and their action potential. Appl. Microbiol. Biotechnol. 81, 1001–1012.

Kai, M., Effmert, U., Piechulla, B., 2016. Bacterial-plant-interactions: approaches to unravel the biological function of bacterial volatiles in the rhizosphere. Front. Microbiol. 7, 108.

Kanchiswamy, C.N., Malnoy, M., Maffei, M.E., 2015. Chemical diversity of microbial volatiles and their potential for plant growth and productivity. Front. Plant Sci. 6, 151.

Kesarwani, M., Hazan, R., He, J., et al., 2011. A quorum-sensing regulated small volatile molecule reduces acute virulence and promotes chronic infection phenotypes. PLoS Pathog. 7, e1002192.

Kim, K.-S., Lee, S., Ryu, C.-M., 2013. Interspecific bacterial sensing through airborne signals modulates locomotion and drug resistance. Nat. Commun. 4, 1809.

Ladyniga, N., Dedyukhina, E.G., Vainshtein, M.B., 2006. A review on microbial synthesis of hydrocarbons. Process Biochem. 41, 1001–1014.

Lee, J.H., Lee, J., 2010. Indole as an intercellular signal in microbial communities. FEMS Microbiol. Rev. 34 (426), 44.

Lee, B., Farag, M.A., Park, H.B., Kloepper, J.W., Lee, S.H., Ryu, C.M., 2012. Induced resistance by a long-chain bacterial volatile: elicitation of plant systemic defense by a C13 volatile produced by *Paenibacillus polymyxa*. PLoS One 7, e48744. Available from: https://doi.org/10.1371/journal.pone.0048744.

Lemfack, M.C., Nickel, J., Dunkel, M., Preissner, R., Piechulla, B., 2013. mVOC: a database of microbial volatiles. Nucleic Acids Res. 42, D744–D748.

Lemfack, M.C., Nickel, J., Dunkel, M., et al., 2014. mVOC: a database of microbial volatiles. Nucleic Acids Res. 42, D744–D748.

Lemfack, M.C., Ravella, S.R., Lorenz, N., Kai, M., Jung, K., et al., 2016. Novel volatiles of skin-borne bacteria inhibit the growth of Gram-positive bacteria and affect quorum-sensing controlled phenotypes of Gram-negative bacteria. Syst. Appl. Microbiol. 39, 503–515.

Létoffé, S., Audrain, B., Bernier, S.P., Delepierre, M., Ghigo, J.-M., 2014. Aerial exposure to the bacterial volatile compound trimethylamine modifies antibiotic resistance of physically separated bacteria by raising culture medium pH. mBio 5, . Available from: https://doi.org/10.1128/mBio.00944-13e00944-13m.

Lirk, P., Bodrogi, F., Raifer, H., et al., 2003. Elective haemodialysis increases exhaled isoprene. Nephrol. Dial. Transpl. 18, 937–941.

Liu, X.-M., Zhang, H., 2015. The effects of bacterial volatile emissions on plant abiotic stress tolerance. Front. Plant Sci. 6, 774. Available from: https://doi.org/10.3389/fpls.2015.00774.

Marilley, L., Casey, M.G., 2004. Flavors of cheese products: metabolic pathways, analytical tools, and identification of producing strains. Int. J. Food Microbiol. 90, 139–159.

Mattila, J.T., Thomas, A.C., 2014. Nitric oxide synthase: non-canonical expression patterns. Front. Immunol. 5, 478.

Meldau, D.G., Meldau, S., Hoang, L.H., Underberg, S., Wünsche, H., Baldwin, I.T., 2013. Dimethyl disulfide produced by the naturally associated bacterium *Bacillus sp* B55 promotes *Nicotiana attenuata* growth by enhancing sulfur nutrition. Plant Cell 25, 2731–2747.

Meruva, N.K., Penn, J.M., Farthing, D.E., 2004. Rapid identification of microbial VOCs from tobacco molds using closed-loop stripping and gas chromatography/time-of-flight mass spectrometry. J. Ind. Microbiol. Biot. 31, 482–488.

Montes Vidal, D., von Rymon-Lipinski, A.-L., Ravella, S., Groenhagen, U., Herrmann, J., et al., 2017. Long-chain alkyl cyanides: unprecedented volatile compounds released by *Pseudomonas* and *Micromonospora* bacteria. Angew. Chem. Int. Ed. Engl. 56, 4342–4346.

Muckschel, B., Simon, O., Klebensberger, J., et al., 2012. Ethylene glycol metabolism by *Pseudomonas putida*. Appl. Env. Microb. 78, 8531–8539.

Mueller, R.S., Beyhan, S., Saini, S.G., et al., 2009. Indole acts as an extracellular cue regulating gene expression in *Vibrio cholerae*. J. Bacteriol. 191, 3504–3516.

Netzker, T., Shepherdson, E.M.F., Zambri, M.P., Elliot, M.A., 2020. Bacterial volatile compounds: functions in communication, cooperation, and competition. Annu. Rev. Microbiol. 74 (1). Available from: https://doi.org/10.1146/annurev-micro-011320-015542.

Nijland, R., Burgess, J.G., 2010. Bacterial olfaction. Biotechnol. J. 5, 974–977.

Ojuederie, O., Olanrewaju, O., Babalola, O., 2019. Plant growth promoting rhizobacterial mitigation of drought stress in crop plants: implications for sustainable agriculture. Agronomy 9 (11), 712. Available from: https://doi.org/10.3390/agronomy9110712.

Ortíz-Castro, R., Contreras-Cornejo, H.A., Macías-Rodríguez, L., LópezBucio, J., 2009. The role of microbial signals in plant growth and development. Plant Signal. Behav. 4, 701–712.

Penuelas, J., Asensio, D., Tholl, D., et al., 2014. Biogenic volatile emissions from the soil. Plant Cell Environ. 37, 1866–1891.

Plyuta, V., Lipasova, V., Popova, A., Koksharova, O., Kuznetsov, A., et al., 2016. Influence of volatile organic compounds emitted by *Pseudomonas* and *Serratia* strains on *Agrobacterium tumefaciens* biofilms. APMIS 124, 586–594.

Que, Y.A., Hazan, R., Strobel, B., et al., 2013. A quorum-sensing small volatile molecule promotes antibiotic tolerance in bacteria. PLoS One 8, e80140.

Raza, W., Ling, N., Liu, D., et al., 2016. Volatile organic compounds produced by *Pseudomonas fluorescens* WR-1 restrict the growth and virulence traits of *Ralstonia solanacearum*. Microbiol. Res. 192, 103–113.

Ryu, C.-M., 2015. Bacterial volatiles as airborne signals for plants and bacteria. In: Lugtenberg, B. (Ed.), Principles of Plant-Microbe Interactions. Springer, Berne.

Ryu, C.M., Farag, M.A., Hu, C.H., et al., 2003. Bacterial volatiles promote growth in Arabidopsis. Proc. Natl. Acad. Sci. U. S. A. 100, 4927–4932.

Sánchez-López, Á.M., Baslam, M., De Diego, N., et al., 2016. Volatile compounds emitted by diverse phytopathogenic microorganisms promote plant growth and flowering through cytokinin action. Plant Cell Environ. 39, 2592–2608.

Scholler, C.E., Gurtler, H., Pedersen, R., et al., 2002. Volatile metabolites from *Actinomycetes*. J. Agr. Food Chem. 50, 2615–2621.

Schulz, S., Dickschat, J.S., 2007. Bacterial volatiles: the smell of small organisms. Nat. Prod. Rep. 24, 81442.

Sharifi, R., Lee, S.M., Ryu, C.M., 2018. Microbe-induced plant volatiles. New Phytol. 218. Available from: https://doi.org/10.1111/nph.14955.

Son, J.S., Lee, S.J., Jun, S.Y., Yoon, S.J., Kang, S.H., Paik, H.R., et al., 2010. Antibacterial and biofilm removal activity of a podoviridae *Staphylococcus aureus* bacteriophage SAP-2 and a derived recombinant cell-wall-degrading enzyme. Appl. Microbiol. Biotechnol. 86, 1439–1449.

Song, G.C., Ryu, C.M., 2013. Two volatile organic compounds trigger plant self-defense against a bacterial pathogen and a sucking insect in cucumber under open field conditions. Int. J. Mol. Sci. 14, 9803–9819.

Stefels, J., 2000. Physiological aspects of the production and conversion of DMSP in marine algae and higher plants. J. Sea Res. 43, 183–197.

Tahir, H.A., Gu, Q., Wu, H., et al., 2017. Plant growth promotion by volatile organic compounds produced by *Bacillus subtilis* SYST2. Front. Microbiol. 8. Available from: https://doi.org/10.3389/fmicb.2017.00171.

Tellez, M.R., Schrader, K.K., Kobaisy, M., 2001. Volatile components of the *cyanobacterium, Oscillatoria perornata* (Skuja). J. Agr. Food Chem. 49, 5989–5992.

Velázquez-Becerra, C., Macías-Rodríguez, L.I., López-Bucio, J., Altamirano-Hernández, J., Flores-Cortez, I., Valencia-Cantero, E., 2010. A volatile organic compound analysis from *Arthrobacter agilis* identifies dimethylhexadecylamine, an amino-containing lipid modulating bacterial growth and *Medicago sativa* morphogenesis in vitro. Plant Soil 339, 329–340. Available from: https://doi.org/10.1007/s11104-010-0583-z.

Velázquez-Becerra, C., Macías-Rodríguez, L.I., López-Bucio, J., AltamiranoHernández, J., Flores-Cortez, I., Valencia-Cantero, E., 2011. A volatile organic compound analysis from *Arthrobacter agilis* identifies dimethylhexadecylamine, an amino-containing lipid modulating bacterial growth and *Medicago sativa* morphogenesis in vitro. Plant Soil 339, 329–340.

Vespermann, A., Kai, M., Piechulla, B., 2007. Rhizobacterial volatiles affect the growth of fungi and *Arabidopsis thaliana*. Appl. Env. Microbiol. 73, 5639–5641. Available from: https://doi.org/10.1128/AEM.01078-07.

Wang, J., Zhou, C, Xiao, X., et al., 2017. Enhanced iron and selenium uptake in plants by volatile emissions of *Bacillus amyloliquefaciens* (BF06). Appl. Sci. 7.

Wenke, K., Wanke, D., Kilian, J., Berendzen, K., Harter, K., Piechulla, B., 2012. Volatiles of two growth-inhibiting rhizobacteria commonly engages AtWRKY18 function. Plant J. 70, 445–459.

Whiteson, K.L., Meinardi, S., Lim, Y.W., et al., 2014. Breath gas metabolites and bacterial metagenomes from cystic fibrosis airways indicate active pH neutral 2, 3-butanedione fermentation. ISME J. 8, 1247–1258.

Wu, Y.-N., Feng, Y.-L., Paré, P.W., et al., 2016. Beneficial soil microbe promotes seed germination, plant growth, and photosynthesis in herbal crop *Codonopsis pilosula*. Crop. Pasture Sci. 67, 91–98.

Yu, Q., Tang, C., Kuo, J., 2000. A critical review on methods to measure apoplastic pH in plants. Plant Soil 219, 29–40.

Zhang, H., Xie, X., Kim, M.S., Kornyeyev, D.A., Holaday, S., Pare, P.W., 2008. Soil bacteria augment Arabidopsis photosynthesis by decreasing glucose sensing and abscisic acid levels in planta. Plant J. 56, 264–273.

Zhang, H., Sun, Y., Xie, X., Kim, M.S., Dowd, S.E., Pare, P.W., 2009. A soil bacterium regulates plant acquisition of iron via deficiency-inducible mechanisms. Plant J. 58, 568–577.

Zhou, J.-Y., Li, X., Zheng, J.-Y., Dai, C.-C., 2016. Volatiles released by endophytic *Pseudomonas fluorescens* promoting the growth and volatile oil accumulation in *Atractylodes lancea*. Plant Physiol. Biochem. 101, 132–140.

Zhu, J.J., Bean, H.D., Kuo, Y.M., et al., 2010. Fast detection of volatile organic compounds from bacterial cultures by secondary electrospray ionization-mass spectrometry. J. Clin. Microbiol. 48, 4426–4431.

Zhu, J., Hill, J.E., 2013. Detection of *Escherichia coli* via VOC profiling using secondary electrospray ionization-mass spectrometry (SESIMS). Food Microbiol. 34, 412–417.

Further reading

Chen, J., Tang, J., Shi, H., Tang, C., Zhang, R., 2016. Characteristics of volatile organic compounds produced from five pathogenic bacteria by headspace-solid phase micro-extraction/gas chromatography-mass spectrometry. J. Basic Microbiol. 57 (3), 228–237. Available from: https://doi.org/10.1002/jobm.201600505.

Hamilton-Kemp, T., Newman, M., Collins, R., et al., 2005. Production of the long-chain alcohols octanol, decanol, and dodecanol by *Escherichia coli*. Curr. Microbiol. 51, 82–86.

Kai, M., Piechulla, B., 2009. Plant growth promotion due to rhizobacterial volatiles an effect of CO2. FEBS Lett. 583, 3473–3477. Available from: https://doi.org/10.1016/j.febslet.2009.09.053.

Lorenz, M.C., Fink, G.R., 2002. Life and death in a macrophage: role of the glyoxylate cycle in virulence. Eukaryot. Cell 1, 657–662.

Ryu, C.M., Farag, M.A., Hu, C.H., et al., 2004. Bacterial volatiles induce systemic resistance in Arabidopsis. Plant Physiol. 134, 1017–1026.

Sharifi, R., Ryu, C.-M., 2018. Revisiting bacterial volatile-mediated plant growth promotion: lessons from the past and objectives for the future. Ann. Bot. 122 (3), 349–358. Available from: https://doi.org/10.1093/aob/mcy108.

Chapter 17

Importance of microbial secondary metabolites in health care applications

Ruchira Mitra[1,2], Jing Han[1,3], Hua Xiang[1,3] and Surojit Bera[4]

[1]*State Key Laboratory of Microbial Resources, Institute of Microbiology, Chinese Academy of Sciences, Beijing, P.R. China,* [2]*International College, University of Chinese Academy of Sciences, Beijing, P.R. China,* [3]*College of Life Science, University of Chinese Academy of Sciences, Beijing, P.R. China,* [4]*Department of Microbiology, School of Bioengineering and Biosciences, Lovely Professional University, Jalandhar, India*

Chapter Outline

17.1 Introduction

An increase in global average life expectancy in recent decades is mainly attributed to improvement in health care sectors. The availability of proper treatment and supply of medicinal and supportive drugs has ensured a better life to most of the world's population. The current health care industry is mostly driven by chemically synthesized drugs. Almost

Volatiles and Metabolites of Microbes. DOI: https://doi.org/10.1016/B978-0-12-824523-1.00011-0

more than 60% of the marketed drugs are of synthetic origin (Bade et al., 2010). Implementation of stricter regulation in the drug approval process has further pushed the pharmaceutical companies toward synthetic drugs. Synthetic drugs contain highly purified and processed molecules, which effectively reduce toxicity. Furthermore, the manufacture of synthetic drugs is much faster and more cost-effective, which speeds up commercialization. Despite all these advantages, consumers are switching their interests from synthetic drugs to natural drugs. This is because a major concern of synthetic drugs is the unacceptable side effects to the human body. Moreover, the pharmaceutical residues cause severe damage to our ecosystem when discharged to the environment (Sirés and Brillas, 2012). Thus, with this growing awareness, consumers are looking for more naturally derived medicinal drugs and health-promoting supplements.

Nature has provided a wide range of bioactive molecules in plants, animals, and microorganisms. From time immemorial, plants were considered the most common sources of medicines and health supplements. However, the discovery of penicillin from *Penicillium notatum* by Alexander Fleming in 1928 brought microorganisms into the limelight as a potent source for biomolecules (Fleming, 1944). Since then, researchers are continuously exploring microorganisms from various sources to isolate bioactive components with potential applications. Several microbe-derived bioactive compounds have immense health benefits and can serve as a potential candidate in health care applications. A microbial mode of synthesis is also an effective approach in terms of time, cost, space, and most importantly environment (Bera et al., 2016). However, a major challenge is most of the microbe-derived bioactive molecules are synthesized at a low concentration. A promising solution to this problem is the metabolic engineering of the microbial strains to facilitate the enhanced synthesis of targeted products. Furthermore, advancement in recombinant DNA technology has allowed the successful development of recombinant proteins such as insulin and human growth hormones using microbial expression systems (Graumann and Premstaller, 2006). This proves that microbes offer a promising and robust platform for efficient synthesis of natural health care products. "Biologics" is a collective term that represents a set of molecules whose active pharmaceutical component is derived from living organisms (Park et al., 2019). Biologics, including monoclonal antibodies, vaccines, blood factor derivatives, enzymes, and recombinant proteins have complex structures and are produced using various cutting-edge techniques of biotechnology. As of 2017, the global market price of biologics was US$254.9 billion and is estimated to increase by up to US$580 billion by 2026 (Credence Research, 2018). Hence, it is well understood that naturally derived products are gradually gaining importance in the global market and microorganisms held a special mention in this scenario.

17.1.1 Microbial metabolite: a new era in bioactive compounds

Microorganisms synthesize small molecules as an intermediate or end product of any physiological metabolism which is termed as microbial metabolites. Broadly, these metabolites are of two types, primary and secondary. Primary metabolites are essentially involved in the growth, development, and reproduction of the microorganism. This type includes carbohydrate, protein, fat, nucleic acid, hormone, ethanol, and other fermentation end products that are directly involved in its growth. Secondary metabolites are not essential for growth, development, and reproduction of the microorganism but involved in overall maintenance and homeostasis of microorganisms (James, 2017). Secondary metabolites are mostly synthesized during or at the end of the stationary phase and are often required for survival during unfavorable conditions. Examples of secondary metabolites include antibiotics, terpenoids, phenolics, alkaloids, etc. The discovery of penicillin brought a revolutionary change in the world of secondary metabolites. However, previously the screening of microbes for secondary metabolites was mostly confined to search for components with antimicrobial potential. Fortunately with passing years the situation started changing wherein bioactive molecules with other therapeutic activities, such as anticancer, antioxidant, and antidiabetic, were identified and characterized. The total market price of the microbe and microorganism-derived products was approximately US$143.5 billion in 2014 and it has been estimated that this price will increase to US$306 billion by 2020 (Singh et al., 2017). Such an upsurge in the market value indicates the growing demand for microbial metabolites.

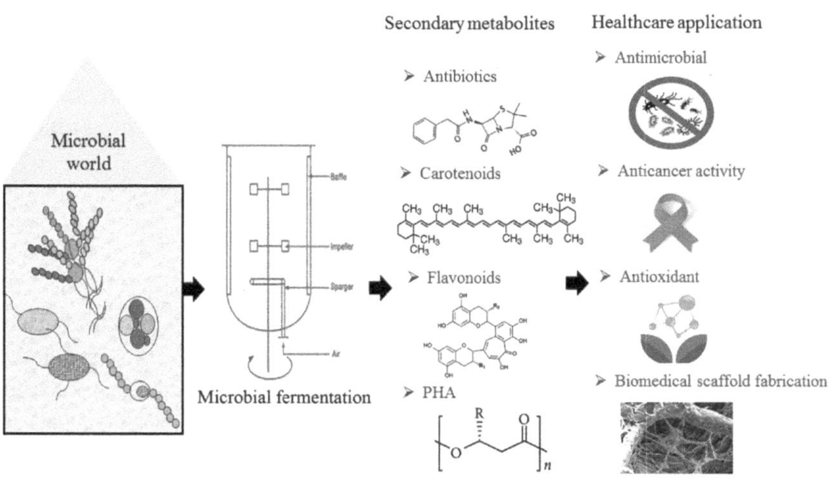

FIGURE 17.1 Schematic diagram depicting fermentation of microorganisms to isolate bioactive compounds with potential health care applications. Microbes producing bioactive molecules from different sources are isolated, identified, and characterized. The bioactive molecules are extracted after fermentation and their application in various health care sectors is studied.

The objective of this chapter is to provide an insight into the microbial metabolites by emphasizing secondary metabolites (Fig. 17.1). Secondary metabolites are synthesized by microorganisms as a response to different kinds of environmental conditions. Some of these secondary metabolites have demonstrated antimicrobial activities, acute and chronic disease prevention, potential biomedical properties, and many more, which can contribute to health care sectors. Thus, in the following chapter, we have summarized some interesting findings from the last few decades on selective secondary metabolites for a better understanding of their health benefiting potential to the readers.

17.2 Antibiotics

Modern medicine is highly dependent on the effective use of antibiotics. The use of antibiotics dates back to ancient times. The Eber's papyrus from 1500 BCE documented the use of moldy bread to treat purulent wounds in China, Greece, Serbia, Egypt, and other ancient civilizations (Haas, 1999). It is now known that the healing effect of the bread is due to the antibiotics produced by the molds present on it. Before the discovery of penicillin, antibiotics were chemically synthesized, which included salvarsan, the first effective treatment for syphilis (Gelpi et al., 2015). However, the discovery of penicillin unveiled the ability of microorganisms to produce natural compounds that can kill other microbes. This finding was a milestone that revolutionized the course of medicine. The term "*antibiotic*" was first proposed by Selman Waksman in 1941 to describe the natural molecules produced by a microorganism that can act against the growth of other organisms (Clardy et al., 2009). Antibiotics are the most common microbial secondary metabolite commercially exploited. There are several classes of antibiotics that have been discovered until now (Table 17.1). Mostly, the classes of antibiotics are categorized based on their mode of action. Some of the important antibiotic classes have been discussed in the following section.

17.2.1 Inhibition of cell wall synthesis by β-lactams and glycopeptide aminoacid

β-Lactam is one of the oldest and most common classes of antibiotics used in prevention of bacterial infections. As the name suggests, this antibiotic contains a β-lactam ring in its core structure (Wanger et al., 2017). β-Lactam antibiotics work by inhibiting the biosynthesis of cell wall. The cell wall is the essential protective structure that covers the cytoplasmic membrane of most bacterial cell. A vital constituent of cell wall is peptidoglycan, a mesh-like polymer of glycan chain cross-linked by peptides. Penicillin-binding proteins (PBPs) play a key role in the morphology of the cell wall. The high molecular weight PBPs are major peptidoglycan synthase as they possess

TABLE 17.1 Different antibiotics from various microbial source and their targets.

Name of antibiotic	Microbial source	Type	Target	References
Fengycin	*Bacillus subtilis*	Antifungal lipopeptide	Filamentous fungi	Vanittanakom et al. (1986)
Polymyxin	*Bacillus polymyxa*	Nonribosomal cyclic lipopeptide antibiotics	Gram-negative bacterial infections	Paulus and Gray (1964)
Bacillopeptins	*B. subtilis* FR-2	Iturin-group antibiotic	Fungi	Kajimura et al. (1995)
Pyrrolnitrin	*Burkholderia cepacia*	Pyrrole antibiotic	Fungi, *Streptomyces antibioticus*	El-Banna and Winkelmann (1998)
Fusacandins	*Fusarium sambucinum*	Papulacandin class antibiotic	Fungi, Gram-positive bacteria such as *Staphylococcus aureus* or *Micrococcus luteus*	Kaneda and Kajimura (2002), Jackson et al. (1995)
Gentamicin	*Micromonospora echinospora*	Aminoglycoside antibiotic complex	Gram-positive organisms such as *S. aureus*, Gram-negative infections	Himabindu, and Jetty (2006)
Myxothiazol	*Myxococcus fulvus*	Competitive inhibitor of ubiquinol	Fungi	Gerth et al. (1980)
Pestiocandin	*Pestalotiopsis humus* FKI-7473	Papulacandin class antibiotic	Multidrug-sensitive yeasts	Sakai et al. (2018)
Stigmatellin	*Stigmatella aurantiaca*	Potent inhibitor of the quinol oxidation (QO) site	Yeast, filamentous fungi, few Gram-positive, for example, *Micrococcus luteus*	Kunze et al. (1984)

(*Continued*)

TABLE 17.1 (Continued)

Name of antibiotic	Microbial source	Type	Target	References
Caprazamycins	*Streptomyces* sp.	Lipo-nucleoside antibiotics	*Mycobacterium tuberculosis*, *Pseudomonas aeruginosa*, various drug resistant strains	Igarashi et al. (2005)
Kakadumycins	*Streptomyces* sp. NRRL 30566	Inhibits RNA synthesis by binding to DNA	Malarial parasite *Plasmodium falciparum*	Castillo et al. (2003)
Thienamycin	*Streptomyces cattleya*	Carbapenem antibiotic (β-lactam family, resistant to bacterial β-lactamase enzymes)	Gram-positive, Gram-negative bacteria	Coulthurst et al. (2005)
Daptomycin	*Streptomyces roseosporus*	Cyclic lipopeptide antibiotic	Gram-positive pathogens	Baltz et al. (2006)
Actinorhodin	*Streptomyces coelicolor*	Polyketide antibiotic	Gram-positive bacteria	Xu et al. (2012)

glycosyltransferase activity for polymerizing glycan chain and transpeptidase activity responsible for their cross-linking (Cho et al., 2014). β-Lactam antibiotic functions by acylating the transpeptidase. Transpeptidase interferes with the transpeptidation process leading to loss of cell integrity and finally cell lysis (Pandey and Cascella, 2019). The low molecular weight PBPs, referred to as PG hydrolase, play an important role in morphogenesis. β-Lactam causes a drug-induced imbalance between cell wall synthesis and hydrolysis, causing cell damage (Cho et al., 2014). They are classified into four main classes: penicillin, cephalosporin, carbapenem, and monobactam. Several bacteria produce β-lactamase, an enzyme capable of hydrolyzing the structural integrity of the lactam ring and inactivating β-lactam antibiotics (Whitehouse et al., 2018). β-Lactamase enzyme resists the activity of penicillin, cephalosporin, carbapenem, and monobactam. This drug resistance leads to the introduction of β-lactamase inhibitors, including clavulanate, sulbactam, and tazobactam (Drawz and Bonomo, 2010). Interestingly the β-lactamase inhibitor does not affect the other changes caused by β-lactam antibiotic such as alteration in PBPs (Dowling, 2004). Hence, β-lactamase inhibitor helps to extend the coverage of β-lactam antibiotic against β-lactamase producing bacteria. Combinations such as amoxicillin–clavulanic acid, piperacillin–tazobactam, and ampicillin–sulbactam are currently used to provide broad spectrum of activity (Bush and Bradford, 2019).

Another important class of antibiotic is glycopeptide, which is actinomycete-derived and contains unique tricyclic or tetracyclic heptapeptide cores that are mostly glycosylated and sometimes also have side chains made with lipophilic fatty acid (Butler et al., 2014). Glycopeptide antibiotic binds to D-alanyl-D-alanine terminus of peptidoglycan precursors, interfering with transpeptidation and transglycosylation reactions of the cell wall synthesis (Binda et al., 2014). The oldest glycopeptide antibiotic, that is, vancomycin, is still used as an essential medicine against clinically significant pathogens, including methicillin-resistant *Staphylococcus aureus, Streptococcus* sp., susceptible *Enterococcus* sp. (Zeng et al., 2016). However, a major concern is that some organisms are resistant to glycopeptide (Lebreton and Cattoir, 2019). This resistance is due to the *van* genes that encodes a ligase responsible for synthesis of low-affinity peptidoglycan precursors with D-Ala-D-Lactate or D-Ala-D-Serine (Alby and Miller, 2018). This results in a 1000-fold decrease in affinity of vancomycin for the binding site. Hydrophobic modification of glycopeptide-appended carbohydrates is a beneficial approach against glycopeptide-resistant pathogens which includes antibiotics such as Ortivancin, Dalbavancin, and TD-6424 (Thibodeaux et al., 2007).

17.2.2 Inhibition of protein biosynthesis by aminoglycosides and chloramphenicol

Aminoglycosides are a broad spectrum antibiotic that acts by inhibiting the protein biosynthesis. Aminoglycosides target the 30S subunit of the bacterial ribosome (Kapoor et al., 2017). Aminoglycosides bind to the A-site on the 16S ribosomal RNA of the 30S ribosomal subunit (Krause et al., 2016). This interaction changes the conformation and leads to misreading of the genetic code on the messenger RNA template. This allows incorporation of incorrect amino acids into polypeptide chain, resulting into erroneous protein synthesis. Aminoglycosides is helpful in treating Gram-negative bacterial infections. However, few anaerobic bacteria have acquired the resistance against aminoglycosides by altering its binding site in ribosomal subunit. The other mechanisms involved in aminoglycoside-resistance includes, reduced antibiotic uptake by the bacteria due to membrane impermeability and aminoglycoside-modifying enzymes, which modifies and inactivates the antibiotic (Mingeot-Leclercq et al., 1999).

Chloramphenicol is a broad spectrum bacteriostatic antibiotic. It has good penetration ability into many body tissues, pleural and ascitic fluids, and placenta as well. Chloramphenicol enters the bacterial cell by passive or facilitated diffusion, binds to 23S r-RNA of 50S ribosomal subunit, and inhibits the protein synthesis by blocking peptidyltransferase activity, binding and movement of ribosomal substrates, and translation termination (Xaplanteri et al., 2003). Chloramphenicol may also bind to the 30S subunit of ribosome and inhibits protein synthesis (Maddison et al., 2008). Chloramphenicol acetyltransferase is a bacterial enzyme which catalyzes the acetyl-S-CoA-dependent acetylation of chloramphenicol at the 3-hydroxyl group (Shaw, 1983). This acetylated chloramphenicol is incapable of binding to the ribosomal subunit and thus, inactivates chloramphenicol.

17.2.3 Inhibition of DNA replication by quinolones

Quinolones are broad spectrum antibiotics, containing a basic bicyclic core structure, used against both Gram-positive and Gram-negative bacteria. It served as an effective treatment for community-acquired and severe nosocomial infections (Pham et al., 2019). Quinolones are favored for their high potency, bioavailability, convenient formulations, high serum concentrations, and lower side effect possibilities (Andersson and MacGowan, 2003). Quinolones function by interfering with DNA replication and leading to DNA synthesis inhibition. During formation of the DNA replication fork in bacteria, DNA gyrase, and topoisomerase IV, the heterotetramer enzymes, generates double-stranded breaks in the strands which are resealed by the enzyme itself. This break and reseal induces negative supercoiling which relaxes the excessive positive supercoiling of the strand during DNA replication. The enzyme-DNA complex, also known as DNA cleavage complex, is the target for quinolone antibiotics. Quinolones noncovalently binds to this DNA cleavage complex which results in accumulation of the DNA replication machinery at the replication forks (Hooper, 1999). This blocks the resealing of the strand and hence interrupts the DNA replication cycle, which eventually inhibits the bacterial growth or kills it. However, resistance to quinolones has already been reported. The most common mechanism of resistance to quinolones is mutation in amino acids of the targeted enzymes which alter the geometry of quinolone-binding site of the enzyme, thus hindering the binding process of the antibiotic (Pham et al., 2019). Other mechanisms include reduced drug uptake due to reduced permeability, protection of topoisomerase target by small pentapeptide-repeat proteins (Qnr proteins), and quinolone inactivation through acetylation carried out by plasmid-encoded aminoglycoside-modifying enzyme variant (Rossolini et al., 2017).

17.3 Anticarcinogenic properties of carotenoids

Carotenoids are natural pigments widely present in plants and microbes (Mitra et al., 2015). They are broadly categorized as carotenes (hydrocarbons) and xanthophylls (oxygenated derivatives of hydrocarbon) (Bera, 2019) (Table 17.2). Carotenoids have several therapeutic potentials including antioxidant, provitamin A activity, and activity against acute and chronic health disorders (Bera et al., 2017). Thus, carotenoids have high promise in pharmaceutical and nutraceutical technology (Bera and Dutta, 2017). Cancer is the second-leading cause of death worldwide, accounting for 9.6 million deaths in 2018 (WHO, 2018). The global cancer burden is expected to increase up to 22.2 million by 2030 (Vineis and Wild, 2014). According to WHO, lung, prostate, colorectal, stomach, and liver cancer are the most common types of cancer in men, while breast, colorectal, lung, cervical, and thyroid cancer are the most common among women. It is believed that the concerted effort of research and medical science would definitely reduce the invasiveness of the disease in the future. The anticancer effect of carotenoid is being well-researched for a long time. In 1991, Ziegler, through prospective and retroprospective studies, suggested that carotenoid intake reduces the risk of different types of cancer (Ziegler, 1991). A prospective cohort study, conducted by Giovannucci et al. (1995), observed that lycopene intake lowered the risk of prostate cancer. Likewise, over the last decades, lots of researchers conducted different clinical trials and epidemiological studies to investigate the effects of carotenoids on cancer.

β-Carotene is a bright red-orange colored carotenoid that is produced by bacteria (Brevibacterium sp., *Flavobacterium* sp.), fungi (Mucor, Phycomyces, *Blakeslea trispora*), and microalgae (*Dunaliella* sp.). In 2000, a comparative study revealed that β-carotene offer better chemopreventive activity than retinoic acid against diethylnitrosamine induced hepatocarcinogenesis in rat (Bishayee et al., 2000). β-Carotene extracted from *Dunaliella salina* possesses potential anticancer activity. Administration of lyophilized powder of *D. salina* showed a significant cancer reduction in 20-methylcholanthrene-induced fibrosarcoma in wister rats. It was suggested that the presence of 9-*cis*-β-carotene contribute to

TABLE 17.2 Different types of carotenoid and their microbial sources.

Organism	Carotenoid type	References
Haematococcus pluvialis	Astaxanthin	Kobayashi et al. (1991)
Chlorella zofingiensis	Astaxanthin	Liu et al. (2014)
Phaffia rhodozyma	Astaxanthin	Johnson et al. (1980)
Mucor circinelloides	β-Carotene	Papp et al. (2013)
Dunaliella salina	β-Carotene	Zhu and Jiang (2008)
Dunaliella bardawil	β-Carotene	Ben-Amotz and Avron (1983)
Rhodotorula glutinis RY-06	β-Carotene	Wang et al. (2008)
Flavobacterium multivorum	β-Carotene	Bhosale and Bernstein (2004)
Brevibacterium sp. KY-4313	β-Carotene	Hsieh et al. (1974)
Kocuria marina DAGII	β-Cryptoxanthin	Mitra et al. (2016)
Bradyrhizobium sp. Strain ORS278	Canthaxanthin	Hannibal et al. (2000)
Dietzia natronolimnaea HS-1	Canthaxanthin	Khodaiyan et al. (2007)
Gordonia jacobaea MV-26	Canthaxanthin	Veiga-Crespo et al. (2005)
Dietzia maris NIT-D	Canthaxanthin	Bera et al. (2015)
Micrococcus roseus	Canthaxanthin	Cooney et al. (1966)
Brevibacterium sp. strain KY-4313	Canthaxanthin	Nelis and De Leenheer (1989)
Haloferax alexandrinus strain TM^T	Canthaxanthin	Asker and Ohta (2002)
Corynebacterium michiganense	Canthaxanthin	Saperstein and Starr (1954)
Chlorella zofingiensis	Canthaxanthin	Li et al. (2006)
Dactylococcus. dissociatus	Canthaxanthin	Grama et al. (2014)
Chlorella protothecoides	Lutein	Shi et al. (1999)
Rhodospirillum rubrum	Lycopene	Wang et al. (2012)
Haloferax mediterranei	Lycopene	Zuo et al. (2018)
Rhodobacter viridis JA737	Neurosporene	Ramaprasad et al. (2013)
Paracoccus zeaxanthinifaciens	Zeaxanthin	Berry et al. (2003)

(Continued)

TABLE 17.2 (Continued)

Organism	Carotenoid type	References
Zeaxanthinibacter enoshimensis TD-ZE3^T	Zeaxanthin	Asker et al. (2007a)
Mesoflavibacter zeaxanthinifaciens TD-ZX30^T	Zeaxanthin	Asker et al. (2007b)
Nubsella zeaxanthinifaciens TDMA-5^T	Zeaxanthin	Asker et al. (2008)
Aquibacter zeaxanthinifaciens CC-AMZ-304^T	Zeaxanthin	Hameed et al. (2014a)
Gramella planctonica CC-AMWZ-3T	Zeaxanthin	Shahina et al. (2014)
Mesoflavibacter aestuarii KYW614^T	Zeaxanthin	Lee et al. (2014)
Gramella oceani CC-AMSZ-T^T	Zeaxanthin	Hameed et al. (2014b)

the anticancer activity (Raja et al., 2007). *D. salina* powder reduced tumor by 83.4% in 7,12-dimethylbenz(*a*)anthracene (DMBA) induced mammary cancer in wister rats (Srinivasan et al., 2017). *D. salina* triggered apoptosis in human oral squamous carcinoma cells (KB) in a dose-dependent manner (Chiu et al., 2017). Ethanol extract of carotenoids of *D. salina* reduced proliferation and induced apoptosis and cell arrest at the G0/G1 phase in human nonsmall cell lung cancer (NSCLC) cells (A549) (Sheu et al., 2008). β-Carotene extracted from *D. salina* EU5891199 showed an effective apoptosis of 32% in human prostate cancer cell line (PC-3) whereas synthetic β-carotene showed 27% (Jayappriyan et al., 2013). Similarly, the extracted β-carotene caused 70% apoptosis compared to 30% by synthetic β-carotene in MDA-MB-231 breast cancer cells (Olmos et al., 2015). Moreover, the cytotoxic effect of the *D. salina* extract against another breast cancer cells (i.e., MCF-7) increased under stressed conditions such as high temperature, high salinity, and nitrogen stress (Singh et al., 2016). This was because the stressed condition led to a higher accumulation of carotenoid. Multidrug resistance (MDR) is one of the major failures in cancer treatment, which causes recurrence of a variety of blood cancers and solid tumors, including breast, ovarian, lung, and lower gastrointestinal tract cancers. In a study by Teng et al. (2016), β-carotene sensitized the multidrug resistant human cervical cancer cell line (KB-vin), and human nonsmall cell lung carcinoma resistant cell line (NCI-H460/MX20) to chemotherapeutic agents such as paclitaxel, doxorubicin, and mitoxantrone. β-Carotene possessed MDR reversal ability and thus, it could be considered as a potential candidate in combinatorial cancer treatment.

Fucoxanthin is a marine carotenoid present in brown algae (*Undaria pinnatifida, Laminaria japonica*) and diatoms (*Phaeodactylum tricornutum* and *Cylindrotheca closterium*) (Zhang et al., 2015). Fucoxanthin extract from *L. japonica* inhibited NSCLC cell growth in nude mice and proliferation of human colon adenocarcinoma cells (WiDr) by cell cycle arrest and apoptosis (Mei et al., 2017; Das et al., 2005). *U. pinnatifida* extract containing fucoxanthin exhibited cytotoxicity against melanoma cell at low concentrations (Wang et al., 2014). A notable study in 2008 revealed that fucoxanthin inhibited the growth of human hepatic carcinoma cell (HepG2) through cell cycle arrest at G_0/G_1 phase (Das et al., 2008). The study also indicated that fucoxanthin decreased the activity of cyclin D/cdk4 in HepG2 cells by decreasing the levels of D-type cyclins, which was the possible reason behind its antitumorgenic activity. Fucoxanthin induced apoptosis and antiproliferative effect on human colon cancer cell line (Caco-2) (Hosokawa et al., 2004). It was reasoned that during uptake of fucoxanthin by Caco-2, it is converted to fucoxanthinol, which has stronger growth inhibition properties than the former. Thus, fucoxanthin metabolites may be responsible for its antiproliferative action on Caco-2 cells. This study was further extended. The effect of fucoxanthin was examined on six human colorectal cell lines (DLD-1, HCT116, SW620, Caco-2, Colo205, WiDr) and twenty surgically resected colorectal cancer tissue specimen. Fucoxanthin and fucoxanthinol both exhibited dose-dependent anticancer effects and the effect of latter was found to be stronger (Takahashi et al., 2015). Yu et al. (2011) reported that tumor proliferation was inhibited in human gastric adenocarcinoma cells (MGC-803) treated with fucoxanthin. It down regulated the expressions of CyclinB1 and survivin and induced cell cycle arrest in the G2/M phase of the cell. The decrease in cyclin B1 was associated with JAK/STAT signal pathway. Fucoxanthin also induced apoptosis in the MGC-803 cells in a dose-dependent manner. In a study by Oliveira-Junior et al. (2016), fucoxanthin was confirmed to be a natural anticancer compound as it exerted antitumoral, antimetastatic and antiangiogenic activities in animal models. This was probably due to the interaction of fucoxanthin with lipid raft in cancer cells leading to the apoptosis and eventually death of the cells (Oliveira-Junior et al., 2016).

Haematococcus pluvialis, a unicellular green microalga, is a rich source of astaxanthin, a ketocarotenoid. *H. pluvialis* extract arrested the proliferation of colon cancer cell (HCT-116) and promoted apoptosis (Palozza et al., 2009). Similarly, this natural astaxanthin demonstrated antiproliferative effect on HepG2 cell line (Nagaraj et al., 2012). *Chlorella zofingiensis* is another potential alga for commercial production of astaxanthin (Ip and Chen, 2005). Interestingly, *C. zofingiensis* also accumulates lutein, a xanthophyll which has potential anticarcinogenic activity against cervical carcinoma, lung cancer, and breast cancer cell lines (Del Campo et al., 2004; Gansukh et al., 2019; Zhang et al., 2018; Swanson et al., 2016). Among xanthophylls, only β-cryptoxanthin serves as a major vitamin A precursor due to the presence of a single unsubstituted β-ionone ring (Mitra et al., 2017a). Iskandar et al. (2016) reported that β-cryptoxanthin supplementation reduced 4-[methyl nitrosamino]-1-[3-pyridyl]-1-butanone (NNK)-induced lung tumorigenesis in A/J mice (Iskandar et al., 2016). This protective effect of β-cryptoxanthin was associated with downregulation of α7-nAChR/PI3K signaling. The study recommended the use of β-cryptoxanthin as a chemotherapeutic agent against lung cancer. A study published in 2013 suggested that β-cryptoxanthin could become a potential therapeutic in the treatment of gastric cancer (Wu et al., 2013). β-Cryptoxanthin suppressed the cell growth, cell migration, cell cycle and expression of associated proteins like cyclin D1 and cyclin E in stomach tumor cell line, BGC-823 (derived from human gastric gland adenocarcinoma). The study also suggested that β-cryptoxanthin would be more effective in the early stage rather than the later stage of carcinogenesis (Wu et al., 2013). However, till now, only a few microbial sources of β-cryptoxanthin including *Kocuria marina* DAGII (Mitra et al., 2017b) and *Flavobacterium lutescens* (Serrato-Joya et al., 2006) have been reported. Thus, it is important to explore more microbial sources for these valuable carotenoids for their widespread application in cancer treatment. However, one of the major drawbacks in microbial carotenoid production is high-priced substrates. Hence, researchers are now focussing on usage of cheap substrates such as agro-industrial wastes to produce carotenoids at reduced cost (Mitra and Dutta, 2018; Bera et al., 2015). This is an effective approach which will lead to sustainable industrial production to meet the increasing global demand for microbial carotenoids.

17.4 Flavonoids

Flavonoids are natural polyphenolic compounds commonly found in the plant kingdom. It is a diversified group divided into several subclasses, for example, chalcones, aurones, flavanones, flavones, flavonols, isoflavonoids, flavandiols, anthocyanins, phlobaphenes, and condensed tannins (Leonard et al., 2006). Synthesis of flavonoid proceeds through phenylpropanoid pathway (Falcone Ferreyra et al., 2012). In the first step, 4-coumaroyl-CoA ligase converts phenylpropanoic acids to the coenzyme A (CoA) esters. Second, chalcone synthase condenses three molecules of malonyl CoA with one molecule of CoA ester to synthesize chalcone. Finally, chalcone isomerase converts chalcone to (2S)-flavanone in a stereospecific manner. (2S)-flavanone is the common precursor for the different flavonoids subclasses (Leonard et al., 2007). For instance, flavanone 3β-hydroxylase catalyzes the formation of (2R, 3R)-*trans*-dihydroflavonols from (2S)-flavanone. Finally, flavonol synthase desaturates dihydroflavonols to flavonols.

Flavonoids have numerous health benefits. They exhibit strong antioxidant, antiinflammatory, antiobesity, and antiplatelet activities (Kozłowska and Szostak-Węgierek, 2017). They are promising bioactive molecules for prevention of chronic diseases such as diabetes, hypertension, and cardiovascular disorder (Panche et al., 2016). Flavonoids play a crucial role in preventing cancer and neurodegenerative disorders like Parkinson's and Alzheimer's diseases (Heim et al., 2002). Flavonoids, such as quercetin and macluraxanthone, inhibit acetylcholinesterase and butyrylcholinsterase in a concentration-dependent manner, which is a key property for drugs against Alzheimer's diseases (Khan et al., 2009). Consumption of dietary quercetin was found to be inversely associated with incidence of prostate, lung, stomach, endometrium, esophagus, and breast cancer (Brusselmans et al., 2005). Flavonoids also possess hepatoprotective properties. Silymarin, a flavonoid consisting of three phytochemical, that is, silybin, silidianin, and silicristin promotes hepatocyte regeneration and inhibits liver fibrogenesis (Feher and Lengyel, 2012). Interestingly, many researchers have reported the antiviral activity of flavonoids. Flavonoids such as baicalein, robustaflavone, and hinokiflavone inhibited HIV-1 reverse transcriptase, catechin inhibited DNA polymerase of HIV-1, demethylated gardenin A and robinetin inhibited HIV-1 proteinase, and quercetin even demonstrated antidengue virus properties (Kumar and Pandey, 2013). Owing to their immense health benefits, flavonoids are gaining attraction as a potential component in drug making. Hence, to combat with its growing demand researchers have targeted engineered microbes for flavonoid production. To date, microbial sources including *Escherichia coli*, *Saccharomyces cerevisiae*, and *Streptomyces* species have produced and modified flavonoids (Pandey et al., 2016). Table 17.3 enlists some flavonoids produced from these microbes.

17.4.1 Engineering of microbial hosts for flavonoid production

E. coli is one of the most experimented microbes for synthesis of valuable products. Using metabolic engineering techniques, synthesis of flavanones in *E. coli* was achieved. An artificial gene cluster containing phenylalanine ammonia lyase (PAL) from yeast *Rhodotorula rubra*, 4-coumarate:coenzyme A ligase (4CL) from the actinomycete *Streptomyces coelicolor* A3(2) and chalcone synthase (CHS) from licorice plant *Glycyrrhiza echinata* was constructed (Hwang et al., 2003). *E. coli* containing the artificial gene cluster synthesized pinocembrin and naringenin flavanone from phenylalanine and tyrosine, respectively. The PAL enzyme exhibited tyrosine ammonia lyase (TAL) activity and thus, both phenylalanine and tyrosine was deaminated by PAL enzyme to yield cinnamic acid and 4-coumaric acid, respectively. Generally, cinnamate-4-hydroxylase (C4H) enzyme hydroxylates cinnamic acid to 4-coumaric acid and catalyzes the ligation of 4-coumaric acid to CoA thioester to generate 4-coumaroyl-CoA. However, in the engineered *E. coli*, the bacterial 4CL ligated CoA thioester to both cinnamic acid and 4-coumaric acid and bypassed the C4H step. Finally, CHS condensed cinnamoyl-CoA or 4-coumaroyl-CoA with malonyl CoA to

TABLE 17.3 Various flavonoids produced by *Escherichia coli*, *Streptomyces* sp., and *Saccharomyces cerevisiae*.

Organism	Flavonoid name	Production (mg/L)	References
Escherichia coli	Apigenin	30	Lee et al. (2015)
E. coli	Genkwanin	41	Lee et al. (2015)
E. coli	Pinocembrin	67.81	Cao et al. (2016)
E. coli	Naringenin	391	Wu et al. (2014)
E. coli	Fisetin	0.3	Stahlhut et al. (2015)
E. coli	Eriodictyol	107	Zhu et al. (2014)
E. coli	Catechin	910.9	Zhao et al. (2015a,b)
E. coli	Apigetrin	16.6	Thuan et al. (2018)
E. coli	7-O-methyl aromadendrin	30	Malla et al. (2012)
E. coli	Phloroglucinol	3800	Cao et al. (2011)
Streptomyces albus	Apigenin	0.384	Marín et al. (2017)
S. albus	Luteolin	0.872	Marín et al. (2017)
S. albus	Eriodictyol	1.256	Marín et al. (2017)
Streptomyces clavuligerus	Naringenin	184	Álvarez-Álvarez et al. (2015)
Streptomyces venezuelae	Apigenin	15.3	Park et al. (2011)
S. venezuelae	Chrysin	30.9	Park et al. (2011)
Saccharomyces cerevisiae	Kaempferol	66.29	Duan et al. (2017)
S. cerevisiae	Naringenin	90	Lyu et al. (2017)
S. cerevisiae	Eriodictyol	200	Amor et al. (2010)

yield pinocembrin chalcone or naringenin chalcone, respectively. The chalcones were then converted to the corresponding flavanones, pinocembrin and naringenin nonenzymatically under alkaline conditions. However, the flavanones formed by the nonenzymatic conversion were in racemic form. Hence, to obtain the natural forms of (2S)-pinocembrin and (2S)-naringenin, chalcone isomerase (CHI) from *Pueraria* plant was further added to the artificial gene cluster (Miyahisa et al., 2005). Moreover, the production of flavanones were increased to 60 mg/L by increasing the intracellular pool of malonyl CoA by overexpressing acetyl-CoA carboxylase from *Corynebacterium glutamicum* and optimizing the culture conditions (Miyahisa et al., 2005). In the next study, flavone synthase I gene from *Petroselinum crispum* was expressed in the engineered *E. coli* strain (Miyahisa et al., 2006). The flavone synthase introduces double bond between C-2 and C-3 of flavanones and converts them to flavones. Thus, flavone apigenin or chrysin was synthesized when tyrosine or phenylalanine was supplemented in the medium with no traces of naringenin or pinocembrin. Similarly, under the successive actions of flavanone 3β-hydroxylase (F3H) and flavonol synthase (FLS), (2S)-flavanones are converted into the corresponding flavonols. F3H catalyzes the stereospecific 3β-hydroxylation of flavanones to the respective (2R, 3R)-dihydroflavonols and FLS introduces double bond between C-2 and C-3 of (2R, 3R)-dihydroflavonols to yield flavonols. Hence, introduction of *F3H* and *FLS* genes from the *Citrus* species into the flavanone-producing recombinant *E. coli* yielded kaempferol from tyrosine or galangin from phenylalanine (Miyahisa et al., 2006). In another study, PAL, C4H, 4CL, and CHS from *Arabidopsis thaliana* was expressed in *E. coli* (Watts et al., 2004). However, C4H was non-functional and blocked the synthesis of naringenin. Hence, 4-coumaric acid was exogenously fed to promote synthesis of naringenin. Leonard et al. (2008) improved the production of flavanones by engineering the central metabolism of flavanone-producing engineered *E. coli* strain E2. Modification of the acetate-acetyl-CoA–malonyl CoA metabolic node by overexpressing the acetyl-CoA carboxylase (ACC) subunits from *Photorhabdus luminescens* in *E. coli* augmented the malonyl CoA precursor availability. This increased the flavanone production up to 576%, compared to the parent *E. coli* strain E2. Furthermore, coexpression of chimeric biotin ligase (BirA) composed of N terminus of BirA from *E. coli* and C terminus of BirA from *P. luminescens* with ACC subunits further improved flavanone production to 1124%, as compared to parent *E. coli* strain E2. Another alternative strategy was designed in the same study which aimed at improving flavanone production by reducing the toxic effects of acetate accumulation. Amplification of the acetate assimilation pathways in *E. coli* combined with overexpression of ACC yielded *E. coli* strain producing 429 mg/L pinocembrin, which is 1.379% higher than E2. Furthermore, introduction of the malonate utilization genes, *matB* encoding malonyl CoA synthetase and *matC* encoding putative dicarboxylate carrier protein from *Rhizobium trifolii* in the parent *E. coli* strain E2 improved pinocembrin production up to 480 mg/L (Leonard et al., 2008). Also, fatty acid synthase genes *fabB* and *fabF* were repressed in the *E. coli* parent strain to inhibit competitive reaction pathways which yielded 710 mg/L pinocembrin (Leonard et al., 2008). Hence, it is well-evident that *E. coli* is an efficient platform for metabolic engineering to produce various subclasses of flavonoids.

Similar to *E. coli*, flavanone-producing *S. cerevisiae* strain was obtained by introducing the *PAL* gene from *Rhodosporidium toruloides*, 4CL gene from *A. thaliana*, and CHS from *Hypericum androsaemum* (Jiang et al., 2005). PAL exhibited TAL activity and hence, the recombinant yeast strain AH22 produced 7 mg/L of naringenin and 0.8 mg/L of pinocembrin. Introduction of the soluble flavone synthase I and membrane-bound flavone synthase II (FSII) in flavanone-producing recombinant yeast strain yielded flavone molecules chrysin, apigenin, and luteolin but the intermediate flavanones pinocembrin, naringenin, and eriodictyol, respectively, were also present in the medium (Leonard et al., 2005). Expression of naringenin biosynthesis genes from *A. thaliana* in *S. cerevisiae* led to low naringenin yield (Koopman et al., 2012). Removal of the tyrosine feedback inhibition, and deletion of the decarboxylase-encoding genes together with increasing copy number of the chalcone synthase gene and heterologous expression of TAL, improved the naringenin concentration by 40-fold. Similarly, metabolic engineering of *Streptomyces* species has been done to produce flavone and flavanone. Heterologous construction of the flavanone biosynthetic pathway in *Streptomyces venezuelae* led to synthesis of naringenin and pinocembrin (Park et al., 2009). Using similar approaches, de novo biosynthesis of apigenin, luteolin and eriodictyol was realized in *Streptomyces albus* (Marín et al., 2017).

17.5 Polyhydroxyalkanoates

Polyhydroxyalkanoates (PHA) are polyesters of hydroxyalkanoates synthesized by microorganisms as intracellular reserves of carbon and energy (Raza et al., 2018). PHA is produced under unbalanced conditions of limited essential nutrients and excessive carbon supply (Valappil et al., 2007). It is a class of microbial secondary metabolite which has gained emerging importance in biomedical applications. They are biodegradable in nature with good biocompatibility, hydrophobicity, enantiomerically pure, and have high molecular weight with low polydispersity (Pillai et al., 2017). Table 17.4 enlists various microorganisms producing PHA. The most commonly produced PHA by microbial fermentation is polyhyrodxybutyrate (PHB), a homopolymer of 3-hydroxybutyrate (3-HB). However, PHB is brittle, highly crystalline, thermally unstable, and possess poor elastic property, and thus offers narrow processing window and limited industrial application (Chen et al., 2002a,b). Poly(3-hydroxybutyrate-*co*-3-hydroxyvalerate) (PHBV) is a copolymer of 3-hydroxybutyrate and 3-hydroxyvalerate (3HV), which presents superior properties compared to PHB. The 3HV units in PHBV reduce the crystallinity and make it more flexible (Ferre-Guell and Winterburn, 2018). In addition to controlled biodegradability and good biocompatibility, PHBV offers advanced mechanical properties and better processability which are some of the essential criteria in biomedical applications (Han et al., 2017). Poly(4-hydroxybutyric acid) (P4HB) is another biopolyester which is suitable for medical applications. It is the only PHA which received FDA approval for clinical use of monofilament suture in general soft tissue

TABLE 17.4 Variety of PHA-producing microorganisms.

Organism	PHA	References
Aliidongia dinghuensis 7M-Z19^T	PHB	Chen et al. (2017)
Bacillus cereus YB-4	P(3HB-*co*-2 mol% 3HV)	Mizuno et al. (2010)
Bacillus thuringiensis	PHB	Singh et al. (2013)
Comamonas sp. EB172	PHB	Zakaria et al. (2010)
Comamonas acidovorans	P4HB	Saito and Doi (1994)
Cupriavidus necator H16	P(3HB-*co*-8 mol% 3HV)	Obruca et al. (2010)
Fodinicurvata sediminis YIM D82^T	PHB	Wang et al. (2009)
Fodinicurvata fenggangensis YIM D812^T	PHB	Wang et al. (2009)
Halogranum amylolyticum	P(3HB-*co*-20.1 mol% 3HV)	Zhao et al. (2015a,b)
Haloferax mediterranei	P(3HB-*co*-8.55 mol% 3HV)	Chen et al. (2019b)
H. mediterranei	P(3HB-*co*-60.3 mol% 3HV)	Han et al. (2015)
H. mediterranei	P(3HB-*co*-21.8%-3HV-*co*-5.1%-4HB)	Koller et al. (2007)
Halomonas bluephagenesis TD01	PHB	Tan et al. (2011)
H. bluephagenesis TY194 (Δ*sdhE*, G7: P_{porin}-*ppc*)	P(3HB-*co*-25 mol% 3HV)	Chen et al. (2019a)
H. bluephagenesis TD68–194	P(3HB-*co*-16 mol% 4HB)	Ye et al. (2020)
Hydrogenophaga pseudoflava	P4HB	Choi et al. (1999)
Paracoccus sp. LL1	PHB	Sawant et al. (2015)
Ralstonia eutropha H16	PHB	Kahar et al. (2004)

approximation and/or ligation in 2007 (Utsunomia et al., 2020). P4HB is strong, significantly more flexible than other synthetic absorbable polymers such as polyglycolide and poly-L-lactide and has tensile strength almost comparable to ultrahigh molecular weight polyethylene (Martin and Williams, 2003). Fermentation is the favored route for P4HB synthesis as chemical synthesis failed to synthesize high molecular weight P4HB, desirable for biomedical applications. However, very few existing microbes such as *Comamonas acidovorans* (Saito and Doi, 1994), *Hydrogenophaga pseudoflava* (Choi et al., 1999) synthesize P4HB from 4HB precursors as most microbes incorporate 3HB and 3HV, producing copolymers P34HB and PHBV4HB (Mitra et al., 2020). Efficient microbial synthesis of P4HB requires metabolic engineering and heterologous construction of 4HB pathway (Fu et al., 2014). The following section explores the application of PHBV and P4HB in biomedical sectors.

17.5.1 Industrial production of PHA

Despite the importance of PHA in biomedical sectors, industrial production of PHA is not well flourished. There are only around 24 companies worldwide involved in PHA production (Chen, 2009). The industrial strains for PHA production include *Alcaligenes latus*, *Bhurkolderia* sp., *Ralstonia eutropha* and recombinant *E. coli* (Chen, 2010). The main reason behind the limited commercialization of PHA is its high production cost. The substrates used for PHA production primarily accounts for this high cost. Hence, researchers are focusing on using various agro-industrial wastes as substrates for PHA synthesis to reduce PHA production cost. Several microbes have the ability to produce PHA using low-cost substrates. For instance, *R. eutropha* utilizes cane molasses to produce 1.63 g/L PHB (Bozorg et al., 2015). *Pannonibacter phragmitetus* ERC8 uses glycerol waste as sole carbon source to produce 1.36 g/L of PHA (Ray et al., 2016). *Pseudomonas aeruginosa* 42A2 can utilize waste-free fatty acids from soybean oil (WFFA) as well as waste frying oil (WFO) to accumulate 66.1% (wt.) and 29.4% (wt.) PHA, respectively (Fernández et al., 2005). *Bacillus subtilis* produces 2.5 g/L PHA from 10% sugarcane molasses (Anjali et al., 2014). *Halomonas campisalis* MCMB-1027 accumulates 47% (wt.) PHBV from 1% (v/v) aqueous extract of bagasse (Kulkarni et al., 2015). It can utilize and convert 62% of the sugar from bagasse into cell mass and PHBV. *Halomonas bluephagenesis* TD01 produced 40 g/L CDW containing 60% PHB in 14-day unsterile and continuous process (Tan et al., 2011). By engineering the TCA cycle, *H. bluephagenesis* was capable to incorporate up to 25 mol% 3HV from glucose under unsterile conditions (Chen et al., 2019a). The strain synthesized 6.3 g/L CDW containing 65% PHBV in shake flask experiments. On the other hand, heterologous pathway construction for 4HB monomer supply and deletion of the succinate semialdehyde dehydrogenase genes (*gabD*) in *H. bluephagenesis* led to incorporation of 16 mol% 4HB from glucose (Ye et al., 2020). This engineered strain produced 48.2 g/L CDW containing 75% P(3HB-*co*-16 mol% 4HB) in 7-L bioreactor under unsterile conditions. Thus, *H. bluephagenesis* is a potential strain for industrial PHA production. Among archaeal sources of PHA, *Haloferax mediterranei* is a promising candidate for PHA production using low-cost substrates. It can efficiently utilize a range of agro-industrial wastes including vinasse, olive mill wastewater, cheese whey, and rice-based ethanol stillage as a sole carbon source for PHBV production (Mitra et al., 2020). It can also efficiently utilize chitin to produce 1 g/L PHBV (Hou et al., 2014). Thus, usage of low-cost substrates is a sustainable approach to curb the PHA production cost. Furthermore, parameters like carbon and nitrogen ratio, salinity, feeding strategy, fermentation apparatus, and mode of fermentation affects the PHA yield (Mitra et al., 2020). Thus, optimization of the fermentation process parameters and culture condition can also improve the productivity which can help in cost reduction. The other developments which might help in lowering of PHA cost is achieving high cell density at short time period with super-high PHA content in the cell dry weight, obtaining large PHA granules for easy PHA recovery, controlled lysis of the PHA-containing cells, continuous fermentation, and controlling PHA molecular weight (Chen, 2009).

17.5.2 Tissue engineering applications of PHA

Tissue engineering is an emerging interdisciplinary field which emphasizes on repairing and regenerating damaged tissue and organs. It includes techniques to utilize various polymers as scaffolding materials to support cell growth, deliver growth factors and other molecules, and promote successful tissue or organ regeneration (Howard et al., 2008). PHBV with 8 mol% 3-HV is an optimum composition to fabricate scaffold due to its low cytotoxicity and high cell compatibility (Zhu et al., 2007). PHBV microsphere as tissue engineering scaffold offers several benefits. PHBV microspheres are biocompatible, small sized, and spherical shaped, and thus they would elicit less inflammatory response which would reduce chances of invasive surgeries (Chen and Tong, 2012). The 3D structure of the microsphere maintains the viability and phenotype of the cell (Zhu et al., 2009). Microspheres enable sustained and controlled release of drug molecules to regulate cell growth and promote vascularization (Zhu et al., 2007). Moreover, since PHBV is biodegradable, it could be degraded at the implant site, leaving space for tissue regeneration (Chen and Tong, 2012). *H. mediterranei* produced tailor-made random PHBV and higher-order PHBV showed faster degradation under implantation condition and excellent biocompatibility compared to bacteria-produced PHBV (Han et al., 2015, 2017). Furthermore, they exhibited good cell attachment and proliferation of rat fibroblast and osteoblast cells. 45S5 bioactive glass-based scaffolds are well-researched for application in bone tissue engineering. However, they have insufficient mechanical strength and low fracture toughness. Vancomycin encapsulated in PHBV microsphere was used to homogeneously coat the 45S5 bioactive glass-based scaffold (Li et al., 2014). This PHBV microsphere coated scaffold released vancomycin in a controlled manner and also improved the mechanical properties of the scaffold. This modified scaffold solved dual purposes and offered potential advantages in bone tissue engineering. Similarly, PHBV microsphere was embedded in poly(L-lactic-*co*-glycolic) acid (PLGA) matrix, another biodegradable polymer with excellent biocompatibility, to formulate a new scaffold model (Huang et al., 2010). This modified scaffold exhibited improved comprehensive strength, thus suggesting potential future applications as load-bearing scaffold in bone tissue engineering. As a neural tissue engineering scaffold, PHBV microspheres supported growth and proliferation of neural cell line (PC12), neuronal maturation, promoted higher degree of axon and dendritic segregation and also supported differentiation of neural progenitor cells into neurons (Chen and Tong, 2012). Bacterial infection after total joint arthroplasty is a challenging complication. Antimicrobial peptides, tachyplesin I, tagged with PHA-granule-associated protein and immobilized on PHBV effectively inhibited the growth of both Gram-negative and Gram-positive bacteria (Xue et al., 2018). It also promoted fibroblast proliferation in vitro and accelerated wound healing in a deep-wound mouse model in vivo. PHBV nanofibers loaded with silver nanoparticle demonstrated high antibacterial activity against *S. aureus* and *Klebsiella pneumonia* with no cell cytotoxicity. This showed the potential use of PHBV as a scaffold in minimizing chances of infection after a total joint arthroplasty (Xing et al., 2010). PHBV nanofibers also hold potential importance in skin tissue engineering applications. As scaffold, PHBV nanofibers favored adhesion and proliferation of human keratinocytes (Sundaramurthi et al., 2013). Expression of loricrin and keratin-1 genes were significantly higher on PHBV nanofibers and genes associated with peripheral blood lymphocyte activation were downregulated. This demonstrated the biocompatibility of PHBV nanofibers for skin tissue engineering applications.

Wound healing is a highly programmed process comprising of four phases, hemostasis, inflammation, proliferation, and remodeling (Guo and DiPietro, 2010). PHBV scaffold loaded with adipose-derived stem cells (ASCs) showed high potential in improving skin wound healing with reduced scarring (Zonari et al., 2015). The PHBV scaffold provided sufficient mechanical property that is required to withstand the stress occurred during wound healing. PHBV scaffold with ASC demonstrated enhanced vascularization and reduced differentiation to myofibroblast. The PHBV scaffold allowed exudate and inflammatory cell infiltration which enhanced the degradation of the PHBV structure. The neo-skin formed on the PHBV scaffold demonstrated the presence of well-organized dermal matrix, sebaceous gland and hair follicles. A novel cerium oxide nanoparticles incorporated electrospun PHBV membrane showed excellent cytocompatibility and cell adhesion

(Augustine et al., 2019). Moreover, it demonstrated significant achievements in in vivo diabetic wound healing, including cell infiltration and granulation tissue formation. Thus, PHBV scaffold model holds high promise in wound healing applications.

Research on P4HB has shown several promising results in various medical conditions. In congenital cardiac defects, patch augmentation of right ventricular outflow tract and pulmonary artery is common. Porous P4HB patch seeded with vascular cells such as endothelial cells, smooth muscle cells, and fibroblasts, implanted to augment pulmonary artery in juvenile sheep model showed tissue generation in organized and functional manner accompanied by formation of cellular and extracellular matrix contents (Stock et al., 2000). This was an optimistic study which showed the feasibility of P4HB use as a patch material in pulmonary circulation. Interestingly, a trileaflet valve scaffold was fabricated using a composite material made of polyglycolic acid (PGA) mesh coated with thin P4HB layer (Hoerstrup et al., 2000). P4HB added favorable mechanical strength to the highly porous polyglycolic acid mesh. This trileaflet heart valve seeded with ovine myofibroblasts and endothelial cells, functioned up to 5 months in sheep model. Fabrication of small caliber vascular graft from PGA/P4HB composite scaffold, seeded with endothelial cells, was feasible (Hoerstrup et al., 2001). These tissue engineered vascular grafts demonstrated in vitro tissue maturation and extracellular matrix formation under "biomimetic" environment. As a potential treatment of intracardiac wall defects, PGA/P4HB composite scaffold seeded with human umbilical cord-derived cells was used to develop tissue engineered occluder membranes (Weber et al., 2011). This implant was biodegradable, autologous, thermoresistant with growth and remodeling capacity. Thus, P4HB-based tissue engineered scaffold has immense potentials in cardiovascular applications.

17.5.3 Medical devices

In 2007, the FDA-approved P4HB suture was the first long-term resorbable suture which reached the market. Since then, P4HB scaffolds have been developed and commercialized. BioFiber Scaffold was the first orthopedic soft tissue scaffold made from P4HB (Williams et al., 2016). This scaffold was suitable for tendon repair, soft tissue reinforcement as well as provided lattice for tissue ingrowth. Phasix Mesh manufactured by knitting P4HB monofilament fibers into a fully resorbable scaffold helped in rapid tissue incorporation (Martin et al., 2013). Phasix Mesh was a durable scaffold which provided mechanical reinforcement of soft tissue and aimed hernia repair applications. GalaFLEX scaffold is a macroporous, P4HB monofilament long-term resorbable implant that can provide immediate soft tissue support and allow robust tissue ingrowth (Williams et al., 2016). GalaFLEX scaffold has shown promise in plastic and reconstructive surgery and cosmetic breast surgeries in humans (Adams et al., 2016). Collectively, development of P4HB-based biomedical devices with improved long-term outcomes and soft tissue reinforcement is still ongoing and has high potential applications in near future.

17.6 Conclusion

In this chapter, we have presented the importance of microbial secondary metabolite to human health care in terms of four selected major categories. Starting with antibiotics, we described the mode of mechanism of several antibiotic classes in preventing microbial infections. The discovery of antibiotics revolutionized the world of microbes and scientists began considering microbes as an efficient cell factory for producing bioactive molecules, beneficial for human health. Second, we emphasized on the anticancer properties of carotenoids. Although carotenoids are well-distributed in plants, microbial synthesis of carotenoids is necessary to fulfill its growing demand. Similarly, the plant-derived flavonoids are full of health-benefiting properties and thus necessitate microbial synthesis by employing various metabolic engineering approaches, and this constitutes the third major category of the chapter. Finally, the fourth category includes PHA, a microbial secondary metabolite offering high promise in biomedical applications.

The health care system is one of the most important pillars for humankind and it is essential to strengthen it so that it can combat against numerous diseases suffered by human beings. Microbes are ubiquitously present in our nature. Their evolving role in synthesis of different biomolecules has fascinated researchers worldwide. Since human beings are continually threatened by new types of health disorders, it is high time to consider microbes as one of the solutions. Although a new era has begun where microbes are being exploited for production of various novel metabolites it is merely a scratch in the infinite pool. Hence, it is important for researchers worldwide to accelerate the pace of mining more and more beneficial microbes for extracting novel metabolites which might be applicable to human race.

Acknowledgments

Authors express their thanks to Lovely Professional University (Punjab, India), Institute of Microbiology, Chinese Academy of Sciences, (Beijing, China), University of Chinese Academy of Sciences (UCAS, China), National Key Research and Development Program (No. 2018YFA0900200), the National Natural Science Foundation of China (No. 31970031), and CAS-TWAS President's Fellowship (to RM).

References

Adams Jr, W.P., Toriumi, D.M., Van Natta, B.W., 2016. Clinical use of GalaFLEX in facial and breast cosmetic plastic surgery. Aesthet. Surg. J. 36 (Suppl. 2), S23–S32.

Alby, K., Miller, M.B., 2018. Mechanisms and detection of antimicrobial resistance. Principles and Practice of Pediatric Infectious Diseases. Elsevier, pp. 1467–1478.

Álvarez-Álvarez, R., Botas, A., Albillos, S.M., Rumbero, A., Martín, J.F., Liras, P., 2015. Molecular genetics of naringenin biosynthesis, a typical plant secondary metabolite produced by *Streptomyces clavuligerus*. Microb. Cell Fact. 14 (1), 1–12.

Amor, I.L.B., Hehn, A., Guedon, E., Ghedira, K., Engasser, J.M., Chekir-Ghedrira, L., et al., 2010. Biotransformation of naringenin to eriodictyol by *Saccharomyces cerevisiea* functionally expressing flavonoid 3′ hydroxylase. Nat. Prod. Commun. 5 (12), 1934578X1000501211.

Andersson, M.I., MacGowan, A.P., 2003. Development of the quinolones. J. Antimicrob. Chemother. 51 (Suppl. 1), 1–11.

Anjali, M., Sukumar, C., Kanakalakshmi, A., Shanthi, K., 2014. Enhancement of growth and production of polyhydroxyalkanoates by *Bacillus subtilis* from agro-industrial waste as carbon substrates. Compos. Interfaces 21 (2), 111–119.

Asker, D., Ohta, Y., 2002. Production of canthaxanthin by *Haloferax alexandrinus* under non-aseptic conditions and a simple, rapid method for its extraction. Appl. Microbiol. Biotechnol. 58 (6), 743–750.

Asker, D., Beppu, T., Ueda, K., 2007a. *Zeaxanthinibacter enoshimensis* gen. nov., sp. nov., a novel zeaxanthin-producing marine bacterium of the family Flavobacteriaceae, isolated from seawater off Enoshima Island, Japan. Int. J. Syst. Evol. Microbiol. 57 (4), 837–843.

Asker, D., Beppu, T., Ueda, K., 2007b. *Mesoflavibacter zeaxanthinifaciens* gen. nov., sp. nov., a novel zeaxanthin-producing marine bacterium of the family Flavobacteriaceae. Syst. Appl. Microbiol. 30 (4), 291–296.

Asker, D., Beppu, T., Ueda, K., 2008. *Nubsella zeaxanthinifaciens* gen. nov., sp. nov., a zeaxanthin-producing bacterium of the family Sphingobacteriaceae isolated from freshwater. Int. J. Syst. Evol. Microbiol. 58 (3), 601–606.

Augustine, R., Hasan, A., Patan, N.K., Dalvi, Y.B., Varghese, R., Antony, A., et al., 2019. Cerium oxide nanoparticle incorporated electrospun poly (3-hydroxybutyrate-co-3-hydroxyvalerate) membranes for diabetic wound healing applications. ACS Biomater. Sci. Eng. 6 (1), 58–70.

Bade, R., Chan, H.F., Reynisson, J., 2010. Characteristics of known drug space. Natural products, their derivatives and synthetic drugs. Eur. J. Med. Chem. 45 (12), 5646–5652.

Baltz, R.H., Brian, P., Miao, V., Wrigley, S.K., 2006. Combinatorial biosynthesis of lipopeptide antibiotics in *Streptomyces roseosporus*. J. Ind. Microbiol. Biotechnol. 33 (2), 66–74.

Ben-Amotz, A., Avron, M., 1983. On the factors which determine massive β-carotene accumulation in the halotolerant alga *Dunaliella bardawil*. Plant Physiol. 72 (3), 593–597.

Bera, S., 2019. Carotenoids: updates on legal statutory and competence for nutraceutical properties. Curr. Res. Nutr. Food Sci. 7 (2), 300.

Bera, S., Dutta, D., 2017. Encapsulation and release of a bacterial carotenoid from hydrogel matrix: characterization, kinetics and antioxidant study. Eng. Life Sci. 17 (7), 739–748.

Bera, S., Bharadwaj, V., Chaudhuri, S., Dutta, D., 2015. Strong antioxidant property of bacterial canthaxanthin obtained by raw coconut water supplementation as an additional nutrient source. In: Proceedings of 2015 International Conference on Bio-Medical Engineering and Environmental Technology, pp. 129–136.

Bera, S., Sharma, V.P., Dutta, S., Dutta, D., 2016. Biological decolorization and detoxification of malachite green from aqueous solution by *Dietzia maris* NIT-D. J. Taiwan Inst. Chem. Eng. 67, 271–284.

Bera, S., Chaudhuri, S., Dutta, D., 2017. Stability and antioxidant activity of bacterial canthaxanthin in aloe vera model system. Res. J. Pharm. Biol. Chem. Sci. 8 (4), 1169–1179.

Berry, A., Janssens, D., Hümbelin, M., Jore, J.P., Hoste, B., Cleenwerck, I., et al., 2003. *Paracoccus zeaxanthinifaciens* sp. nov., a zeaxanthin-producing bacterium. Int. J. Syst. Evol. Microbiol. 53 (1), 231–238.

Bhosale, P., Bernstein, P.S., 2004. β-Carotene production by *Flavobacterium multivorum* in the presence of inorganic salts and urea. J. Ind. Microbiol. Biotechnol. 31 (12), 565–571.

Binda, E., Marinelli, F., Marcone, G.L., 2014. Old and new glycopeptide antibiotics: action and resistance. Antibiotics 3 (4), 572–594.

Credence Research, 2018. Biologics Market By Product (Recombinant Therapeutic Proteins, Blood & Blood Components, Cellular and Gene Therapy, Vaccines), By Disease (Oncology, Autoimmune Diseases, Diabetes, Infectious Disease, Others) - Growth, Future Prospects, and Competitive Analysis, 2018–2026. Credence Research, Inc.

Bishayee, A., Sarkar, A., Chatterjee, M., 2000. Further evidence for chemopreventive potential of β-carotene against experimental carcinogenesis: diethylnitrosamine-initiated and phenobarbital-promoted hepatocarcinogenesis is prevented more effectively by β-carotene than by retinoic acid. Nutr. Cancer 37 (1), 89–98.

Bozorg, A., Vossoughi, M., Kazemi, A., Alemzadeh, I., 2015. Optimal medium composition to enhance poly-β-hydroxybutyrate production by *Ralstonia eutropha* using cane molasses as sole carbon source. Appl. Food Biotechnol. 2 (3), 39–47.

Brusselmans, K., Vrolix, R., Verhoeven, G., Swinnen, J.V., 2005. Induction of cancer cell apoptosis by flavonoids is associated with their ability to inhibit fatty acid synthase activity. J. Biol. Chem. 280 (7), 5636–5645.

Bush, K., Bradford, P.A., 2019. Interplay between β-lactamases and new β-lactamase inhibitors. Nat. Rev. Microbiol. 17 (5), 295.

Butler, M.S., Hansford, K.A., Blaskovich, M.A., Halai, R., Cooper, M.A., 2014. Glycopeptide antibiotics: back to the future. J. Antibiot. 67 (9), 631–644.

Cao, Y., Jiang, X., Zhang, R., Xian, M., 2011. Improved phloroglucinol production by metabolically engineered *Escherichia coli*. Appl. Microbiol. Biotechnol. 91 (6), 1545–1552.

Cao, W., Ma, W., Wang, X., Zhang, B., Cao, X., Chen, K., et al., 2016. Enhanced pinocembrin production in *Escherichia coli* by regulating cinnamic acid metabolism. Sci. Rep. 6, 32640.

Castillo, U., Harper, J.K., Strobel, G.A., Sears, J., Alesi, K., Ford, E., et al., 2003. Kakadumycins, novel antibiotics from Streptomyces sp. NRRL 30566, an endophyte of *Grevillea pteridifolia*. FEMS Microbiol. Lett. 224 (2), 183–190.

Chen, G.Q., 2009. A microbial polyhydroxyalkanoates (PHA) based bio-and materials industry. Chem. Soc. Rev. 38 (8), 2434–2446.

Chen, G.Q., 2010. Industrial production of PHA. Plastics From Bacteria. Springer, Berlin, Heidelberg, pp. 121–132.

Chen, W., Tong, Y.W., 2012. PHBV microspheres as neural tissue engineering scaffold support neuronal cell growth and axon–dendrite polarization. Acta Biomater. 8 (2), 540–548.

Chen, C., Fei, B., Peng, S., Zhuang, Y., Dong, L., Feng, Z., 2002a. Nonisothermal crystallization and melting behavior of poly (3-hydroxybutyrate) and maleated poly (3-hydroxybutyrate). Eur. Polym. J. 38 (8), 1663–1670.

Chen, Y., Yang, G., Chen, Q., 2002b. Solid-state NMR study on the structure and mobility of the noncrystalline region of poly (3-hydroxybutyrate) and poly (3-hydroxybutyrate-co-3-hydroxyvalerate). Polymer 43 (7), 2095–2099.

Chen, M.H., Zhou, X.Y., Ou, F.H., Xia, F., Lv, Y.Y., Qiu, L.H., 2017. *Aliidongia dinghuensis* gen. nov., sp. nov., a poly-β-hydroxybutyrate-producing bacterium isolated from *Pinus massoniana* forest soil. Int. J. Syst. Evol. Microbiol. 67 (2), 212–217.

Chen, Y., Chen, X.Y., Du, H.T., Zhang, X., Ma, Y.M., Chen, J.C., et al., 2019a. Chromosome engineering of the TCA cycle in *Halomonas bluephagenesis* for production of copolymers of 3-hydroxybutyrate and 3-hydroxyvalerate (PHBV). Metab. Eng. 54, 69–82.

Chen, J., Mitra, R., Zhang, S., Zuo, Z., Lin, L., Zhao, D., et al., 2019b. Unusual phosphoenolpyruvate (PEP) synthetase-like protein crucial to enhancement of polyhydroxyalkanoate accumulation in *Haloferax mediterranei* revealed by dissection of PEP-pyruvate interconversion mechanism. Appl. Environ. Microbiol. 85 (19), e00984-19.

Chiu, H.F., Liao, J.Y., Lu, Y.Y., Han, Y.C., Shen, Y.C., Venkatakrishnan, K., et al., 2017. Antiproliferative, anti-inflammatory and pro-apoptotic effects of *Dunaliella salina* on human KB oral carcinoma cells. J. Food Biochem. 41 (3), e12349.

Cho, H., Uehara, T., Bernhardt, T.G., 2014. Beta-lactam antibiotics induce a lethal malfunctioning of the bacterial cell wall synthesis machinery. Cell 159 (6), 1300–1311.

Choi, M.H., Yoon, S.C., Lenz, R.W., 1999. Production of poly (3-hydroxybutyric acid-co-4-hydroxybutyric acid) and poly (4-hydroxybutyric acid) without subsequent degradation by *Hydrogenophaga pseudoflava*. Appl. Environ. Microbiol. 65 (4), 1570–1577.

Clardy, J., Fischbach, M.A., Currie, C.R., 2009. The natural history of antibiotics. Curr. Biol. 19 (11), R437–R441.

Cooney, J.J., Marks, H.W., Smith, A.M., 1966. Isolation and identification of canthaxanthin from *Micrococcus roseus*. J. Bacteriol. 92 (2), 342–345.

Coulthurst, S.J., Barnard, A.M., Salmond, G.P., 2005. Regulation and biosynthesis of carbapenem antibiotics in bacteria. Nat. Rev. Microbiol. 3 (4), 295–306.

Das, S.K., Hashimoto, T., Shimizu, K., Yoshida, T., Sakai, T., Sowa, Y., et al., 2005. Fucoxanthin induces cell cycle arrest at G0/G1 phase in human colon carcinoma cells through up-regulation of p21WAF1/Cip1. Biochim. Biophys. Acta Gen. Subj. 1726 (3), 328–335.

Das, S.K., Hashimoto, T., Kanazawa, K., 2008. Growth inhibition of human hepatic carcinoma HepG2 cells by fucoxanthin is associated with down-regulation of cyclin D. Biochim. Biophys. Acta Gen. Subj. 1780 (4), 743–749.

Del Campo, J.A., Rodriguez, H., Moreno, J., Vargas, M.A., Rivas, J., Guerrero, M.G., 2004. Accumulation of astaxanthin and lutein in *Chlorella zofingiensis* (Chlorophyta). Appl. Microbiol. Biotechnol. 64 (6), 848–854.

Dowling, P.M., 2004. Pharmacologic Principles. Equine Internal Medicine, second ed. pp. 169–233.

Drawz, S.M., Bonomo, R.A., 2010. Three decades of β-lactamase inhibitors. Clin. Microbiol. Rev. 23 (1), 160–201.

Duan, L., Ding, W., Liu, X., Cheng, X., Cai, J., Hua, E., et al., 2017. Biosynthesis and engineering of kaempferol in *Saccharomyces cerevisiae*. Microb. Cell Fact. 16 (1), 1–10.

El-Banna, N., Winkelmann, G., 1998. Pyrrolnitrin from *Burkholderia cepacia*: antibiotic activity against fungi and novel activities against streptomycetes. J. Appl. Microbiol. 85 (1), 69–78.

Falcone Ferreyra, M.L., Rius, S., Casati, P., 2012. Flavonoids: biosynthesis, biological functions, and biotechnological applications. Front. Plant Sci. 3, 222.

Feher, J., Lengyel, G., 2012. Silymarin in the prevention and treatment of liver diseases and primary liver cancer. Curr. Pharm. Biotechnol. 13 (1), 210–217.

Fernández, D., Rodríguez, E., Bassas, M., Viñas, M., Solanas, A.M., Llorens, J., et al., 2005. Agro-industrial oily wastes as substrates for PHA production by the new strain *Pseudomonas aeruginosa* NCIB 40045: effect of culture conditions. Biochem. Eng. J. 26 (2-3), 159–167.

Ferre-Guell, A., Winterburn, J., 2018. Biosynthesis and characterization of polyhydroxyalkanoates with controlled composition and microstructure. Biomacromolecules 19 (3), 996–1005.

Fleming, A., 1944. The discovery of penicillin. Br. Med. Bull. 2 (1), 4–5.
Fu, X.Z., Tan, D., Aibaidula, G., Wu, Q., Chen, J.C., Chen, G.Q., 2014. Development of *Halomonas* TD01 as a host for open production of chemicals. Metab. Eng. 23, 78–91.
Gansukh, E., Mya, K.K., Jung, M., Keum, Y.S., Kim, D.H., Saini, R.K., 2019. Lutein derived from marigold (*Tagetes erecta*) petals triggers ROS generation and activates Bax and caspase-3 mediated apoptosis of human cervical carcinoma (HeLa) cells. Food Chem. Toxicol. 127, 11–18.
Gelpi, A., Gilbertson, A., Tucker, J.D., 2015. Magic bullet: Paul Ehrlich, Salvarsan and the birth of venereology. Sex. Transm. Infect. 91 (1), 68–69.
Gerth, K., Irschik, H., Reichenbach, H., Trowitzsch, W., 1980. Myxothiazol, an antibiotic from *Myxococcus fulvus* (Myxobacterales). J. Antibiot. 33 (12), 1474–1479.
Giovannucci, E., Ascherio, A., Rimm, E.B., Stampfer, M.J., Colditz, G.A., Willett, W.C., 1995. Intake of carotenoids and retino in relation to risk of prostate cancer. J. Natl. Cancer Inst. 87 (23), 1767–1776.
Grama, B.S., Chader, S., Khelifi, D., Agathos, S.N., Jeffryes, C., 2014. Induction of canthaxanthin production in a *Dactylococcus* microalga isolated from the Algerian Sahara. Bioresour. Technol. 151, 297–305.
Graumann, K., Premstaller, A., 2006. Manufacturing of recombinant therapeutic proteins in microbial systems. Biotechnol. J. Healthc. Nutr. Technol. 1 (2), 164–186.
Guo, S.A., DiPietro, L.A., 2010. Factors affecting wound healing. J. Dent. Res. 89 (3), 219–229.
Haas, L.F., 1999. Papyrus of Ebers and Smith. J. Neurol. Neurosurg. Psychiatry 67 (5), 578.
Hameed, A., Shahina, M., Lin, S.Y., Lai, W.A., Hsu, Y.H., Liu, Y.C., et al., 2014a. *Aquibacter zeaxanthinifaciens* gen. nov., sp. nov., a zeaxanthin-producing bacterium of the family Flavobacteriaceae isolated from surface seawater, and emended descriptions of the genera *Aestuariibaculum* and *Gaetbulibacter*. Int. J. Syst. Evol. Microbiol. 64 (1), 138–145.
Hameed, A., Shahina, M., Lin, S.Y., Liu, Y.C., Lai, W.A., Young, C.C., 2014b. *Gramellaoceani* sp. nov., a zeaxanthin-producing bacterium of the family Flavobacteriaceae isolated from marine sediment. Int. J. Syst. Evol. Microbiol. 64 (8), 2675–2681b.
Han, J., Wu, L.P., Hou, J., Zhao, D., Xiang, H., 2015. Biosynthesis, characterization, and hemostasis potential of tailor-made poly (3-hydroxybutyrate-co-3-hydroxyvalerate) produced by *Haloferax mediterranei*. Biomacromolecules 16 (2), 578–588.
Han, J., Wu, L.P., Liu, X.B., Hou, J., Zhao, L.L., Chen, J.Y., et al., 2017. Biodegradation and biocompatibility of haloarchaea-produced poly (3-hydroxybutyrate-co-3-hydroxyvalerate) copolymers. Biomaterials 139, 172–186.
Hannibal, L., Lorquin, J., D'Ortoli, N.A., Garcia, N., Chaintreuil, C., Masson-Boivin, C., et al., 2000. Isolation and characterization of canthaxanthin biosynthesis genes from the photosynthetic bacterium Bradyrhizobium sp. strain ORS278. J. Bacteriol. 182 (13), 3850–3853.
Heim, K.E., Tagliaferro, A.R., Bobilya, D.J., 2002. Flavonoid antioxidants: chemistry, metabolism and structure-activity relationships. J. Nutr. Biochem. 13 (10), 572–584.
Himabindu, M., Jetty, A., 2006. Optimization of nutritional requirements for gentamicin production by *Micromonospora echinospora*. Indian J. Exp. Biol. 44 (10), 842.
Hoerstrup, S.P., Sodian, R., Daebritz, S., Wang, J., Bacha, E.A., Martin, D.P., et al., 2000. Functional living trileaflet heart valves grown in vitro. Circulation 102 (Suppl. 3), Iii-44–Iii-49.
Hoerstrup, S.P., Zünd, G., Sodian, R., Schnell, A.M., Grünenfelder, J., Turina, M.I., 2001. Tissue engineering of small caliber vascular grafts. Eur. J. Cardio-Thoracic Surg. 20 (1), 164–169.

Hooper, D.C., 1999. Mode of action of fluoroquinolones. Drugs 58 (2), 6–10.

Hosokawa, M., Kudo, M., Maeda, H., Kohno, H., Tanaka, T., Miyashita, K., 2004. Fucoxanthin induces apoptosis and enhances the antiproliferative effect of the PPARγ ligand, troglitazone, on colon cancer cells. Biochim. Biophys. Acta Gen. Subj. 1675 (1), 113–119.

Hou, J., Han, J., Cai, L., Zhou, J., Lü, Y., Jin, C., et al., 2014. Characterization of genes for chitin catabolism in *Haloferax mediterranei*. Appl. Microbiol. Biotechnol. 98 (3), 1185–1194.

Howard, D., Buttery, L.D., Shakesheff, K.M., Roberts, S.J., 2008. Tissue engineering: strategies, stem cells and scaffolds. J. Anat. 213 (1), 66–72.

Hsieh, L.K., Lee, T.C., Chichester, C.O., Simpson, K.L., 1974. Biosynthesis of carotenoids in *Brevibacterium* sp. KY-4313. J. Bacteriol. 118, 385–393.

Huang, W., Shi, X., Ren, L., Du, C., Wang, Y., 2010. PHBV microspheres–PLGA matrix composite scaffold for bone tissue engineering. Biomaterials. 31 (15), 4278–4285.

Hwang, E.I., Kaneko, M., Ohnishi, Y., Horinouchi, S., 2003. Production of plant-specific flavanones by *Escherichia coli* containing an artificial gene cluster. Appl. Environ. Microbiol. 69 (5), 2699–2706.

Igarashi, M., Takahashi, Y., Shitara, T., Nakamura, H., Naganawa, H., Miyake, T., et al., 2005. Caprazamycins, novel lipo-nucleoside antibiotics, from *Streptomyces* sp. J. Antibiot. 58 (5), 327–337.

Ip, P.F., Chen, F., 2005. Production of astaxanthin by the green microalga *Chlorella zofingiensis* in the dark. Process. Biochem. 40 (2), 733–738.

Iskandar, A.R., Miao, B., Li, X., Hu, K.Q., Liu, C., Wang, X.D., 2016. β-Cryptoxanthin reduced lung tumor multiplicity and inhibited lung cancer cell motility by downregulating nicotinic acetylcholine receptor α7 signaling. Cancer Prev. Res. 9 (11), 875–886.

Jackson, M., Frost, D.J., Karwowski, J.P., Humphrey, P.E., Dahod, S.K., Choi, W.S., et al., 1995. Fusacandins A and B; novel antifungal antibiotics of the papulacandin class from *Fusarium sambucinum*. J. Antibiot. 48 (7), 608–613.

James, K.D., 2017. Animal metabolites: from amphibians, reptiles, aves/birds, and invertebrates. Pharmacognosy. Academic Press, pp. 401–411.

Jayappriyan, K.R., Rajkumar, R., Venkatakrishnan, V., Nagaraj, S., Rengasamy, R., 2013. In vitro anticancer activity of natural β-carotene from *Dunaliella salina* EU5891199 in PC-3 cells. Biomed. Prevent. Nutr. 3 (2), 99–105.

Jiang, H., Wood, K.V., Morgan, J.A., 2005. Metabolic engineering of the phenylpropanoid pathway in *Saccharomyces cerevisiae*. Appl. Environ. Microbiol. 71 (6), 2962–2969.

Johnson, E.A., Villa, T.G., Lewis, M.J., 1980. *Phaffia rhodozyma* as an astaxanthin source in salmonid diets. Aquaculture 20 (2), 123–134.

Kahar, P., Tsuge, T., Taguchi, K., Doi, Y., 2004. High yield production of polyhydroxyalkanoates from soybean oil by *Ralstonia eutropha* and its recombinant strain. Polym. Degrad. Stab. 83 (1), 79–86.

Kajimura, Y., Sugiyama, M., Kaneda, M., 1995. Bacillopeptins, new cyclic lipopeptide antibiotics from *Bacillus subtilis* FR-2. J. Antibiot. 48 (10), 1095–1103.

Kaneda, M., Kajimura, Y., 2002. New antifungal antibiotics, bacillopeptins and fusaricidins. Yakugaku zasshi: J. Pharm. Soc. Jpn. 122 (9), 651–671.

Kapoor, G., Saigal, S., Elongavan, A., 2017. Action and resistance mechanisms of antibiotics: a guide for clinicians. J. Anaesthesiol. Clin. Pharmacol. 33 (3), 300.

Khan, M.T.H., Orhan, I., Şenol, F.S., Kartal, M., Şener, B., Dvorská, M., et al., 2009. Cholinesterase inhibitory activities of some flavonoid derivatives and chosen xanthone and their molecular docking studies. Chem. Biol. Interact. 181 (3), 383–389.

Khodaiyan, F., Razavi, S.H., Emam-Djomeh, Z., Mousavi, S.M.A., Hejazi, M.A., 2007. Effect of culture conditions on canthaxanthin production by *Dietzia natronolimnaea* HS-1. J. Microbiol. Biotechnol. 17 (2), 195.

Kobayashi, M., Kakizono, T., Nagai, S., 1991. Astaxanthin production by a green alga, *Haematococcus pluvialis* accompanied with morphological changes in acetate media. J. Ferment. Bioeng. 71 (5), 335–339.

Koller, M., Hesse, P., Bona, R., Kutschera, C., Atlić, A., Braunegg, G., 2007. Biosynthesis of high quality polyhydroxyalkanoate co-and terpolyesters for potential medical application by the archaeon *Haloferax mediterranei*, Macromol. Symp., 253. WILEY-VCH Verlag, Weinheim, pp. 33–39.

Koopman, F., Beekwilder, J., Crimi, B., van Houwelingen, A., Hall, R.D., Bosch, D., et al., 2012. De novo production of the flavonoid naringenin in engineered *Saccharomyces cerevisiae*. Microb. Cell Fact. 11 (1), 155.

Kozłowska, A., Szostak-Węgierek, D., 2017. Flavonoids–food sources, health benefits, and mechanisms involved. In: Mérillon, J.-M., Ramawat, K.G. (Eds.), Bioactive Molecules in Food. Springer, Cham, pp. 1–27.

Krause, K.M., Serio, A.W., Kane, T.R., Connolly, L.E., 2016. Aminoglycosides: an overview. Cold Spring Harb. Perspect. Med. 6 (6), a027029.

Kulkarni, S.O., Kanekar, P.P., Jog, J.P., Sarnaik, S.S., Nilegaonkar, S.S., 2015. Production of copolymer, poly (hydroxybutyrate-co-hydroxyvalerate) by *Halomonas campisalis* MCM B-1027 using agro-wastes. Int. J. Biol. Macromol. 72, 784–789.

Kumar, S., Pandey, A.K., 2013. Chemistry and biological activities of flavonoids: an overview. Sci. World J. 2013.

Kunze, B., Kemmer, T., Höfle, G., Reichenbach, H., 1984. Stigmatellin, a new antibiotic from *Stigmatella aurantiaca* (myxobacterales). J. Antibiot. 37 (5), 454–461.

Lebreton, F., Cattoir, V., 2019. Resistance to glycopeptide antibiotics. Bacterial Resistance to Antibiotics: From Molecules to Manpp. pp. 51–80.

Lee, J.H., Hwang, Y.M., Baik, K.S., Choi, K.S., Ka, J.O., Seong, C.N., 2014. *Mesoflavibacter aestuarii* sp. nov., a zeaxanthin-producing marine bacterium isolated from seawater. Int. J. Syst. Evol. Microbiol. 64 (6), 1932–1937.

Lee, H., Kim, B.G., Kim, M., Ahn, J.H., 2015. Biosynthesis of two flavones, apigenin and genkwanin in *Escherichia coli*. J. Microbiol. Biotechnol. 25 (9), 1442–1448.

Leonard, E., Yan, Y., Lim, K.H., Koffas, M.A., 2005. Investigation of two distinct flavone synthases for plant-specific flavone biosynthesis in *Saccharomyces cerevisiae*. Appl. Environ. Microbiol. 71 (12), 8241–8248.

Leonard, E., Yan, Y., Koffas, M.A., 2006. Functional expression of a P450 flavonoid hydroxylase for the biosynthesis of plant-specific hydroxylated flavonols in *Escherichia coli*. Metab. Eng. 8 (2), 172–181.

Leonard, E., Lim, K.H., Saw, P.N., Koffas, M.A., 2007. Engineering central metabolic pathways for high-level flavonoid production in *Escherichia coli*. Appl. Environ. Microbiol. 73 (12), 3877–3886.

Leonard, E., Yan, Y., Fowler, Z.L., Li, Z., Lim, C.G., Lim, K.H., et al., 2008. Strain improvement of recombinant *Escherichia coli* for efficient production of plant flavonoids. Mol. Biopharm. 5 (2), 257–265.

Li, H.B., Fan, K.W., Chen, F., 2006. Isolation and purification of canthaxanthin from the microalga *Chlorella zofingiensis* by high-speed counter-current chromatography. J. Sep. Sci. 29 (5), 699–703.

Li, W., Ding, Y., Rai, R., Roether, J.A., Schubert, D.W., Boccaccini, A.R., 2014. Preparation and characterization of PHBV microsphere/45S5 bioactive glass composite scaffolds with vancomycin releasing function. Mater. Sci. Eng. C 41, 320–328.

Liu, J., Sun, Z., Gerken, H., Liu, Z., Jiang, Y., Chen, F., 2014. *Chlorella zofingiensis* as an alternative microalgal producer of astaxanthin: biology and industrial potential. Mar. Drugs 12 (6), 3487–3515.

Lyu, X., Ng, K.R., Lee, J.L., Mark, R., Chen, W.N., 2017. Enhancement of naringenin biosynthesis from tyrosine by metabolic engineering of *Saccharomyces cerevisiae*. J. Agric. Food Chem. 65 (31), 6638–6646.

Maddison, J.E., Watson, A.D., Elliott, J., 2008. Antibacterial drugs. Small Anim. Clin. Pharmacol. 2, 148–168.

Malla, S., Koffas, M.A., Kazlauskas, R.J., Kim, B.G., 2012. Production of 7-O-methyl aromadendrin, a medicinally valuable flavonoid, in *Escherichia coli*. Appl. Environ. Microbiol. 78 (3), 684–694.

Marín, L., Gutiérrez-del-Río, I., Yagüe, P., Manteca, Á., Villar, C.J., Lombó, F., 2017. De novo biosynthesis of apigenin, luteolin, and eriodictyol in the actinomycete *Streptomyces albus* and production improvement by feeding and spore conditioning. Front. Microbiol. 8, 921.

Martin, D.P., Williams, S.F., 2003. Medical applications of poly-4-hydroxybutyrate: a strong flexible absorbable biomaterial. Biochem. Eng. J. 16 (2), 97–105.

Martin, D.P., Badhwar, A., Shah, D.V., Rizk, S., Eldridge, S.N., Gagne, D.H., et al., 2013. Characterization of poly-4-hydroxybutyrate mesh for hernia repair applications. J. Surg. Res. 184 (2), 766–773.

Mei, C., Zhou, S., Zhu, L., Ming, J., Zeng, F., Xu, R., 2017. Antitumor effects of Laminaria extract fucoxanthin on lung cancer. Mar. Drugs 15 (2), 39.

Mingeot-Leclercq, M.P., Glupczynski, Y., Tulkens, P.M., 1999. Aminoglycosides: activity and resistance. Antimicrob. Agents Chemother. 43 (4), 727–737.

Mitra, R., Dutta, D., 2018. Growth profiling, kinetics and substrate utilization of low-cost dairy waste for production of β-cryptoxanthin by *Kocuria marina* DAGII. R. Soc. Open Sci. 5 (7), 172318.

Mitra, R., Samanta, A.K., Chaudhuri, S., Dutta, D., 2015. Growth kinetics and carotenoid production of a newly isolated bacterium on glucose as a carbon source during batch fermentation. In: Int. Conf. on Advances in Biotechnology (BioTech) Proceedings Singapore: Global Science and Technology Forum.

Mitra, R., Samanta, A.K., Chaudhuri, S., Dutta, D., 2016. Impact of carbon source on β-cryptoxanthin production by *Kocuria marina* DAGII: a classical approach. Mater. Today Proc. 3 (10), 3269–3275.

Mitra, R., Samanta, A.K., Chaudhuri, S., Dutta, D., 2017a. Effect of selected physico-chemical factors on bacterial β-cryptoxanthin degradation: stability and kinetic study. J. Food Process. Eng. 40 (2), e12379.

Mitra, R., Chaudhuri, S., Dutta, D., 2017b. Modelling the growth kinetics of *Kocuria marina* DAGII as a function of single and binary substrate during batch production of β-Cryptoxanthin. Bioprocess. Biosyst. Eng. 40 (1), 99–113.

Mitra, R., Xu, T., Xiang, H., Han, J., 2020. Current developments on polyhydroxyalkanoates synthesis by using halophiles as a promising cell factory. Microb. Cell Fact. 19, 1–30.

Miyahisa, I., Kaneko, M., Funa, N., Kawasaki, H., Kojima, H., Ohnishi, Y., et al., 2005. Efficient production of (2S)-flavanones by *Escherichia coli* containing an artificial biosynthetic gene cluster. Appl. Microbiol. Biotechnol. 68 (4), 498–504.

Miyahisa, I., Funa, N., Ohnishi, Y., Martens, S., Moriguchi, T., Horinouchi, S., 2006. Combinatorial biosynthesis of flavones and flavonols in *Escherichia coli*. Appl. Microbiol. Biotechnol. 71 (1), 53–58.

Mizuno, K., Ohta, A., Hyakutake, M., Ichinomiya, Y., Tsuge, T., 2010. Isolation of polyhydroxyalkanoate-producing bacteria from a polluted soil and characterization of the isolated strain Bacillus cereus YB-4. Polym. Degrad. Stab. 95 (8), 1335–1339.

Nagaraj, S., Rajaram, M.G., Arulmurugan, P., Baskaraboopathy, A., Karuppasamy, K., Jayappriyan, K.R., et al., 2012. Antiproliferative potential of astaxanthin-rich alga *Haematococcus pluvialis* Flotow on human hepatic cancer (HepG2) cell line. Biomed. Prevent. Nutr. 2 (3), 149–153.

Nelis, H.J., De Leenheer, A.P., 1989. Reinvestigation of *Brevibacterium* sp. strain KY-4313 as a source of canthaxanthin. Appl. Environ. Microbiol. 55 (10), 2505–2510.

Obruca, S., Marova, I., Snajdar, O., Mravcova, L., Svoboda, Z., 2010. Production of poly (3-hydroxybutyrate-co-3-hydroxyvalerate) by *Cupriavidus necator* from waste rapeseed oil using propanol as a precursor of 3-hydroxyvalerate. Biotechnol. Lett. 32 (12), 1925–1932.

Oliveira-Junior, R.G., Thiéry, V., Sergent, O., Chevanne, M., Picot, L., 2016. Could fucoxanthin interaction with lipid rafts mediate its cytotoxicity in cancer cells? J. Oceanogr. Mar. Res. 4 (1), 1000144.

Olmos, J., Gómez, R., Rubio, V., 2015. Apoptosis comparison effects between synthetic and natural β-carotene from *Dunaliella salina* on MDA-MB-231 breast cancer cells. J. Microb. Biochem. Technol. 7, 051–056.

Palozza, P., Torelli, C., Boninsegna, A., Simone, R., Catalano, A., Mele, M.C., et al., 2009. Growth-inhibitory effects of the astaxanthin-rich alga *Haematococcus pluvialis* in human colon cancer cells. Cancer Lett. 283 (1), 108–117.

Panche, A.N., Diwan, A.D., Chandra, S.R., 2016. Flavonoids: an overview. J. Nutr. Sci. 5.

Pandey, N., Cascella, M., 2019. Beta lactam antibiotics. StatPearls [Internet]. StatPearls Publishing.

Pandey, R.P., Parajuli, P., Koffas, M.A., Sohng, J.K., 2016. Microbial production of natural and non-natural flavonoids: pathway engineering, directed evolution and systems/synthetic biology. Biotechnol. Adv. 34 (5), 634–662.

Papp, T., Csernetics, Á., Nagy, G., Bencsik, O., Iturriaga, E.A., Eslava, A.P., et al., 2013. Canthaxanthin production with modified *Mucor circinelloides* strains. Appl. Microbiol. Biotechnol. 97 (11), 4937–4950.

Park, S.R., Yoon, J.A., Paik, J.H., Park, J.W., Jung, W.S., Ban, Y.H., et al., 2009. Engineering of plant-specific phenylpropanoids biosynthesis in *Streptomyces venezuelae*. J. Biotechnol. 141 (3-4), 181–188.

Park, S.R., Ahn, M.S., Han, A.R., Park, J.W., Yoon, Y.J., 2011. Enhanced flavonoid production in *Streptomyces venezuelae* via metabolic engineering. J. Microbiol. Biotechnol. 21 (11), 1143–1146.

Park, S.R., Yoon, Y.J., Pham, J.V., Yilma, M.A., Feliz, A., Majid, M.T., et al., 2019. A review of the microbial production of bioactive natural products and biologics. Front. Microbiol. 10, 1404.

Paulus, H., Gray, E., 1964. The biosynthesis of polymyxin B by growing cultures of *Bacillus polymyxa*. J. Biol. Chem. 239, 865–871.

Pham, T.D., Ziora, Z.M., Blaskovich, M.A., 2019. Quinolone antibiotics. Med. Chem. Comm. 10 (10), 1719–1739.

Pillai, A.B., Kumar, A.J., Thulasi, K., Kumarapillai, H., 2017. Evaluation of short-chain-length polyhydroxyalkanoate accumulation in *Bacillus aryabhattai*. Braz. J. Microbiol. 48 (3), 451–460.

Raja, R., Hemaiswarya, S., Balasubramanyam, D., Rengasamy, R., 2007. Protective effect of *Dunaliella salina* (Volvocales, Chlorophyta) against experimentally induced fibrosarcoma on wistar rats. Microbiol. Res. 162 (2), 177–184.

Ramaprasad, E.V.V., Sasikala, C., Ramana, C.V., 2013. Neurosporene is the major carotenoid accumulated by *Rhodobacter viridis* JA737. Biotechnol. Lett. 35 (7), 1093–1097.

Ray, S., Prajapati, V., Patel, K., Trivedi, U., 2016. Optimization and characterization of PHA from isolate *Pannonibacter phragmitetus* ERC8 using glycerol waste. Int. J. Biol. Macromol. 86, 741–749.

Raza, Z.A., Abid, S., Banat, I.M., 2018. Polyhydroxyalkanoates: characteristics, production, recent developments and applications. Int. Biodeterior. Biodegrad. 126, 45–56.

Rossolini, G.M., Arena, F., Giani, T., 2017. Mechanisms of antibacterial resistance. Infect. Dis. 2, 1181–1196 (Fourth Edition).

Saito, Y., Doi, Y., 1994. Microbial synthesis and properties of poly (3-hydroxybutyrate-co-4-hydroxybutyrate) in *Comamonas acidovorans*. Int. J. Biol. Macromol. 16 (2), 99–104.

Sakai, K., Suga, T., Iwatsuki, M., Chinen, T., Nonaka, K., Usui, T., et al., 2018. Pestiocandin, a new papulacandin class antibiotic isolated from *Pestalotiopsis humus*. J. Antibiot. 71 (12), 1031–1035.

Saperstein, S., Starr, M.P., 1954. The ketonic carotenoid canthaxanthin isolated from a colour mutant of *Corynebacterium michiganense*. Biochem. J. 57 (2), 273–275.

Sawant, S.S., Salunke, B.K., Kim, B.S., 2015. Degradation of corn stover by fungal cellulase cocktail for production of polyhydroxyalkanoates by moderate halophile *Paracoccus* sp. LL1. Bioresour. Technol. 194, 247–255.

Serrato-Joya, O., Jiménez-Islas, H., Botello-Álvarez, E., Rico-Martínez, R., Navarrete-Bolaños, J.L., 2006. Production of β-cryptoxanthin, a provitamin-A precursor, by *Flavobacterium lutescens*. J. Food Sci. 71 (6), E314–E319.

Shahina, M., Hameed, A., Lin, S.Y., Lee, R.J., Lee, M.R., Young, C.C., 2014. *Gramella planctonica* sp. nov., a zeaxanthin-producing bacterium isolated from surface seawater, and emended descriptions of *Gramella aestuarii* and *Gramella echinicola*. Antonie van leeuwenhoek 105 (4), 771–779.

Shaw, W.V., 1983. Chloramphenicol acetyltransferase: enzymology and molecular biology. Crit. Rev. Biochem. 14 (1), 1–46.

Sheu, M.J., Huang, G.J., Wu, C.H., Chen, J.S., Chang, H.Y., Chang, S.J., et al., 2008. Ethanol extract of *Dunaliella salina* induces cell cycle arrest and apoptosis in A549 human non-small cell lung cancer cells. In Vivo 22 (3), 369–378.

Shi, X.M., Liu, H.J., Zhang, X.W., Chen, F., 1999. Production of biomass and lutein by *Chlorella protothecoides* at various glucose concentrations in heterotrophic cultures. Process. Biochem. 34 (4), 341–347.

Singh, M., Kumar, P., Patel, S.K., Kalia, V.C., 2013. Production of polyhydroxyalkanoate co-polymer by *Bacillus thuringiensis*. Indian J. Microbiol. 53 (1), 77–83.

Singh, P., Baranwal, M., Reddy, S.M., 2016. Antioxidant and cytotoxic activity of carotenes produced by *Dunaliella salina under stress*. Pharm. Microbiol. 54 (10), 2269–2275.

Singh, R., Kumar, M., Mittal, A., Mehta, P.K., 2017. Microbial metabolites in nutrition, healthcare and agriculture. 3 Biotech 7 (1), 15.

Sirés, I., Brillas, E., 2012. Remediation of water pollution caused by pharmaceutical residues based on electrochemical separation and degradation technologies: a review. Environ. Int. 40, 212–229.

Srinivasan, R., Chaitanyakumar, A., Mageswari, A., Gomathi, A., Kumar, J.P., Jayasindu, M., et al., 2017. Oral administration of lyophilized *Dunaliella salina*, a carotenoid-rich marine alga, reduces tumor progression in mammary cancer induced rats. Food Funct. 8 (12), 4517–4527.

Stahlhut, S.G., Siedler, S., Malla, S., Harrison, S.J., Maury, J., Neves, A.R., et al., 2015. Assembly of a novel biosynthetic pathway for production of the plant flavonoid fisetin in *Escherichia coli*. Metab. Eng. 31, 84–93.

Stock, U.A., Sakamoto, T., Hatsuoka, S., Martin, D.P., Nagashima, M., Moran, A.M., et al., 2000. Patch augmentation of the pulmonary artery with bioabsorbable polymers and autologous cell seeding. J. Thorac. Cardiovasc. Surg. 120 (6), 1158–1167.

Sundaramurthi, D., Krishnan, U.M., Sethuraman, S., 2013. Biocompatibility of poly (3-hydroxybutyrate-co-3-hydroxyvalerate)(PHBV) nanofibers for skin tissue engineering. J. Biomed. Nanotechnol. 9 (8), 1383–1392.

Swanson, H.M., Smith Jr, J.R., Gong, X., Rubin, L.P., 2016. Lutein, but not other carotenoids, selectively inhibits breast cancer cell growth through several molecular mechanisms. FASEB J. 30 (1 Suppl.), 34-2.

Takahashi, K., Hosokawa, M., Kasajima, H., Hatanaka, K., Kudo, K., Shimoyama, N., et al., 2015. Anticancer effects of fucoxanthin and fucoxanthinol on colorectal cancer cell lines and colorectal cancer tissues. Oncol. Lett. 10 (3), 1463–1467.

Tan, D., Xue, Y.S., Aibaidula, G., Chen, G.Q., 2011. Unsterile and continuous production of polyhydroxybutyrate by *Halomonas* TD01. Bioresour. Technol. 102 (17), 8130–8136.

Teng, Y.N., Sheu, M.J., Hsieh, Y.W., Wang, R.Y., Chiang, Y.C., Hung, C.C., 2016. β-Carotene reverses multidrug resistant cancer cells by selectively modulating human P-glycoprotein function. Phytomedicine 23 (3), 316–323.

Thibodeaux, C.J., Liu, H.W., Thorson, J.S., 2007. Complementary routes to natural product glycodiversification: pathway engineering and glycorandomization, Comprehensive Glycoscience, From Chemistry to Systems Biology, 3. pp. 373–396.

Thuan, N.H., Chaudhary, A.K., Van Cuong, D., Cuong, N.X., 2018. Engineering co-culture system for production of apigetrin in *Escherichia coli*. J. Ind. Microbiol. Biotechnol. 45 (3), 175–185.

Utsunomia, C., Ren, Q., Zinn, M., 2020. Poly (4-hydroxybutyrate): current state and perspectives. Front. Bioeng. Biotech. 8, 257.

Valappil, S.P., Peiris, D., Langley, G.J., Herniman, J.M., Boccaccini, A.R., Bucke, C., et al., 2007. Polyhydroxyalkanoate (PHA) biosynthesis from structurally unrelated carbon sources by a newly characterized *Bacillus* spp. J. Biotechnol. 127 (3), 475–487.

Vanittanakom, N., Loeffler, W., Koch, U., Jung, G., 1986. Fengycin-a novel antifungal lipopeptide antibiotic produced by *Bacillus subtilis* F-29-3. J. Antibiot. 39 (7), 888–901.

Veiga-Crespo, P., Blasco, L., Rosa dos Santos, F., Poza, M., Villa, T.G., 2005. Influence of culture conditions of Gordonia jacobaea MV-26 on canthaxanthin production. Int. Microbiol. 8 (1), 55–58.

Vineis, P., Wild, C.P., 2014. Global cancer patterns: causes and prevention. Lancet 383 (9916), 549–557.

Wang, S.L., Chen, G.T., Qi, G.H., Yang, Z.P., Li, B., 2008. Formulation of culture medium for superoxide dismutase production by Rhodotorula glutinis RY-06 strain with high yield of beta-carotene. 2008 2nd International Conference on Bioinformatics and Biomedical Engineering. IEEE, pp. 959–961.

Wang, Y.X., Liu, J.H., Zhang, X.X., Chen, Y.G., Wang, Z.G., Chen, Y., et al., 2009. *Fodinicurvata sediminis* gen. nov., sp. nov. and *Fodinicurvata fenggangensis* sp. nov., poly-β-hydroxybutyrate-producing bacteria in the family Rhodospirillaceae. Int. J. Syst. Evol. Microbiol. 59 (10), 2575–2581.

Wang, G.S., Grammel, H., Abou-Aisha, K., Sägesser, R., Ghosh, R., 2012. High-level production of the industrial product lycopene by the photosynthetic bacterium *Rhodospirillum rubrum*. Appl. Environ. Microbiol. 78 (20), 7205–7215.

Wang, S.K., Li, Y., White, W.L., Lu, J., 2014. Extracts from New Zealand *Undaria pinnatifida* containing fucoxanthin as potential functional biomaterials against cancer in vitro. J. Funct. Biomater. 5 (2), 29–42.

Wanger, A., Chavez, V., Huang, R.S.P., Wahed, A., Actor, J.K., Dasgupta, A., 2017. Antibiotics, antimicrobial resistance, antibiotic susceptibility testing, and therapeutic drug monitoring for selected drugs. Microbiol. Mol. Diagn. Pathol. 119–153. 10.1016.B978-0-12-805351-5.00007-7.

Watts, K.T., Lee, P.C., Schmidt-Dannert, C., 2004. Exploring recombinant flavonoid biosynthesis in metabolically engineered *Escherichia coli*. Chembiochem. 5 (4), 500–507.

Weber, B., Schoenauer, R., Papadopulos, F., Modregger, P., Peter, S., Stampanoni, M., et al., 2011. Engineering of living autologous human umbilical cord cell-based septal occluder membranes using composite PGA-P4HB matrices. Biomaterials 32 (36), 9630–9641.

Whitehouse, C.A., Zhao, S., Tate, H., 2018. Antimicrobial resistance in Campylobacter species: mechanisms and genomic epidemiology, Advances in Applied Microbiology, 103. Academic Press, pp. 1–47.

WHO, 2008. https://www.who.int/news-room/fact-sheets/detail/cancer.

Williams, S.F., Martin, D.P., Moses, A.C., 2016. The history of GalaFLEX P4HB scaffold. Aesthet. Surg. J. 36 (Suppl. 2), S33–S42.

Wu, C., Han, L., Riaz, H., Wang, S., Cai, K., Yang, L., 2013. The chemopreventive effect of β-cryptoxanthin from mandarin on human stomach cells (BGC-823). Food Chem. 136 (3), 1122–1129.

Wu, J., Yu, O., Du, G., Zhou, J., Chen, J., 2014. Fine-tuning of the fatty acid pathway by synthetic antisense RNA for enhanced (2S)-naringenin production from L-tyrosine in *Escherichia coli*. Appl. Environ. Microb. 80 (23), 7283–7292.

Xaplanteri, M.A., Andreou, A., Dinos, G.P., Kalpaxis, D.L., 2003. Effect of polyamines on the inhibition of peptidyltransferase by antibiotics: revisiting the mechanism of chloramphenicol action. Nucleic Acids Res. 31 (17), 5074–5083.

Xing, Z.C., Chae, W.P., Baek, J.Y., Choi, M.J., Jung, Y., Kang, I.K., 2010. In vitro assessment of antibacterial activity and cytocompatibility of silver-containing PHBV nanofibrous scaffolds for tissue engineering. Biomacromolecules 11 (5), 1248–1253.

Xu, Y., Willems, A., Au-yeung, C., Tahlan, K., Nodwell, J.R., 2012. A two-step mechanism for the activation of actinorhodin export and resistance in *Streptomyces coelicolor*. MBio 3 (5).

Xue, Q., Liu, X.B., Lao, Y.H., Wu, L.P., Wang, D., Zuo, Z.Q., et al., 2018. Anti-infective biomaterials with surface-decorated tachyplesin I. Biomaterials 178, 351–362.

Ye, J., Hu, D., Yin, J., Huang, W., Xiang, R., Zhang, L., et al., 2020. Stimulus response-based fine-tuning of polyhydroxyalkanoate pathway in *Halomonas*. Metab. Eng. 57, 85–95.

Yu, R.X., Hu, X.M., Xu, S.Q., Jiang, Z.J., Yang, W., 2011. Effects of fucoxanthin on proliferation and apoptosis in human gastric adenocarcinoma MGC-803 cells via JAK/STAT signal pathway. Eur. J. Pharmacol. 657 (1), 10–19.

Zakaria, M.R., Tabatabaei, M., Ghazali, F.M., Abd-Aziz, S., Shirai, Y., Hassan, M.A., 2010. Polyhydroxyalkanoate production from anaerobically treated palm oil mill effluent by new bacterial strain *Comamonas* sp. EB172. World J. Microbiol. Biotechnol. 26 (5), 767–774.

Zeng, D., Debabov, D., Hartsell, T.L., Cano, R.J., Adams, S., Schuyler, J.A., et al., 2016. Approved glycopeptide antibacterial drugs: mechanism of action and resistance. Cold Spring Harb. Perspect. Med. 6 (12), a026989.

Zhang, H., Tang, Y., Zhang, Y., Zhang, S., Qu, J., Wang, X., et al., 2015. Fucoxanthin: a promising medicinal and nutritional ingredient. Evid. Based Complement. Alternat. Med. 2015.

Zhang, W.L., Zhao, Y.N., Shi, Z.Z., Cong, D., Bai, Y.S., 2018. Lutein inhibits cell growth and activates apoptosis via the PI3K/AKT/mTOR signaling pathway in A549 human non-small-cell lung cancer cells. J. Environ. Pathol. Toxicol. Oncol. 37 (4).

Zhao, S., Jones, J.A., Lachance, D.M., Bhan, N., Khalidi, O., Venkataraman, S., et al., 2015a. Improvement of catechin production in *Escherichia coli* through combinatorial metabolic engineering. Metab. Eng. 28, 43–53.

Zhao, Y.X., Rao, Z.M., Xue, Y.F., Gong, P., Ji, Y.Z., Ma, Y.H., 2015b. Poly (3-hydroxybutyrate-co-3-hydroxyvalerate) production by Haloarchaeon *Halogranum amylolyticum*. Appl. Microbiol. Biotechnol. 99 (18), 7639–7649.

Zhu, Y.H., Jiang, J.G., 2008. Continuous cultivation of *Dunaliella salina* in photobioreactor for the production of β-carotene. Eur. Food Res. Technol. 227 (3), 953–959.

Zhu, X.H., Wang, C.H., Tong, Y.W., 2007. Growing tissue-like constructs with Hep3B/HepG2 liver cells on PHBV microspheres of different sizes. J. Biomed. Mater. Res. Part B 82 (1), 7–16.

Zhu, X.H., Wang, C.H., Tong, Y.W., 2009. In vitro characterization of hepatocyte growth factor release from PHBV/PLGA microsphere scaffold. J. Biomed. Mater. Res. Part A 89 (2), 411–423.

Zhu, S., Wu, J., Du, G., Zhou, J., Chen, J., 2014. Efficient synthesis of eriodictyol from L-tyrosine in *Escherichia coli*. Appl. Environ. Microbiol. 80 (10), 3072–3080.

Ziegler, R.G., 1991. Vegetables, fruits, and carotenoids and the risk of cancer. Am. J. Clin. Nutr. 53 (1), 251S–259S.

Zonari, A., Martins, T.M., Paula, A.C.C., Boeloni, J.N., Novikoff, S., Marques, A.P., et al., 2015. Polyhydroxybutyrate-co-hydroxyvalerate structures loaded with adipose stem cells promote skin healing with reduced scarring. Acta Biomater. 17, 170–181.

Zuo, Z.Q., Xue, Q., Zhou, J., Zhao, D.H., Han, J., Xiang, H., 2018. Engineering *Haloferax mediterranei* as an efficient platform for high level production of lycopene. Front. Microbiol. 9, 2893.

Chapter 18

Role of fungal metabolites as biopesticides: an emerging trend in sustainable agriculture

Tuyelee Das[1], Champa Keya Tudu[1], Samapika Nandy[1], Devendra Kumar Pandey[2] and Abhijit Dey[1]

[1]*Department of Life Sciences, Presidency University, Kolkata, India,* [2]*Department of Biotechnology, Lovely Faculty of Technology and Sciences, Lovely Professional University, Jalandhar, India*

Chapter Outline

18.1 Introduction

18.1.1 Background and definitions of pesticides, biopesticides, and fungal metabolites

Plant pests are the living organisms which hamper the growth and yield of the plant that includes insects, weeds, gastropod mollusks, rodents, etc. (Agrios, 2003). To combat the pest attack and to prevent crop loss, farmers have relied upon pesticides for a long time. Pesticides are chemical compounds, which are used to control, discourage, and prevent pests in the agricultural field that actually works upon different stages of growth and development of pests. However, the negative impacts of chemical pesticides

Volatiles and Metabolites of Microbes. DOI: https://doi.org/10.1016/B978-0-12-824523-1.00014-6

were highlighted largely during the last 50 years, which have pointed out ecosystem destruction, chemical toxicity at different food chain, hyperaccumulation of toxic chemicals in higher rank of consumers, development of commercial pesticide resistance among pests, and acute soil and water toxicity that delivered huge impacts on agriculture and human health (Gerwick and Sparks, 2014). It was noted that unprecedented use of chemical pesticides contributed to the development of chronic human diseases and disorders related to the respiratory system, reproduction system, and neurodegenerative diseases (Hernandez et al., 2010) as well as cancer and disease related to metabolism toxicity (Hakeem et al., 2016; Kumar et al., 2012; Mostafalou and Abdollahi, 2017). Hence, there was an urgency to opt for more sustainable approaches like biopesticides for long. The use of hazard free biopesticides for crop protection is a novel biocontrol strategy for pest management, which also fulfils the basic criteria of sustainable agricultural practice. There is a population boom worldwide, which exponentially conferred huge "demand-supply" pressure on agricultural output. Under this context, application of top-notch tools to ensure crop protection is a critical step to meet our accelerated food demand. Pest infestation can create the havoc on crop plants and potentially reduce the annual crop yield. According to UC Agriculture and Natural Resources scientists, 10%–40% of all major crops (Wheat, rice, maize, soybean, and potato) has reportedly exhibited reduction in gross productivity pest attacks and other plant diseases (https://californiaagtoday.com/pests-diseases-cause-worldwide-damage-crops/).

Biopesticides are naturally derived compounds or organisms; mostly the living microorganisms or secondary metabolites, which control pests by nontoxic mechanisms without incurring any damage to the ecosystem. Nevertheless, the gradual increase in world population, rapid decline in biodiversity, and extensive expansion of agricultural sector needs the most eco-friendly, sustainable, and reliable approaches to ensure crop yield at its best (Sinha and Biswas, 2011). There are many advantages of biopesticides such as good shelf-life, easy decomposition, environmentally friendly, zero side effects, broad application range, increased crop yield, etc. Due to the increasing popularity of biopesticides worldwide over synthetic pesticides, the present market value of biopesticides is estimated at 4.3 billion USD in 2020 and is expected to grow by 14.7% up to 2025 with annual turnover of 8.5 billion USD (www.marketsandmarkets.com/Market-Reports/biopesticides-267.html). Biopesticides are divided into various classes, that is, (1) microbial pesticides (entomopathogenic bacteria, fungi, viruses, protozoa, nematodes, and their metabolites) and (2) botanical pesticides (Fisher and Garczynski, 2012; Sarwar, 2015). Among different biopesticides, the fungi and fungal secondary metabolites are extensively studied for their interesting contribution in crop yield enhancement synced with suppression of pest attack. The overall impact is comparable with different chemical pesticides. Fungal secondary metabolites are low-molecular-weight compounds, divided into three main classes: nonribosomal peptides, polyketides, and terpenes, which have not been involved in the growth of fungus (Boruta, 2018). These metabolites are beneficial and harmful as well. Harmful metabolites are known as toxins that can destroy or suppress certain insects, pathogens, and weeds referred to as pests. Various commercial products based on fungi (*Trichoderma, Gliocladium, Ampelomyces*) are available in the market for controlling pests (Vinale et al., 2008a,b). These fungi produced secondary metabolites do act as biocontrol agents. These secondary metabolites not only protect the crop plants from pests and pathogen attacks but also improve crop yields and overall plant growth. In Fig. 18.1 we have highlighted different modes of biopesticide action and Fig. 18.2 depicts the versatile role of fungal metabolites.

In the following section, we have provided a comprehensive idea about the mechanism by which the fungal strains and its secondary metabolites actually prevent pest attack without compromising the crop yield. In Fig. 18.3, we have added chemical structures of fungal metabolites derived from ChemSpider software.

18.2 Fungal secondary metabolites and their mechanism as biopesticide

There is a number of fungi and different secondary metabolites that act as biopesticides. Commercially available fungal biopesticides are based on

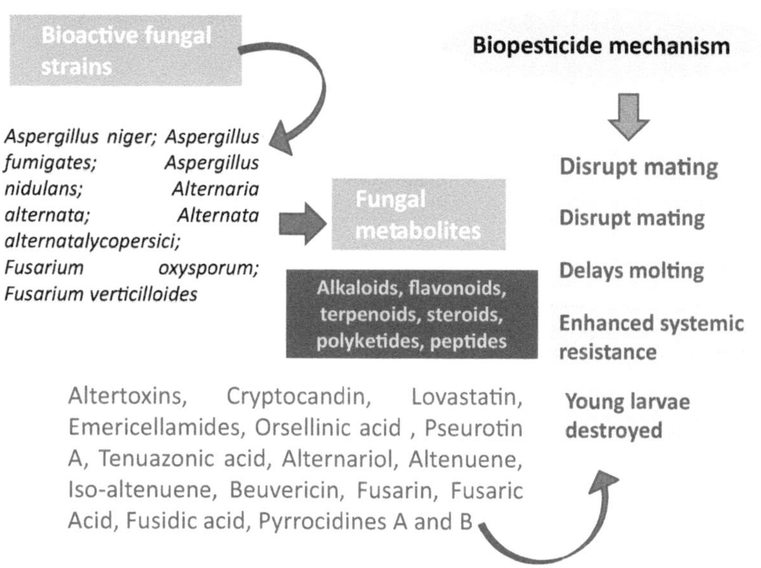

FIGURE 18.1 Bioactive fungi and their natural products and their mode of action as biopesticides.

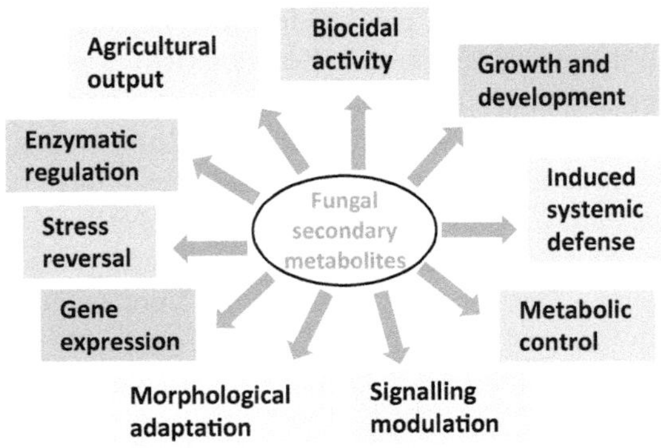

FIGURE 18.2 Multifaceted role of fugal secondary metabolites as biopesticides.

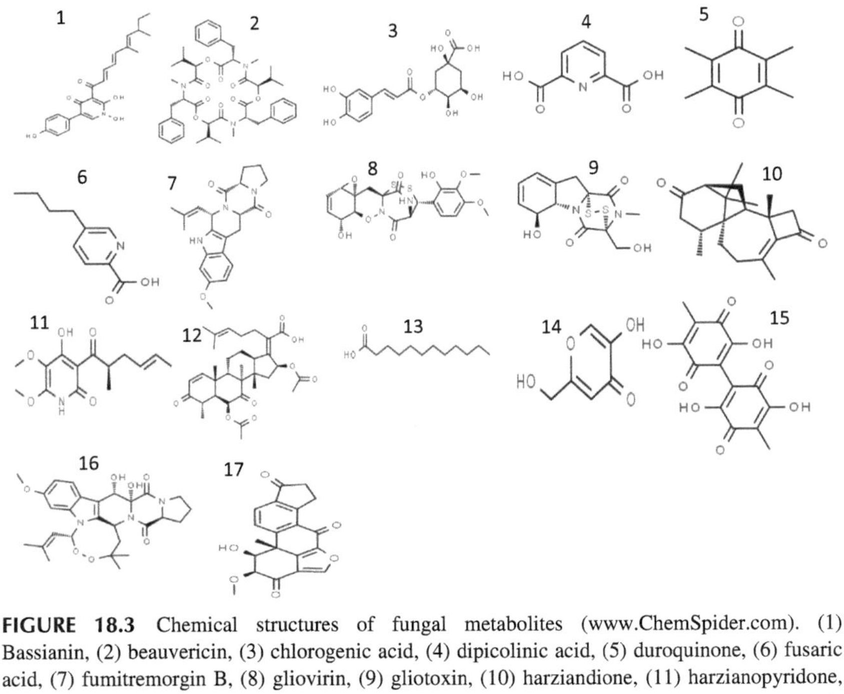

FIGURE 18.3 Chemical structures of fungal metabolites (www.ChemSpider.com). (1) Bassianin, (2) beauvericin, (3) chlorogenic acid, (4) dipicolinic acid, (5) duroquinone, (6) fusaric acid, (7) fumitremorgin B, (8) gliovirin, (9) gliotoxin, (10) harziandione, (11) harzianopyridone, (12) helvolic acid, (13) lauric acid, (14) kojic acid, (15) oosporein, (16) rugulosin, (17) verruculogen.

different strains of *Beauveria, Hirsutella, Gliocladium, Metarhizium anisopliae, Lecanicillium lecanii*, and *Trichoderma* (Vinale et al., 2008a,b). *Trichoderma* spp. is a free-living ascomycota fungi used as biopesticides in India increased crop yield in wheat, rice, potato, tomato, chilly, etc. (Kumar et al., 2017). Available commercial product of *Trichoderma* spp. is active against fungi causing wilt, root rot, and wood decay and foliar fungal diseases. Soil-born fungi *Gliocladium catenulatum*–based commercial products are active against seed, root, stem rot, and wilt disease. Fungal metabolites are produced by the mevalonic acid pathway, shikimic acid pathway, polyketide biosynthetic pathway, or by acid-derived pathways. In the following we have discussed the broad spectrum role of plant fungi and its metabolites. Table 18.1 presents different fungal metabolites used as biopesticides.

18.2.1 Plant insect control metabolites by fungi

18.2.1.1 *Beauveria bassiana*

Beauveria bassiana conidia and fungal metabolites reduced the reproduction rate of *Aphis gossypii* (greenfly). This study reported that *A. gossypii* has the capacity to detect fungal metabolites and then avoid making any interaction with the leaf, detected by choice assays (Gurulingappa et al., 2011). *B. bassiana* was at first produced conidia on the upper side of leaves where insects prefer to sit so that an insect could detect presence of certain kind of metabolites by consuming the leaves. Afterwards, on detection of the secondary metabolites the insects may avoid these plants for oviposition and feeding. *B. bassiana* also started its action into an insect by forming appressoria, proteins, and enzymes after attachment with insect and breaking cuticle. Mostly this fungus destroys the insect's immune system and kills it (Lewis et al., 2009). Cysteine-rich protein hydrophobins (Hyd1 and Hyd2) appeared on the fungi surface and that acts as a virulence factor that helps fungi to attach on the outer side of the cuticular layer of pests. Cytochrome P450, catalases, esterases, long-chain alcohols, and aldehyde dehydrogenases degrade the lipid layer and proteins of insect cells after attachment (Ortiz-Urquiza and Keyhani, 2013; Zhang et al., 2011). Secondary metabolites from *B. bassiana* have an important role in insect growth and immune system suppression (Feng et al., 2015).

18.2.1.2 *Hirsutella thompsonii*

Hirsutellin A was reported present in the *Hirsutella thompsonii* filtrate (Mazet and Vey, 1995). Hirsutellin A is an extracellular fungal toxic protein that has insecticidal activity (Boucias et al., 1998). Omoto and Clayton study showed by increasing Hirsutellin A concentration in leaf bioassay simultaneously decreased egg number of *Phyllocoptruta oleivora*, a citrus rust mite (Omoto and McCoy, 1998).

18.2.1.3 *Metarhizium anisopliae*

M. anisopliae is an essential hyphomycete entomopathogenic fungus present in the soil and insects, and found in different geographical regions throughout the world (Tulloch, 1976; Roberts and St. Leger, 2004). This fungus can

TABLE 18.1 Different fungal metabolites used as biopesticides.

Fungal source (fungal group)	Name of the metabolites	Plant	Pathogen/disease	Mechanism	References
Acremonium implicatum (Ascomycota)	NMP	*Solanum lycopersicum*	*Meloidogyne incognita*	Reduced tomato root galls by 66.7% in pot experiment, inhibited egg hatching of pathogen	Tian et al. (2014)
Aspergillus flavus (Ascomycota)	Duroquinone, naphthelene, lauric acid	NMP	*Hyblaea puera, Atteva fabriciella, Eligma narcissus*	Percent larval mortality via bioassay	Senthilkumar et al. (2014)
Aspergillus fumigatus (Ascomycota)	Fumitremorgin B, 12β-hydroxy-13α-methoxyverruculogen TR-2, verruculogen, helvolic acid	*Melia azedarach*	*Spodoptera frugiperda*	Weight reduction, increase larval mortality	Li et al. (2012)
Aspergillus oryzae (Ascomycota)	NMP	*Rhizophora mucronata*	*Spodoptera litura*	Acetylcholinesterase (Ache) inhibition	Abraham et al. (2015)
	Kojic acid	*Cornus alba*	*M. incognita*	Sequence analysis of the ITS rDNA, calmodulin (CaM), and β-tubulin (BenA) genes, inhibit egg hatchability	Kim et al. (2016)
Aspergillus niger (Ascomycota)	NMP	*Acacia arabica*	*S. litura*	Decrease in consumption rate accounts for decrease in growth rate	Kaur et al. (2016)

Beauvaria bassiana, Metarhizium anisopliae (Ascomycota)	BBF1 and BBF2, MAF1 and MAF2	NMP	*Dysdercus cingulatus*	BBF2 interfere with the digestive process, pest undergoes starvation (weight reduction), metabolites bind to hydrophobic site, interfere with protease enzymes	Sahayaraj and Tomson (2010)
Cladosporium sp. (Ascomycota)	NMP	*Tinospora cordifolia*	*S. litura*	Reduced larval survival, adult emergence, longevity, and reproductive potential of pathogen	Thakur et al. (2013)
Claviceps purpurea (Ascomycota)	PF-2, GA	*Achnatherum inebrians*	*Aphis gossypii*	Crude fermentation extracts of isolates GA and PF-2 compounds, reduce growth rate	Shi et al. (2013)
Cladosporium velox	Chlorogenic acid	*T. cordifolia*	*S. litura*	Resistance to insect by targeting the alpha glucosidases, improvement of crop production	Singh et al. (2016)
Coniothyrium minitans (Ascomycota)	Macrosphelide A	NMP	*Sclerotinia sclerotiorum, Sclerotium cepivorum*	Isolation of macrosphelide A reduced growth of fungi and bacteria, at IG_{50} values indicate inhibition of ascomycetes	McQuilken et al. (2003)
Fusarium oxysporum (Ascomycota)	Bikaverin, fusaric acid	NMP	*Bursaphelenchus xylophilus*	Synergistic interaction, reduce environmental pollution	Lee et al. (2007)
Gliocladium virens (Ascomycota)	Gliovirin, gliotoxin	NMP	*Pythium ultimum*	Antibiotic production	Howell et al. (1993)

(*Continued*)

TABLE 18.1 (Continued)

Fungal source (fungal group)	Name of the metabolites	Plant	Pathogen/disease	Mechanism	References
Lecanicillium lecanii, B. bassiana	Bassianin, beauvericin, bassionalide, beauveroiolide, bassacridin, oosporein, tenellin	NMP	*A. gossypii*	Reduced insect survival and reproduction, with conidia pathogenesity demonstrated	Gurulingappa et al. (2011)
Metarhizium flavoviride	Viridoxins A, Viridoxins B	NMP	*Leptinotarsa decemlineata*	Inhibit mycelial extension, loss of insect toxic properties	Gupta et al. (1993)
Nigrospora sp. (Ascomycota)	NMP	*T. cordifolia*	*S. litura*	Decreased ECI and ECD to insect biomass	Thakur et al. (2012)
Nomuraea rileyi (Ascomycota)	Phosphorothioic acid, pyrrolo[1,2-a] pyrazine-1,4-dione, hexahydro-3-(2-methylpropyl)-	NMP	*S. litura*	Reduced reactive oxygen species, oxidative stress, membrane disruption, and protein unfolding	Namasivayam et al. (2018)
Paecilomyces redivivus	4-(4′carboxy-2′-ethyl-hydroxypentyl)-5,6,-dihydro-6-methylcyclobuta[b]-pyridine-3,6-dicarboxylic acid,	NMP	*M. incognita*	NMP	Liu et al. (2009)

Phialocephala scopiformis (Ascomycota)	Rugulosin	*Picea glauca*	*Choristoneura fumiferana*	Growth rate of the insect pest is reduced, exposing larvae to increased disease and predation by birds and reducing the damage to trees	Walker et al. (2016)
Phomopsis phaseoli (Ascomycota)	3-Hydroxypropionic acid	*Betula pendula, Betula pubescens*	*M. incognita*	NMP	Schwarz et al. (2004)
Talaromyces pinophilus (Ascomycota)	Ferrirubin, 3-*O*-methylfunicone, herquline B	*Vicia faba*	*Acyrthosiphon pisum*	Isolation of F36CF organic extract	Vinale et al. (2017)
Trichoderma harzianum (Ascomycota)	NMP	*Solanum tuberosum, Piper nigrum, Malus domestica, S. lycopersicum, Cucumis sativus, Daucus carota*	*F. oxysporum, Alternaria alternate, S. sclerotiorum, Fusarium solani, Geotrichum candidum*	Chitinase, β-1-3 glucanase and β-1-4 glucanase production, antibiotics, competitions, inorganic plant nutrient resolution, disabling pathogen enzymes, and encouraging endurance	Tozlu et al. (2018)
T. harzianum (Ascomycota)	T22azaphilone, T39butenolide	NMP	*Rhizoctonia solani*	Antibiosis assays of secondary metabolites	Vinale et al. (2006)
	6-Pentyl-α-pyrone	NMP	*R. solani, Fusaarium oxysporum*	Inhibit germination	Scarselletti and Faull (1994)
T. harzianum, Trichoderma atroviride (Ascomycota)	6-n-pentyl-6H-pyran-2-one (6PP), gliotoxin, viridin, harzianopyridone, harziandione, peptaibols	*Lycopersicum esculentum, Brassica napus, Pisum sativum*	*R. solani, P. ultimum*	Detection of over expression of pathogenesis-related (PR) proteins	Vinale et al. (2008a,b)

(*Continued*)

TABLE 18.1 (Continued)

Fungal source (fungal group)	Name of the metabolites	Plant	Pathogen/disease	Mechanism	References
Trichoderma viride (Ascomycota)	NMP	Pigeon pea	*Fusarium udum*	Seed and soil treatment, control seed and seedling diseases	Bhattacharjee and Dey (2014)
	NMP	Rose	*Botrytis cinerea*	Seed and soil treatment, control seed and seedling diseases	Bhattacharjee and Dey (2014)
Verticillium biguttatum (Ascomycota)	Bigutol, methylbigutol	NMP	*R. solani*	Inhibited mycelial extension	Morris et al. (1995)
Verticillium lecanii (Ascomycota)	Lignin	Citrus fruits	*Penicillium digitatum*	Permeability barriers prevent pathogen spread, strong cell rigidity, inhibit pathogen growth and fungal spread	Benhamou (2004)
	Dipicolinic acid	NMP	*Calliphora erythrocephala*	Loss of pathogenecity	Claydon and Grove (1982)

be used as a potential biopesticide against several insects. This has been categorized as a safe and ecofriendly biocontrol agent (Butt et al., 2001a,b). Soil fungi *M. anisopliae* attacks a wide range of insects by infiltrating the cuticle layer of any part of the insect body. Fungus conidia first attached to an insect body than started releasing enzymes (chitinases) that degrade polysaccharides, increase osmotic pressure, and change lipid composition of insect cells so that *M. anisopliae* can easily grow on the insect. After sometimes, appressoria appears from the fungus that helps the hyphae to go inside the insect integument. MCL1 protein appears (for Metarhizium collagen-like protein) in the fungi cell after induction that has an antiadhesive role, which gives protection against phagocytosis (Wang and St. Leger, 2006; Zheng et al., 2011). The mode of action of *Metarhizium* is by increasing virulence factor by stimulating more enzymes. Hernandez et al. (2010) showed overexpression of the cat1 gene increases catalase in *M. anisopliae* that leads to better tolerance to exogenous H_2O_2. Virulence factors of *M. anisopliae* also increased by more conidial germination. In *M. anisopliae* AaIT gene responsible for insect-specific neurotoxin production and showed neurotoxin reduced wing movement of *Aedes aegypti* (Wang and St. Leger, 2007).

18.2.1.4 Gliocladium fimbriatum

Gliocladium fimbriatum fungus is a good alternative to chemical pesticides against Fusarium wilt plant disease (Fitrianingsih et al., 2019).

18.2.1.5 Verticillium lecanii

Verticillium lecanii are useful against aphid control. The mode of action of this fungus is the same as *M. anisopliae* and *B. bassiana*. The study has proved that *V. lecanii* has insecticidal effects against *Calliphora erythrocephala* (blowfly). Dipicolinic acid (pyridine-2,6-dicarboxylic acid) was isolated from the acidic extract and C25 compounds were isolated from neutral extracts of *V. lecanii* (Claydon and Grove, 1982). *V. lecanii* has been reported to produce certain metabolites bassionalide, vertilecanin A, vertilecanin A methyl ester, vertilecanin C, decenedioic acid, and 10-hydroxy-8-decenoic acid (Suzuki et al., 1977; Soman et al., 2001).

18.2.2 Plant insect control metabolites by fungal metabolites

18.2.2.1 Beauvericin

Beauvericin is a cyclohexadepsipeptide mycotoxin produced by *B. bassiana*, *Fusarium* spp., and *Paeeilomycesfumoso-roseus* (Grove and Pople, 1980; Logrieco et al., 1998; Munkvold et al., 1998), reported beauvericin has insecticidal activity against *C. erythrocephala Aedes aegypti* (Grove and Pople, 1980), *Lygus* spp. (Leland and McGuire, 2004), *Spodoptera frugiperda* (Fornelli et al., 2004), and *Schizaphis graminum* (Jestoi, 2008). It was reported that mixed application of *B. bassiana* and *Nicotiana tabacum* added in a field experiment at a robusta coffee plantation reduced coffee berry borer (*Hypothenemus hampei*) attack by 1.54% and 0.33%. Beauvericin is involved in the parasitization process to insect larvae and makes the insects less active (Haryuni et al., 2019). The particular mode of action of beauvericin is not understood. However, beauvericin increased cell membrane permeability and acts as an ionophore, which leads to calcium-mediated apoptotic pathways in cells (Kouti et al., 2003; Wang and Xu, 2012).

18.2.2.2 Destruxins

Destruxins are cyclic depsipeptidic mycotoxin mostly produced by *M. anisopliae* showed insecticide activity against lepidopteran insects. Destruxins A and B were discovered in 1960 from the in vitro study of *M. anisopliae* toxin production (Kodaira, 1961). Destruxins disrupt the balance between calcium and hydrogen ions in insect cells and suppress vacuolar-type ATPase (Muroi et al., 1994). Destruxin A is a cytotoxic and pathogenetic compound that damages insects' innate immunity (Hu et al., 2007; Pal et al., 2007). Additionally, destruxins also inhibits protein and nucleic acid synthesis. Destruxins A, B, and E showed antifeedant properties in cabbage when treated with destruxins resulting in feeding reduction of *Phaedon cochleariae* (leaf beetle) and *Plutella xylostella* (cabbage mot) larvae. Antifeeding and insecticidal activities of destruxins are dose dependent. All the larvae died at 100 ppm without damaging any leaves, however, at 300 ppm treatment 90% larvae died with few damaged leaves. LC_{50} values of destruxins A and E against *P. cochleariae* is (30 and 17 ppm), which is lower than the *P. xylostella* (79 and 58 ppm) value (Amiri et al., 1999). Additionally, Destruxin E showed antifeeding activity in *Empoasca vitis* (leafhoppers) (Maniania, 1994). Dextruxins (Destruxin A-760 and Destruxin A-724) showed pathogenicity against one of the prominent olive pest *Saissetia oleae* and given protection from *S. oleae* in both in vitro and field study with 163 and 190 ppm treatment on the third larval instars stage (M.M, 2018). Though the mechanism of destruxins is not clear, one recent study demonstrated that BmTudor-sn protein in *Bombyx mori* (silkworm) cell (Bm12) has a binding site for destruxins A at the 100 μM level that binds in Leu704 amino acid. This study could be made clear the idea of the molecular mechanism of Destruxin A for the production of new biopesticides (Wang et al., 2019). Destruxin A also induced the upregulation of heat shock protein BmHSCP in Bm12 cells (Zhang et al., 2017).

18.2.2.3 Viridoxins

Viridoxin A and B are secondary metabolites isolated from *Metarhizium flavoviride* that showed insecticidal activity against *Leptinotarsa decemlineata*, the Colorado potato beetle (Gupta et al., 1993).

18.2.2.4 Gliotoxin

Gliotoxin is an epipolythiodiketopiperazine that was first time isolated from *G. fimbriatum*, Aside from it also isolated from *Gliocladium virens* the study revealed that gliotoxin was active against *Pythium ultimum* and *Rhizoctonia solani* pathogen. Cell wall–degrading enzymes improve gliotoxin antipathogenic activity (Whilhite et al., 1994; Lorito et al., 1996).

18.2.2.5 Gliovirin

Another toxic metabolite gliovirin was isolated from *G. virens*, which has shown activity against *P. ultimum* affecting the cotton seedling (Howell and Stipanovic, 1983)p.

18.2.2.6 Enalin A analogues

Enalin A analogues are like ergosterol peroxide from *Verticillium* spp. isolated from roots of *Rehmannia glutinosa*, and they revealed antibiotic activity against *Rhizoctonia* sp., *Septoria* spp., and *Fusarium* spp. These metabolites also inhibit *Verticillium* spp. itself. Ergosterol peroxide inhibits pathogens at a very low concentrations of 0.97 μg/mL (You et al., 2009).

18.2.2.7 Oosporein

Oosporein, which was first isolated from *Oospora colorans* and was later isolated from mycoparasite *Verticillium psalliotae*, revealed inhibitory action against *Phytophthora infestans*, a late blight pathogen affecting tomato (Niu, 2017; Wainwright et al., 1986).

18.2.2.8 Bigutol

Bigutol and methylbigutol originally obtained from *Verticillium biguttutum* showed antifungal activity against the growth of *R. solani* with 138 mg/mL concentration, which suggests these metabolites could be served as a potent biopesticide (Morris et al., 1995).

18.2.3 Special metabolites class and selective fungal genus and their role as biopesticide

18.2.3.1 Trichoderma

Trichoderma is used as biopesticide agents and showed different modes of action based on inter- or intraspecies competition (Chet, 1987; Ghisalberti and Sivasithamparam, 1991; Lorito et al., 1996). *Trichoderma* colonizes with the root of the plant by competing with pathogens (Sivan and Chet, 1989) and it produces low-molecular-weight secondary metabolites (alkyl pyrones, diketopiperazines, isonitriles, polyketides, and petaibols). Moreover, the study of the host-pathogen and *Trichoderma* interaction revealed antibiotic activity against various phytopathogenic fungi (Gajera et al., 2013). *Trichoderma* also produces cellulases and lytic enzymes (proteases, b-1,3-glucanases, chitinases), which degrade a pathogen's cell wall (Harman et al., 1993).

Secondary metabolites from *Trichoderma* spp. were evaluated for their antibacterial potential against phytopathogenic bacteria (*Ralstonia solanacearum, Xanthomonas compestris*, and *Meloidogyne incognita*). Out of the different growth media used, solid wheat media showed the highest mortality rate and inhibition of egg hatching capability for *M. incognita*. *Trichoderma* fungal metabolites are isolated by ethyl acetate solvent (Khan et al., 2020). *R. solanacearum* and *X. compestris* are the two most prominent plant pathogens of potato, tomato, and pepper (Camesano, 2015; Jones et al., 2005; Scherf et al., 2010). One in vitro study highlight the successful usage of *Trichoderma harzianum* (ET 4 and ET 14) as biopesticides showed hyperparasitic effects in plant pathogens *Fusarium oxysporum* from *Solanum tuberosum* and *Piper nigrum, Fusarium solani* from *Cucumis sativus, Alternaria alternate* from *Malus domestica* and *Solanum lycopersicum, Sclerotinia sclerotiorum* from *C. sativus*, and *Geotrichum candidum* from *Daucus carota* without causing any human health-related problems (Tozlu et al., 2018). *Trichoderma* also possess strong nematocidal properties. *T. harzianum* (T22 and T39) strains produced secondary metabolites T22azaphilone, T39butenolide, harzianolide, and harzianopyridone as observed by liquid chromatography/mass spectrometry-based studies.

T22azaphilone showed weak antifungal activity against *Gaeumannomyces graminis* var. *tritici*. However, T39 butenolide and harzianolide metabolites completely inhibit *G. graminis* var. *tritici*. Harzianopyridone metabolites inhibit growth of *R. solani*, *G. graminis* var. *tritici*, and *P. ultimum* (Vinale et al., 2006). *Trichoderma* spp. secondary metabolites 1,8-dihydroxy-3-methyl-anthraquinone, T22azaphilone, T39butenolide, harzianopyridone, harzianolide, and 1-hydroxy-3-methyl-anthraquinone has the capacity to increase the expression of pathogenesis-related (PR) proteins that helps in plant defense. Harzianolide and 6PP increase the growth of tomato and canola seedlings as observed through in vitro studies. T22azaphilone, 1-hydroxy-3-methyl-anthraquinone, 1,8-dihydroxy-3-methyl-anthraquinone, of *T. harzianum* A6, showed low activity against *G. graminis* var. *tritici*. However, harzianolide (200 mg) inhibited *G. graminis* var. *tritici* effectively (Vinale et al., 2008a,b). Cork oak has been attacked by brine shrimp species *Artemia salina*.

18.2.3.2 Anthraquinones

Anthraquinones are quinone derivative fungal metabolite, produced by several species of *Trichoderma* (*Trichoderma virens*, *Trichoderma polysporum*, *Trichoderma atroviride*, *Trichoderma viride*, *T. harzianum*), *Phoma foveata*, *Ascochyta pisi*, and *Digitalis* spp. (Stefano et al., 1999). Anthraquinone includes pachybasin, chrysophanol, emodin, and trichodermaol (Adachi and Lu, 1983; Betina and Kubela, 1987; Donnelly and Sharidan, 1986; Slater et al., 1967). Anthraquinone is causing postintestinal distress in birds. Birds after ingestion of anthraquinone-rich fruit develop irritation in the gut; hence, they avoid the particular food next time (Avery et al., 1997; Deliberto et al., 2016). Chrysophanol produced from *T. viride* (Stefano et al., 1999), *T. harzianum* (Liu et al., 2007), Aspergillus, Penicillium, Paecilomyces, Paraconiothyrium, Drechslera, and Phaeospheria has ecological importance (Yusuf et al., 2019). The first time chrysophanaol was isolated from *Penicillium islandicum* (Ghosh et al., 1978). Chrysophanol is produced in fungi via the polymalonate pathway (PMA). Liu et al. showed that chrysophanol isolated from *T. harzianum* ETS-323 enhanced the growth of *Brassica oleracea* var. capitata and initiate defense in the host by induced resistance against *Botrytis cinerea* (Liu et al., 2016). Chrysophanol treatment to cabbage induced defense genes ascorbate peroxidase (apx), b-1,3-glucanase (b-1,3-glucanase), pathogenesis-related protein 1 (PR-1), glutathione S-transferase (GST), and deoxycytidine deaminase (DCD) (Liu et al., 2016).

18.2.3.3 Peptaibols

Peptaibols are a class of linear short type peptides made up of nonstandard amino acids (isovaline, hydroxyproline, and ethylnorvaline) that at the first time was isolated from *T. viride* (Reusser, 1967). Other peptaibols producing species are plant-pathogenic fungi like *Acremonium* (Sharman et al., 1996), *Paecilomyces* (Rossi et al., 1987), *Emericellopsis* (Berg et al., 1996), and *Trichoderma* sp. (Daniel and Rodrigues Filho, 2007). *Trichoderma citrinoviride* produced peptaibols secondary metabolites in the liquid culture, which was screened and analyzed by TLC, HPLC, MALDI-TOF MS, and nano-ESI-QTOF MS, and they showed strong antifungal activity against *Biscogniauxia mediterranea* and *Artemisia salin* (Maddau et al., 2009).

18.2.3.4 Bassionalide

Bassianolide, an ionophore-induced metabolite that acts against the corn earworm larvae (Champlin and Grula, 1979). In sweet potato whitefly, the application of *V. lecanii* fungal metabolite extract (100 mg/L) reduced manyfold the egg hatching capability. Another metabolite toxin (V3450 and Vp28) successfully acted as repellent to whitefly (*Bemisia tabaci*) and spraying adults with 1000 mg/L toxin resulted in antifeedant effects (Logan and Birkett, 2007). Additionally, *V. lecanii* is also active against citrus fruit pathogen *Penicillium digitatum* which are responsible for causing the green mold. Metabolite production and accumulation were directly associated with pathogen inhibition and acts as a mechanical barrier (Benhamou, 2004).

18.2.4 Nematicidal metabolites produced by fungi

Fungi directly or indirectly paralyzes nematodes and are seen as significant biocontrol agents against nematode-mediated crop loss. Fungal secondary metabolites produced by *Pleurotus ostreatus* and *Syncephalastrum racemosum* fungi were tested as potent nematicides. *P. ostreatus* produces droplets that contain toxins and ostreatin, which in contact with nematodes changes their body shape, shrinks their head part, and makes them immobilized (Barron and Thorn, 1987). Avermectin is a nematicidal metabolite present in *S. racemosum* fungus and is active against plant pathogenic root-knot nematode. The combined application of *S. racemosum* and avermectin in the greenhouse condition was found to be more effective than individual treatment. Combined treatments also increased cucumber yield and significantly reduced the number of active nematodes (Huang et al., 2014). *Paecilomyces redivivus* that produces different nematosidal metabolites along with (4′carboxy-2′-ethyl-hydroxypentyl)-5,6,-dihydro-6-methylcyclobuta[b]-pyridine-3,6-dicarboxylic acid was found to be active against *M. incognita* at 50.86 mg/L concentration (Liu et al., 2009). Tetradecalactone metabolites, caryospomycins A, B, and C were active against *Bursaphelenchus xylophilus* (pine wilt nematode) and they were obtained from *Caryospora callicarpa* (Dong et al., 2007).

18.2.5 Plant weed controlling metabolites produced by fungi

Alternaria alternate toxin (AAL) toxin, tenuazonic, tentoxin, maculosins, and zinniol have been isolated from *Alternaria* sp. AAL and is toxic to many weeds like jimsonweed, black nightshade, prickly sida, and hemp sesbania. The study of the mode of action of this toxin has revealed sharp increase in electrolyte leakage and substantial chlorosis in weed plants (Abbas et al., 1995a,b). In *Lemna pausicostata* L., cellular electrolyte leakage and chlorophyll loss occurred after 72 hours treatment with 20−40 nM (Abbas et al., 1995a,b). *Chenopodium album* L., which almost grows everywhere except in the desert, could be controlled by *Ascochyta caulina* toxin ascaulitoxin. Fungal culture filtrate of this fungi contains ascaulitoxin, its aglycone derivatives, and trans-4-aminoproline toxins. Combined application of these toxins with spores of *A. caulina* enhanced biocontrol capacity against *C. album* under greenhouse conditions (Vurro et al., 2001).

18.3 Future prospects and conclusions

Many secondary metabolites have been isolated and screened from various beneficial fungi. Most of the fungi are utilized as biocontrol agents in the agriculture sector. Fungi represent a novel source of metabolites that could be used as an efficient biocontrol tool to make biopesticides. *Trichoderma* is one of the most important and explored species of fungi, which has been employed for its significant effects on suppressing plant diseases. Additionally, *V. lecanii* and *G. fimbriatum* produces metabolites like gliotoxin, bigutol, oosporein, bassionalide, and gliovirin, which are novel compounds applicable in plant disease management. The production of natural compounds depend on the species and strain of the selected fungus. Moreover, the biosynthesis pathways, culture condition, growth pattern, isolation procedures, and extraction conditions are important parameters. The mechanisms of biopesticides are different among several categories of pests since fungi-based biopesticides kill insects by disrupting insect body coat and invade an insect's body by hyphae, which causes immune system destruction of the insects. Fungi-based nematocides are useful in making the nematodes immobilized and paralyzed. On the other hand, weed control through production of several mycotoxins could be achieved by electrolyte leakage and chlorosis, although the majority of studies have not revealed clearly the exact mode of metabolite action. Metabolite applications in field experiments are very limited although the available commercial products, and natural product-based start-ups are rapidly filling the gap in scientific knowledge to strategically provide a better alternative to chemical pesticides. Unfortunately, funding for biopesticide-based research is still lacking in many developing countries where the application of ecofriendly pesticide is much more needed. Despite many challenges, the fungal pesticide could have a vivid future in India. Already, *B. bassiana, Beauveria brongniartii, M. anisopliae, V. lecanii, H. thompsonii*, and *Paecilomyces fumosoroseus* based biopestisides were registered as bioinsecticides and bionematicides. Sophistication in advanced and cost-effective biotechnological techniques for the quality improvement and identification of novel metabolites are crucial. In the future, the industrial-scale isolation and wide application from these strains on a large scale will mark a new era in chemical pesticide-free sustainable agriculture.

References

Abbas, H.K., Duke, S.O., Paul, R.N., Riley, R.T., Tanaka, T., 1995a. AAL-toxin, a potent natural herbicide which disrupts sphingolipid metabolism of plants. Pestic. Sci. 43, 181–187.

Abbas, H.K., Tanaka, T., Duke, S.O., Boyette, C.D., 1995b. Susceptibility of various crop and weed species to AAL-toxin, a natural herbicide. Weed Technol. 125–130.

Abraham, S., Basukriadi, A., Pawiroharsono, S., Sjamsuridzal, W., 2015. Insecticidal activity of ethyl acetate extracts from culture filtrates of mangrove fungal endophytes. Mycobiology 43, 137–149.

Adachi, Y., Lu, B.C.-Y., 1983. Supercritical fluid extraction with carbon dioxide and ethylene. Fluid Phase Equilib. 14, 147–156.

Agrios, G.N., 2003. DISEASES I plant pathology, principles. Encycl. Appl. Plant Sci. 203–212.

Amiri, B., Ibrahim, L., Butt, T.M., 1999. Antifeedant properties of destruxins and their potential use with the entomogenous fungus *Metarhizium anisopliae* for improved control of crucifer pests. Biocontrol Sci. Technol. 9, 487–498.

Avery, M.L., Humphrey, J.S., Decker, D.G., 1997. Feeding deterrence of anthraquinone, anthracene, and anthrone to rice-eating birds. J. Wildl. Manag. 1359–1365.

Barron, G.L., Thorn, R.G., 1987. Destruction of nematodes by species of Pleurotus. Can. J. Bot. 65 (4), 774–778.

Benhamou, N., 2004. Potential of the mycoparasite, *Verticillium lecanii*, to protect citrus fruit against *Penicillium digitatum*, the causal agent of green mold: a comparison with the effect of chitosan. Phytopathology 94 (7), 693–705.

Berg, A., Ritzau, M., Ihn, W., Schlegel, B., Fleck, W.F., Heinze, S., et al., 1996. Isolation and structure of bergofungin, a new antifungal peptaibol from *Emericellopsis donezkii* HKI 0059. J. Antibiot. 49, 817–820.

Betina, V., Kubela, Ŝ., 1987. Uncoupling effect on fungal hydroxyanthraquinones on mitochondrial oxidative phosphorylation. Chem. Biol. Interact. 62, 179–186.

Bhattacharjee, R., Dey, U., 2014. An overview of fungal and bacterial biopesticides to control plant pathogens/diseases. Afr. J. Microbiol. Res. 8, 1749–1762.

Boruta, T., 2018. Uncovering the repertoire of fungal secondary metabolites: from Fleming's laboratory to the International Space Station. Bioengineered 9 (1), 12–16.

Boucias, D.G., Farmerie, W.G., Pendland, J.C., 1998. Cloning and sequencing of cDNA of the insecticidal toxin hirsutellin A. J. Invertebr. Pathol. 72 (3), 258–261.

Butt, T.M., Jackson, C.W., Magan, N., 2001a. Fungi as Biocontrol Agents: Progress, Problems and Potential. CAB International, Wallingford.

Butt, T.M., Jackson, C., Magan, N., 2001b. Introductionfungal biological control agents: problems and potential. In: Butt, T.M., Jackson, C., Magan, N. (Eds.), Fungi as Biocontrol Agents: Progress, Problems and Potential. CAB International, Wallingford, pp. 1–9. , Mode of action of biopestisides.

Camesano, T.A., 2015. Nanotechnology to Aid Chemical and Biological Defense. Springer, New York.

Champlin, F.R., Grula, E.A., 1979. Noninvolvement of Beauvericin in the entomopathogenicity of *Beauveria bassiana*. Appl. Environ. Microbiol. 37, 1122–1126.

Chet, I., 1987. Trichoderma – application, mode of action, and potential as biocontrol agent of soilborne plant pathogenic fungi. In: Chet, I. (Ed.), Innovative Approaches to Plant Disease Control. John Wiley & Sons, New York, pp. 137–160.

Claydon, N., Grove, J.F., 1982. Insecticidal secondary metabolic products from the entomogenous fungus *Verticillium lecanii*. J. Invertebr. Pathol 40 (3), 413–418.

Daniel, J.F.D.S., Rodrigues Filho, E., 2007. Peptaibols of Trichoderma. Nat. Prod. Rep. 24, 1128–1141.

Deliberto, S.T., Werner, S.J., Collins, F., Deliberto, S.T., Services, W., Wildlife, N., et al., 2016. Review of anthraquinone applications for pest. Pest. Manag. Sci. 72, 1813–1825.

Dong, J., Zhu, Y., Song, H., Li, R., He, H., Liu, H., et al., 2007. Nematicidal resorcylides from the aquatic fungus *Caryospora callicarpa* YMF1. 01026. J. Chem. Ecol. 33 (5), 1115–1126.

Donnelly, D.M.X., Sharidan, M.H., 1986. Anthraquinones from *Trichoderma polysporum*. Phytochemistry 25, 2303–2304.

Feng, P., Shang, Y., Cen, K., Wang, C., 2015. Fungal biosynthesis of the bibenzoquinone oosporein to evade insect immunity. Proc. Natl. Acad. Sci. U. S. A. 112, 11365–11370.

Fisher, T.W., Garczynski, S.F., 2012. Isolation, culture, preservation, and identification of entomopathogenic bacteria of the *Bacilli*. Manual of Techniques in Invertebrate Pathology. Academic Press, London, UK, pp. 75–99.

Fitrianingsih, A., Martanto, E.A., Abbas, B., 2019. The effectiveness of fungi *Gliocladium fimbriatum* and *Trichoderma viride* to control fusarium wilt disease of tomatoes (*Lycopersicum esculentum*). Indian J. Agric. Res. 53 (1), 57–61.

Fornelli, F., Minervini, F., Logrieco, A., 2004. Cytotoxicity of fungal metabolites to lepidopteran (Spodoptera frugiperda) cell line (SF-9). J. Invertebr. Pathol. 85 (2), 74–79.

Gajera, H., Domadiya, R., Patel, S., Kapopara, M., Golakiya, B., 2013. Molecular mechanism of Trichoderma as bio-control agents against phytopathogen system - a review. Curr. Res. Microbiol. Biotechnol. Curr. Res. Microbiol. Biotechnol. 1, 133–142.

Gerwick, B.C., Sparks, T.C., 2014. Natural products for pest control: an analysis of their role, value and future. Pest. Manag. Sci. 70, 1169–1185.

Ghisalberti, E.L., Sivasithamparam, K., 1991. Antifungal antibiotics produced by Trichoderma spp. Soil Biol. Biochem. 23, 1011–1020.

Ghosh, A.C., Manmade, A., Kobbe, B., Townsend, J.M., Demain, A.L., 1978. Production of luteoskyrin and isolation of a new metabolite, pibasterol, from *Penicillium islandicum* Sopp. Appl. Environ. Microbiol. 35, 563–566.

Grove, J.F., Pople, M., 1980. The insecticidal activity of beauvericin and the enniatin complex. Mycopathologia 70, 103–105.

Gupta, S., Krasnoff, S.B., Renwick, J.A.A., Roberts, D.W., Steiner, J.R., Clardy, J., 1993. Viridoxins A and B: novel toxins from the fungus *Metarhizium flavoviride*. J. Org. Chem. 58 (5), 1062–1067.

Gurulingappa, P., McGee, P.A., Sword, G., 2011. Endophytic *Lecanicillium lecanii* and *Beauveria bassiana* reduce the survival and fecundity of *Aphis gossypii* following contact with conidia and secondary metabolites. Crop. Prot. 30, 349–353.

Hakeem, K.R., Akhtar, M.S., Abdullah, S.N.A., 2016. Effects of pesticides on environment. Plant Soil Microbes Vol. 1 Implic. Crop. Sci. 1–366.

Harman, G.E., Hayes, C.K., Lorito, M., Broadway, R.M., Di Pietro, A., Peterbauer, C., et al., 1993. Chitinolytic enzymes of *Trichoderma harzianum*: purification of chitobiosidase and endochitinase. Phytopathology 83, 313–318.

Haryuni, Koernia Dewi, T.S., Suprapti, E., Rahman, S.F., Gozan, M., 2019. The effect of *Beauveria bassiana* on the effectiveness of *Nicotiana tabacum* extract as biopesticide against *Hypothenemus hampei* to robusta coffee. Int. J. Technol. 10, 159–166.

Hernandez, C.E.M., Guerrero, I.E.P., Hernandez, G.A.G., Solis, E.S., Guzman, J.C.T., 2010. Catalase overexpression reduces the germination time and increases the pathogenicity of the fungus *Metarhizium anisopliae*. Appl. Microbiol. Biotechnol. 87 (3), 1033–1044.

Howell, C.R., Stipanovic, R.D., 1983. Gliovirin, a new antibiotic from *Gliocladium virens*, and its role in the biological control of *Pythium ultimum*. Can. J. Microbiol. 29 (3), 321–324.

Howell, C.R., Stipanovic, R.D., Lumsden, R.D., 1993. Antibiotic production by strains of *Gliocladium virens* and its relation to the biocontrol of cotton seedling diseases. Biocontrol Sci. Technol. 3, 435–441.

Hu, Q.B., Ren, S.X., An, X.C., Qian, M.H., 2007. Insecticidal activity influence of destruxins on the pathogenicity of *Paecilomyces javanicus* against *Spodoptera litura*. J. Appl. Entomol. 131 (4), 262–268.

Huang, W.K., Sun, J.H., Cui, J.K., Wang, G.F., Kong, L.A., Peng, H., et al., 2014. Efficacy evaluation of fungus *Syncephalastrum racemosum* and nematicide avermectin against the root-knot nematode *Meloidogyne incognita* on cucumber. PLoS One 9 (2), e89717.

Jestoi, M., 2008. Emerging Fusarium-mycotoxins fusaproliferin, beauvericin, enniatins, and moniliformin—a review. Crit. Rev. Food Sci. Nutr. 48 (1), 21–49.

Jones, J.T., Furlanetto, C., Kikuchi, T., 2005. Horizontal gene transfer from bacteria and fungi as a driving force in the evolution of plant parasitism in nematodes. Nematology 7 (5), 641–646.

Kaur, T., Kaur, J., Kaur, A., Kaur, S., 2016. Larvicidal and growth inhibitory effects of endophytic *Aspergillus niger* on a polyphagous pest, *Spodoptera litura*. Phytoparasitica 44, 465–476.

Khan, R.A.A., Najeeb, S., Hussain, S., Xie, B., Li, Y., 2020. Bioactive secondary metabolites from *Trichoderma* spp. against phytopathogenic fungi. Microorganisms 8, 817.

Kim, T.Y., Jang, J.Y., Jeon, S.J., Lee, H.W., Bae, C.H., Yeo, J.H., et al., 2016. Nematicidal activity of kojic acid produced by *Aspergillus oryzae* against *Meloidogyne incognita*. J. Microbiol. Biotechnol. 26, 1383–1391.

Kumar, N., Pathera, A.K., Saini, P., Kumar, M., 2012. Harmful effects of pesticides on human health. Ann. Agri Bio Res. 17, 125–127.

Kodaira, Y., 1961. Toxic substances to insects, produced by *Aspergillus ochraceus* and Oopsra destructor. Agric. Biol. Chem. 25 (3), 261–262.

Kouti, K., Lemmens, M., Gruber, L.R., 2003. Beauvericin induced channels in ventricular myocytes and liposomes. Biochim Biophys Acta 1609, 203–210.

Kumar, G., Maharshi, A., Patel, J., Mukherjee, A., Singh, H.B., Sarma, B.K., 2017. Trichoderma: a potential fungal antagonist to control plant diseases. SATSA Mukhapatra-Annual Tech. Issue 21, 206–218.

Lee, S., Sunog, N.D., Kim, J.C., 2007. Nematicidal activity of bikaverin and fusaric acid isolated from *Fusarium oxysporum* against pine wood nematode *Bursaphelenchus xylophilus*. Plant Pathol. J. 23, 318321.

Leland, J.E., McGuire, M.R., 2004. Strain selection of a fungal entomopathogen, *Beauveria bassiana*, for control of plant bugs (*Lygus* spp.). In Proceedings of the 2004 Beltwide Cotton Conferences. National Cotton Council, Memphis, TN, pp. 1287–1496.

Lewis, M.W., Robalino, I.V., Keyhani, N.O., 2009. Uptake of the fluorescent probe FM4-64 by hyphae and haemolymph-derived in vivo hyphal bodies of the entomopathogenic fungus *Beauveria bassiana*. Microbiology 155, 3110–3120.

Li, X.J., Zhang, Q., Zhang, A.L., Gao, J.M., 2012. Metabolites from *Aspergillus fumigatus*, an endophytic fungus associated with *Melia azedarach*, and their antifungal, antifeedant, and toxic activities. J. Agric. Food Chem. 60, 3424–3431.

Liu, S.Y., Lo, C.T., Chen, C., Liu, M.Y., Chen, J.H., Peng, K.C., 2007. Efficient isolation of anthraquinone-derivatives from *Trichoderma harzianum* ETS 323. J. Biochem. Biophys. Methods 70, 391–395.

Liu, S.Y., Liao, C.K., Lo, C.T., Yang, H.H., Lin, K.C., Peng, K.C., 2016. Chrysophanol is involved in the biofertilization and biocontrol activities of *Trichoderma*. Physiol. Mol. Plant Pathol. 96, 1–7.

Liu, Y.J., Zhai, C.Y., Liu, Y., Zhang, K.Q., 2009. Nematicidal activity of *Paecilomyces* spp. and isolation of a novel active compound. J. Microbiol. 47 (3), 248–252.

Logan, J.G., Birkett, M.A., 2007. Juvenile hormonal activity of mono- and polyenehomobenzenes in *Drosophila virilis*, *Spodoptera frugiperda* and *Araschnia levana*. Pest. Manag. Sci. 63, 809–814.

Logrieco, A., Moretti, A., Castella, G., Kostecki, M., Golinski, P., Ritieni, A., et al., 1998. Beauvericin production by Fusarium species. Appl. Environ. Microbiol. 64, 3084–3088.

Lorito, M., Farkas, V., Rebuffat, S., Bodo, B., Kubicek, C.P., 1996. Cell wall synthesis is a major target of mycoparasitic antagonism by *Trichoderma harzianum*. J. Bacteriol. 178, 6382–6385.

Maddau, L., Cabras, A., Franceschini, A., Linaldeddu, B.T., Crobu, S., Roggio, T., et al., 2009. Occurrence and characterization of peptaibols from *Trichoderma citrinoviride*, an endophytic fungus of cork oak, using electrospray ionization quadrupole time-of-flight mass spectrometry. Microbiology 155, 3371–3381.

Maniania, N.K., 1994. Contact toxicity of the mycotoxin destruxin E to *Empoasca vitis* (Göthe) (Hom., Cicadellidae). J. Appl. Entomol. 117, 135–143.

Mazet, I., Vey, A., 1995. Hirsutellin A, a toxic protein produced in vitro by *Hirsutella thompsonii*. Microbiology 141 (6), 1343–1348.

McQuilken, M.P., Gemmell, J., Hill, R.A., Whipps, J.M., 2003. Production of macrosphelide A by the mycoparasite *Coniothyrium minitans*. FEMS Microbiol. Lett. 219, 27–31.

Morris, R.A.C., Ewing, D.F., Whipps, J.M., Coley-Smith, J.R., 1995. Antifungal hydroxymethylphenols from the mycoparasite *Verticillium biguttatum*. Phytochemistry 39, 1043–1048.

Mostafalou, S., Abdollahi, M., 2017. Pesticides: an update of human exposure and toxicity. Arch. Toxic. 91, 549–599.

Munkvold, G., Stahr, H.M., Logrieco, A., Moretti, A., Ritieni, A., 1998. Occurrence of fusaproliferin and beauvericin in *Fusarium*-contaminated livestock feed in Iowa. Appl. Environ. Microbiol. 64, 3923–3926.

Muroi, M., Shiragami, N., Takatsuki, A., 1994. Destruxin B, a specific and readily reversible inhibitor of vacuolar-type H + -translocating ATPase. Biochem. Biophys. Res. Commun. 205, 1358–1365.

Namasivayam, S.K.R., Bharani, R.A., Karunamoorthy, K., 2018. Insecticidal fungal metabolites fabricated chitosan nanocomposite (IM-CNC) preparation for the enhanced larvicidal activity - an effective strategy for green pesticide against economic important insect pests. Int. J. Biol. Macromol. 120, 921–944.

Niu, X., 2017. Perspectives in Sustainable Nematode Management Through Pochonia chlamydosporia Applications for Root and Rhizosphere Health, Perspectives in Sustainable Nematode Management Through *Pochonia chlamydosporia* Applications for Root and Rhizosphere Health.

Omoto, C., McCoy, C.W., 1998. Toxicity of purified fungal toxin hirsutellin A to the citrus rust mite phyllocoptruta oleivora (ash.). J. Invertebr. Pathol. 72 (3), 319–322.

Ortiz-Urquiza, A., Keyhani, N.O., 2013. Action on the surface: entomopathogenic fungi versus the insect cuticle. Insects 4, 357–374.

Pal, S., Leger, R.J.S., Wu, L.P., 2007. Fungal peptide Destruxin A plays a specific role in suppressing the innate immune response in *Drosophila melanogaster*. J. Biol. Chem. 282 (12), 8969–8977.

Reusser, F., 1967. Biosynthesis of antibiotic U-22,324, a cyclic polypeptide. J. Biol. Chem. 242, 243–247.

Roberts, D.W., St. Leger, R.J., 2004. Metarhizium spp., cosmopolitan insect-pathogenic fungi: mycological aspects. Adv. Appl. Microbiol. 54, 1–70.

Rossi, C., Tuttobello, L., Ricci, M., Casinovi, C.G., Radios, L., 1987. Leucinostatin D, a novel peptide antibiotic from *Paecilomyces marquandii*. J. Antibiot. Res. 40, 130–133.

Sabbour, M.M., 2018. Toxicity of two extracted destruxin against *Palpita unionalis* (Lepidoptera: Crambidae) under laboratory and field conditions. Am. J. Chem. 3, 1–5.

Sahayaraj, K., Tomson, M., 2010. Impact of two pathogenic fungal crude metabolites on mortality, biology and enzymes of *Dysdercus cingulatus* (Fab.) (Hemiptera: Pyrrhocoridae). J. Biopestic. 3, 163.

Sarwar, M., 2015. The dangers of pesticides associated with public health and preventing of the risks. Int. J. Bioinform. Biomed. Eng. 1 (2), 130–136.

Scarselletti, R., Faull, J.L., 1994. In vitro activity of 6-pentyl-α-pyrone, a metabolite of *Trichoderma harzianum*, in the inhibition of *Rhizoctonia solani* and *Fusarium oxysporum* f. sp. lycopersici. Mycol. Res. 98, 1207–1209.

Scherf, J.M., Milling, A., Allen, C., 2010. Moderate temperature fluctuations rapidly reduce the viability of *Ralstonia solanacearum* race 3, biovar 2, in infected geranium, tomato, and potato plants. Appl. Environ. Microbiol. 2010 (76), 7061–7067.

Schwarz, M., Köpcke, B., Weber, R.W., Sterner, O., Anke, H., 2004. 3-Hydroxypropionic acid as a nematicidal principle in endophytic fungi. Phytochemistry 65, 2239–2245.

Senthilkumar, N., Murugesan, S., Babu, D.S., 2014. Metabolite profiling of the extracts of endophytic fungi of entomopathogenic significance, *Aspergillus flavus* and *Nigrospora sphaerica* isolated from tropical tree species of India, *Tectona grandis* L. J. Agric. Life Sci. 1, 108–114.

Sharman, G.J., Try, A.C., Williams, D.H., Ainsworth, A.M., Beneyto, R., Gibson, T.M., et al., 1996. Structural elucidation of XR586, a peptaibol-like antibiotic from *Acremonium persicinum*. Biochem. J. 320, 723–728.

Shi, Y., Zhang, X., Lou, K., 2013. Isolation, characterization, and insecticidal activity of an endophyte of drunken horse grass, *Achnatherum inebrians*. J. Insect Sci. 13, 151.

Singh, B., Kaur, T., Kaur, S., Manhas, R.K., Kaur, A., 2016. Insecticidal potential of an endophytic *Cladosporium velox* against *Spodoptera litura* mediated through inhibition of alpha glycosidases. Pestic. Biochem. Phys. 131, 46–52.

Sinha, B., Biswas, I., 2011. Potential of bio-pesticides in Indian agriculture vis-a-vis rural development. SSRN Electron. J.

Sivan, A., Chet, I., 1989. Te possible role of competition between *Trichoderma harzianum* and *Fusarium oxysporium* on rhizosphere colonization. Phytopathology 79, 198–203.

Slater, G.P., Haskins, R.H., Hogge, L.R., Nesbitt, L.R., 1967. Metabolic products from a *Trichoderma viride* Pers. ex Fries. Can. J. Chem. 45 (1), 92–96.

Soman, A.G., Gloer, J.B., Angawi, R.F., Wicklow, D.T., Dowd, P.F., 2001. Vertilecanins: new phenopicolinic acid analogues from *Verticillium lecanii*. J. Nat. Prod. 64, 189–192.

Stefano, S.D., Nicoletti, R., Beato, O., 1999. Pachibasina e crisofanolo, antrachinoni prodotti dal fungo *Trichoderma aureoviride*. Il Tabacco 7, 21–24.

Suzuki, A., Kanaoka, M., Isogai, A., Murakoshi, S., Ichinoe, M., Tamura, S., 1977. Bassianolide, a new insecticidal cyclodepsipeptide from *Beauveria bassiana* and *Verticillium lecanii*. Tetrahedron Lett. 25, 2167–2170.

Thakur, A., Kaur, S., Kaur, A., Singh, V., 2012. Detrimental effects of endophytic fungus *Nigrospora* sp. on survival and development of *Spodoptera litura*. Biocontrol Sci. Technol. 22, 151–161.

Thakur, A., Kaur, S., Kaur, A., Singh, V., 2013. Enhanced resistance to *Spodoptera litura* in endophyte infected cauliflower plants. Environ. Entomol. 42, 240–246.

Tian, X., Yao, Y., Chen, G., Mao, Z., Wang, X., Xie, B., 2014. Suppression of *Meloidogyne incognita* by the endophytic fungus *Acremonium implicatum* from tomato root galls. Int. J. Pest. Manage 60, 239–245.

Tozlu, E., Tekiner, N., Kotan, R., 2018. Screening of *Trichoderma harzianum* Rifai (1969) isolates of domestic plant origin against different fungal plant pathogens for use as biopesticide. Fresenius Environ. Bull. 27, 4232–4238.

Tulloch, M., 1976. The genus *Metarhizium*. Trans. Br. Mycol. Soc. 66, 407–411.

Vinale, F., Marra, R., Scala, F., Ghisalberti, E.L., Lorito, M., Sivasithamparam, K., 2006. Major secondary metabolites produced by two commercial *Trichoderma* strains active against different phytopathogens. Lett. Appl. Microbiol. 43, 143–148.

Vinale, F., Sivasithamparam, K., Ghisalberti, E.L., Marra, R., Woo, S.L., Lorito, M., 2008a. Trichoderma-plant-pathogen interactions. Soil Biol. Biochem. 40 (1), 1–10.

Vinale, F., Sivasithamparam, K., Ghisalberti, E.L., Marra, R., Barbetti, M.J., Li, H., et al., 2008b. A novel role for *Trichoderma* secondary metabolites in the interactions with plants. Physiol. Mol. Plant Pathol. 72, 80–86.

Vinale, F., Nicoletti, R., Lacatena, F., Marra, R., Sacco, A., Lombardi, N., et al., 2017. Secondary metabolites from the endophytic fungus *Talaromyces pinophilus*. Nat. Prod. Res. 31, 1778–1785.

Vurro, M., Zonno, M.C., Evidente, A., Andolfi, A., Montemurro, P., 2001. Enhancement of efficacy of *Ascochyta caulina* to control *Chenopodium album* by use of phytotoxins and reduced rates of herbicides. Biol. Control. 21 (2), 182–190.

Wainwright, M., Betts, R.P., Teale, D.M., 1986. Antibiotic activity of oosporein from *Verticillium psalliotae*. Trans. Br. Mycol. Soc. 86, 168–170.

Walker, A.K., Frasz, S.L., Seifert, K.A., Miller, J.D., Mondo, S.J., LaButti, K., et al., 2016. Full genome of *Phialocephala scopiformis* DAOMC 229536, a fungal endophyte of spruce producing the potent anti-insectan compound rugulosin. Genome Announc. 4, e01768–15.

Wang, C., St. Leger, R.J., 2006. A collagenous protective coat enables *Metarhizium anisopliae* to evade insect immune responses. Proc. Natl. Acad. Sci. U. S. A 103, 6647–6652.

Wang, C., St. Leger, R.J., 2007. A scorpion neurotoxin increases the potency of a fungal insecticide. Nat. Biotechnol. 2007 (25), 1455–1456.

Wang, J., Hu, W., Hu, Q., 2019. BmTudor-sn is a binding protein of destruxin A in silkworm Bm12 cells. Toxins 11, 67.

Wang, Q., Xu, L., 2012. Beauvericin, a bioactive compound produced by fungi: a short review. Molecules 17 (3), 2367–2377.

Whilhite, S.E., Lumsden, R.D., Straney, D.C., 1994. Mutational analysis of gliotoxin production by the biocontrol fungus *Gliocladium virens* in relation to suppression of Pythium damping-off. Phytopathology 84 (8), 816–821.

You, F., Han, T., Wu, J. z, Huang, B. k, Qin, L. p, 2009. Antifungal secondary metabolites from endophytic *Verticillium* sp. Biochem. Syst. Ecol. 37, 162–165.

Zhang, S., Xia, Y.X., Kim, B., Keyhani, N.O., 2011. Two hydrophobins are involved in fungal spore coat rodlet layer assembly and each play distinct roles in surface interactions, development and pathogenesis in the entomopathogenic fungus, *Beauveria bassiana*. Mol. Microbiol. 80, 811–826.

Yusuf, M.A., Singh, B.N., Sudheer, S., Kharwar, R.N., Siddiqui, S., Abdel-Azeem, A.M., et al., 2019. Chrysophanol: a natural anthraquinone with multifaceted biotherapeutic potential. Biomolecules 9 (2), 68.

Zhang, H., Hu, W., Xiao, M., Ou, S., Hu, Q., 2017. Destruxin A induces and binds HSPs in *Bombyx mori* Bm12 cells. J. Agric. Food Chem. 65, 9849–9853.

Zheng, P., Xia, Y., Xiao, G., Xiong, C., Hu, X., Zhang, S., et al., 2011. Genome sequence of the insect pathogenic fungus *Cordyceps militaris*, a valued traditional chinese medicine. Genome Biol. 12, 1–21.

Chapter 19

Endophytes producing active constituents in *Centella asiatica* with a special emphasis on asiaticoside and madecassoside: a review update

Shreya Sikdar Mitra[1], Anuradha Mukherjee[2], Potshanghbam Nongdam[3], Devendra Kumar Pandey[4] and Abhijit Dey[1]

[1]*Department of Life Sciences, Presidency University, Kolkata, India,* [2]*MMHS, Joynagar, India,* [3]*Department of Biotechnology, Manipur University, Imphal, India,* [4]*Department of Biotechnology, Lovely Faculty of Technology and Sciences, Lovely Professional University, Jalandhar, India*

Chapter Outline

Volatiles and Metabolites of Microbes. DOI: https://doi.org/10.1016/B978-0-12-824523-1.00013-4

19.1 Introduction

A diverse array of organic compounds often referred to as the phytochemicals or secondary metabolites are procured from the vascular plants. Various roles are played by these molecules in the entire duration of a plantâ€™s life, stretching across from configurational roles to the protective ones. Plants producing such chemicals of medicinal significance are habitually entitled as the medicinal plants. Such plants providing various traditional phytochemicals are conventionally deployed by the pharmaceutical companies for production of diverse range of formulations. Modern times have seen the steep ascent in the use of botanicals all over the globe. According to the World Health Organization (WHO) approximately 80% of the total global population are found dependent on different phytochemicals (Ekor, 2014). The vital role played by the medicinal plant communities are mainly due to the occurrence of discrete chemical compounds that are found to generate several beneficial physiological impacts on the human body. Extensive range of bioactive chemical components of different plants include saponin, alkaloids, flavanoids, tannins, sterols, and phenols. At times, drugs obtained from the medicinal plants are favored over other available drugs for their uncomplicated and broad spectrum activity. Additionally, adverse side effects associated with these drugs is marginal in contrast to the chemotherapeutic drugs. Medicinal plants heterogeneity is found bountiful in India that is a repository of huge floral diversity. Since prehistoric times, use of phytochemical-based drugs is found ubiquitously in the customary age-old medicinals like Ayurveda, Siddha, and Unani. Globally around 426 biomes with a variety of habitat diversity have emerged as the richest centers for plant genetic resources (Ravishankar and Shukla, 2007). Among 18,665 flowering species, only around 3000 plants have been utilized for the multifarious formulations in traditional systems of medicines like Ayurveda, Unani, and Siddha (Schippmann et al., 2006). Herbal treatments mainly comprise of plants or raw plant extracts containing an assorted range of components that are commonly believed to have a collaborative effect. The neoteric reemergence of common interest in herbal methods can be chiefly assigned to the following causes shown in Fig. 19.1.

But this extensive exploitation of plant resources for the purpose of harnessing miscellaneous scopes of medicinally important chemicals serve as potential threat for the plant species driving them on the verge of extinction. Investigating alternative techniques in isolation of the phytochemicals without pestering the

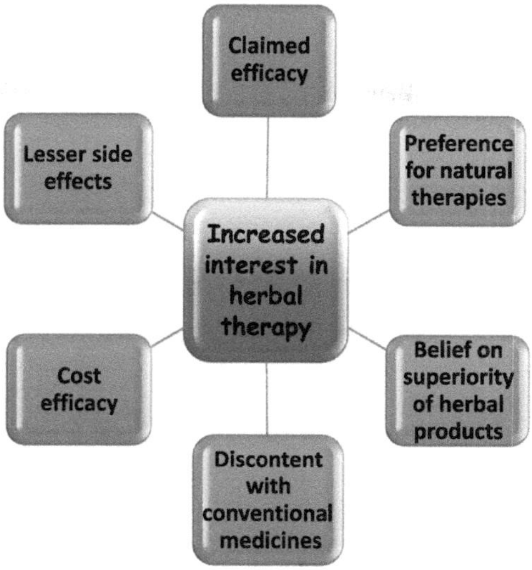

FIGURE 19.1 The causes behind soaring demand of herbal medicines.

plant communities is the current need. Fungal endophytic colonizers of the plants is one such inevitable tool in this field.

19.1.1 Endophytes

In the year 1866, de Bary for the first time made use of the term "endophyte" (Gr. *endon*, within, *phyton*, plant). The microbial community involving fungi, bacteria, or viruses that are mainly found to reside inside plant cells come under the broad purview of the term endophyte. Literal meaning of the word endophyte stands for "in the plant" (endon Gr. = within, phyton = plant). The use of this terminology is as broad as its literal definition encompasses around the gamut of potential hosts and occupants, such as bacteria, fungi, plants, algae, and insects in plants (Kobayashi and Palumbo, 2000; Stone et al., 2000; Marler et al., 1999; Feller and Mathis, 1997; Peters, 1991). Any organ of the host can be inhabited. Similarly diversified is the term "endophyte" used for the flexible strategies of the life history of symbiosis, encompassing facultatively saprobic to parasitic, and exploitive to mutualistic. These sets of organisms apparently are found inside the host tissue without apparently causing any disease.

19.1.2 Centella asiatica

Centella asiatica (L.) Urban, synonym *Hydrocotyle asiatica* L. (Eng. Indian pennywort, French *Hydrocotyle asiatique*, German *Asiatischer Wassernabel*) ordinarily known as Gotu kola or Tiger Grass, Asiatic pennywort, Indian pennywort, or Spade leaf, is the member of the Umbelliferae (Apiaceae) family. It is a nonwoody, slender, weakly aromatic, and usually a creeping tropical perennial herb valued for its medicinal aspects. It is indigenous to the South East Asian countries such as India, Indonesia, China, Malaysia, and Sri Lanka, as well as in the African continent in regions like South Africa and Madagascar and also found in the warmer regions of Oceania and Central America (Jamil et al., 2007). It is mainly endemic to the hotter provinces of both the hemispheres. In the undomesticated and uncultivated conditions this plant is found to survive in humid and shady sites up to an elevation of 7000 ft. occasionally and can be frequently found to survive in various aqueous habitats such as the streams, ponds, river banks, and irrigated fields. It is also found to grow by the stone walls or rocky regions at a height of approximately 2000 ft. in India and Sri Lanka (Chandrika and Kumara, 2015). It can grow up to 40 cm tall. The plant is reputed for its sizable medical significance that can be mainly assigned to the discrete groups of bioactive molecules found in the plant.

19.2 Historical background

In conventional fables and folklore medicine there is a mention of the plant as the cure of a variety of diseases and has been recognized for different therapeutic properties. Different forms of existing literature source reveals that Gotu kola had medicinal application in the Indian province since eternity. There has been notable use of this plant in the Indian Ayurvedic system and is well perpetuated for boosting lastingness in individuals. As time progressed, its utilitarian values were perceived as it was promoted for skin therapy topically and internally. Thus, it acted as a healing agent against leprosy, lupus, and eczema. There is the mention about the plant "Manduka pami" in the ancient Hindu text *Susruta Samhita*, which in modern times is believed perhaps to be *C. asiatica*. In the 19th century, the American eclectics who were quite accustomed to different types of medicinal properties associated with the plants that were used in the treatment of leprosy also noted the application of close relatives of *C. asiatica*. In France, during the 1800s the herb extract was believed to be a drug molecule. It has also been documented that the ultimate cure of Dr. Boiteau in 1852, distressed with leprosy for a long time, could be attributed to the Gotu kola treatment used while suffering from leprosy for several years. At the Centennial International Exhibition of Melbourne in 1888 the display of the Gotu kola juice as a medicine by the group of doctors from Brisbane was another mentionable event (Sayasinha et al., 1999). In China, Gotu kola was famed as an antiaging agent, and reports made reference to a Chinese herbalist, Li Ching Yun, who was reported to have survived miraculously for 256 years, surviving 23 wives, believed to have happened due to a consistent usage of Gotu kola (Sayasinha et al., 1999).

19.3 Bioactive compounds

Different varieties of secondary metabolic molecules are found as the active ingredients of *C. asiatica* mainly incorporating the pentacyclic triterpenoids saponins and sapogenins accumulatively recognized as centellosides, which are acclaimed for their hallmarked antimicrobial and defensive functions against the pathogens. Nearly around 25 triterpenoid compounds have been reported from *C. asiatica* in the existing literature sources, although identical, synonyms and disputed findings are there too (Gallego et al., 2017). These broad category of terpenoid saponins include mainly the molecules, namely asiaticoside, madecassoside, brahmoside, thankuniside, sceffoleoside, and centellose, and among them the primary and the profusely occurring sapogenins are the asiaticoside, madecassoside asiatic, brahmic, centellic, and madecassic acids. The dominant biological activity showing molecules among these are asiatic acid, madecassic acid, asiaticoside, and madecassoside depicted in Table 19.1. However disparity in the quantity and content of triterpene components are found in accordance with the local and the different types of associated environmental features. Other groups of active secondary metabolites found in this plant are volatile oils, tannins, flavonoids, phytosterol, resins, mucilages, fatty acids, free amino acids, and sugars (Bylka et al., 2013; James and Dubery, 2009).

19.4 Biosynthesis pathway of the centellosides

Centellosides belong to the group of the triterpene saponins and sapogenins found to possess an ursane or oleane skeleton, illustrated in Table 19.1. Triterpenes are 30 carbon atoms bearing terpenoids that emerge by the conjugation of two farnesyl diphosphate (FPP) units to forming a transitional molecule named squalene. Fig. 19.2 provides the depiction of the entire biosynthetic pathway of this group of pentacyclic triterpenoids. Head-to-tail fusion of dimethylallyl diphosphate (DMAPP) and two isopentyl diphosphate (IPP) molecules gives rise to FPP (Croteau et al., 2000). The familiar secondary metabolite forming pathways such as cytosolic mevalonate pathway or the plastidial methylerythritol phosphate pathway (MEP) forming pyruvate and phosphoglyceraldeyde lead to the synthesis of the precursors IPP and DMAPP (Croteau et al., 2000; Augustin et al., 2011). Farnesyl diphosphate synthase (FPS) accelerates the fusion of two molecules of IPP with DMAPP (obtained via the mevalonate pathway) to give FPP (Fig. 19.3), the routine antecedent pathway maximum of the sesquiterpenes produced by plants (Augustin et al., 2011). The head-to-head union of two molecules of FPP to obtain squalene is regulated by another catalyst, namely squalene synthase (SQS) that is a single microsomal polypeptide. Conversion of squalene to the common transitional product in the biosynthesis of steroids and triterpenoids, namely (3S) squalene 2,3-epoxide, is again assisted by the action of the

TABLE 19.1 Dominant bioactive compounds in *C. asiatica.*

Name of the compound	Molecular formula and average mass	Class of the compound	Structures of the compound (from https://www.chemspider.com)
Asiaticoside	$C_{48}H_{78}O_{19}$, 959.122 Da	Triterpenoid saponin	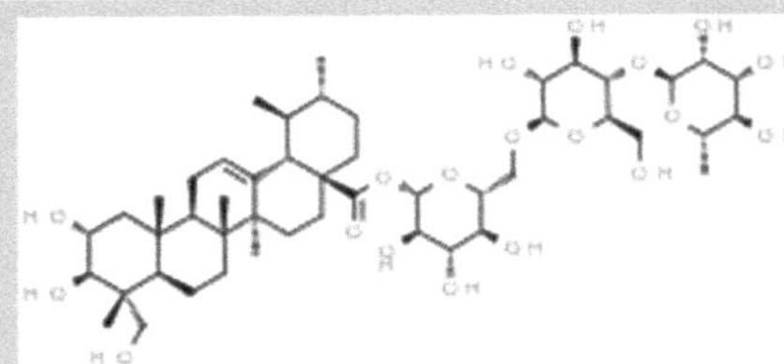
Madecassoside	$C_{48}H_{78}O_{20}$, 975.121 Da	Triterpenoid saponin	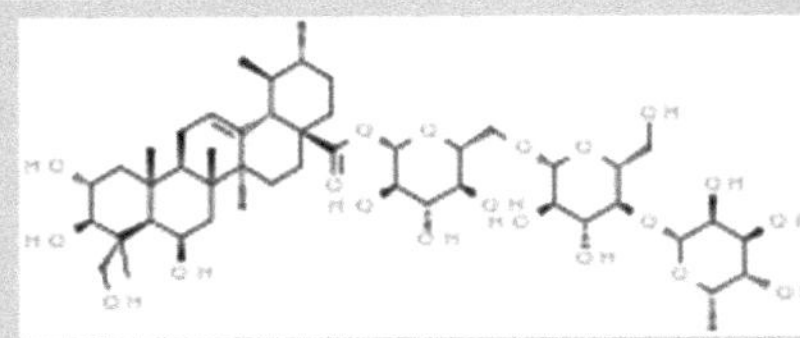
Asiatic acid	$C_{30}H_{48}O$, 488.699 Da	Aglycone of pentacyclic triterpenoid	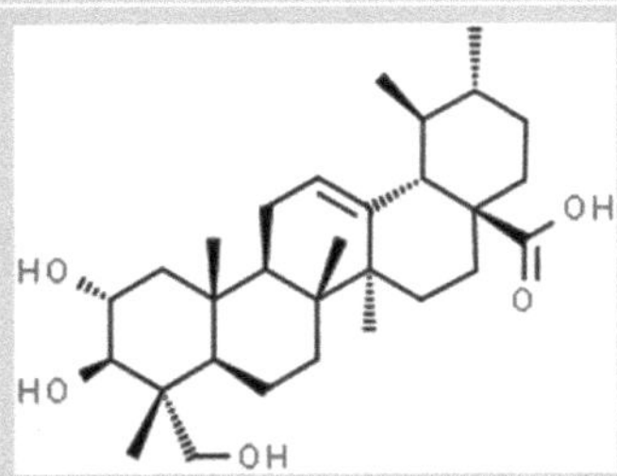
Madecassic acid	$C_{30}H_{48}O$, 504.698 Da	Aglycone of pentacyclic triterpenoid	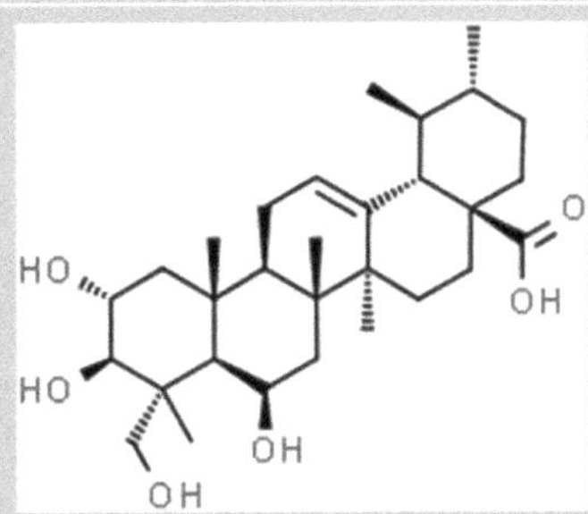

microsomal squalene epoxidase that needs the factors O_2 and NADPH (Christou and Klee, 2004). There is another group of enzymes namely the 2,3-oxidosqualene cyclases (OSCs) are enzymes that regulate the cyclization of squalene to cyclic triterpene alcohols with diverse structural modifications

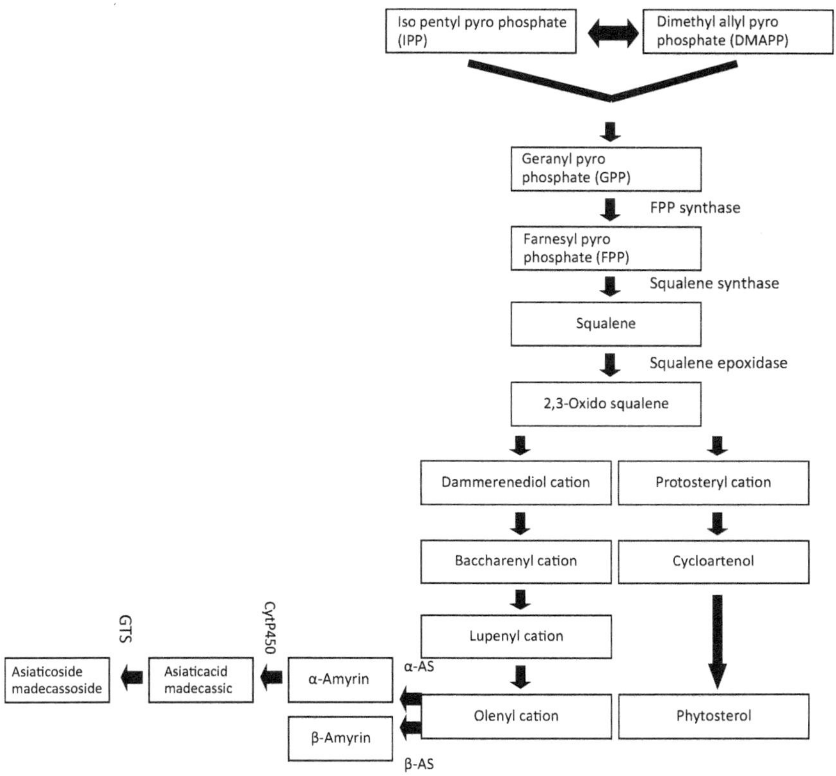

FIGURE 19.2 Biosynthetic pathway of centelloside production.

that results in the generation of molecules with different types of ring structures (Augustin et al., 2011). OSC is responsible for controlling the cyclization of 2,3-oxidosqualene to the protosteryl cation or dammarenyl cation, which is the bifurcating point in the pathway of synthesis of either the sterols and triterpenoid saponins. In between there are formation of different types of intermediate products of various types of structural conformations like protosteryl cation, dammarenyl cation, cycloartenol, lupenyl cation, and olenyl cation, all depicted in Fig. 19.2, which are utilized for the synthesis of various significant groups. The cycloartenol cation acts as the progenitor for the synthesis of the valued metabolite phytosterols that are often found as the dominant structural membrane constituents (Augustin et al., 2011). Another significant group of enzymes are the α/β-amyrin synthases (α/β ASs), which aid in the conversion of the oleanyl cation that emerge from the lupenyl cation to α-amyrin (ursane skeleton) or β-amyrin (oleane skeleton) (Christou and Klee, 2004). The endmost biosynthetic steps of the centelloside pathway have still not been worked out completely but involve various sets, but most likely enzymes and reactions commonly occur in plants (Mugford and Osbourn, 2012). Following cyclization, further diversity is produced through the modification of the products by oxidation, glycosylation, hydroxylation, and other substitutions controlled primarily by cytochrome P_{450}-dependent monooxygenases, glycosyltransferases, and other enzymes. Scant knowledge exists about the enzymes needed for these chemical transformations. In the case of saponins, the oligosaccharide chains are likely produced by the sequential addition of single sugar residues to the aglycone, but triterpenoid glycosylation is not yet elucidated (Augustin et al., 2011). UDP-glucosyltransferases (UGTs) assist in the process of glycosylation of the asiatic acid and madecassic acid to yield the final desired products asiaticoside and madecassoside (Augustin et al., 2011).

19.5 Salient medicinal and pharmacological uses

Diverse reports on the biological activities of *C. asiatica* are available. Broadly the various biologically useful activities of this plant are enumerated in the following paragraphs. Fig. 19.3 summarizes the medicinal uses of *C. asiatica*.

19.5.1 Memory booster and cognitive function improvement

Notable consequences of the water extract of the herb are found on the learning and memory retention of individuals due to declining levels of various

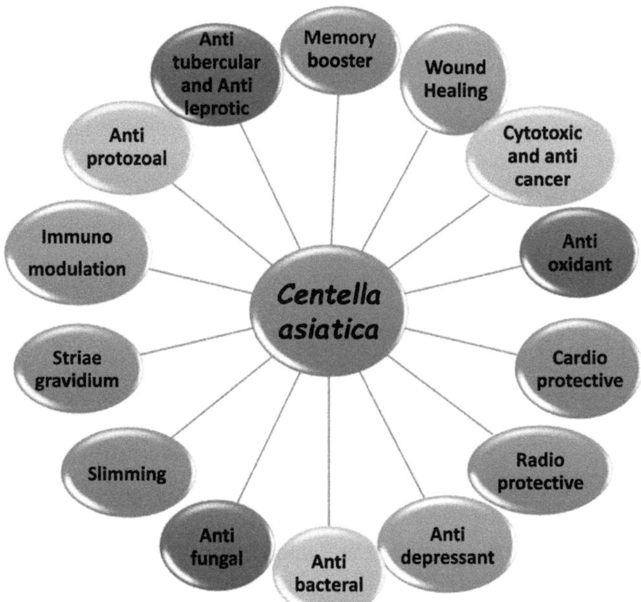

FIGURE 19.3 Summarized medicinal uses of *Centella asiatica*.

neurotransmitters such as dopamine, norepinephrine, and 5-HT and their metabolites in the brain (Nalini et al., 1992). Various set of chemicals like brahmic acid, brahminoside, isobrahmic acid, and brahmoside occur in *C. asiatica* that one way or the other contribute to this property of the herb. Also various other aspects of the herb include psychotropic, sedative, and anticonvulsant properties. In addition, it has functionality in anxiety, mental disorders, and dementia. One such polyherbal amalgamation, Mentat, which contains synergistic action of the different herbs, is found to cause an enhancement of attention, concentration, and memory in children showing various kinds of learning disabilities (Saha et al., 2002).

19.5.2 Wound healing activity

The plant is also associated with the property of healing the wound through an increase in collagen content, DNA, and protein in the cell along with proliferation of the fibroblasts and production of extracellular matrix found in curing of the wounds. An extract known by the name Madecassol, containing madecassic acid, asiaticoside, and asiatic acid as the principal components, is found to have a role in the process of grafting wounds (Srivastava et al., 1997).

19.5.3 Cytotoxic and antitumor activities

Asiatic acid was found as the bioactive compound of the plant possessing antiinflammatory and antiapoptotic properties contributing to its ability of controlling tumor growth, cytotoxicity, and subsequent associated malignancy. The crude extract from *C. asiatica*, along with its semipurified fractions, when administered orally in mice with solid Ehrlich Ascites tumor, induced apoptosis, which was associated with acceleration of the longevity of the affected mice (Babu and Paddikkala, 1994). The efficacy of asiatic acid was noted against skin cancer by Park et al. (2005) and lung cancer by Wu et al. (2017).

19.5.4 Antioxidant activity

Free radical and reactive oxygen species (ROS) generated through various ongoing chemical reactions inside the cells contributes to oxidative stress. All organisms have developed several types of mechanisms to combat this free radical and protect the organisms from the damaging oxidative stresses. In Sprague-Dawley rats, *C. asiatica* powder as well as the plant extract were evaluated for retardation of oxidative stress and the obtained results showed notable drop in the generation of ROS and oxidative stress in rats (Hussin et al., 2007). Steam distillation extraction of essential oil of *C. asiatica* showed exceptional antioxidant nature for lipid-containing food. Its activity was comparable enough with the butylhydroxyanisole (BHA), a synthetic antioxidant (Raza et al., 2009).

Abundance of flavonoid, β-carotene, polyphenol, tannin, vitamin C, etc. found in *C. asiatica* were found to be responsible for its notable antioxidant property (Chandrika and Kumara, 2015). Constant supplementation for 14 days of crude *C. asiatica* methanolic extract in lymphoma-bearing mice resulted in an elevation of the antioxidant enzymes levels and a fall in the ascorbic acid levels (Jayashree et al., 2003).

19.5.5 Cardio-protective property

In the rats showing ischemia-reperfusion, stimulated myocardial infraction was found to be obstructed by the whole plant alcoholic extract administration leading to cardio-protective functions (Pragada et al., 2004).

19.5.6 Radio protective activity

C. asiatica was found to display another salient feature of radioprotection during clinical radiotherapy. It was found to have potency to control different behavioral changes arising as the effect of radiation (Shobi and Goel, 2001).

19.5.7 Antidepressant property

The majority of the triterpene group of molecules found associated with the plant was found to alleviate depression in individuals by causing noteworthy diminishment of the corticosterone level in serum imparting its antidepressant feature (Chen et al., 2005).

19.5.8 Immunomodulatory activity

Constituent triterpenoid saponin and isolated pectin from *C. asiatica* showed immune system revitalizing activities (Singh et al., 2010). In addition, initial immunomodulatory functions were shown by the methanol extracts from the plants (Jayathirtha and Mishra, 2004).

19.5.9 Antibacterial and antifungal activities

The plant showed activity against diverse bacteria that are Gram-positive (*Bacillus subtilis* and *Staphylococcus aureus*) and Gram-negative (*Pseudomonas aeruginosa, Escherichia coli*, and *Shigella sonnei*), and against a few fungi, such as *Candida albicans* and *Aspergillus niger* (Prakash et al., 2017).

19.5.10 Antiprotozoal activity

C. asiatica was found associated with antiprotozoal activity, and complying with the property, a whole plant alcoholic extract showed effective antiprotozoal activity against the protozoan *Entamoeba histolytica* (Dhar et al., 1968).

19.5.11 Antitubercular and antileprotic activities

The primary active molecule of the plant asiaticoside has been found effective in the treatment of leprosy and various types of tuberculosis (Singh et al., 2010).

19.5.12 Slimming role

A pronounced rise in the cyclic adenosine monophosphate content was found associated with successive increase in the nonesterified fatty acids content in human adipocytes through administering of the *C. asiatica* extracts controlling fat deposition and aiding in the process of slimming (Singh et al., 2010).

19.5.13 Striae gravidarum

This is a condition well known as pregnancy stretch marks and are scars on the skin in the abdominal region due to massive dilation of the uterus along with weight gain during pregnancy. Application of an ointment that included *Centella* extract, α-tocopherol, and collagen-elastin hydrolysates resulted in reduction of the number of women developing stretch marks (Young and Jewell, 1996).

19.6 Elicitation

Elicitation is an effective and potential technique used in the biotechnological field for upscaling the generation of plant-derived secondary metabolites. Elicitors are such components that are responsible for triggering the plant immune system, boosting the secondary metabolism pathways inside the plants for the purpose of shielding the plant cells as well as the whole plant. They can be broadly categorized as two groups: (1) abiotic and (2) biotic. As evident from the titles one indicates the abiotic ones, primarily the inorganic group of compounds, and the other the biological or the living ones mainly encompassing the microbial category. In this section we will be mainly discussing the biotic elicitors on our plant of interest. The maximum amount of the biotic elicitors are accredited by the cell membrane-bound particular receptors that covey the signal to the cell via activation of the signal transduction system, initiating various responses associated with the plant defense system such as phytoalexin production, etc. Numerous factors are responsible for controlling the response of the plants toward the elicitors. But this technique is a unique one, which is sustainable and can influence the generation of the bioactive molecules not bothering the available plant resources.

19.7 Fungal endophytic elicitation for bioactive compound synthesis

Scientists assert the recognition of around 100,000 species of fungi to date; however the number may exceed over 1 million. Bioactive compounds, especially those derived from the plant-associated microbes such as endophytes, remains still an unexplored domain (Staniek et al., 2008). Therefore, there is an immense opportunity to explore novel bioactive compounds from such microbes among zillions of plants occupants across various niches and ecosystems. Over the past few decades, the endophytic fungi have been known as a prolific storehouse of natural bioactive constituents with significant promise in the pharmaceutical industry (Mousa and Raizada, 2013). Fascinatingly a number of these secondary metabolites are similar to the ones produced by their hosts and could serve as the replacements for varied and notable plant-based compounds (Mousa and Raizada, 2013). The existing pattern of plant endophytic relations cause stress and switch on different biochemical pathway, inducing greater secondary metabolite production in response to the imposed stress. Despite scores of research on bioactive molecules from fungal endophytes, endophyte research is still in the nascent stage since scant success has been attained in commercial exploitation of such bioactive molecules. Therefore, to recognize the tangible efficiency of endophytic fungi, a complete understanding of ecological, chemical, biochemical, physiological, biotechnological, and bioinformatical aspects is needed. The present review predominantly discusses few fungal colonizers as reported in the plant *C. asiatica* with number of possible future domains that still remain open for exploration.

19.8 Endophytes associated with *C. asiatica* and their utilization

Extensive distribution of endophytic fungi within higher plants has been recognized as a promising source in the manufacture of various bioactive ingredients. These groups of endophytic colonizers have the capability of synthesizing diverse categories of secondary metabolites possessing different sorts of therapeutic activities. The process of fungal elicitation for secondary metabolite generation is definitely an important step for protection of the endangered plants along with obtaining a substantial amount of the desired product. In this literature, few such endophytic fungal elicitors in the case of *C. asatica* are discussed in the following sections.

Prasad et al. (2013) studied the impact of the fungi acting as elicitor molecules on the asiaticoside and biomass production through the use of *C. asiatica* multiple shoot cultures in a dose, and culture age dependent manner. Addition of the fungal culture filtrate (CF) *Trichoderma harzianum* in the 3% v/v in the growth medium on day 10 of a 35-day culture cycle and saw an upshot in the dry weight culture yield of asiaticoside, which was almost 2.53 and 2.35 times higher compared to the used control shoots. There was also a significant increase in the biomass accumulation about 1.24-fold greater than that in the control used. On the contrary, induction of shoots fortified with mycelial extract (ME) of the fungus *Colletotrichum lindemuthianum* (1.5% v/v; added on day 0) although assisted in highest biomass cumulatively of all the elicitation treatments but led to a drop in the asiaticoside content of 1.10 mg/g dry weight, which was almost 3.5 and 8.7 times lower than in the control or *T. harzianum* CF-treated shoots, respectively. Administration of mycelial extract of another fungus *Fusarium oxysporum* (0.5%−1.5% v/v), in general, proved a shoot growth hindrance if added on 0 day of the culture cycle with GI = 4.85−8.45 compared to 11.11 in the case of control along with stunted yield of asiaticoside only about 0.18−0.42 mg dry weight culture^{-1}. Though there was minimal enhancement in the shoot biomass accumulation (GI = 5.68−11.94) over the untreated control, when on day 30, *F. oxysporum* ME (0.5%−1.5% v/v) was fortified in the medium but the asiaticoside production (0.18−0.94 mg dry weight culture^{-1}) was less. Hence the result from this study was first of its type to illustrate the possible application of the fungus *T. harzianum* CF in upscaling the triterpenoid pathway in the medicinal plant *C. asiatica*.

Gupta et al. (2018) reported an elevated production of asiaticoside due to the endophytic relationship shared with the fungus *Colletotrichum gloeosporioides*. A total of 13 endophytic fungi were isolated chiefly from the plant leaves which were found related with the asaticoside production. The discrete authentication of this asiaticoside-producing endophyte was done by employing the technique of internal transcribed spacer-based rDNA sequencing.

LC-MS was used for the chemical certification of the existence of asiaticoside in *C. gloeosporioides* ethyl acetate extract. The synthesis of asiaticoside was calculated in respect to incubation time and subculture generation showed the presence of 62.29 ± 3.36 μg/100 mL of asiaticoside using *C. gloeosporioides* on the 15th day in the first subculture generation, which was followed by a decline in the subsequent generations. Similar type of correlation was also found for the yield and the growth curve of *C. gloeosporioides*. There was a positive consistency found between the synthesis of asiaticoside and the obtained yield.

Jisha et al. (2018) reported an enhancement in asiaticoside production by the use of one such model symbiotic fungal elicitor *Piriformospora indica*. *P. indica*, similar to endophytic fungi, showed colonization with *C. asiatica*. It resulted in different effects like escalated secondary metabolite production, increased production of plant biomass, and an acceleration in the transcriptional level expression of miRNA related with a stress-combating mechanism, perhaps as the result of the endosymbiosis due to the evoking of the plant defense system.

19.8.1 Isolation of endophytic strains

Isolation of endophytic fungi is an initial step of great significance. If the isolation method does not occur properly, the entire process will be voided. Isolation mainly involves surface sterilization of the material, maceration of plant tissues, and culture of the endophytes in an appropriate medium followed by incubation. The sterilization of the plant tissue for endophyte isolation is a step demanding paramount focus since the absence of proper sterilization will generate subsequent microbial contamination of the material, which might give a false notion of the endophytic fungi, and the entire efficiency of the whole experiment will be at stake with misleading outcomes. Washing the plant material with distilled water was followed by the explant surface sterilization method through already available and established protocols with slight modifications. The explants of *C. asiatica* generally are placed under running tap water for washing followed by Tween-20 wash (3–4 times), dipped in 0.1% Bavistin (20 minutes) with constant shaking and then rinsing by sterile water almost 4–5 times. Often further surface sterilization by the use of 70% ethanol for 30 seconds, followed by dipping in 2% sodium hypochlorite for a minute, and again treated with 70% ethanol for 30 seconds and sterile water wash for 3–4 times is also practiced. Finally, the maceration of the sterilized leaf segments is done by cutting with a surgical blade and placing the material in potato dextrose agar (PDA) medium along with 50 mg/L streptomycin for avoiding any sort of bacterial contaminations. Parafilm-sealed petri dishes are placed for incubation at $28^{\circ}C \pm 2^{\circ}C$ in the incubator. The cultures are carefully observed almost every day to see if there is any growth of the endophytic fungal colonies in the leaf segments. Subculturing of the isolated hyphal tips are done on the PDA medium. The morphological characterizations of isolated strains are mainly done using a compound microscope. Slides are normally prepared in lactophenol cotton blue for spore observation and other characteristic features (Gupta et al., 2018). Various pace of growth patterns are observed in the case of different isolated strains. Lastly the obtained pure culture needs to be preserved at $-70^{\circ}C/-130^{\circ}C$ for long-term use. At this extreme temperature, all sorts of biochemical and physiological activities of the respective organisms get halted.

19.8.2 Extraction of fungal metabolites

Following the isolation of the endophytic fungi and maintaining the pure culture, the succeeding move is to harness the chief metabolic products from the microbial species. All literature sources mention normalized and optimized methods of extraction. At the beginning inoculation of the culture of the endophytic fungal strain is mainly done in the potato dextrose broth (PDB) medium following incubation at 28°C with constant agitation until turbidity is reached in the culture. The extraction of the metabolite is primarily done using solvent extraction principle mainly by the use of organic

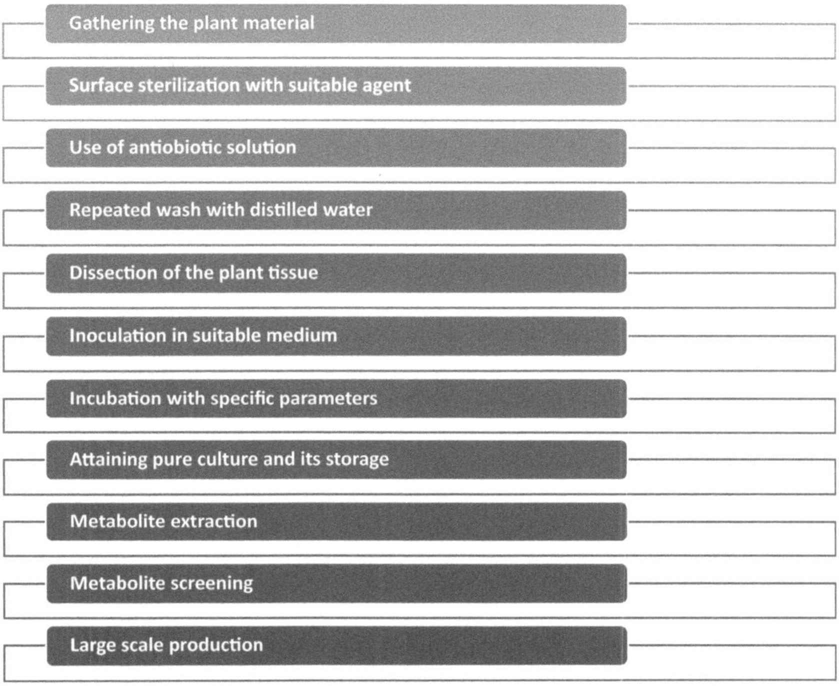

FIGURE 19.4 Flow chart of the discussed process.

solvents such as the methanol or ethyl acetate. The solvent is generally evaporated using a rotary evaporator followed by further dissolution in some other suitable solvents like methanol (Gupta et al., 2018). Ethyl acetate, formic acid, and isopropanol can also be used for the purpose of extraction. Maximum absorbance of the required compound can be checked using the UV spectrophotometer. Confirmation and quantification of the obtained metabolite is mainly done through the preparative analysis of HPLC, LC-MS, and GC-MS that serve as the ultimate confirmatory step for the desired bioactive molecule. Industrial fermentation and various other biotechnological techniques can be worked for the purpose of large-scale synthesis of the interested compound per the needs. The detailed steps of the entire process from isolation to extraction is summarized in a flow chart in Fig. 19.4.

19.9 Limitations in pentacyclic triterpenoid synthesis in *C. asiatica*

Employing fungal endophytes to serve industry-oriented and commercialized purposes has its own set of constraints. Lack of inclusive knowledge in this domain on ecological parameters governing the endophytic interlink age and activities is one such vital shortcoming. The yield and the performance of the secondary metabolite production in the scale-up processes are reliant on a diverse set of parameters. Also several endophytes at times are found instead acting as pathogens with prospective amount of toxicity. Although there are reports on the existence of a great diversity of fungal and even bacterial colonizers associated with *C. asiatica* (Rakotoniriana et al., 2008), marginal among them have been shown to have direct correlations beneficial for the production of the triterpenoids. In fact, research in this domain of the effect of the varied endophytic fungal as well as bacterial members on the secondary metabolite production of *C. asiatica* is a long way to go. Also another point to be mentioned is the complexity of the biosynthetic pathways, which often is related to low yields and thus is commercially unrealistic. Low yields as well as drops in the secondary metabolite quantity on repeated rounds of subculturing of the plant tissue is yet another hurdle to overcome.

19.10 Future avenues for endophyte-mediated centelloside production

The endless hunt for novel drug molecules by the pharmaceutical sector has stressed the procuring bioactive natural compounds from diversified sources. Endophyte-mediated bioactive compound synthesis and their application as alternative therapeutics is one of the contemporary areas of research of utmost gravity as it has its own set of utilities discussed already in the literature. The production of the asiaticoside, madecassoside, and other triterpenoid molecules are mainly obtained from the plant *C. asiatica*. Hence, such endophytic fungi-induced metabolite generation is no doubt of great significance for the purpose of conservation of this plant having immense beneficial roles. The number of reported endophytic colonizers, though, are not many, so obviously there are several impending tasks left on the discovery of more numbers of such colonizers. Putting substantial stress on the bacterial species as the potent candidates for this endophytic relationship is also essential. There are reports highlighting the role of bacteria in the generation of such bioactive compounds, in particular, in the stages of their metabolic pathways while colonizing as endophytes inside the host medicinal plants. Since many reports were retrieved from the database, all mainly focused on the fungal strains associated with different plant parts. The bacterial endophytic association with the plant rarely was found, and barely noted in reports such as Rafat et al. (2012). Immense effort has to be put into the identification and isolation of such associating bacteria.

Bioactive compounds are the specific set of compounds mainly produced through different metabolic pathways in the living system with diversified pharmaceutical consequences and having comparatively reduced side effects. Hence pharmacological parameters of the endophytic metabolites will definitely turn out to be a convenient device in the path of sustainable drug discovery programs. For proper profit optimization and commercialization of such natural resources, friendly endophytic strains along with discrete sets of policies and rules are needed to be set up, along with ample and supportive funding.

19.11 Conclusion

Deploying the variety of endophytic assets for the production of a range of bioactive molecules, such as the earlier-discussed asiaticoside and madecassoside, is a new avenue that can be optimized for meeting the huge demands of these molecules across the various strata through the globe. In parallel, the use of these substitute sources will definitely help to alleviate the vulnerability of the phytoresources arising from their overexploitation. Preliminarily the potential plants for the purpose needs to be meticulously scanned following the endophyte isolation as per the ethnomedical comprehension. Criticality over the plant selection process will guarantee the deserving endophytic candidates compared to an unplanned mass screening. In-depth characterization and segregation of the required microbial isolates along with building unique analytical studies of the particular metabolites will aid in the process of unraveling pioneer bioactive molecules. Basic clarity about the host-endophytic interaction needs to be understood for proper appreciation of the metabolic contribution by the endophyte. All these will affirm the existence of the bioactive compounds and their derived products from the microbial metabolites. Sizable comprehension of multifarious factors associated with the plant-microbe interaction has to be present. Hence more subsequent exploration has to be carried out for the discovery of such endophytic colonizers associated with medicinal plants like *C. asiatica*, which can shelter the plant community along with harnessing the desired molecule.

Conflicts of interest

None.

References

Augustin, J.M., Kuzina, V., Andersen, S.B., Bak, S., 2011. Molecular activities, biosynthesis and evolution of triterpenoid saponins. Phytochemistry 72 (6), 435–457.

Babu, T.D., Paddikkala, J., 1994. DNA fragmentation in Ehrlich Ascites tumour cells by extract of herbal plant *Centella asiatica* (L.). Amala Res. Bull. 14 (4).

Bylka, W., Znajdek-Awiżeń, P., Studzińska-Sroka, E., Brzezińska, M., 2013. Centella asiatica in cosmetology. Postepy Dermatol. Alergol. 30 (1), 46.

Chandrika, U.G., Kumara, P.A.P., 2015. Gotu kola (*Centella asiatica*): nutritional properties and plausible health benefits, Advances in Food and Nutrition Research, Vol. 76. Academic Press, pp. 125–157.

Chen, Y., Han, T., Rui, Y., Yin, M., Qin, L., Zheng, H., 2005. Effects of total triterpenes of *Centella asiatica* on the corticosterone levels in serum and contents of monoamine in depression rat brain. Zhong yao cai = Zhongyaocai = J. Chin. Med. Mater. 28 (6), 492–496.

Christou, P., Klee, H., 2004. Handbook of Plant Biotechnology. John Wiley and Sons, No. LC-0239.

Croteau, R., Kutchan, T.M., Lewis, N.G., 2000. Natural products (secondary metabolites). Biochem. Mol. Biol. Plants 24, 1250–1319.

Dhar, M.L., Dhar, M.M., Dhawan, B.N., Mehrotra, B.N., Ray, C., 1968. Screening of Indian plants for biological activity: I. Indian J. Exp. Biol. 6 (4), 232.

Ekor, M., 2014. The growing use of herbal medicines: issues relating to adverse reactions and challenges in monitoring safety. Front. Pharm. 4, 177.

Feller, I.C., Mathis, W.N., 1997. Primary herbivory by wood-boring insects along an architectural gradient of rhizophora mangle. Biotropica 29 (4), 440–451.

Gallego, A., Ramirez-Estrada, K., Vidal-Limon, H.R., Hidalgo, D., Lalaleo, L., Khan Gupta, S., et al., 2017. Foliar endophytic diversity of *Centella asiatica* (L.) Urban in relation to different seasons and leaf age. Int. J. Curr. Microbiol. Appl. Sci. 6 (6), 468–477.

Gupta, S., Bhatt, P., Chaturvedi, P., 2018. Determination and quantification of asiaticoside in endophytic fungus from *Centella asiatica* (L.) Urban. World J. Microbiol. Biotechnol. 34 (8), 111.

Hussin, M., Abdul-Hamid, A., Mohamad, S., Saari, N., Ismail, M., Bejo, M.H., 2007. Protective effect of *Centella asiatica* extract and powder on oxidative stress in rats. Food Chem. 100 (2), 535–541.

James, J.T., Dubery, I.A., 2009. Pentacyclic triterpenoids from the medicinal herb, *Centella asiatica* (L.) Urban. Molecules 14 (10), 3922–3941.

Jamil, S.S., Nizami, Q., Salam, M., 2007. *Centella asiatica* (Linn.) Urban—a review.

Jayashree, G., Muraleedhara, G.K., Sudarslal, S., Jacob, V.B., 2003. Anti-oxidant activity of *Centella asiatica* on lymphoma-bearing mice. Fitoterapia 74 (5), 431–434.

Jayathirtha, M.G., Mishra, S.H., 2004. Preliminary immunomodulatory activities of methanol extracts of *Eclipta alba* and *Centella asiatica*. Phytomedicine 11 (4), 361–365.

Jisha, S., Gouri, P.R., Anith, K.N., Sabu, K.K., 2018. *Piriformospora indica* cell wall extract as the best elicitor for asiaticoside production in *Centella asiatica* (L.) Urban, evidenced by morphological, physiological and molecular analyses. Plant Physiol. Biochem. 125, 106–115.

Kobayashi, D.Y., Palumbo, J.D., 2000. Bacterial endophytes and their effects on plants and uses in agriculture. Microbial Endophytes. CRC Press, pp. 213–250.

Marler, M.J., Zabinski, C.A., Callaway, R.M., 1999. Mycorrhizae indirectly enhance competitive effects of an invasive forb on a native bunchgrass. Ecology 80 (4), 1180–1186.

Mousa, W.K., Raizada, M.N., 2013. The diversity of anti-microbial secondary metabolites produced by fungal endophytes: an interdisciplinary perspective. Front. Microbiol. 4, 65.

Mugford, S.T., Osbourn, A., 2012. Saponin Synthesis and Function 28. Isoprenoid Synthesis in Plants and Microorganisms: New Concepts and Experimental Approaches, p. 405.

Nalini, K., Aroor, A.R., Rao, A., Karanth, K.S., 1992. Effect of *Centella asiatica* fresh leaf aqueous extract on learning and memory and biogenic amine turnover in albino rats. Fitoterapia 63 (3), 231–238.

Park, B.C., Bosire, K.O., Lee, E.S., Lee, Y.S., Kim, J.A., 2005. Asiatic acid induces apoptosis in SK-MEL-2 human melanoma cells. Cancer Lett. 28 (1), 81–90.

Peters, A.F., 1991. Field and culture studies of *Streblonema macrocystis* sp. nov. (Ectocapales, Phaeophyceae) from Chile, a sexual endophyte of giant kelp. Phycologia 30 (4), 365–377.

Pragada, R.R., Veeravalli, K.K., Chowdary, K.P.R., Routhu, K.V., 2004. Cardioprotective activity of *Hydrocotyle asiatica* L. in ischemia-reperfusion induced myocardial infarction in rats. J. Ethnopharmacol. 93 (1), 105–108.

Prakash, V., Jaiswal, N.I.S.H.I.T.A., Srivastava, M.R.I.N.A.L., 2017. A review on medicinal properties of *Centella asiatica*. Asian J. Pharm. Clin. Res. 10 (10), 69.

Prasad, A., Mathur, A., Kalra, A., Gupta, M.M., Lal, R.K., Mathur, A.K., 2013. Fungal elicitor-mediated enhancement in growth and asiaticoside content of *Centella asiatica* L. shoot cultures. Plant Growth Regul. 69 (3), 265–273.

Rafat, A., Philip, K., Muniandy, S., 2012. A novel source of bioactive compounds: endophytic bacteria isolated from *Centella asiatica*. J. Pure Appl. Microbiol. 6 (1), 1.

Rakotoniriana, E.F., Munaut, F., Decock, C., Randriamampionona, D., Andriambololoniaina, M., Rakotomalala, T., et al., 2008. Endophytic fungi from leaves of *Centella asiatica*: occurrence and potential interactions within leaves. Antonie van Leeuwenhoek 93 (1-2), 27–36.

Ravishankar, B., Shukla, V.J., 2007. Indian systems of medicine: a brief profile. Afr. J. Tradit. Complement. Altern. Med. 4 (3), 319–337.

Raza, S.A., Rehman, A., Adnan, A., Qureshi, F., 2009. Comparison of antioxidant activity of essential oil of *Centella asiatica* and Butylated hydroxyanisole in sunflower oil at ambient conditions. Biharean Biol. 3 (1), 71–75.

Saha, A., Bhatia, B.D., Kulkarni, K.S., 2002. Evaluation of the efficacy of Mentat in children with learning disability: a placebo-controlled double-blind clinical trial. Neurosci. Today 6 (3), 184–188.

Sayasinha, P., Warnasuriya, D., Dissanayake, H., 1999. History, medicinal and aromatic plant series: vol.1. Information Services Centre Industrial Technology Institute, Colombo, p. 1.

Schippmann, U.W.E., Leaman, D., Cunningham, A.B., 2006. A comparison of cultivation and wild collection of medicinal and aromatic plants under sustainability aspects. Frontis 75–95.

Shobi, V., Goel, H.C., 2001. Protection against radiation-induced conditioned taste aversion by *Centella asiatica*. Physiol. Behav. 73 (1-2), 19–23.

Singh, S., Gautam, A., Sharma, A., Batra, A., 2010. *Centella asiatica* (L.): a plant with immense medicinal potential but threatened. Int. J. Pharm. Sci. Rev. Res. 4 (2), 9–17.

Srivastava, R., Shukla, Y.N., Kumar, S., 1997. Chemistry and pharmacology of *Centella asiatica*: a review. J. Med. Arom. Plant Sci. 19, 1049–1056.

Staniek, A., Woerdenbag, H.J., Kayser, O., 2008. Endophytes: exploiting biodiversity for the improvement of natural product-based drug discovery. J. Plant. Interact. 3 (2), 75–93.

Stone, J.K., Bacon, C.W., White Jr, J.F., 2000. An overview of endophytic microbes: endophytism defined. Microbial Endophytes. CRC Press, pp. 17–44.

Wu, T., Geng, J., Guo, W., Gao, J., Zhu, X., 2017. Asiatic acid inhibits lung cancer cell growth in vitro and in vivo by destroying mitochondria. Acta Pharm. Sin. B 7 (1), 65–72.

Young, G., Jewell, D., 1996. Creams for preventing stretch marks in pregnancy. Cochrane Database Syst. Rev. (1).

Chapter 20

Endophytes producing bioactive compounds from *Piper* spp.: a review on utilization, bottlenecks, and future perspectives

Shreya Sikdar Mitra[1], Protha Biswas[1], Anuradha Mukherjee[2], Potshanghbam Nongdam[3], Devendra Kumar Pandey[4] and Abhijit Dey[1]

[1]*Department of Life Sciences, Presidency University, Kolkata, India,* [2]*MMHS, Joynagar, India,* [3]*Department of Biotechnology, Manipur University, Imphal, India,* [4]*Department of Biotechnology, Lovely Faculty of Technology and Sciences, Lovely Professional University, Jalandhar, India*

Chapter Outline

Volatiles and Metabolites of Microbes. DOI: https://doi.org/10.1016/B978-0-12-824523-1.00017-1

20.1 Introduction

There is an ever-intensifying surge for the unveiling of novel and beneficial compounds that can provide aid to various facets of human conditions and ailments. With dwindling interest in common pharmaceutical treatments, drug resistance in bacteria due to antibiotic misuse, appearance of incurable diseases, overuse of chemotherapeutic sources, cost efficiency, etc. are a few of the listed causes for the shift of attention to the newly emerging alternative sources of treatment. Among these alternative methods, herbal or plant-based therapies are gaining a universal approval due to the multiple benefits attached with them. But along with these issues the environmental degradation and loss of biodiversity are a few indelible and haunting factors that in no way can be avoided. Along with the production of the desired active compounds, the preservation of the plant diversity can in no way be ignored. As a result, contemporary technological advancements are taking place so that in an efficient manner both the purposes can be met, that is, the generation of the desired bioactive molecules without causing any harm to the depleting biotic resources. Endophyte-mediated production of secondary bioactive molecules is one such method.

20.1.1 Endophytes

The word "endophyte" (Gr. *endon*, within, *phyton*, plant) was coined by de Bary (1866) for the first time. The microorganisms (bacteria, fungi, or viruses) that are mainly found to reside inside plant cells come under the wide range of the term endophyte. Literal meaning of the word endophyte stands for "in the plant" (endon Gr. = inside or within; phyton = plant).

The usage of this term is as vast as its literal definition encompasses the gamut of effective hosts and occupants, for example, bacteria (Kobayashi and Palumbo, 2000), fungi (Stone et al., 2000), plants (Marler et al., 1999), insects in plants (Feller and Mathis, 1997), and algae within algae (Peters, 1991). There is no specificity for the organ of the host in which the endophyte can associate. This group of varied and flexible microorganisms are found in diversified habitats starting from the ice-laden poles to the warmer tropical belts along with geothermal soils, deserts, oceans, rainforests, estuarine ecosystems, and several more. Similar to this is the application of the term "endophyte" for diverse life histories with symbiotic associations, encompassing parasitic to facultative, and to saprobic, and mutualistic to exploitive. These set of organisms apparently are found inside the host tissue without leading to any disease occurrences. These microbial life forms entwined with the shelter-giving plant are relatively unexplored, underexamined, and promising reservoir of novel natural products for application in agriculture, industry, and medicine. It is remarkable to note that out of approximately 300,000 plant species on earth, almost each plant serves as the host to one or many endophytes. Only a few of these host plants have been well documented and deciphered with respect to endophyte biology. Consequently, the scope to find unrevealed and enthralling endophytic microbial community amidst the multitude of vegetal communities in different settings and ecosystems is extraordinary. The maximum of existing literature sources mainly focus on the examples, methods, and rationale of profuse endophytes isolated and investigated over the course of many years.

20.1.2 Reasons for choice of microbial diversity

There are ample reasons for selecting the microbial communities for the secondary metabolite generation and the subsequent biological functions. At first, this tiny group of organisms have made their way ever since life originated on the earth because they evolved with diverse metabolic paths and mechanisms for generating bounty of secondary metabolites that have over time acted as their armors by helping to survive within the wide environmental challenges they are inflicted upon. As a result, the discovery of a single microbe can be associated with the prospective yield of bountiful, totally unexplored compounds having bioactivity for diverse medicinal causes. It has been estimated by Demain et al. (2000) that fewer than about 16% described fungal species have been isolated, cultured, and investigated. These reported species in all possibilities account for less than 5% of the total fungal species awaiting detailed investigation. There are several feasible benefits of using microbial sources for the synthesis of the desired bioactive molecule that are briefly depicted in Fig. 20.1.

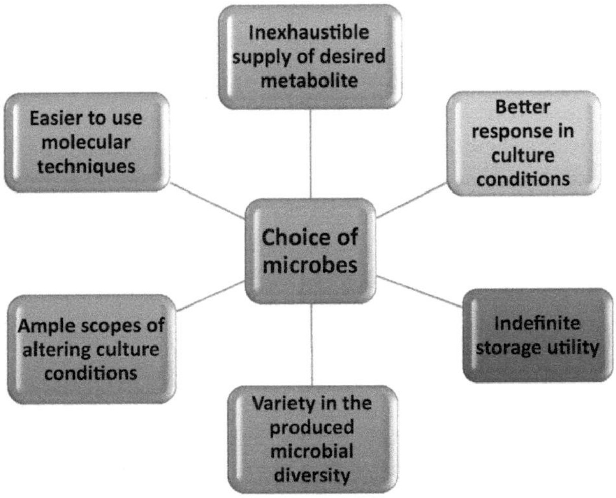

FIGURE 20.1 The reasons for selection of microbial resources for secondary metabolite production.

20.2 Piperaceae family

Piperaceae, frequently referred to as the pepper family in the order Piperales, provide significant monetary benefit due to its important member *Piper nigrum*, the commercial source of white and black pepper. The family consists of around five genera, of which two *Piper* (around 2000 species) and *Peperomia* (around 1600 species) are the two principal ones. This is an economically significant family, being the origin of a multitude of secondary metabolic bioactive compounds possessing immense medicinal potency along with different other salient usages. In a majority of the *Piper* species, the predominant bioactive molecules are groups of different alkaloids and amides (Scott et al., 2005). Maximum plant species of this family are found to grow as herbs, vines, shrubs, and trees, and mainly have distribution across tropical and subtropical regions. A strong and pungent flavor is found to be associated with the leaves of the members of the family Piperaceae that are normally found to grow singly. The flowers are lacking the calyx and the corolla, and are compactly arranged in the form of dense spikes. An essential oil obtained from distillation of peppercorns is used in the process of making meat sauces. *Piper cubeba* is of considerable significance in Southeast Asia, and is the source of cubeb, which has different uses as medicines and as flavoring agents in cigarettes and bitters. In the Orient, it is a wide practice to chew the betel pepper, *Piper betle*, leaves with betel nut (*Areca catechu*) slices and lime, and is popular as a mild stimulant. A traditional drink of Fiji and other Pacific Islands, variously popular as kawakawa, aiva, kava, and yagona, is prepared from the root of another important species *Piper methysticum;* it is also known for its sedative and narcotic properties. *Peperomia* species are mostly grown as low herbs; however a few grow as epiphytes on trees. Many species grown in the soil are also propagated as house plants for their beautiful foliage.

20.2.1 Ancient background and usage of the genus *Piper*

In the prehistoric traditional Indian medicine such as Ayurveda, both types of pepper had found their usage. They were mainly used for the purpose of controlling diabetes, an inducer of the central nervous system (CNS), digestive tonics, antispasmodics, aphrodisiacs, blood purifiers, and antipyretics. Since early times, pepper has been included in various customary formulas for the purpose of boosting the efficacy of different other active secondary molecules such as curcumin and vasicine. The first written evidence about *Piper longum* was given by Hippocrates, who focused more on the medicinal uses of the plant than on its uses as the spice. In the 6th or 5th century BCE, long pepper reached Greece; long before the European discovery of the New World, long pepper was a salient and reputed spice. The history of black pepper is somewhat associated with (and often mistaken with) that of long pepper, although Theophrastus differentiated the two in his initial work of plant studies. The Romans were familiar with both and often times mentioned both of them as just piper. Pliny even out of mistake had a belief that both dried black pepper and long pepper had the common origin. In the beginning of 12th century round or black pepper started to compete with long pepper in Europe and substituted it by the 14th century. At present, long pepper is gradually becoming rare in general commerce.

20.2.2 Distribution of the family

The genus *Piper* is pantropical in its distribution but mainly predominant in the warm tropics and the subtropics, especially common in central Asia, specifically in India, and also in South and Central America. It is thought to be endemic to India as they are found commonly growing in the wild, distributed across Himachal Pradesh, Arunachal Pradesh, Khasi, and Jayantia hills, and the different neighboring regions (Srivastava et al., 2000) but widely grown in the southern belts of the Kerala region (Parthasarathy et al., 2006). The scientific name *Piper* and the common name "pepper" were obtained from the Sanskrit word "pippali," referring to the long pepper *P. longum. Piper* is the representative genera of the family Piperaceae.

20.2.3 A few noteworthy members of the genera *Piper*

20.2.3.1 *Piper nigrum*

P. nigrum is about a 9 m (30 ft.) long woody climber endemic to southern India and to Sri Lanka and grown mainly in majority of tropical areas having a definite soil moisture and warmer temperatures. The pungency of *Piper* is characterized by the presence of a chemical named chavicine that is an isomeric form of the piperine molecule. Also present are the alkaloids piperine and piperidine. Popular across the globe as the king of spices for its pungent smell due to the occurrence of the active chemicals. Peppercorn of *P. nigrum* as a whole or some of its active constituents have usage in various food items. White pepper is obtained from the same species, but black pepper is prepared by short cooking and drying of the unripe fruits; white pepper is mainly made up of the dried, naked, ripe seeds (Srivastava et al., 2000). Several *P. nigrum* parts, including the obtained secondary metabolites, are being used as drug, preservative, and insecticidal and larvicidal agents. The numerous biological role of the secondary metabolites of this species include antiapoptotic, antibacterial, antifungal, antidepressant, antidiarrheal, antipyretic, antimutagenic, antitumor, antimetastatic, antioxidative, antiinflammatory, antispasmodic, gastric ailments curer, hepatoprotective, insecticidal, and larvicidal activities. Other roles of this species involve reducing oxidative stresses, bioavailability of different active compounds, increased bioavailability of vaccine, protection against diabetes, and numerous other biological activities (Scott et al., 2008).

20.2.3.2 *Piper longum*

Ordinarily called Indian long pepper, or *pipli*, this a flowering vine from the family Piperaceae, mainly grown for its fruits that are normally dried and are used as a spice for the purpose of seasoning dishes. It has similarity in taste with its near kin *P. nigrum* from which black, white, and green pepper are obtained. Though often used since medieval times in spice mixes such as "strong powder," long pepper in recent times has been a very insignificant ingredient in European dishes, but is still used in vegetable pickles in India and Nepal, in spice mixtures in North Africa, and in cooking in Indonesia and Malaysia. It is popularly sold in Indian grocery stores, where it is usually marked as *pippali*. *Pippali* is the one of the chief spice constituents of the Indian metropolis of Lucknow and Nihari, the national dishes of Pakistan. It is also a primary and commonly used constituent in several Ayurvedic formulations (Kumar et al., 2011).

20.2.3.3 *Piper betle*

P. betle is a vine from the family Piperaceae that has economic importance for its leaves. An evergreen perennial plant characterized by its white catkin and famous glossy heart-shaped leaves. The betel plant had originated in the areas of South and South East Asia. Betel leaf is regularly taken in Asia and other parts of the world by many Asian emigrants, as betel *quid* or as *paan*, with *A. catechu* nut and/or tobacco. In India and Sri Lanka, a sheaf of betel leaves as a part of the custom is presented as the sign of regard and to mark the starting of some pious occasions. It is also used in cooking, mostly as raw, for its peppery taste. Around 300 years ago in China it was used for different medicinal purposes (Ahuja and Ahuja, 2011).

20.3 Bioactive compounds

20.3.1 Alkaloids and amides A

A wide range of alkaloids and other related molecules as the amides were isolated from different species of the genus *Piper*. One of the most profound one is piperine. Piperine [1-[5-[1,3-benzodioxol-5-yl]-1-oxo-2,4, pentadienyl] piperidine] is the primary pungent alkaloid found in the fruits of *P. nigrum*, together with chavicine, that is found as its stereo-isomer that slowly is converted back to piperine during storage, leading to the gradual decrease and finally loss of the characteristic pungent odor. Few others, *viz.* piperettine, piperderidine, methyl piperine, asarinine, iperonaline, piperlongumine, piperundecalidine, piperlongumi-nine, pregumidiene, piperlongumine, pellitorine, brachystamide, brachystamide-A, pipercide, brachystine, longamide, tetrahydropiperine, terahydropiperlongu-mine, dehydropipernonaline piperidine, trimethoxy cinnamoyl-piperidine, and refractomide A are also recorded in the genus *Piper*. Different plant parts are found to be involved with the production of these compounds (Das et al., 1996; Kirtikar and Basu, 1935; Shankaracharya et al., 1997; Sharma et al., 1983; Tabuneng et al., 1983).

20.3.2 Esters

From the fruits of the *P. longum*, an ester named as tridecyl-dihydro-pcoumaarate, eicosanyl-(E)-p-coumarate and Z-12-octandecenoic−glycerol-monoester was isolated (Tabuneng et al., 1983).

20.3.3 Volatile oils

Along with the alkaloids, another important class of molecules is the volatile oils. Generally these oils are found as the amalgam of different essential oils. The three significant constituents are (apart from the volatile piperine) pentadecane, caryophyllene (both around 17.8%), and bisaboline (11%). Others are terpinoline, thujine, p-cymene, zingiberine, dihydrocarveol, and p-methoxy acetophenone (Das et al., 1996; Kirtikar and Basu, 1935; Shankaracharya et al., 1997; Sharma et al., 1983; Tabuneng et al., 1983). Long pepper is found to have a lesser amount of essential oil than its allies (about 1%), that contain sesquiterpene hydrocarbons and ethers (β-caryophyllene, β-caryophyllene oxide, bisabolene, each 10% to 20%, and α-zingiberene, 5%), and 18% pentadecane, 7% tridecane, and 6% heptadecane as saturated aliphatic hydrocarbons.

20.3.4 Lignans

The primary lignans obtained from the fruits of *P. longum* are sesamin, pulviatilol, and fargesin (Kirtikar and Basu, 1935; Rastogi et al., 1993).

20.4 Biosynthetic pathway

The principal bioactive compounds from the genus *Piper* as seen in Table 20.1 are mainly alkaloids and related amides. Being the alkaloids, they are mainly nitrogen-containing molecules and as a result are primarily derived from various amino acids. In this case they are mainly formed from the precursor amino acid L-lysine and then several chemical modifications are involved in the synthesis of the desired molecule piperine. Some variations are found at times but the common biosynthetic pathway of piperine production is depicted below in Fig. 20.2.

20.5 Notable pharmacological and medicinal uses

There are several reports on the widespread medicinal utilities of the two important plants of Piperaceae family, namely *P. nigrum* and *P. longum*. They have been noted for the anticarcinogenic, immunomodulatory, antimicrobial

TABLE 20.1 Some dominant bioactive compounds in the genus *Piper*.

Name of the active compound	Class of the compound	Molecular formula and average mass	Structures of the compound
Piperine	Alkaloid	$C_{17}H_{19}NO_3$, 285.338 Da	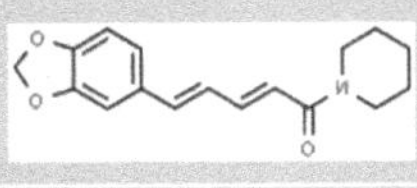
Piperidine	Alkaloid	$C_5H_{11}N$, 85.147 Da	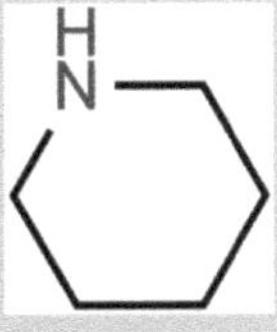
Chavicine	Alkaloid	$C_{17}H_{19}NO_3$, 285.338 Da	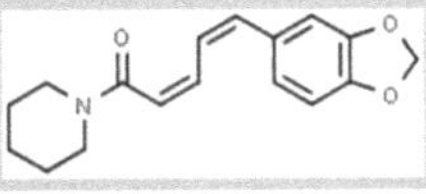

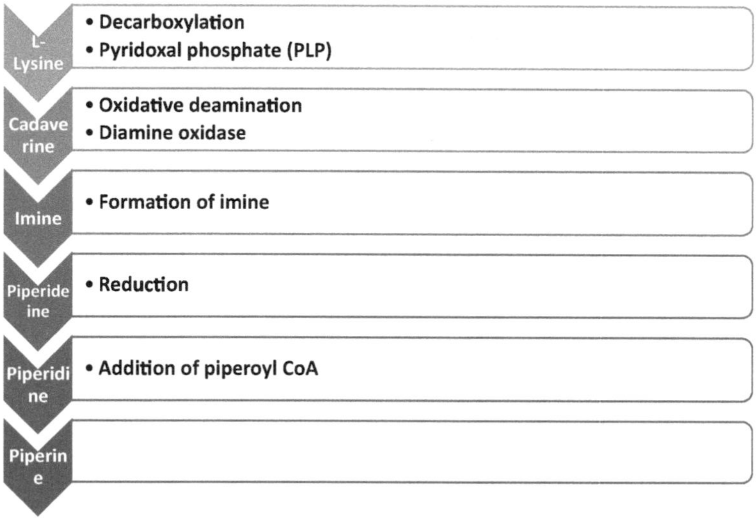

FIGURE 20.2 Biosynthetic pathway of piperine production.

(Yang et al., 2002), hepatoprotective, antiinflammatory (Darshan and Doreswamy, 2004), antiasthmatic, and antiulcer (Bai and Xu, 2000) activities. *Zingiber officinale* (ginger) also contains piperine which is attributed to some of these medicinal properties like antioxidant (Reddy and Lokesh, 1992) and antiinflammatory (Jiang, 2005). It also possess antioxidant and biotransformative properties and has shown the ability of absorbing few drug molecules such as sulphadiazine, tetracyline, rifampicin, and phenytoin (Wu, 2007). Some of the noted importance are mentioned below and a few are summarized in Fig. 20.3.

20.5.1 Antioxidant activity

A number of spices consisting of (*P. nigrum*, *P. longum*, and *Z. officinale*), herbs (*Cyperus rotundus* and *Plumbago zeylanica*), and salts that form a mixture commonly termed as the *Amrita Bindu* were examined for any antioxidant property. The test revealed that the antioxidant efficacy of the spices and herbs were in the following order: *P. nigrum* > *P. longum* > *C. rotundus* > *P. zeylanca* > *Z. officinale*, indicating on the high potentiality of *Piper* species for antioxidant status. Also, long pepper *P. longum* was found to display significant efficiency as an antioxidant by reducing oxidative stresses created through free-radicals. Roots extracted in petroleum ether and piperine from roots of *P. longum* were found to diminish the lipid peroxide levels and maintain glutathione content, highlighting on the antioxidant status of the species against the free-radical generated oxidative stresses (Natarajan et al., 2006).

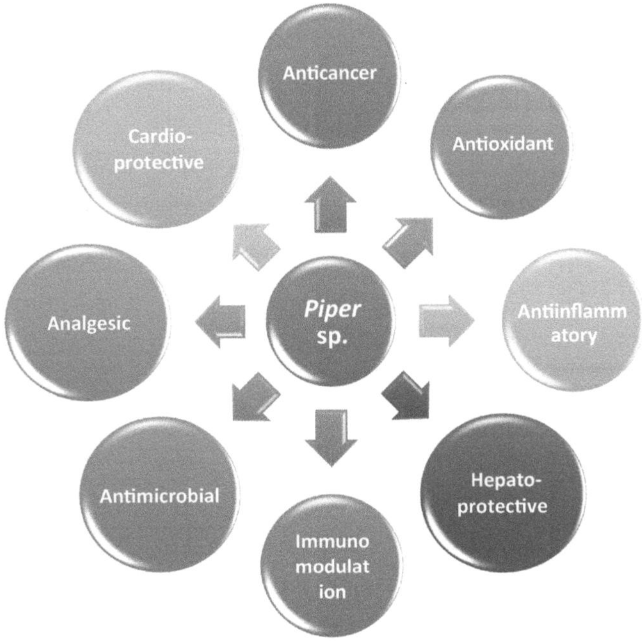

FIGURE 20.3 Summarized health benefits of the genus *Piper* from family Piperaceae.

20.5.2 Antiinflammatory activity

A significant amount of antiinflammatory activity was observed by the fruit decoction of *P. longum* by the use of rat edema stimulated by carrageenan (Sharma and Singh., 1980).

20.5.3 Hepatoprotective activity

Liver regeneration property was considerably improved through the use of fruit extract from *Piper* by reducing fibrosis, but no protection was offered from the cirrhotic damages or changes in the liver architectures in rodents in experiments. Ethanol extract of *P. longum* reduced the chances of carbon tetrachloride (CCl_4) induced liver fibrosis. Piperine was found to exhibit safety against CCl_4 and *tert*-butyl hydroperoxide mediated hepatotoxicity by lowering lipid peroxidation both in vitro and in vivo, enzymatic leakage, and by obstructing the reduction of glutathione (GSH) and total thiols in the treated mice. But a lower hepato-protective efficiency was shown by piperine in comparison to silymarin, a noted hepatoprotective drug (Koul and Kapil., 1993).

20.5.4 Immunomodulatory activity

Specific and nonspecific immune-stimulatory properties of *P. longum* fruits were investigated by different techniques such as the phagocytic index, macrophage migration index, and hemagglutination titer in mice. A popular formulation used in Ayurveda, commonly known as *Pippali rasayana* comprising of long peppers, was examined in *Giardia lamblia* infected mice and was found to stimulate macrophages and showing an increase in the values of accelerated macrophage migration index and phagocytic index in infected animals hinting at the immune-stimulatory activity of the formulation (Tripathi et al., 1999).

20.5.5 Analgesic activity

Roots from the *P. longum* displayed opioid-type analgesia in a rat tail-flick study and for nonsteroidal antiinflammatory drug (NSAID)-type analgesia employing acetic-acid writhing methodology using pentazocine and ibuprofen as the controls. *P. longum* root powder in an aqueous suspension was orally administered to mice and rats. The study established that the roots from *P. longum* behaved as feeble opioid but displayed efficiency as the NSAID type of analgesic activity (Vedhanayaki et al., 2003).

20.5.6 Antimicrobial activity

Ethyl acetate and petroleum ether extracts of *P. longum* were found to display different antimicrobial activities toward variety of microbial species (Ali et al., 2007).

20.5.7 Larvicidal activity

P. longum derived ethanol extracts sowed potency against early fourth instar larvae of *Aedes aegypti* mosquitoes in larvicidal bioassay (Yang et al., 2002).

20.5.8 Some other uses

Ethanol extract of the *P. longum* fruit contains piperine, piperlonguminine, and pipernonaline that are found as the major antihyperlipidaemic components. Significant effect was found following its application in vivo, which was compared to the commercially used drug simvastatin (Jin et al., 2009). Restricting the acyl CoA diacylglycerol acyltransferase has come up as an efficient therapeutic measure against a cure for obesity. Compounds with piperidine groups are found to be efficient acyl CoA diacylglycerol acyltransferase inhibitors (Lee et al., 2005). Toward six fungal species, namely *Rhizoctonia solani, Pyricularia oryzae, Puccinia recondita, Phytophthora infestans, Botrytis cinerea*, and *Erysiphe graminis*, any considerable fungicidal effect of *P. longum* was examined using control synthetic fungicides (dichlofluanid, chlorothalonil, and mancozeb) and four commercially available compounds, namely piperine, piperlongumine, eugenol, and piperettine. Though there was a varied response against the pathogens, there was indication on the fungicidal activity of the phytochemicals derived from *P.*

longum (Lee et al., 2001). The alkaloids piperidine and piperine obtained from the ethanol extraction of *P. longum* fruit has been found to function as the monoamine oxidase inhibitor (MAOI) that is used as therapy for depression. Hence they can have efficiency as an antidepression agent (Lee et al., 2008).

20.6 Elicitation

Elicitation is an effective and potential technique used in the field of biotechnology as a scale-up process for the synthesis of various bioactive plant secondary metabolites. Elicitors are all such molecules that act as a stimulant for the plant immune system, in turn enhancing the secondary metabolism pathways inside the plants with the aim of guarding the plant cells as well the whole plant against foreign invasion. Elicitors are broadly classified into two major groups: (1) abiotic and (2) biotic. As suggested by the titles, one indicates the abiotic ones, primarily the inorganic group of compounds, and the other the biological or the living ones mainly encompassing the microbial category. This literature mainly focuses on the effects of the biotic elicitors on our plant of interest. Maximum of the biotic elicitors are accredited by the cell membrane–bound particular receptors that conveys the signal to the cell via activation of the signal transduction system, initiating various responses associated with the plant defense system such as the phytoalexin production, etc. Numerous factors are responsible for controlling the response of the plants toward the elicitors, but the technique is unique in itself as it generates the desired bioactive molecules without having any impact on the available floral wealth.

20.6.1 Endophytic elicitation for bioactive compound synthesis

Endophytes are the secret gems of microbial diversity. Because they are found to exist asymptomatically they have received lesser focus in comparison with their pathogenic counterparts. Thus they demonstrate an undermined and lesser-exploited source in the search for novel molecules from different hidden groups of microbes. The diversified studies of this microbes point to them as voracious producers of various exploitable compounds both for agrochemical and medicinal causes. The finding of novel, potent bioactive molecules is doubtlessly of great significance as the demand for newer molecules along with saving the available resource has become the new focus. In relation with this has been the discovery of various groups of elicitor compounds that can serve a dual purpose. Endophytes being one such biotic elicitors as a result are gaining focus in the present scenario. The endophytic research domain opened up with the preliminary reports of the synthesis of one of the highly demanding molecule paclitaxel obtained from an endophyte of Northwest Pacific yew (Stierle et al., 1993). After this initial discovery, this research area spread its wings along with the reports on the

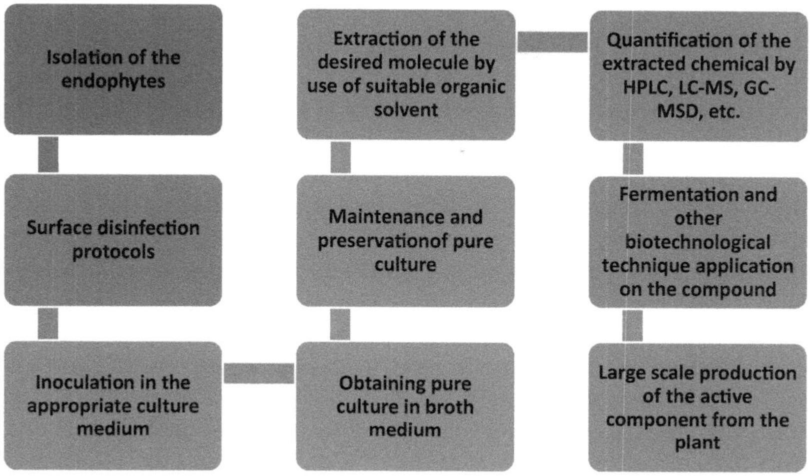

FIGURE 20.4 Flow chart of process of secondary metabolite isolation from the endophytes.

isolation of an array of other important anticancer therapeutics from fungal endophytes, including camptothecin, vincristine, podophyllotoxin, and so forth (Kharwar et al., 2011). Although scientists assert the recognition of 100,000 fungal species to date, there could be the possibility of over 1 million extant species of fungi. Bioactive compounds, especially obtained from the plant-associated microbes such as endophytes, still remain an unexplored domain (Staniek et al., 2008). Fig. 20.4 presents a flow chart of the process of secondary metabolite isolation from the endophytes. The present review predominantly discusses on few fungal colonizers as reported in the plant genera *Piper* with the number of possible future domains that still remain open for exploration.

20.6.2 Endophytes associated with the genus *Piper* and their utilization

Endophytes have now been widely recognized as an efficient and acceptable source for the production of necessary bioactive molecules. As already discussed, piperine and other associated molecules have wide-scale medicinal and multitude of other benefits. The compounds of significance are reported to be produced by endophytes associated with different genera of *Piper*. Literature reveals the presence of both fungal and bacterial endophytes from the genus *Piper*. Krishna and Yashwanti (2019) reported the presence of the endophytic fungi *Colletotrichum gloeosporioides* obtained from *P. nigrum* and extracted piperine alkaloid by the method of submerged fermentation of the isolated fungal colonizer. Chithra et al. (2014) performed an experiment with the aim of recognizing endophytic fungal members that can produce piperine from *P. nigrum*. Through screening of different endophytic fungi, the isolate that was identified was *C. gloeosporioides* had the ability of producing piperine that was confirmed by high-performance liquid chromatography (HPLC) and liquid chromatography-mass spectrometry (LC-MS) and was expected to have great industrial efficiency. Verma et al. (2011) in liquid culture synthesized piperine from the endophytic fungal partner *Periconia* sp. was the pioneer record of the alternative source for this compound besides *P. longum*. The fungus-derived highly bioactive piperine showed considerable antimycobacterial activity against two bacterial species *viz. Mycobacterium tuberculosis* and *Mycobacterium smegmetis*. Single-crystal X-ray crystallography was used for crystallizing the molecule and significant potency of piperine was also obtained, and the molecules were found to have resemblance with the origin. Jasim et al. (2013) reported in their study, isolation, and identification of two bacteria *viz. Klebsiella* sp. (PnB 10) and *Enterobacter* sp. (PnB 11) displaying outstanding growth stimulating characters. The results also provided substantial evidence on the utilities of the endophytic bacterial isolates found in the study, and also their promising growth-enhancing ability in *P. nigrum*. Aravind et al. (2009) made an attempt to segregate and identify endophytic bacterial diversity associated with black pepper (*P. nigrum*) opposing the fungal attack by *Phytophthora capsici* resulting in foot rot disease. About 74 bacterial endophytes were isolated, characterized, and investigated against *P. capsici*. Six genera were found from *Bacillus* spp. (22 strains), *Pseudomonas* spp. (20 strains), *Arthrobacter* spp. (15 strains), *Micrococcus* spp. (7 strains), *Serratia* and *Curtobacterium* sp. (1 strain each), and eight unidentified strains were obtained from root and stem internal tissues. Three isolates *viz.* IISRBP 17, IISRBP 25, and IISRBP 35 were found to have efficiency against *Phytophthora* inhibition in multilevel screening assays that accounted for 70% disease inhibition during greenhouse trials. Black pepper associated *Pseudomonas aeruginosa*, *Pseudomonas putida*, and *Bacillus megaterium* were identified as effective in opposing endophytes toward the biocontrol of *Phytophthora* foot rot disease in black pepper. This report provided proof for the endophytic bacterial diversity in the stem and roots of black pepper, with utility as biocontrol agent against *P. capsici* infection.

20.7 Isolation, extraction, screening, and confirmation of endophytes producing piperine and allied compounds

20.7.1 Isolation of the endophytic fungi

Generally the associated endophytic fungi are obtained from the living, mature, and healthy plant parts collected from suitable sources. Following the isolation, unanimously the next salient step is to carry out the surface sterilization procedures following the standardized sterilization protocols. Normally the plant materials, at first, are washed under the running tap water for around 10 minutes following immersion in NaOCl (2.5% available chlorine) for 10 minutes, and further treatment with 70% ethanol for around 1 minute. Finally, the samples are rinsed with sterile distilled water a number of times (Aravind et al., 2009) and the final washed samples are spread, plated on arginine glycerol agar (composition: agar 20 g/L; glycerol 20 g/L; l-arginine 2.5 g/L; NaCl 1 g/L; $CaCO_3$ 0.1 g/L; $FeSO_4 \cdot 7H_2O$ 0.1 g/L; $MgSO_4 \cdot 7H_2O$ 0.1 g/L) as control. The cutting of the plant samples aseptically are done into about 1 cm long segments, and the cut segments are then placed on agar-containing medium such as on arginine glycerol agar with 50 g/mL nalidixic acid. The plates are normally left for incubation at 28°C for 5 days and any type of fungal growth are reported. The fungal isolates found are later purified and kept on PDA (composition: potato infusion 200 g/L; dextrose 20 g/L; agar 20 g/L) medium.

20.7.2 Isolation of the endophytic bacteria

Screening of the bacterial endophytes is dependent to a great extent on the correct selection of the culture medium. In maximum cases, nutrient agar is selected for bacterial culture. At times some other types of agar like the beef extract peptone agar, LB agar, King's B agar, soy agar, arginine glycerol agar, starch caesin agar, caesin agar, etc., are also deployed. As liquid medium for culture purposes, Luria broth and nutrient broth are used. For checking any type of fungal contamination, antifungal agents like nalidixic acid, nystatin, etc. are used. Often supplementation with 1%–3% sugars such as glucose, sucrose, and fructose is done as a carbon source to the growing organisms. Also, amino acid or ammonium salt supplementation is also done as the source of nitrogen. Another critical factor for proper culturing of microorganisms is the pH of the medium as all the bacterial population are not dependent on the same type of pH. Some may be acidophilic, some basophilic, and some are totally neutral.

20.7.3 Extraction of the piperine and associated functional molecules

A variety of metabolite extraction procedures are used for the purpose of extracting the bioactive metabolites from the endophytic colonizers. The commonly employed organic solvents for the extraction purpose in an optimized manner involves methanol, ethyl acetate, dimethyl sulphoxide, etc. In addition, solvent extraction procedures including acidified isopropanol are also mentioned in some reports.

20.7.4 Screening and confirmation for the production of piperine using HPLC

The processes and techniques outlined by Verma et al. (2011) or Chithra et al. (2014), widely used for the purpose of screening piperine production, are generally used with some slight changes. At first, after isolation of the endophytic fungal members, inoculation is done into 200 mL of potato dextrose broth and incubated in a shaker at 30°C ± 2°C for 30 days. After incubation, filtration of the fermentation broth are carried, employing filter paper and cheese cloth to separate any fungal biomass. The culture filtrates are generally extracted two times using equal volume of ethyl acetate, and a rotary evaporator was used to concentrate the organic phase at 50°C under vacuum to achieve the crude extract powder. Isolated endophytic fungi extracted in ethyl acetate are analyzed by HPLC with piperine as the standard. Many former studies give a clear hint at the HPLC analyses being the most authentic and brisk techniques for the purpose of detecting piperine and its analogous compounds (Bajad et al., 2002). The crude extract thus obtained has to be resuspended in methanol and are analyzed for the available HPLC system normally using C 18 column (150 mm × 2.00 mm × 5 m). A mixture of methanol:water (70:30) is employed as the mobile phase that is delivered normally at a flow rate of 0.3 mL/min with detection at a wavelength of 344 nm and the recording of the chromatogram is carried out. The presence of piperine is confirmed by comparison of the sample peaks with the respective standard piperine peaks at the comparable retention time (Singh et al., 2012).

20.8 Loopholes in the piperine and related bioactive molecules production

Deploying the endophytic colonizers to serve industry-oriented and commercialized purposes has its own set of constraints. Lack of inclusive knowledge in this domain on ecological parameters governing the endophytic interlinking age and activities is one such vital shortcoming. The yield and the performance of the secondary metabolite production in the scale-up processes are reliant on a diverse set of parameters, which impact the overall process. Several endophytes at times are found to instead be acting as pathogens with the prospective amount of toxicity. Low yield, as well as a drop in the secondary metabolite quantity on repeated rounds of subculturing of the plant tissue, is another shortcoming to be taken care of. Over the last several years, the endophytes have been deciphered as the generous origin of various active natural metabolites with prospective advantages in the pharma industry (Mousa and Raizada, 2013). Fascinatingly, many of these secondary bioactive chemicals are similar to the ones obtained from their respective hosts, and could be used as the replacements for varied and notable plant-based compounds (Mousa and Raizada, 2013). The existing pattern of plant endophytic relations causes stress and switches on different biochemical pathways, inducing greater response and secondary metabolite generation due to the imposed stress. Despite scores of records of bioactive compounds from fungal and bacterial endophytes, endophytic exploration is still in the nascent stage as scanty positive results have been attained in the commercialization of secondary metabolite synthesis. Therefore, in order to recognize the tangible efficiency of the endophytes, a complete understanding of all relevant parameters is desired.

20.9 Future perspectives

Because the diverse scope of the various medicinal plants in human healthcare sectors is already known and validated, the associated group of colonizers attached with them will also be valuable in this field and can serve as the potential source of therapeutic compounds along with the set of the host plants. The group of endophytes are already popular for multiple chemical metabolites synthesized by their metabolic system, and has attracted the focus of researchers from the domain of biochemistry, microbiology, and other allied sciences to utilize them for manifold causes in the medicinal field. Investigating these alternative sources of beneficial bioactive molecules are the thriving areas of research by the pharmaceutical industry, both in developed and developing nations. Though a handful of secondary metabolite generating endophytes are isolated from the genus *Piper*, the number may be abundant for isolation of this major bioactive compound from this group of endophytes on a large-scale basis and in the coming future. For this purpose, accelerated screening of the endophytes producing the desired compounds has to be carried out. The discovery of a greater number of the endophytes will no doubt serve as a solution to mitigate the shortage of supply of the required material. Additionally, contemporary biotechnological and other research-oriented technologies have to be taken up in this field of endophyte-mediated production of useful secondary metabolites.

20.10 Conclusion

The diversified pharmacological utilities of the different secondary metabolites isolated from the genus *Piper* focus on their potential to be used in the future for the purpose of drug development programs. Many investigations have previously reported on the quantification of these compounds, making use of various strategies such as MS, HPLC, etc. Focusing on the diverse endophytic sources of such compounds along with research on various aspects of the host plants is reaching new heights. More inquiry needs to be done in this domain to produce satisfactory levels of the required metabolites sufficient for large-scale commercial uses of these compounds. Greater exploration has to be carried out to find a sizable amount of endophytes that can help in substantial production of the desired compounds along with increasing the cost efficiency and protecting plant resources from further exploitation for this purpose.

Acknowledgments

The first author would like to express her gratitude to Mr. Souvik Mitra for his constant support and guidance.

Conflict of interest

None.

References

Ahuja, S.C., Ahuja, U., 2011. Betel leaf and betel nut in folklore on cultivation. Asian Agrihist. 16 (1), 75–81.

Ali, M.A., Alam, N.M., Yeasmin, M.S., Khan, A.M., Sayeed, M.A., Rao, V.B., 2007. Antimicrobial screening of different extracts of *Piper longum* Linn. Res. J. Agri. Biol. Sci. 3 (60), 852–857.

Aravind, R., Kumar, A., Eapen, S.J., Ramana, K.V., 2009. Endophytic bacterial flora in root and stem tissues of black pepper (*Piper nigrum* L.) genotype: isolation, identification and evaluation against *Phytophthora capsici*. Lett. Appl. Microbiol. 48 (1), 58–64.

Bai, Y.F., Xu, H., 2000. Protective action of piperine against experimental gastric ulcer. Acta Pharmacol. Sin. 21 (4), 357–359.

Bajad, S., Johri, R.K., Singh, K., Bedi, K.L., 2002. Simple high performance liquid chromatography method for the simultaneous determination of ketoconazoleand piperine in rat plasma and hepatocyte culture. J. Chromatogr. A 949 (1–2), 43–47.

Chithra, S., Jasim, B., Sachidanandan, P., Jyothis, M., Radhakrishnan, E.K., 2014. Piperine production by endophytic fungus *Colletotrichum gloeosporioides* isolated from *Piper nigrum*. Phytomedicine 21 (4), 534–540.

Darshan, S., Doreswamy, R., 2004. Patented anti inflammatory plant drug development from traditional medicine. Phytother. Res. Int. J. Devoted Pharmacol. Toxicol. Eval. Nat. Prod. Deriv. 18 (5), 343–357.

Das, B., Kashinatham, A., Srinivas, K.V.N.S., 1996. Alkamides and other constituents of *Piper longum*. Planta Med. 62 (06), 582.

de Bary, A., 1866. Morphologie und Physiologie Pilze, Flechten, und myxomyceten. Hofmeister's Handbook of Physiological Botany, Vol.2. Leipzig.

Demain, A.L., 2000. New leads for pharmaceutical and agrochemical industries. In: Wrigley, S. K., Hayes, M.A., Thomas, R., Chrystal, E.J.T., Nicholson, N. (Eds.), Biodiversity. The Royal Society of Chemistry, Cambridge, UK, pp. 3–16.

Feller, I.C., Mathis, W.N., 1997. Primary herbivory by wood-boring insects along an architectural gradient of rhizophora mangle. Biotropica 29 (4), 440–451.

Jasim, B., Jimtha, C.J., Jyothis, M., Radhakrishnan, E.K., 2013. Plant growth promoting potential of endophytic bacteria isolated from *Piper nigrum*. Plant Growth Regul. 71 (1), 1–11.

Jiang, H., 2005. Modern tools to study traditional medicinal plants: ginger and turmeric. PhD Thesis. The University of Arizona.

Jin, Z., Borjihan, G., Zhao, R., Sun, Z., Hammond, G.B., Uryu, T., 2009. Antihyperlipidemic compounds from the fruit of *Piper longum* L. Phytother. Res. Int. J. Devoted Pharmacol. Toxicol. Eval. Nat. Prod. Deriv. 23 (8), 1194–1196.

Kharwar, R.N., Mishra, A., Gond, S.K., Stierle, A., Stierle, D., 2011. Anticancer compounds derived from fungal endophytes: their importance and future challenges. Nat. Prod. Rep. 28 (7), 1208–1228.

Kirtikar, K., Basu, B., 1935. Indian Medicinal Plants. Lalit Mohan Basu Publisher, Allahabad, India.

Kobayashi, D.Y., Palumbo, J.D., 2000. Bacterial endophytes and their effects on plants and uses in agriculture. Microbial Endophytes. CRC Press, pp. 213–250.

Koul, I.B., Kapil, A., 1993. Evaluation of the liver protective potential of piperine, an active principle of black and long peppers. Planta Med. 59 (05), 413–417.

Krishna, T., Yashwanti, K., 2019. Piperine production from endophytic fungi of *Piper nigrum* L. and its *in silico* approach for anti-inflammatory and anti-mycobacterial potential. Int. Res. J. Biol. Sci. 8 (12), 8–15.

Kumar, S., Kamboj, J., Sharma, S., 2011. Overview for various aspects of the health benefits of *Piper longum* linn. fruit. J. Acupunct. Meridian Stud. 4 (2), 134–140.

Lee, S.E., Park, B.S., Kim, M.K., Choi, W.S., Kim, H.T., Cho, K.Y., et al., 2001. Fungicidal Activity of Pipernonaline, a Piperidine Alkaloid Derived From Long Pepper. Piper longum.

Lee, S.A., Hong, S.S., Han, X.H., Hwang, J.S., Oh, G.J., Lee, K.S., et al., 2005. Piperine from the fruits of *Piper longum* with inhibitory effect on monoamine oxidase and antidepressant-like activity. Chem. Pharm. Bull. 53 (7), 832–835.

Lee, S.A., Hwang, J.S., Han, X.H., Lee, C., Lee, M.H., Choe, S.G., et al., 2008. Methylpiperate derivatives from *Piper longum* and their inhibition of monoamine oxidase. Arch. Pharm. Res. 31 (6), 679.

Marler, M.J., Zabinski, C.A., Callaway, R.M., 1999. Mycorrhizae indirectly enhance competitive effects of an invasive forb on a native bunchgrass. Ecology 80 (4), 1180–1186.

Mousa, W.K., Raizada, M.N., 2013. The diversity of anti-microbial secondary metabolites produced by fungal endophytes: an interdisciplinary perspective. Front. Microbiol. 4, 65.

Natarajan, K.S., Narasimhan, M., Shanmugasundaram, K.R., Shanmugasundaram, E.R.B., 2006. Antioxidant activity of a salt–spice–herbal mixture against free radical induction. J. Ethnopharmacol. 105 (1-2), 76–83.

Parthasarathy, U., Saji, K.V., Jayarajan, K., Parthasarathy, V.A., 2006. Biodiversity of Piper in South India–application of GIS and cluster analysis. Curr. Sci. 652–658.

Peters, A.F., 1991. Field and culture studies of *Streblonema macrocystis* new species (Ectocarpales, Phaeophyceae) from Chile, a sexual endophyte of giant kelp. Phycologia 30, 365–377.

Rastogi, R.P., Mehrotra, B.N., Pastogi, R.P., 1993. Compendium of Indian Medicinal Plants, 2. Central Drug Research Institute. Lucknow and Publications & Information Directorate, New Delhi, India, p. 10.

Reddy, A.C.P., Lokesh, B.R., 1992. Studies on spice principles as antioxidants in the inhibition of lipid peroxidation of rat liver microsomes. Mol. Cell. Biochem. 111 (1–2), 117–124.

Scott, I.M., Gagnon, N., Lesage, L., Philogene, B.J.R., Arnason, J.T., 2005. Efficacy of botanical insecticides from Piper species (Piperaceae) extracts for control of European chafer (Coleoptera: Scarabaeidae). J. Econ. Entomol. 98 (3), 845–855.

Scott, I.M., Jensen, H.R., Philogène, B.J., Arnason, J.T., 2008. A review of Piper spp. (Piperaceae) phytochemistry, insecticidal activity and mode of action. Phytochem. Rev. 7 (1), 65.

Shankaracharya, N.B., Jaganmohan Rao, L., Pura Naik, J., Nagalakshmi, S., 1997. Characterisation of chemical constituents of Indian long pepper (*Piper longum* L.). J. Food Sci. Technol. 34 (1), 73–75.

Sharma, A.K., Singh, R.H., 1980. Screening of anti-inflammatory activity of certain indigenous drugs on carrageenin induced hind paw oedema in rats. Bull. Med. Res. 11, 262–271.

Sharma, R.K., Rathore, Y.K.S., Kumar, S., 1983. Chemical examination of dried fruits of Pippali (*Piper longum* Linn.). J. Sci. Res. Plant Med. 4 (4), 63.

Singh, S., Solanki, U.B., Jain, S., Dhanwani, R.K., Kumar, S., 2012. Quantitative analysis of piperine in an ayurvedic formu-lation using reverse phase high performance liquid chromatography. Asian J. Pharm. Med. Sci. 2 (1), 11–15.

Srivastava, S., Gupta, M.M., Tripathi, A.K., Kumar, S., 2000. 1,3-Benzodioxole-5-(2,4,8-triene-methyl nonaoate) and 1,3-benzodioxole-5-(2,4,8-triene-isobutyl nonaoate) from *Piper mullesua*. Indian J. Chem. 39B, 946–949.

Staniek, A., Woerdenbag, H.J., Kayser, O., 2008. Endophytes: exploiting biodiversity for the improvement of natural product-based drug discovery. J. Plant Interact. 3 (2), 75–93.

Stierle, A., Strobel, G., Stierle, D., 1993. Taxol and taxane production by *Taxomyces andreanae*, an endophytic fungus of Pacific yew. Science 260 (5105), 214–216.

Stone, J.K., Bacon, C.W., White Jr., J.F., 2000. An overview of endophytic microbes: endophytism defined. Microbial Endophytes. CRC Press, pp. 17–44.

Tabuneng, W., Bando, H., Amiya, T., 1983. Studies on the constituents of the crude drug "Piperis Longi Fructus." On the alkaloids of fruits of *Piper longum* L. Chem. Pharm. Bull. 31 (10), 3562–3565.

Tripathi, D.M., Gupta, N., Lakshmi, V., Saxena, K.C., Agrawal, A.K., 1999. Antigiardial and immunostimulatory effect of *Piper longum* on giardiasis due to *Giardia lamblia*. Phytother. Res. Int. J. Devoted Pharmacol. Toxicol. Eval. Nat. Prod. Deriv. 13 (7), 561–565.

Vedhanayaki, G., Shastri, G.V., Kuruvilla, A., 2003. Analgesic activity of *Piper longum* Linn. root. Indian J. Exp. Biol. 41 (6), 649–651.

Verma, V.C., Lobkovsky, E., Gange, A.C., Singh, S.K., Prakash, S., 2011. Piperine production by endophytic fungus Periconia sp. isolated from *Piper longum* L. J. Antibiot. 64 (6), 427–431.

Wu, H., 2007. Isolation and Characterization of Natural Products From Ginger and *Allium ursinum* (Doctoral dissertation). Rutgers University-Graduate School-New Brunswick.

Yang, Y.C., Lee, S.G., Lee, H.K., Kim, M.K., Lee, S.H., Lee, H.S., 2002. A piperidine amide extracted from *Piper longum* L. fruit shows activity against *Aedes aegypti* mosquito larvae. J. Agric. Food Chem. 50 (13), 3765–3767.

Chapter 21

Recent advances and future prospects of indole alkaloids producing endophytes from *Catharanthus roseus*

Prabhjot Kaur[1], Abhijit Dey[2], Vijay Kumar[1], Padmanabh Dwivedi[3], R.M. Banik[4], Ranjit Singh[5] and Devendra Kumar Pandey[1]

[1]*Department of Biotechnology, Lovely Faculty of Technology and Sciences, Lovely Professional University, Jalandhar, India,* [2]*Department of Life Sciences, Presidency University, Kolkata, India,* [3]*Department of Plant Physiology, Institute of Agriculture Sciences, BHU, Varanasi, India,* [4]*School of Biochemical Engineering, IIT (BHU), Varanasi, India,* [5]*Department of Chemistry, Saroop Rani Government College for Women, Amritsar, India*

Chapter Outline

Conflict of interest

There is no conflict of interest.

Volatiles and Metabolites of Microbes. DOI: https://doi.org/10.1016/B978-0-12-824523-1.00018-3

21.1 Introduction

Catharanthus roseus L. (G) Don (Apocynaceae) (formerly *Vinca rosea* L.), also known as Madagascar periwinkle (Fig. 21.1), is an important medicinal plant, well known as anticancer, antidiabetic, and antimalarial, which is known for use in Indian folklore and by traditional medicinal herbalists for two centuries. *C. roseus*, an evergreen herb (also known as "Sadabahar" in Hindi language) is widely grown as an ornamental plant (characteristic as summer blooming flower). *Catharanthus* is a perennial tropical/subtropical medicinal plant, consisting of seven species from the Madagascar region viz. *C. coriaceus*, *C. lanceus*, *C. longifolius*, *C. ovalis*, *C. roseus*, *C. scitulus*, and *C. trichophyllus*, and only one species (*C. pusillus*) from India. *C. roseus* can be grown under any environmental conditions (best growth and maximum yield of secondary metabolites reported at 25°C–45°C) (Prabhu and Rajeswari, 2017).

21.1.1 Classification

Domain: Eukarya
- Kingdom: Plantae
- Subkingdom: Tracheobionta
- Super Division: Spermatophyta
- Division: Magnoliophyta

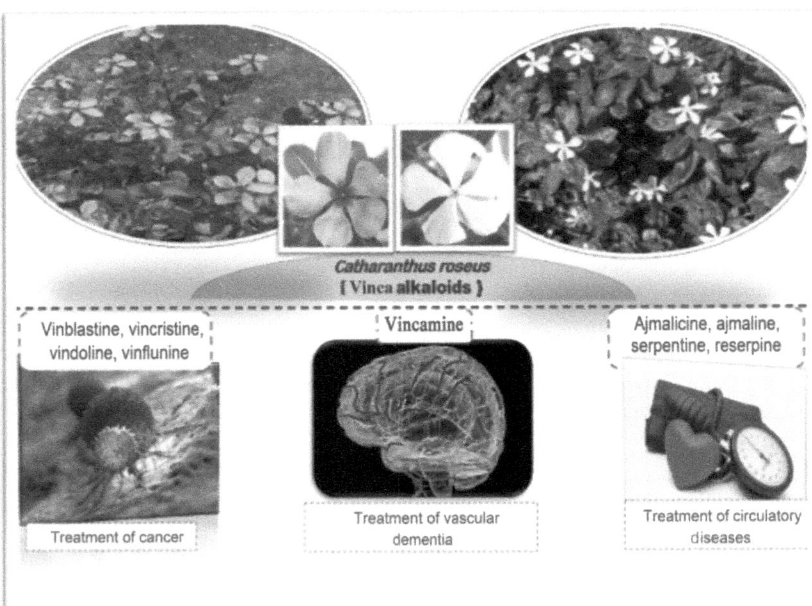

FIGURE 21.1 Medicinally important terpenoid indole alkaloids (TIAs) from *Catharanthus roseus* plant.

Class: Magnoliopsida
Subclass: Asteridae
Super Order: Gentiananae
Order: Gentianales
Family: Apocynaceae
Subfamily: Rauvolfioideae
Tribe: Vinceae
Genus: *Catharanthus*
Species name: *roseus*

21.1.2 Major alkaloids

C. roseus is nothing less than a chemical factory, producing more than 130 terpenoid indole alkaloids (TIAs), 25 of which are dimeric in nature, and it is considered to be a model plant to study TIA biosynthesis. The medicinally important TIAs are vinblastine (VB), vincristine (VC), vindoline, vinflunine, vincamine, ajmalicine, ajmaline, catharanthine serpentine, and reserpine (Fig. 21.2). Among TIAs, VB and VC are the widely recognized antimitotic vinca alkaloids, which were the first natural drugs used clinically as anticancer agents and are only found in *C. roseus* (Alam et al., 2017). Vinca leukoblastine, Velbe, Leurocristine, and Oncovin are some marketed drugs of VB and VC, respectively (Heijden et al., 2004). These compounds are present in small amounts in the Madagascar periwinkle, *C. roseus* leaves.

VB, VC, and vinorelbine (semisynthetic derivative of VB) drugs are used as an antimitotic drug that inhibits tubulin polymerization and prevents the formation of the mitotic spindle, which leads to the mitotic arrest and eventual cell death. These compounds have been reported in the treatment of cancer patients with Hodgkin's and non-Hodgkin's lymphomas, other lymphomas (breast, renal, testicular, thyroid, and brain tumors), and leukemias (Gajalakshmi et al., 2013). Other significant vinca alkaloids, viz. vindoline, vinflunine, vincamine, ajmalicine, ajmaline, serpentine, and reserpine, are also of great pharmaceutical interest. Ajmalicine is one of the most popular antihypertensive drugs obtained from the root barks of *C. roseus*. It is used in the treatment of circulatory diseases and also shows antimicrobial, cytotoxic, central depressant, and antioxidant activities. Reserpine has been used as an antipsychotic and antihypertensive drug, whereas serpentine is used as a sedative (Ambrin et al., 2020; Pan et al., 2015).

Due to the presence of these valuable alkaloids, this plant is widely used in traditional medicines worldwide. The whole plant of *C. roseus* extract has historically been used in different countries such as Australia, Brazil, England, Europe, India, Jamaica, Kenya, Malaysia, South Africa, Taiwan, Thailand, West Indies, etc. for the treatment of diabetes mellitus.

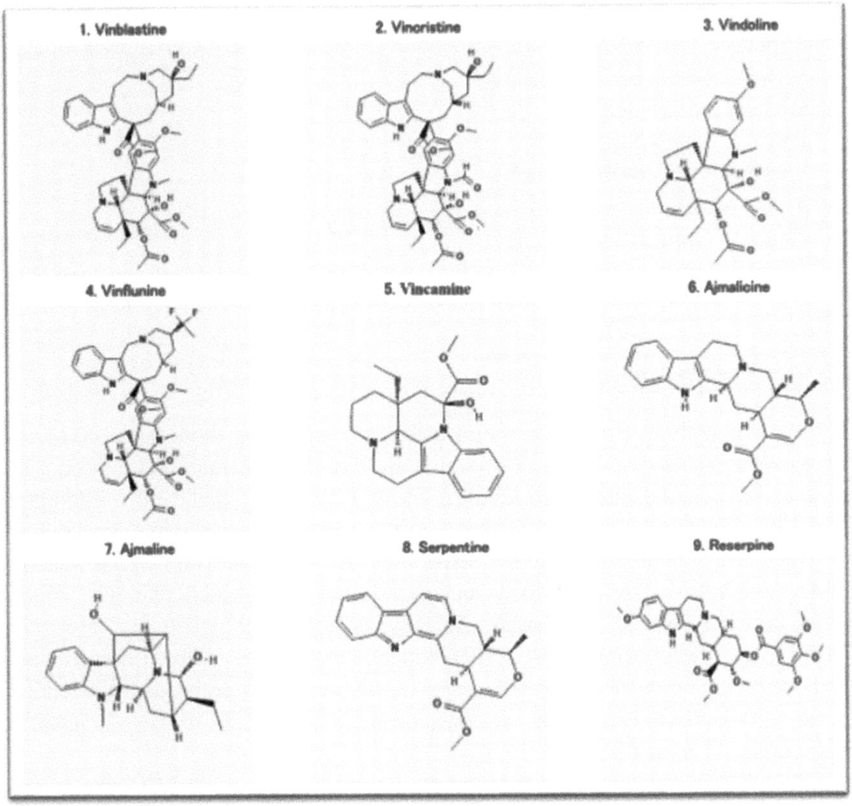

FIGURE 21.2 Molecular structures of significant terpenoid indole alkaloids (TIAs).

In India, a hot extract of the whole plant material has been used to treat cancer and Hodgkin's disease, while in China and North Vietnam, the hot extract was used as a menstrual regulator. In Peru, a hot extract of the whole plant is taken for the treatment of heart diseases (Nisar et al., 2016).

21.2 Biosynthetic pathway of indole alkaloids

C. roseus is considered a model plant to investigate biosynthesis of TIAs. TIA pathway is a very complex pathway (at least 35 intermediates are known) consisting of several multisteps and regulatory genes. Overview of key genes, enzymes, and biosynthetic steps for the production of catharanthine, vindoline, VB, and VC are presented in three steps in Fig. 21.3A–C. Vindoline and catharanthine are the precursors needed for the biosynthesis of significant dimeric alkaloids: VB and VC.

21.3 Source of endophyte(s) for indole alkaloids

Endophytic microbes (colonizing healthy medicinal plant's tissues) have the ability to produce common bioactive compounds from similar precursors. Endophytes are mostly bacteria or fungi inhabiting the host plants asymptomatically, which is advantageous in the healthy growth of plants and helps them to adapt in all environments, acting as remediators of stress conditions. Relationships (obligate or facultative) of endophytic microbes or mutualistic associations are always beneficial to both the partners. Several researchers reported cross-talk and chemotactic responses between root exudates and endophytic microbes (Khare et al., 2018). There is strong evidence of the plant endosymbiotic associations and their significant impact on the production of bioactive compounds in medicinal plants (Kaur et al., 2020a).

C. roseus is only responsible for synthesizing the TIAs, and out of these alkaloids, two bisindole alkaloids viz. VB and VC (valuable antitumor compounds) are most promising. But the *C. roseus* plant is well known to produce lower amounts of these significant alkaloids. Different endophytes viz. fungal and bacterial strains are reported for the production of these alkaloids. Such fungal endophytes include *Aspergillus japonicas* (Singh et al., 2020), *Choanephora infundibulifera* (Singh et al., 2020; Pandey et al., 2016), *Curvularia* sp. (Singh et al., 2020; Parthasarathy et al., 2020; Pandey et al., 2016), *Eutypella* sp. (Kuriakose et al., 2016), *Fusarium oxysporum* (Kumar et al., 2013), *Nigrospora sphaerica* (Ayob et al., 2017), and *Talaromyces radicus* (Palem et al., 2016). In comparison, very few reports on the bacterial endophytes for the production of vinca alkaloids are present. These bacterial endophytes include *Microbacterium* sp. (Anjum and Chandra, 2019), *Micrococcus* sp. (Tiwari et al., 2013), *Staphylococcus sciuri* (Tiwari et al., 2013), and *Pseudomonas* sp. (Singh et al., 2020; Jaleel et al., 2009).

These bacterial and fungal endophytes isolated from *C. roseus* plant significantly influenced the yield of vinca alkaloids (Table 21.1). Singh et al. (2020) and Pandey et al. (2016) reported about 200%–400% increased yield of serpentine and vindoline from fungal strains viz. *Curvularia* sp. (CATDLF5), *C. infundibulifera* (CATDLF6), and 100%–200% increased amount of ajmalicine from *Aspergillus japonicus* (CATDRF2). On the other hand, Tiwari et al. (2013), Jaleel et al. (2009), and Anjum and Chandra (2019) also found a significant amount of serpentine, ajmalicine, and vindoline from bacterial isolates in *C. roseus*. For VB production, several studies reported the significant amount 182 μg/L (Parthasarathy et al., 2020); 76 μg/L (Kumar et al., 2013); 70 μg/L (Palem et al., 2016); and 0.868 μg/mL (Ayob et al., 2017). For endophytic production of VC, significant amounts viz. 670 μg/L, 67 μg/L and 53 μg/L were reported by Palem et al. (2016), Kumar et al. (2013) and Kuriakose et al. (2016), respectively.

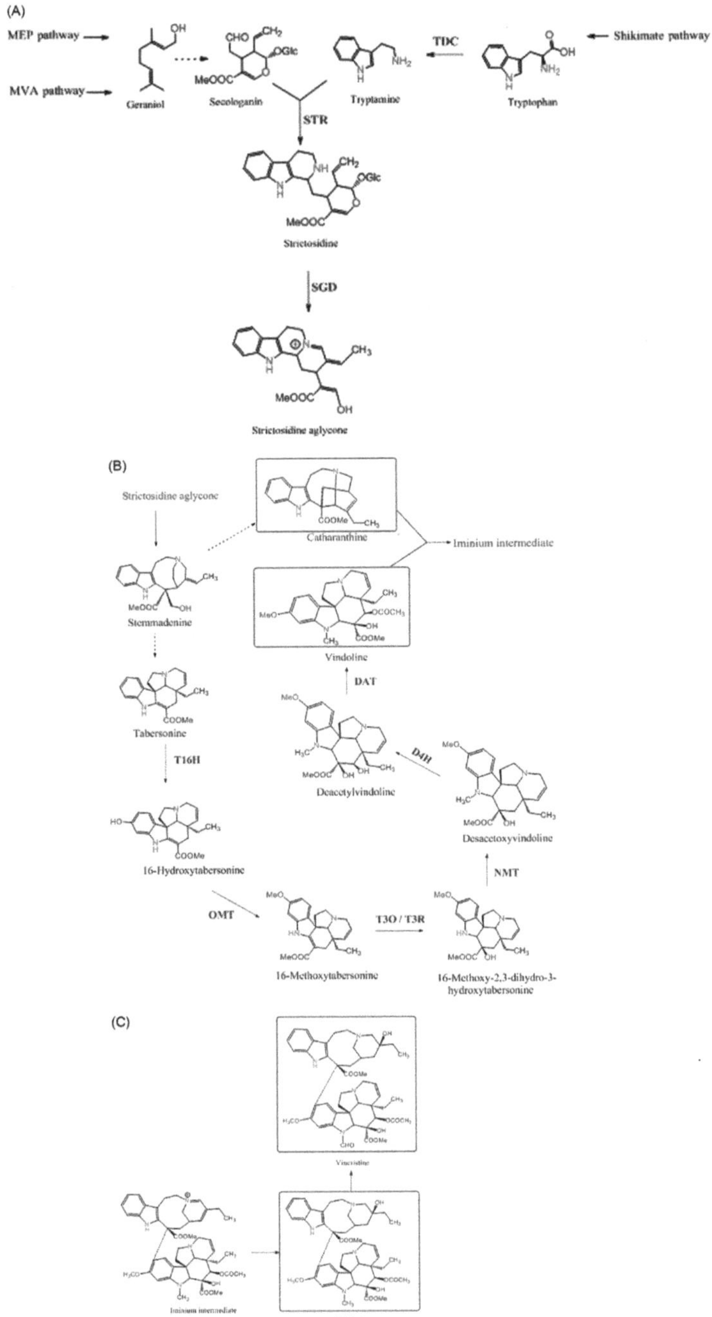

FIGURE 21.3 (A) Biosynthesis of *Strictosidine aglycone* (precursor of catharanthine and vindoline); (B) biosynthesis of TIAs: vindoline and catharanthine (precursors of vinblastine and vincristine); and (C) biosynthesis of TIAs: vinblastine and vincristine.

TABLE 21.1 Isolation of different endophyte(s) from *Catharanthus roseus* L. and quantification of indole alkaloids.

Sl. No.	Name (s) of vinca alkaloids	Type (s) of endophytes	Endophytic organisms	Content obtained	References
1.	Serpentine	Fungal strains	*Curvularia* sp. (CATDLF5) and *Choanephora infundibulifera* (CATDLF6)	Yield increased by 211.7%–337.6%	Singh et al. (2020)
	Ajmalicine	Both fungal and bacterial strains	*Aspergillus japonicus* (CATDRF2) and *Pseudomonas* sp. (CATDS5)	123.4%–203.8%	
2.	Vinblastine	Fungal strain	*Curvularia verruculosa*	182 μg/L	Parthasarathy et al. (2020)
3.	Vindoline	Bacterial strain	*Microbacterium* sp.	82 μg/L	Anjum and Chandra (2019)
4.	Vinblastine	Fungal strain	*Nigrospora sphaerica*	0.868 μg/mL	Ayob et al. (2017)
5.	Vincristine	Fungal strain	*Eutypella* sp.	53 μg/L	Kuriakose et al. (2016)
6.	Vindoline	Fungal strains	*Curvularia* sp. (CATDLF5) and *C. infundibulifera* (CATDLF6)	Yield increased by 229%–403%	Pandey et al. (2016)
7.	Vincristine	Fungal strain	*Talaromyces radicus*-CrP20	670 μg/L	Palem et al. (2016)
	Vinblastine			70 μg/L	
8.	Vinblastine	Fungal strain	*Fusarium oxysporum* (AA-CRL-6)	76 μg/L	Kumar et al. (2013)
	Vincristine			67 μg/L	

(*Continued*)

TABLE 21.1 (Continued)

Sl. No.	Name (s) of vinca alkaloids	Type (s) of endophytes	Endophytic organisms	Content obtained	References
9.	Vindoline	Bacterial strains	*Staphylococcus sciuri* (V1) and *Micrococcus* sp. (V3)	Yield increased by: V1 - 38.36%, V3 - 68.51%	Tiwari et al. (2013)
	Ajmalicine			V3 - 46.34%	
	Serpentine			V3 - 54.74%	
10.	Vindoline, ajmalicine, serpentine, and catharanthine	Bacterial strain	*Pseudomonas fluorescens*	Yield increased significantly	Jaleel et al. (2009)

21.4 Extraction and quantification of indole alkaloids from endophyte(s)/*Catharanthus roseus*

Extraction yield of desired compound and bioactivity is largely affected by the selection of extraction solvents and techniques, and therefore extraction of bioactive compounds from endophytes/cultured tissues/plant material is the most important step (Kaur et al., 2019; Jeong and Lim, 2018; Pandey and Kaur, 2018; Paeizi et al., 2018). In the 1970s, the first patent for the extraction and isolation of VB by using organic solvents was published (Jones, 1976). Later, several conventional and nonconventional extraction techniques were used, aiming to obtain the maximum yield of alkaloids (Table 21.2).

Most commonly ethyl acetate, chloroform, and methanol are used by researchers for the extraction of indole alkaloids from fungal extracts (Parthasarathy et al., 2020), different plant parts of *C. roseus* (Kidd et al., 2019; Jeong and Lim, 2018), and cultured tissues (Moon et al., 2018). In addition, some researchers soak plant material in the acidic aqueous phase (3% HCl, formic acid, and H_2SO_4) at pH 3 for extraction and purification of alkaloids (Abouzeid et al., 2019; Mall et al., 2019; Fouad and Hafez, 2018; Paeizi et al., 2018; Liu et al., 2016; Zhang et al., 2014). Alternatively, green and cost-effective extraction methods such as ultrasonication (Jeong and Lim, 2018; Liu et al., 2017), enzyme-assisted negative pressure cavitation extraction (Luo et al., 2014), and supercritical fluid extraction (Falcão et al., 2017) methods were optimized.

21.5 Biological activities associated with endophyte (s) isolated from *Catharanthus roseus* L

Vinca alkaloids are well known for their significant pharmacological properties, and thus several purified endophytes from *C. roseus* have been explored for their biological potential (Table 21.3). In these studies, different fungal and bacterial strains were screened for cytotoxic activity (Parthasarathy et al., 2020; Ali et al., 2019; Dhayanithy et al., 2019; Palem et al., 2016), antagonistic activity (Myo et al., 2020), antimicrobial activity (Ali et al., 2019; Sudharshana et al., 2019), antimycotoxigenic activity (Sudharshana et al., 2019), anticancer activity (Akther et al., 2019; Ayob et al., 2017; Muñoz-Rojas, 2017; Kuriakose et al., 2016), antifungal activity (Yan et al., 2018), antibacterial activity (Ranjan and Jadeja, 2017; Chandrakar and Gupta, 2017), and antidiabetic activity (Isavella Rosaline and Agastian, 2013). Indole alkaloids have historically been used in different developing and developed countries to combat diseases related to cancer, heart, diabetes mellitus, etc. Further, detailed in vitro and in vivo pharmacological examinations are required for the treatment of diseases.

TABLE 21.2 Extraction and quantification of indole alkaloids from endophyte(s)/*Catharanthus roseus*.

Sl. No.	Extraction material	Target alkaloid	Extraction solvent	Extraction method	Quantification method	References
1.	Fungal extract	Vinblastine	Ethyl acetate	Soaking	HPTLC and LC-MS	Parthasarathy et al. (2020)
2.	Leaves	Vincamine, catharanthine, and vindoline	Acidic aqueous solution and washing with chloroform	Soaking	HPLC	Abouzeid et al. (2019)
3.	Leaves	Catharanthine, vindoline, and serpentine	Chloroform	Soaking	UPLC-MS-DAD	Kidd et al. (2019)
4.	Leaves	Vincristine, vinblastine, catharanthine, and vindoline	Methanol, acidic aqueous (3% HCl) and washing with hexane and chloroform	Soaking	HPLC	Mall et al. (2019)
5.	Cultured tissues	Total alkaloids	Methanol, acid water (pH 3) and petroleum ether, washing with chloroform	Soaking	Spectro-photometer	Fouad and Hafez (2018)
6.	Plant parts	Ajmalicine, serpentine, catharanthine, vinblastine, vincristine, and vindoline	70% methanol and ethyl acetate	Ultrasonication	UPLC-ESI-Q-TOF	Jeong and Lim (2018)
7.	Plant parts	Ajmaline, yohimbine, vindesine, ajmalicine, serpentine, vincristine, vinblastine, vindoline, and reserpine	Ethanol	Ultrasonication	UHPLC–ESI–MS/MS	Kumar et al. (2018)

8.	Cultured tissues	Vindoline, vinblastine, vincristine, and catharanthine	Methanol	Bath sonication	HPLC	Moon et al. (2018)
9.	Cultured tissues	Vindoline, catharanthine, vincristine, vinblastine, and ajmalicine	Methanol, H_2SO_4, ethyl acetate	Soaking	HPLC	Paeizi et al. (2018)
10.	Flowers	Catharanthine, vincristine, and vinblastine	Methanol	Soaking	LC-MS	Schweizer et al. (2018)
11.	Plant sample	Vinblastine	Carbon dioxide and ethanol (2% w/w)	Supercritical fluid extraction	HPLC	Falcão et al. (2017)
12.	Plant sample	Vindoline, catharanthine, serpentine, and vinblastine	Methanol	Ultrasonication	ESI/MS/MS and TOF-MS	Liu et al. (2017)
13.	Cultured tissues	Vincristine and vinblastine	Methanol	Refluxing	HPTLC	Maqsood and Abdul (2017)
14.	Hairy roots	Vinblastine, vincristine, and catharanthine	Methanol	Bath sonication	HPLC	Hanafy et al. (2016)
15.	Plant sample	Vincristine, vinblastine, vindoline, and catharanthine	2% formic acidic water:methanol	Ultrasonication	HPLC	Liu et al. (2016)
16.	Plant sample	Ajmalicine, serpentine, catharanthine, vindoline, vincristine, and vinblastine	Methanol	Bath sonication	HPLC	Pan et al. (2016)
17.	Plant parts	Vinblastine, vincristine, vindoline, and catharanthine	0.1% H_2SO_4 methanol	Ultrasonication	LC-ESI-MS/MS	Zhang et al. (2014)
18.	Plant parts	Vindoline, catharanthine, vincristine, and vinblastine	80% ethanol	Enzyme-assisted negative pressure cavitation extraction	HPLC	Luo et al. (2014)

TABLE 21.3 Biological activities associated with endophyte(s) isolated from *Catharanthus roseus* L.

Sl. No.	Endophyte (s)	Identified alkaloid	Biological activity	Tested against	References
1.	Fungal strain: *Curvularia verruculosa*	Vinblastine	Cytotoxicity	HeLa cell line	Parthasarathy et al. (2020)
2.	Bacterial strain: *Bacillus amyloliquefaciens* (DSM7)	–	Antagonistic activity	Dual culture and agar well diffusion methods	Myo et al. (2020)
3.	Fungal strains: *Aspergillus fumigatus* and *Fusarium oxysporum*	Total alkaloids	Cytotoxicity and antimicrobial activity	Human breast cancer (MCF-7) and liver cancer (HEPG-2) cell lines using SRB assay method	Ali et al. (2019)
4.	Fungal strain: *Chaetomium nigricolor*	–	Cytotoxicity and free radical scavenging potential	HeLa and MCF-7 cells	Dhayanithy et al. (2019)
5.	Fungal strain: *Alternaria alternata*	–	Antimicrobial and antimycotoxigenic activities	Bacteria, yeast, and fungi with MICs (7.8–250 μg/mL)	Sudharshana et al. (2019)
6.	Fungal strain: *Botryosphaeria rhodina*	–	Anticancer activity	Lung (A549) cancer cell lines	Akther et al. (2019)
7.	Fungal strain: *Diaporthe* sp.	–	Antifungal activity	Plant pathogenic test fungi and oomycetes	Yan et al. (2018)
8.	Fungal strain: *Nigrospora sphaerica*	Vinblastine	Anticancer activity	Breast cell line cancer (MDA-MB 231)	Ayob et al. (2017)
					Muñoz-Rojas (2017)

9.	Bacterial strain: *Micrococcus yunnanensis* (rsk5)	–	Antibacterial activity	Human pathogen *Staphylococcus aureus*	Ranjan and Jadeja (2017)
10.	Endophytic actinomycetes	–	Antibacterial activity	Human pathogenic bacteria	Chandrakar and Gupta (2017)
11.	Fungal strain: *A. alternata*	–	Acetyl-cholinestearse inhibitory activity	Ellman's method	Bhagat et al. (2016)
12.	Fungal strain: *Eutypella* sp.	Vincristine	Anticancer activity	Human squamous carcinoma cells - A431	Kuriakose et al. (2016)
13.	Fungal strain: *Talaromyces radicus*	Vincristine and Vinblastine	Cytotoxic activity	HeLa cells	Palem et al. (2016)
14.	Endophytic actinomycetes	–	Antidiabetic activity	Rat	Isavella Rosaline and Agastian (2013)

Fungal endophytes investigated for different biological activities include *Aspergillus fumigatus* (Ali et al., 2019), *Alternaria alternata* (Sudharshana et al., 2019; Bhagat et al., 2016), *Botryosphaeria rhodina* (Akther et al., 2019), *F. oxysporum* (Ali et al., 2019), *Chaetomium nigricolor* (Dhayanithy et al., 2019), *Curvularia verruculosa* (Parthasarathy et al., 2020), *Diaporthe* sp. (Yan et al., 2018), *Eutypella* sp. (Kuriakose et al., 2016), *N. sphaerica* (Ayob et al., 2017; Muñoz-Rojas, 2017), and *T. radicus* (Palem et al., 2016). Very few bacterial isolates viz. *Bacillus amyloliquefaciens* (Myo et al., 2020), *Micrococcus yunnanensis* (Ranjan and Jadeja, 2017), and some Actinomycetes (Chandrakar and Gupta, 2017; Isavella Rosaline and Agastian, 2013) were analyzed for these studies.

21.6 Endophyte (s) and other biotechnological strategies for the enhancement of indole alkaloids production in *Catharanthus roseus* L

Around 500 kg dried *C. roseus* plant material is required for the production of 1 g VB, and due to the limited plant-based production of these alkaloids, and high demand of marketed pharmaceuticals, these alkaloids are very costly. Thus, modern biotechnological strategies are best suited for the sustainable production of these valuable indole alkaloids (Kaur et al., 2020b; Verma and Chandel, 2019). Many researchers have worked on alternative methods involving in the high production of indole alkaloids. Several biotechnological methods such as metabolic engineering, use of bioreactors for large-scale production, use of key genes/precursors, abiotic or biotic elicitors, plant tissue culture, and genetic transformation have been used by a lot of researchers.

Few studies linked to the enhanced production of vinca alkaloids are discussed here (Table 21.4). Kidd et al. (2019) reported the 27-fold increase in vindoline production with the application of exogenous secologanin in in vitro cultures. Several studies reported the significant increase of vinca alkaloids by regulating the TIAs biosynthetic genes in cell suspension cultures (with precursor feeding or medium elicited with methyl jasmonate) (Ambrin et al., 2020; Sharma et al., 2019; Liang et al., 2018; Sharma et al., 2018; Fouad and Hafez, 2018; Rizvi et al., 2016; Verma et al., 2015; Suttipanta et al., 2011; Wang et al., 2010). These studies reported the several-fold increase of VB (Sharma et al., 2018, 2019), vindoline (Liang et al., 2018; Sharma et al., 2018; Liu et al., 2011, 2017), catharanthine (Sharma et al., 2018; Liang et al., 2018; Liu et al., 2011; Wang et al., 2010), serpentine (Liu et al., 2017; Suttipanta et al., 2011), and ajmalicine (Ambrin et al., 2020; Suttipanta et al., 2011) (Table 21.4). Hairy root and cell cultures are most commonly used for the overexpression of candidate genes (ORCA2, ORCA3, CrMPK3, DAT, STR, TDS, ZCT1) involved in the biosynthetic pathway of TIAs (Tang and Pan, 2017).

Few studies also reported the use of endophytic fungal strains (Tang et al., 2011), fungal elicitor (*Aspergillus flavus*), inoculation with Arbuscular

TABLE 21.4 Endophyte (s) and other biotechnological strategies for the enhancement of indole alkaloids production in *Catharanthus roseus* L.

Sl. No.	Name (s) of vinca alkaloids	Biotechnological strategy	Type (s) of culture	Yield	References
1.	Ajmalicine	RNA-mediated gene silencing of terpenoid indole alkaloid (TIA) pathway and elicited with Methyl Jasmonate	Suspension cell culture	Significant increase	Ambrin et al. (2020)
2.	Vinblastine	TIAs pathway precursors (Tryptamine and Tryptophan) feeding	Multiple shoot cultures	300 mg/L each of Tryptamine and Tryptophan yielded 0.0277% and 0.0180%, respectively	Sharma et al. (2019)
3.	Catharanthine	Application of exogenous secologanin	In vitro cultures	11-fold increase	Kidd et al. (2019)
	Vindoline			27-fold increase	
4.	Vinblastine	Seedlings treated with *Bacillus pumilus*	Plant growth-promoting rhizobacteria (PGPR) suspension	0.1810%	Al-Zahrany et al. (2019)
	Vincristine			0.1400%	
5.	Catharanthine	Elicitation by UV-C	Cell suspension cultures	7.6-fold increase	Moon et al. (2018)
	Vincristine			3.6-fold increase	
	Vinblastine			4.2-fold increase	
	Vindoline			2-fold increase	

(*Continued*)

TABLE 21.4 (Continued)

Sl. No.	Name (s) of vinca alkaloids	Biotechnological strategy	Type (s) of culture	Yield	References
6.	Vindoline, catharanthine, vincristine, vinblastine, and ajmalicine	Elicitation by Methyl jasmonate and Silver nitrate	Shoot cultures	Significant increase	Paeizi et al. (2018)
7.	Vindoline	Regulation of TIAs biosynthetic genes with Fungal elicitor: (*Aspergillus flavus*)	Cambial meristematic cells (CMCs)	1.45-fold increase	Liang et al. (2018)
	Catharanthine			3.29-fold increase	
	Ajmaline			2.14-fold increase	
8.	Vindoline and catharanthine	Overexpression of candidate genes of TIAs pathway using *Agrobacterium tumefaciens* (LBA1119)	Callus cultures	9-fold increase	Sharma et al. (2018)
	Vinblastine			5-fold increase	
9.	Total alkaloids	Overexpression of CrMPK3 gene and effect of cobalt nanoparticles	Suspension cultures	2-fold increase	Fouad and Hafez (2018)
10.	Vindoline and serpentine	Elicitation of seedlings with Methyl Jasmonate and silencing of CR1 candidate gene	–	Significant increase	Liu et al. (2017)
11.	Vinblastine	Elicitation with yeast extract	Protoplast cultures	22.74%	Maqsood and Abdul (2017)
	Vincristine			48.49%	

12.	Vinblastine, vincristine, and catharanthine	*Agrobacterium rhizogenes*-mediated transformation	Hairy root culture	Significant increase	Hanafy et al. (2016)
13.	Alkaloids	Silencing of transcription gene (ZCT1)	Hairy root culture	Significant increase	Rizvi et al. (2016)
14.	Ajmalicine	Fed-batch cultivation and elicitation with jasmonic acid, methyl jasmonate and KCl	Hairy root culture	123.2 mg/L (4-fold increase)	Thakore et al. (2015)
15.	Vincamine and total alkaloids	Overexpression of rol gene and use of bioreactors	Suspension cultures	0.005% and 2.7%	Verma et al. (2015)
16.	Vindoline, vinblastine, catharanthine, ajmalicine, and serpentine	Arbuscular mycorrhizal fungi (AMF) inoculation	–	Significant increase	Andrade et al. (2013)
17.	Vindoline and vinblastine	Elicitors of hydroxylase, peroxidase, acetyltransferase	Suspension cultures	0.42 and 0.81 mg/g	Guo et al. (2013)
18.	Catharanthine and vindoline	Overexpression of candidate gene (ORCA2)	Hairy root culture	2.03-fold and 3.67-fold increase	Liu et al. (2011)
19.	Total alkaloid	Inoculation of endophytic fungus and its elicitors	Suspension cultures	48% and 32% higher yield	Tang et al. (2011)
20.	Serpentine and ajmalicine	Overexpression of *CrWRKY1* through *A. rhizogenes* (R1000) and elicitation with jasmonate, gibberellic acid, and ethylene	–	291.5 and 15.4 μg/g	Suttipanta et al. (2011)

(*Continued*)

TABLE 21.4 (Continued)

Sl. No.	Name (s) of vinca alkaloids	Biotechnological strategy	Type (s) of culture	Yield	References
21.	Catharanthine	Overexpression of candidate genes (geraniol 10-hydroxylase and ORCA3)	Hairy root culture	6.5-fold increase	Wang et al. (2010)
22.	Total alkaloid and vincristine, vinblastine, catharanthine, and vindoline	Arbuscular mycorrhizal fungi (AMF) inoculation (*Glomus mosseae, Glomus aggregatum, Glomus fasciculatum*)	–	8.19% Total alkaloid content	Ratti et al. (2010)

mycorrhizal fungi (AMF), *Agrobacterium rhizogenes*, and plant growth-promoting rhizobacteria (PGPR) and increased production of the total alkaloid content, VB and VC (Al-Zahrany et al., 2019; Hanafy et al., 2016; Andrade et al., 2013), vindoline, catharanthine, and ajmalicine. Guo et al. (2013) investigated the effects of hydroxylase, peroxidase, acetyltransferase elicitors in suspension cultures and found an increased amount of vindoline (0.42 mg/g) and VB (0.81 mg/g). Alternatively, Maqsood and Abdul (2017) found elicited amount of VB (22.74%) and VC (48.49%) by treating the protoplast cultures with yeast extract.

21.7 Future perspectives and conclusion

The coexistence of bacterial and fungal endophytes with their host plants belonging to various genera has contributed largely to the alteration of secondary metabolite synthetic pathways. The process of adaptation of endophytes under the host's cellular microenvironments triggers special enzyme synthesis, genetic manipulation, and integration of the host genome into their genetic make-up as part of constant evolutionary symbiosis. The biosynthesis pathway of indole alkaloids along with key enzymes can improvise the fermentation techniques to optimize alkaloid production in bacterial and fungal endophytes. To ensure large-scale production, one has to be critical about the choice of endophytes, their host specificity, standardization of inoculation, constant mycelial growth, and stable alkaloid production, which need to be optimized at every single step. Another issue is the cost-effectiveness of the total synthesis process, for which limiting the chances of contamination, determination of the most productive phase of synthesis and optimized isolation, and purification and quantification are mandatory. Various extraction techniques such as maceration, MAE and UAE, etc. for alkaloid have been discussed. Several analytical studies have been reported regarding the quantification of alkaloids through different chromatographic and spectroscopic techniques. Biotechnological strategies for enhanced alkaloid production are summarized in this context, and the factors that play conclusive roles were critically summarized in order to be recommended for successful industrial application of endophyte-mediated alkaloid harvesting procedures.

Acknowledgments

The authors are thankful to Lovely Professional University for providing infrastructure and Presidency University for FRPDF grant.

References

Abouzeid, S., Hijazin, T., Lewerenz, L., Hänsch, R., Selmar, D., 2019. The genuine localization of indole alkaloids in Vinca minor and *Catharanthus roseus*. Phytochemistry 168, 112110.

Akther, T., Mathipi, V., Kumar, N.S., Davoodbasha, M., Srinivasan, H., 2019. Fungal-mediated synthesis of pharmaceutically active silver nanoparticles and anti-cancer property against A549 cells through apoptosis. Environ. Sci. Pollut. Res. 26 (13), 13649–13657.

Alam, M.M., Naeem, M., Khan, M.M.A., Uddin, M., 2017. Vincristine and vinblastine anticancer catharanthus alkaloids: pharmacological applications and strategies for yield improvement. Catharanthus roseus. Springer, Cham, pp. 277–307.

Ali, E.A.M., Abdel-Rahman, T.M.A., Ebrahim, H.A., 2019. Cytotoxicity and anti-microbial activity of alkaloids extracted from *Catharanthus roseus* associated endophytic fungi. Malaysian J. Microbiol. 15 (1), 16–23.

Al-Zahrany, H.M., Elbeshehy, E.K.F., Aldhebiani, A.Y., Almaghrabi, O., Al-Jaddawi, A.A., Alzahrani, A.H., 2019. Effect of *Catharanthus mosaic virus* (CatMV) on antineoplastic alkaloids from periwinkle (*Catharanthus roseus* L.) cultured in Mecca region and resistance induction by plant growth-promoting rhizobacteria. Biotechnol. Biotechnol. Equip. 33 (1), 1392–1401.

Ambrin, G., Ali, H.M., Ahmad, A., 2020. Metabolic regulation analysis of ajmalicine biosynthesis pathway in *Catharanthus roseus* (L.) G. Don suspension culture using nanosensor. Processes 8 (5), 589.

Andrade, S.A.L., Malik, S., Sawaya, A.C.H.F., Bottcher, A., Mazzafera, P., 2013. Association with arbuscular mycorrhizal fungi influences alkaloid synthesis and accumulation in *Catharanthus roseus* and Nicotiana tabacum plants. Acta Physiol. Plant 35 (3), 867–880.

Anjum, N., Chandra, R., 2019. Endophytic bacteria of *Catharanthus roseus* as an alternative source of vindoline and application of response surface methodology to enhance its production. Arch. Biol. Sci. 71 (1), 27–38.

Ayob, F.W., Simarani, K., Zainal Abidin, N., Mohamad, J., 2017. First report on a novel *Nigrospora sphaerica* isolated from *Catharanthus roseus* plant with anticarcinogenic properties. Microb. Biotechnol. 10 (4), 926–932.

Bhagat, J., Kaur, A., Kaur, R., Yadav, A.K., Sharma, V., Chadha, B.S., 2016. Cholinesterase inhibitor (Altenuene) from an endophytic fungus *Alternaria alternata*: optimization, purification and characterization. J. Appl. Microbiol. 121 (4), 1015–1025.

Chandrakar, S., Gupta, A.K., 2017. Antibiotic potential of endophytic actinomycetes of medicinal herbs against human pathogenic bacteria. Proc. Natl. Acad. Sci. India Sect. B: Biol. Sci. 87 (3), 905–915.

Dhayanithy, G., Subban, K., Chelliah, J., 2019. Diversity and biological activities of endophytic fungi associated with *Catharanthus roseus*. BMC Microbiol. 19 (1), 22.

Falcão, M.A., Scopel, R., Almeida, R.N., do Espirito Santo, A.T., Franceschini, G., Garcez, J.J., et al., 2017. Supercritical fluid extraction of vinblastine from *Catharanthus roseus*. J. Supercrit. Fluids 129, 9–15.

Fouad, A.S., Hafez, R.M., 2018. Effect of cobalt nanoparticles and cobalt ions on alkaloids production and expression of CrMPK3 gene in *Catharanthus roseus* suspension cultures. Cell Mol. Biol. 64 (12), 62–69.

Gajalakshmi, S., Vijayalakshmi, S., Devi, R.V., 2013. Pharmacological activities of *Catharanthus roseus*: a perspective review. Int. J. Pharm. Bio. Sci. 4 (2), 431–439.

Guo, Z.G., Liu, Y., Gong, M.Z., Chen, W., Li, W.Y., 2013. Regulation of vinblastine biosynthesis in cell suspension cultures of *Catharanthus roseus*. Plant Cell Tissue Organ. Cult. (PCTOC) 112 (1), 43–54.

Hanafy, M.S., Matter, M.A., Asker, M.S., Rady, M.R., 2016. Production of indole alkaloids in hairy root cultures of *Catharanthus roseus* L. and their anti-microbial activity. South Afr. J. Bot. 105, 9–18.

Heijden, R., Jacobs, D., Snoeijer, W., Hallard, D., Verpoorte, R., 2004. The Catharanthus alkaloids: pharmacognosy and biotechnology. Curr. Med. Chem. 11 (5), 607–628.

Isavella Rosaline, S., Agastian, P., 2013. Glucosidase inhibitory activity and in vitro free radical scavenging activity of ethyl acetate extracts of endophytes isolated from the root of *Catharanthus roseus*. L. Int. J. Pharm. Bio. Sci. 4 (2), B1246–B1258.

Jaleel, C.A., Gopi, R., Azooz, M.M., Panneerselvam, R., 2009. *Pseudomonas fluorescens* as a physiological modulator in the enhancement of medicinally important alkaloids of *Catharanthus roseus*. Acta Pharm. Sci. 51 (2).

Jeong, W.T., Lim, H.B., 2018. A UPLC-ESI-Q-TOF method for rapid and reliable identification and quantification of major indole alkaloids in *Catharanthus roseus*. J. Chromatogr. B 1080, 27–36.

Jones, W.E., 1976. U.S. Patent No. 3,932,417. U.S. Patent and Trademark Office, Washington, DC.

Kaur, P., Gupta, R.C., Dey, A., Pandey, D.K., 2019. Simultaneous quantification of oleanolic acid, ursolic acid, betulinic acid and lupeol in different populations of five Swertia species by using HPTLC-densitometry: comparison of different extraction methods and solvent selection. Ind. Crop. Prod. 130, 537–546.

Kaur, P., Kumar, V., Singh, R., Dwivedi, P., Dey, A., Pandey, D.K., 2020a. Biotechnological strategies for production of camptothecin from fungal and bacterial endophytes. South Afr. J. Bot.

Kaur, P., Gupta, R.C., Dey, A., Malik, T., Pandey, D.K., 2020b. Optimization of salicylic acid and chitosan treatment for bitter secoiridoid and xanthone glycosides production in shoot cultures of Swertia paniculata using response surface methodology and artificial neural network. BMC Plant Biol. 20, 1–13.

Khare, E., Mishra, J., Arora, N.K., 2018. Multifaceted interactions between endophytes and plant: developments and prospects. Front. Microbiol. 9, 2732.

Kidd, T., Easson, M.L., Qu, Y., De Luca, V., 2019. Inter-organ transport of secologanin allows assembly of monoterpenoid indole alkaloids in a *Catharanthus roseus* mutant. Phytochemistry 159, 119–126.

Kumar, A., Patil, D., Rajamohanan, P.R., Ahmad, A., 2013. Isolation, purification and characterization of vinblastine and vincristine from endophytic fungus *Fusarium oxysporum* isolated from *Catharanthus roseus*. PLoS One 8 (9), e71805.

Kumar, S., Singh, A., Kumar, B., Singh, B., Bahadur, L., Lal, M., 2018. Simultaneous quantitative determination of bioactive terpene indole alkaloids in ethanolic extracts of *Catharanthus roseus* (L.) G. Don by ultra high performance liquid chromatography–tandem mass spectrometry. J. Pharm. Biomed. Anal. 151, 32–41.

Kuriakose, G.C., Palem, P.P.C., Jayabaskaran, C., 2016. Fungal vincristine from Eutypella spp - CrP14 isolated from *Catharanthus roseus* induces apoptosis in human squamous carcinoma cell line - A431. BMC Complement. Altern. Med. 16 (1), 302.

Liang, C., Chen, C., Zhou, P., Xu, L., Zhu, J., Liang, J., et al., 2018. Effect of *Aspergillus flavus* fungal elicitor on the production of terpenoid indole alkaloids in *Catharanthus roseus* cambial meristematic cells. Molecules 23 (12), 3276.

Liu, D.H., Ren, W.W., Cui, L.J., Zhang, L.D., Sun, X.F., Tang, K.X., 2011. Enhanced accumulation of catharanthine and vindoline in *Catharanthus roseus* hairy roots by overexpression of transcriptional factor ORCA2. Afr. J. Biotechnol. 10 (17), 3260–3268.

Liu, Z., Wu, H.L., Li, Y., Gu, H.W., Yin, X.L., Xie, L.X., et al., 2016. Rapid and simultaneous determination of five vinca alkaloids in *Catharanthus roseus* and human serum using trilinear component modeling of liquid chromatography–diode array detection data. J. Chromatogr. B 1026, 114–123.

Liu, J., Gao, F., Ren, J., Lu, X., Ren, G., Wang, R., 2017. A novel AP2/ERF transcription factor CR1 regulates the accumulation of vindoline and serpentine in *Catharanthus roseus*. Front. Plant Sci. 8, 2082.

Luo, M., Yang, L.Q., Yao, X.H., Mu, F.S., Zhang, D.Y., Song, Z.Y., et al., 2014. Optimization of enzyme-assisted negative pressure cavitation extraction of five main indole alkaloids from *Catharanthus roseus* leaves and its pilot-scale application. Sep. Purif. Technol. 125, 66–73.

Mall, M., Verma, R.K., Gupta, M.M., Shasany, A.K., Khanuja, S.P.S., Shukla, A.K., 2019. Influence of seasonal and ontogenic parameters on the pattern of key terpenoid indole alkaloids biosynthesized in the leaves of *Catharanthus roseus*. South Afr. J. Bot. 123, 98–104.

Maqsood, M., Abdul, M., 2017. Yeast extract elicitation increases vinblastine and vincristine yield in protoplast derived tissues and plantlets in *Catharanthus roseus*. Revista Brasileira de Farmacognosia 27 (5), 549–556.

Moon, S.H., Pandurangan, M., Kim, D.H., Venkatesh, J., Patel, R.V., Mistry, B.M., 2018. A rich source of potential bioactive compounds with anti-cancer activities by *Catharanthus roseus* cambium meristematic stem cell cultures. J. Ethnopharmacol. 217, 107–117.

Muñoz-Rojas, J., 2017. Importance of producing economic compounds to combat cancer. Microb. Biotechnol. 10 (4), 683–684.

Myo, E.M., Maung, C.E.H., Mya, K.M., Khai, A.A., 2020. Characterization of bacterial endophytes from Myanmar medicinal plants for anti-microbial activity against human and plant pathogens. Braz. J. Pharm. Sci. 56.

Nisar, A., Mamat, A.S., Hatim, M.I., Aslam, M.S., Syarhabil, M., 2016. An updated review on *Catharanthus roseus*: phytochemical and pharmacological analysis. Ind. Res. J. Pharm. Sci. 3 (2), 631–653.

Paeizi, M., Karimi, F., Razavi, K., 2018. Changes in medicinal alkaloids production and expression of related regulatory and biosynthetic genes in response to silver nitrate combined with methyl jasmonate in *Catharanthus roseus* in vitro propagated shoots. Plant Physiol. Biochem. 132, 623–632.

Palem, P.P., Kuriakose, G.C., Jayabaskaran, C., 2016. Correction: an endophytic fungus, *Talaromyces radicus*, isolated from *Catharanthus roseus*, produces vincristine and vinblastine, which induce apoptotic cell death. PLoS One 11 (4), e0153111.

Pan, Q., Mustafa, N.R., Tang, K., Choi, Y.H., Verpoorte, R., 2015. Monoterpenoid indole alkaloids biosynthesis and its regulation in *Catharanthus roseus*: a literature review from genes to metabolites. Phytochem. Rev. 15 (2), 221–250.

Pan, Q., Saiman, M.Z., Mustafa, N.R., Verpoorte, R., Tang, K., 2016. A simple and rapid HPLC-DAD method for simultaneously monitoring the accumulation of alkaloids and precursors in different parts and different developmental stages of *Catharanthus roseus* plants. J. Chromatogr. B 1014, 10–16.

Pandey, D.K., Kaur, P., 2018. Optimization of extraction parameters of pentacyclic triterpenoids from Swertia chirata stem using response surface methodology. 3 Biotech. 8 (3), 152.

Pandey, S.S., Singh, S., Babu, C.V., Shanker, K., Srivastava, N.K., Shukla, A.K., et al., 2016. Fungal endophytes of *Catharanthus roseus* enhance vindoline content by modulating structural and regulatory genes related to terpenoid indole alkaloid biosynthesis. Sci. Rep. 6, 26583.

Parthasarathy, R., Shanmuganathan, R., Pugazhendhi, A., 2020. Vinblastine production by the endophytic fungus *Curvularia verruculosa* from the leaves of *Catharanthus roseus* and its in vitro cytotoxicity against HeLa cell line. Anal. Biochem. 593, 113530.

Prabhu, D.S., Rajeswari, V.D., 2017. *Catharanthus roseus*: the cancer-fighting medicine. Catharanthus roseus. Springer, Cham, pp. 121–151.

Ranjan, R., Jadeja, V., 2017. Isolation, characterization and chromatography based purification of anti-bacterial compound isolated from rare endophytic actinomycetes *Micrococcus yunnanensis*. J. Pharm. Anal. 7 (5), 343–347.

Ratti, N., Verma, H.N., Gautam, S.P., 2010. Effect of Glomus species on physiology and biochemistry of *Catharanthus roseus*. Indian J. Microbiol. 50 (3), 355–360.

Rizvi, N.F., Weaver, J.D., Cram, E.J., Lee-Parsons, C.W., 2016. Silencing the transcriptional repressor, ZCT1, illustrates the tight regulation of terpenoid indole alkaloid biosynthesis in *Catharanthus roseus* hairy roots. PLoS One 11 (7), e0159712.

Schweizer, F., Colinas, M., Pollier, J., Van Moerkercke, A., Bossche, R.V., De Clercq, R., et al., 2018. An engineered combinatorial module of transcription factors boosts production of monoterpenoid indole alkaloids in *Catharanthus roseus*. Metab. Eng. 48, 150–162.

Sharma, A., Verma, P., Mathur, A., Mathur, A.K., 2018. Overexpression of tryptophan decarboxylase and strictosidine synthase enhanced terpenoid indole alkaloid pathway activity and antineoplastic vinblastine biosynthesis in *Catharanthus roseus*. Protoplasma 255 (5), 1281–1294.

Sharma, A., Mathur, A.K., Ganpathy, J., Joshi, B., Patel, P., 2019. Effect of abiotic elicitation and pathway precursors feeding over terpenoid indole alkaloids production in multiple shoot and callus cultures of *Catharanthus roseus*. Biologia 74 (5), 543–553.

Singh, S., Pandey, S.S., Shanker, K., Kalra, A., 2020. Endophytes enhance the production of root alkaloids ajmalicine and serpentine by modulating the terpenoid indole alkaloid pathway in *Catharanthus roseus* roots. J. Appl. Microbiol. 128 (4), 1128–1142.

Sudharshana, T.N., Venkatesh, H.N., Nayana, B., Manjunath, K., Mohana, D.C., 2019. Antimicrobial and anti-mycotoxigenic activities of endophytic *Alternaria alternata* isolated from *Catharanthus roseus* (L.) G. Don.: molecular characterisation and bioactive compound isolation. Mycology 10 (1), 40–48.

Suttipanta, N., Pattanaik, S., Kulshrestha, M., Patra, B., Singh, S.K., Yuan, L., 2011. The transcription factor CrWRKY1 positively regulates the terpenoid indole alkaloid biosynthesis in *Catharanthus roseus*. Plant Physiol. 157 (4), 2081–2093.

Tang, K., Pan, Q., 2017. Strategies for enhancing alkaloids yield in *Catharanthus roseus* via metabolic engineering approaches. Catharanthus roseus. pp. 1–16.

Tang, Z., Rao, L., Peng, G., Zhou, M., Shi, G., Liang, Y., 2011. Effects of endophytic fungus and its elicitors on cell status and alkaloid synthesis in cell suspension cultures of *Catharanthus roseus*. J. Med. Plants Res. 5 (11), 2192–2200.

Thakore, D., Srivastava, A.K., Sinha, A.K., 2015. Model based fed batch cultivation and elicitation for the overproduction of ajmalicine from hairy roots of *Catharanthus roseus*. Biochem. Eng. J. 97, 73–80.

Tiwari, R., Awasthi, A., Mall, M., Shukla, A.K., Srinivas, K.S., Syamasundar, K.V., et al., 2013. Bacterial endophyte-mediated enhancement of in planta content of key terpenoid indole alkaloids and growth parameters of *Catharanthus roseus*. Ind. Crop. Prod. 43, 306–310.

Verma, M.L., Chandel, A. (Eds.), 2019. Biotechnological Production of Bioactive Compounds. Elsevier.

Verma, P., Sharma, A., Khan, S.A., Shanker, K., Mathur, A.K., 2015. Over-expression of *Catharanthus roseus* tryptophan decarboxylase and strictosidine synthase in rol gene integrated transgenic cell suspensions of Vinca minor. Protoplasma 252 (1), 373–381.

Wang, C.T., Liu, H., Gao, X.S., Zhang, H.X., 2010. Overexpression of G10H and ORCA3 in the hairy roots of *Catharanthus roseus* improves catharanthine production. Plant Cell Rep. 29 (8), 887–894.

Yan, D.H., Song, X., Li, H., Luo, T., Dou, G., Strobel, G., 2018. Anti-fungal activities of volatile secondary metabolites of four *Diaporthe* strains isolated from *Catharanthus roseus*. J. Fungi 4 (2), 65.

Zhang, L., Qing-Hui, G.A.I., Yuan-Gang, Z.U., Lei, Y.A.N.G., Yu-Liang, M.A., Yang, L.I.U., 2014. Simultaneous quantitative determination of five alkaloids in *Catharanthus roseus* by HPLC-ESI-MS/MS. Chin. J. Nat. Med. 12 (10), 786–793.

Index

Note: Page numbers followed by "*f*" and "*t*" refer to figures and tables, respectively.

B

D

G

H

I

N

O

P

FSC
www.fsc.org
MIX
Papier | Fördert gute Waldnutzung
FSC® C083411

Kolibri 360 GmbH
Ferdinand-Jühlke-Straße 7
99095 Erfurt, Deutschland
produktsicherheit@kolibri360.de